SOLID-STATE ELECTRONICS: PRINCIPLES AND APPLICATIONS

TIMOTHY J. MALONEY

D0207177

Dirk Publishing Company

Solid-State Electronics: Principles and Applications
Maloney, Timothy J.

Library of Congress Catalog Card Number: 93–91075

Cover design by Patricia Yantz Maloney

For information, write:
Dirk Publishing Co.
Box 217
Lambertville, MI 48144

Phone or Fax (313) 854 5888

Printed in the United States of America

10 9 8 7 6 5 4 3 2 1

ISBN   0-9639857-4-4

CONTENTS

PREFACE

A Word to Students

The electronic revolution continues sweeping us along, for better or worse. For certain, it has brought many improvements in the human condition: Safe, reliable tools that help us to be productive (look at the infrared stress-testing system on page 40 and on page 7 of the color section); marvelous medical equipment that preserves and enhances the quality of life (the angioplasty equipment shown on pages 190 and 224, for example); wonderful opportunities for entertainment (check the animated character on pages 74–75, or the baseball scoreboard on page 504).

These are a few of the positive effects. Because we're in the electronics-promotion business, we will naturally place our emphasis on such positives. You will see many examples of the bright aspects of electronics technology as you work through this book.

But there's a dark side. For example, many of the uses that we have found for our commercial entertainment and news media are questionable, to put it mildly. Just around the corner is the "information highway". (See the communication satellite on page 232 and on page 7 of the color section.) If this information highway is developed in a socially responsible manner, it may prove to be one of the blessings of the electronic age. But if it's mishandled, it could make the media mistakes of the last four decades seem trivial by comparison.

So the large question of for better or for worse is quite unsettled.

In any case, our task here is to understand the technical basis for this rush of activity. The foundation of the electronic revolution is our basic knowledge of electricity and magnetism, which you have studied previously. But that knowledge has been available for over 100 years, while the revolution got rolling only in the last few decades. Clearly, the basic *electrical* knowledge, by itself, wasn't enough. What was required was the development of the solid-state amplifying device, the transistor, to begin the miniaturization process. That is the identifiable event that ignited the revolution.

Therefore, we will begin our labors in Chapter 1 by looking at solid-state structures. This will lead us to an understanding of the simplest solid-state electronic device, the one-junction diode. The diode idea leads to the two-junction transistor in Chapter 4. From there, our world opens up.

Features of This Text

Objectives: Each chapter begins with a list of numbered objectives. These are specific skills that you will learn as you study the material covered in the chapter. A blue check-mark beside the objective's number appears in the outside column at the location(s) within the chapter where you should have achieved that objective. As you are studying, when you encounter this check-mark ask yourself if you really can do what objective number 5 calls for. Keep working at it until you can.

⑤ ✔

Examples: When trying to understand a new idea, everyone finds it helpful to see an example. In this text, examples are numerous. Most of them call for a logical progression of thought. You must first reach conclusion **a**, then reason from there to conclusion **b**, which then leads to **c**, like real-life thinking in solving real electronic problems. Seldom will we have use for the quick, one-liner type of example favored by most texts. For a typical instance, look at Example 2-3 on page 25.

Get This: As we present our explanations, certain crucial statements will occur, that you must understand. Such crucial statements are marked in the side column by the icon shown here. Check page 14 for examples. When you encounter a **Get This**, satisfy yourself that you really do get it. If not, bring it up to your instructor in class.

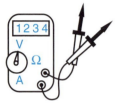

Self-Check: Questions and problems are presented every few pages, marked by the icon shown in the side column. Answer them to practice what you have learned during those pages. The correct answers appear in the back of the book, beginning on page 510. If you are unable to answer a particular question, or if you don't see how a particular problem is solved, ask your instructor for an explanation. See page 17 for a **Self-Check** example.

Troubleshooting: This book gives you a lot of information on troubleshooting, since it is an important skill for an electronics technician. When troubleshooting methods are presented, the DMM icon appears in the side column, as shown here. See page 30 for an example. It also appears when you are asked a troubleshooting question at the end of a chapter.

Chapter Articles: Each chapter contains an article with a photograph, describing in some detail an application of analog electronics. Chapter 1's article, describing automotive dynamometer testing, is on page 10. Read these articles to get a feel for some of the interesting job responsibilities and opportunities in the electronics field.

In addition to the articles, there is a photograph accompanied by a short description on the first page of each chapter. See page 1. And there are other occasional photos throughout the text, with brief descriptions. See page 19.

Other Features:

● An outline of topics, by sections, is given at the beginning of each chapter.

● Also appearing at the chapter beginning is a list of new words and terms that will be introduced in that chapter. When the new word first appears, it is printed in bold type, **like this**. For example, page 14 shows the first use of the term **forward bias**.

● At the end of a chapter, all of its important equations and formulas are brought together. You may find this handy when doing your homework assignments.

● A Summary of the chapter's main ideas follows the collection of formulas. Use the summary to help yourself review for quizzes and exams. For the Formulas and Summary of Chapter 4, look on page 104.

● The mathematics level is intermediate algebra. It is assumed that a hand-held scientific calculator will be used for numerical calculations. When the new (perhaps) calculator functions of $\boxed{\log}$ and $\boxed{10^x}$ are needed for dealing with decibels in Chapter 10, exact keystroke instructions are provided.

● There are extensive Chapter Questions and Problems at each chapter-end. Your instructor will assign some of these for homework. Of course, you can answer as many extras as you want. The more you practice, the more you learn.

YOUR SAFETY

When working on solid-state electronic circuits, follow the same safety practices that apply to basic electric circuits.

● Keep your body insulated from earth ground, with proper clothing and shoes.
● Make sure the insulation is intact on your tool-handles. Make sure your meter test-probes have no cracks in the plastic.
● Remove all metal jewelry from your hands. Put watches, rings and bracelets into your pocket.
● If you are working on a supposed de-energized circuit with the power turned off, be certain that all high-voltage capacitors are discharged.
● Learn the location of the main power-off switch for your work area. Be prepared to turn it off fast if a person is endangered.
● Get training in cardiopulmonary resuscitation, CPR.
● Wear eye-protection when cutting or stripping wire, and when soldering.
● Provide proper ventilation when using printed-circuit chemicals.

Acknowledgments

I really don't know how other technical authors avoid the large number of errors that are forever creeping into a book. My only defense is my good friend Dan Metzger. As always, his careful review and advice were invaluable.

Deep gratitude to my dear wife Pat for her photography that appears throughout. But mostly for her steadying courage in getting us through 1993 – not a vintage year.

Tim Maloney

CHAPTER 1

SOLID-STATE IDEAS

In this visual flight simulator, the learner wears a goggled headset that electronically duplicates the scene through an aircraft's window. The trainer, seated in the rear, manipulates controls to simulate differing flight conditions. He then observes and records the learner's responses.

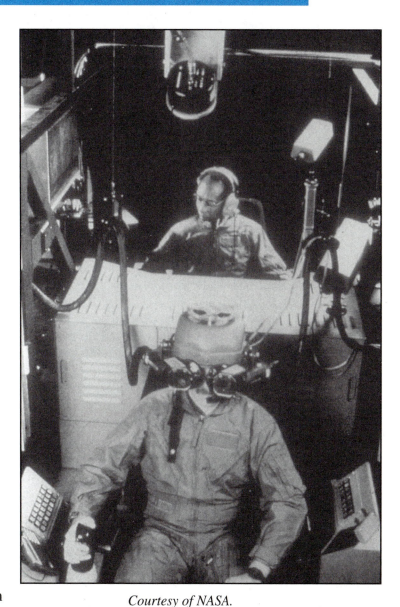

Courtesy of NASA.

OUTLINE

NEW TERMS TO WATCH FOR

valence	*p-n* junction
covalent bonding	depletion region
n-type semiconductor	reverse bias
p-type semiconductor	forward bias

To feel comfortable with solid-state electronic circuits, you need some understanding of the internal operation of solid-state devices. For us, the three most important solid-state devices are the junction diode, the bipolar junction transistor, and the field-effect transistor. So that we can explain the internal operation of these three devices later, let us first study the atomic events in solid silicon.

After studying this chapter, you should be able to:
1. Describe the process of silicon covalent bonding.
2. State the conduction/insulation properties of pure silicon.
3. Name the elements that are used to make *n*-type semiconductors, and state the common characteristic of those elements. Do the same for *p*-type.
4. Describe the mechanism of free-electron current flow in an *n*-type semiconductor.
5. Describe the mechanism of hole-current flow in a *p*-type semiconductor.
6. Describe the events that make a depletion region form around an unbiased *p-n* junction.
7. Explain why a reverse-biased *p-n* junction blocks current.
8. Explain why a forward-biased *p-n* junction conducts current.
9. State the forward conduction voltage for a silicon *p-n* junction.

1-1 PURE SILICON

Silicon is the element that has 14 protons and 14 electrons. Of the 14 electrons, 2 are in the innermost electron orbital shell, called the *K* shell. The *K* shell of any atom is filled by just 2 electrons. A filled shell means that the electrons in that shell are very tightly held by the nucleus. They cannot be involved in any chemical or electrical action. The 2 *K*-shell electrons are shown on the innermost ring in Fig. 1-1, which is our atomic model of silicon.

Figure 1-1
Atomic model of an isolated silicon atom. The *K* shell and the *L* shell are filled. But the *M* shell contains only 4 electrons, out of the 8 that are needed to fill it.

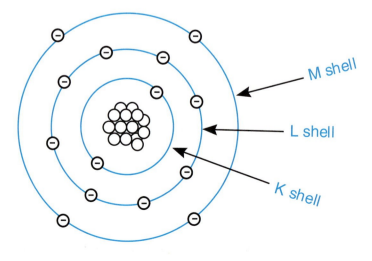

Of silicon's remaining 12 electrons, 8 are in the next orbital shell, the *L* shell. It takes 8 electrons to fill the *L* shell of any atom. So the 8 electrons shown in the second ring of Fig. 1-1 are also held so tightly that they cannot be involved in any chemical or electrical activity.

The last 4 electrons of silicon are in the *M* shell, as Fig. 1-1 shows. An *M* shell behaves like an *L* shell. It takes 8 electrons to fill an *M* shell. Therefore these final 4 electrons of silicon are not tightly held within a filled shell. They can be involved in chemical activity. Such electrons are called **valence** electrons.

In nature, there are many chemical activities, or reactions, that the 4 silicon valence electrons can get involved in. But the one reaction that is important to us is the reaction that bonds great numbers of silicon atoms together in a solid block. Let us explain this silicon bonding by looking at an imaginary situation in which just two silicon atoms come together.

In Fig. 1-2, the nucleus and the filled *K* and *L* shells of each atom are shown lumped together in the center. It is all right to show them lumped like this, because these are the parts of the atoms that can't interact anyway.

When the two silicon atoms come together in this figure, the valence electrons have their orbits shifted. To diagram this shifting, we show the valence (*M*) shell orbit of the left atom shifted slightly to the right of its normal position in Fig. 1-2. And the valence-shell orbit of the right atom is shifted slightly to the left. In other words, the two valence electron shells overlap each other.

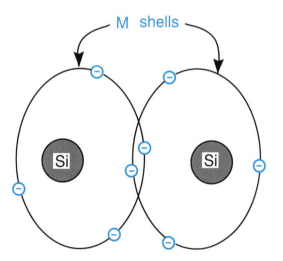

**Figure 1-2
Each atom feels that
the other atom's
valence electrons
have filled its *own*
M shell.**

Because of this overlap, an important change occurs. The atom on the left now acts as if it has 8 electrons in its *M* shell. This is because it still has its own 4 valence electrons, and now it also has part-possession of the 4 valence electrons from the right atom. Of course, the same thing can be said about the atom on the right. It still has its own 4 valence electrons and now it also has part-possession of the 4 valence electrons from the left atom.

With both atoms acting as if they have their *M* shells filled, they form a bonded pair. This means that they are locked together and will not separate unless they are forced to do so. This type of atomic bonding is called **covalent bonding**. The word covalent means that the atoms have co-possession of each other's valence electrons.

We are imagining the situation in Fig. 1-2 in order to explain the idea of covalent bonding. The bonding of just two silicon atoms doesn't really occur in nature.

If a great number of silicon atoms are brought together, every atom forms a covalent bond with its four closest neighbors. Figure 1-3 shows how to visualize this. The center atom's own valence electrons are the ones that are drawn in the corners of the square.

Figure 1-3
Structure of a silicon crystal.
The squares are an aid to
visualization.

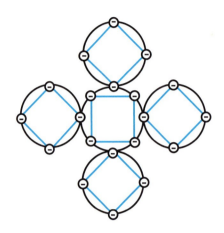

The drawing conveys the idea that the center atom has its *M* shell filled by covalently bonding one electron from each of four neighboring atoms. Of course, each of the four neighboring atoms is also covalently bonding one electron from the center atom. In addition, each neighboring atom is covalently bonding one electron from its other three neighbors, which do not appear in Fig. 1-3. In this way, every silicon atom in an entire large collection has a filled *M* shell.

A large collection of silicon atoms will covalently bond together to form a solid. Because each atom has a filled *M* shell, this solid is chemically and electrically inactive. It is not a good conductor of electric current at room temperature.

So a solid block of pure silicon is an insulator. It is not even a very good semiconductor, by itself.

1-2 DOPING THE SILICON

Semiconductor manufacturers do not leave silicon in a pure state. Instead, they add impurity atoms scattered every so often throughout the solid. There are two kinds of impurity elements. They are:

[1] Elements whose atoms have 5 valence electrons. (5-valent atoms)
[2] Elements whose atoms have 3 valence electrons. (3-valent atoms)
The commonly used impurity elements are listed in Table 1-1.

N-Type Semiconductor

If one of the 5-valent elements, say arsenic, is scattered as an impurity throughout the silicon solid, we call the resulting material *n*-doped silicon, or an ***n-type semiconductor*** . The letter *n* is used because each arsenic atom has an extra electron, carrying a *negative* charge. This extra electron does not fit into the covalent bonding structure of the silicon. The situation is shown in Fig. 1-4.

Because the fifth arsenic electron is not part of the stable covalent bonding, it can break loose from the arsenic atom. Once broken loose, it is able to move about from one silicon atom to the next if an electromotive force, a voltage, is present.

4 ✓

Of course, there is a huge total number of silicon atoms in the solid. If the arsenic doping density is only, say, one in every million silicon atoms, there will still be a great number of free electrons in the solid. Therefore it can support a considerable amount of current. The solid material has become a semiconductor.

One impurity atom per million silicon atoms is typical.

Elements with 5 valence electrons	Elements with 3 valence electrons
Phosphorus (P)	Boron (B)
Arsenic (As)	Aluminum (Al)
Antimony (Sb)	Indium (In)

**Table 1-1
Impurity elements that are used to dope silicon.**

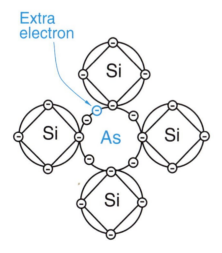

**Figure 1-4
Arsenic is a possible impurity atom in *n*-type material. After bonding with 4 silicon atoms, this arsenic atom has a free electron in its outer shell.**

P-Type Semiconductor

The name *n*-type refers to mobile negative charge carriers, electrons. So it is no surprise that the name *p*-type refers to positive charge carriers, or holes. A ***p*-type semiconductor** has 3-valent atoms, like boron, scattered throughout the silicon. The situation is pictured in Fig. 1-5.

Because the boron atom's *M* shell is not filled, it has a hole where an electron ought to be, if complete bonding were to be achieved. If an electromotive force is present, it may make a neighboring silicon atom transfer one of its bonded electrons into this shell, thereby filling it. Then the positive charge-carrying hole will have moved into the neighboring silicon atom.

Because there are a great number of boron atoms in the solid, a considerable amount of current can flow in this manner. The material has become a *p*-type semiconductor.

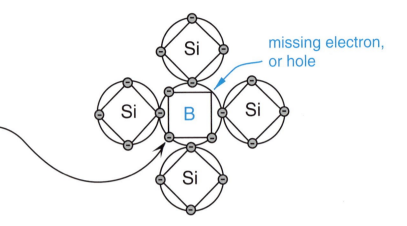

Figure 1-5
Boron is a commonly used impurity atom in *p*-type material. After bonding with 4 silicon atoms, this boron atom has a shortage of 1 electron in its outer shell. A shortage of 1 electron is the same as the presence of 1 hole.

Figure 1-6 points out the internal difference between electron-flow current in an *n*-type semiconductor and hole-flow current in a *p*-type semiconductor. It is important to realize that from an external viewpoint, there is no difference between these two kinds of current. That is, outside the body of the semiconductor solid, you cannot tell the difference between electron flow in one direction and hole flow in the other direction.

Figure 1-6
From an internal atomic vantage point, we can distinguish between free-electron current and hole current. But in the external circuit conductors, it is impossible to tell the difference.

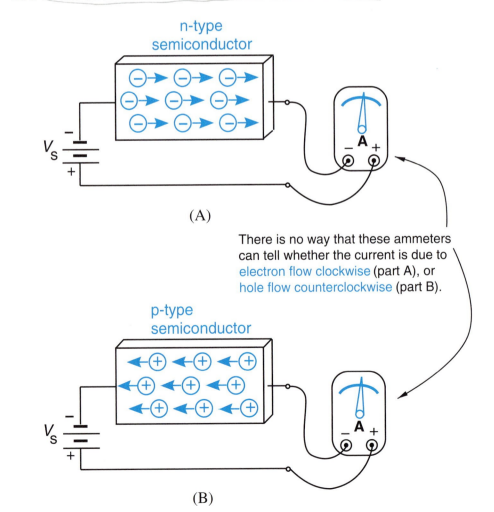

SELF-CHECK FOR SECTIONS 1-1 AND 1-2

1. Silicon has _____ total electrons per atom.
2. Silicon has _____ valence electrons.
3. How many electrons are required to fill the *L* shell or the *M* shell of an atom?
4. When two or more atoms share, or co-possess, each other's outer electrons, a _____ bond is formed.
5. (T-F) Pure solid silicon is a good conductor.
6. (T-F) Pure solid silicon is a semiconductor at room temperatures.
7. (T-F) In doped silicon, roughly one out of every ten atoms is an impurity atom.
8. What do the elements arsenic, antimony and phosphorus have in common?
9. What do the elements aluminum, boron and indium have in common?
10. What kind of atom must be scattered throughout silicon as an impurity in order to create an *n*-type semiconductor?
11. Repeat Question 10 for a *p*-type semiconductor?
12. In an *n*-type semiconductor, there is a(n) _____ associated with each impurity atom.
13. In a *p*-type semiconductor, there is a(n) _____ associated with each impurity atom.
14. (T-F) Before you can insert an ammeter into a circuit that contains a semiconductor, you must first find out what type of semiconductor (*n*-type or *p*-type) you are dealing with.

1-3 FORMING A *P-N* JUNCTION

All alone, doped silicon semiconductor solids are not especially useful electrically. True, they can provide greater or lesser resistance depending on their doping density. Silicon that is heavily doped with impurity atoms has less resistance (is a better conductor) than silicon that is lightly doped with impurity atoms. But there are cheaper and more temperature-stable methods for providing greater or lesser resistance. These methods were described in your introduction to electricity course.

If we deliberately want *a resistance that is highly temperature-dependent, simple doped silicon can give us that.*

However, doped silicon semiconductors become very useful when opposite types are joined together. Figure 1-7(A) shows an *n*-type semiconductor about to be joined to a *p*-type semiconductor. In this diagram, we choose to represent the excess charge-carrying electrons in the *n*-type material as a circle with the letter **e** inside, like this 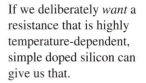. We represent the charge-carrying holes in the *p*-type material by an empty circle, like this ○. We are doing this because we want to reserve **+** and **−** signs for use at a later point in the coming discussion.

In Fig. 1-7(B), the two semiconductors have just at this moment been joined together, establishing a ***p-n* junction**. In both semiconductors, there is some random motion of charge carriers from atom to atom, even without an electromotive force (voltage) being applied. This occurs because at all temperatures above absolute zero, the free electrons in the *n*-type material do not hold completely steady positions within their orbits. Instead, they have a tendency to vibrate around their ideal orbital paths.

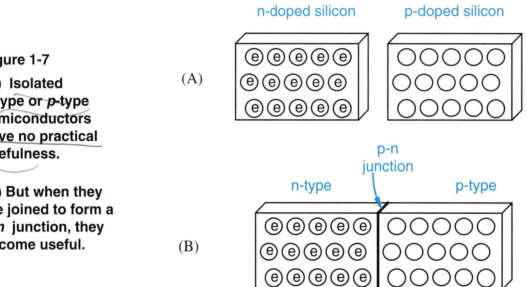

Figure 1-7

(A) Isolated *n*-type or *p*-type semiconductors have no practical usefulness.

(B) But when they are joined to form a *p-n* junction, they become useful.

As the unbonded free arsenic electrons vibrate, they occasionally break away from the parent arsenic atom and move randomly to a neighboring silicon atom. In Fig. 1-7(B) they are just as likely to move to the left, deeper into the *n*-type material, as they are to move to the right, crossing the junction into the *p*-type material. Both directions are equally likely because there is no applied voltage driving the electrons in any particular direction.

Free electrons that move left will continue to vibrate in their new residence atoms. At a later time they may make another random jump. Again, a jump to the right or a jump to the left will be equally likely.

Similar remarks can be made about those free electrons that do happen to move across the *p-n* junction and take up residence in a silicon atom within the *p*-type material. At a later time, as the electron continues to vibrate, it may move deeper into the *p*-type material, or it may randomly move back across the junction to land in a new silicon atom in the *n*-type material.

But notice that we have described these random jumps as always landing in a *silicon* atom. This is proper, because it is quite unlikely that any random jump will cause an electron to land in an impurity atom. Why? Simply because there are so few impurity atoms, compared to the great number of silicon atoms.

But by normal random luck, a free electron will once in a while land in a boron impurity atom in the *p*-type material. When that happens, the electron has landed in a hole. It therefore fills the outer shell of the boron atom, completing the covalent bonding of that boron atom. Since bonded electrons are much less likely to vibrate loose than unbonded electrons, the recently free electron can now be considered to be captured.

Get This

Whenever a free electron from the *n*-side of a junction happens to fall into a hole on the *p*-side, it gets stuck there. It is no longer free to wander about.

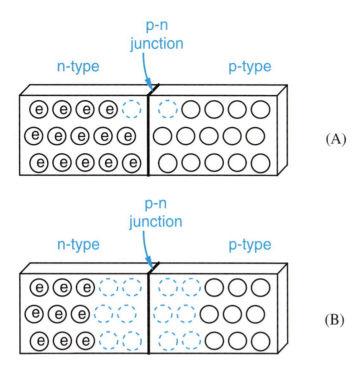

n-type

p-n
junction

p-type

(A)

n-type

p-n
junction

p-type

(B)

Figure 1-8

By random wandering,
free electrons from the
n-side fall into holes on the
p-side. This wipes out both
the free electron and the hole.

(A) A short time after the
junction has been formed.

(B) Later.

One single capture is shown pictorially in Fig. 1-8(A). Note that this figure shows only 14 free electrons on the *n*-side, and 14 holes on the *p*-side. Compare to Fig. 1-7(B), which showed the junction region at the moment that it was formed, having 15 free electrons and 15 holes in the near vicinity. Clearly, the junction region now has fewer charge carriers than it started with.

One individual capture is not significant, of course. But there are large numbers of free electrons wandering around the region of the *p-n* junction. As the nanoseconds tick off, more and more free electrons will randomly fall into holes. At some point in time the situation will be as shown in Fig. 1-8(B). In that figure, 6 of the original 15 free electrons have been captured. Therefore 6 of the original 15 holes have been eliminated. Think of the number 6 as representing a much larger number, perhaps 6 billion (6×10^9).

Look very carefully at what has happened to the arsenic atoms and to the boron atoms in the vicinity of the junction Each arsenic atom on the left side has become net positively (+) charged, because it has lost an electron. Each boron atom on the right side has become net negatively (−) charged, because it has gained an extra electron. Figure 1-8(B) has been redrawn in Fig. 1-9 to show this new situation regarding charge distribution.

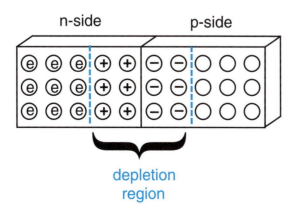

n-side

p-side

depletion
region

Figure 1-9

Under the condition of zero
applied voltage, random
wandering of free electrons
eventually causes a buildup
of **+** charge on the n-side
of the junction and **−** charge
on the *p*-side.

The charge redistribution shown in Fig. 1-9 puts a halt to the random wandering of free electrons near the junction. If any free electron from deep in the *n*-type material happens to wander close to the junction region, it will be repelled by the outer-shell negative charge concentration on the right side of the junction. Therefore that electron will quickly go back into the deep part of the *n*-type material.

Since there are no longer any free electrons on the *n*-side near the junction and no longer any holes on the *p*-side near the junction, we say that the region around the junction is depleted of charge carriers. In Fig. 1-9, the region between the dashed lines is called the **depletion region**.

6 ✔

AUTOMOTIVE DYNAMOMETER TESTING

The U.S. Environmental Protection Agency requires car manufacturers to estimate the city and highway mileage for each of their vehicle models. The mileage testing must be done under carefully controlled driving conditions. These conditions cannot be reliably achieved in actual on-the-road driving, due to uncontrollable variables like wind speed and slightly different driving techniques by human test drivers. However, identical driving conditions can be achieved with a laboratory dynamometer.

As shown in this photo, the car is driven forward until its drive-axle wheels are resting between two floor-mounted rollers. The rollers are mechanically connected to a large dc generator underneath the floor plates. Then, by varying the electric resistive load on the dc generator, the mechanical torque demand on the car's engine and drive train can be varied in a precisely controlled manner.

A standard computer program, the same for all manufacturers, is used to vary the car's wheel speed and torque through the controlled test cycle. The engine's throttle position and the car's brakes are controlled by mechanical actuators attached to the throttle and brake linkages in the engine compartment. With the vehicle's overall acceleration and braking sequence under computer/dynamometer control, the driver never touches the pedals . In this way, the variable human factor is eliminated and fair comparisons can be made. *Courtesy of General Motors Corp.*

SELF-CHECK FOR SECTION 1-3

15. (T–F) A piece of *n*-type doped silicon, isolated by itself, has an overall net negative (−) charge.
16. Describe the overall net charge contained by a piece of *p*-type doped silicon, isolated by itself. Explain this answer.
17. (T–F) In a newly created *p-n* junction, most random jumps that are made by free electrons cause the electrons to land in holes.
18. In an *n*-type semiconductor region, what is the overall net charge on an arsenic atom that has lost a free electron?
19. In an *n*-type semiconductor region, what is the overall net charge on the neighboring silicon atom that has just received the free electron.
20. The answer to Question 17 is False. Explain why that statement is false.

1-4 APPLYING VOLTAGE TO THE
P-N JUNCTION

Now that we understand the idea of the depletion region, we can explain how a *p-n* junction behaves when an external voltage is applied.

Junction Blocking

Figure 1-10(A) shows the *p-n* junction at the very moment that the switch closes, applying an external voltage that is positive (**+**) on the *n*-type material, and negative (**−**) on the *p*-type material. With the circuit complete, the free electrons in the *n*-type material are attracted to and held by the **+** terminal of the voltage source. Therefore the *n*-type material loses even more of its free electrons, making the left side of the depletion region *wider*. The same argument applies to the *p*-type material. Its holes are attracted to and held by the **−** terminal of the source, causing the depletion region to widen on the right side as well.

During the short time that the charge carriers are moving toward the terminals of the voltage source, an ammeter in an external wire will measure a small value of current. Figure 1-10(A) shows this short-lived current.

Eventually, the circuit stabilizes as indicated in Fig. 1-10(B). The final depletion region is much wider than it was to begin with. There are virtually no charge carriers remaining near the *p-n* junction. Therefore the *p-n* junction cannot support any current flow. It acts like an open circuit. The overall circuit current drops to zero, as shown by the ammeter in Fig. 1-10(B).

7

Summarizing the circuit action in Fig. 1-10(B), we can say:

A *p-n* junction will **block** current when the external voltage is applied with the **+** terminal to the *n*-side and the **−** terminal to the *p*-side.

Get This

Figure 1-10

Applying a reverse voltage to a *p-n* junction.

(A) Initially, a small amount of current flows.

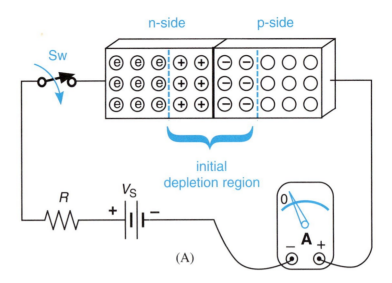

(B) Within a very short time (a few nanoseconds) the depletion region has widened to the point where all current stops.

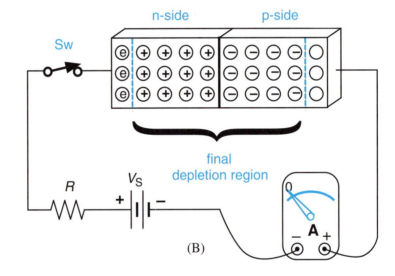

Applying the external voltage as shown above is called **reverse-biasing** the *p-n* junction. Figure 1-11 shows several circuits with reverse bias.

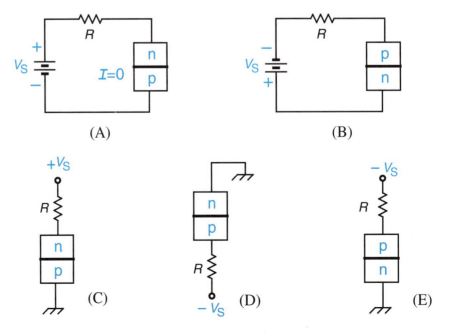

Figure 1-11 Examples of reverse-biasing a *p-n* junction. In every case, *I* = 0.

Junction Conducting

Figure 1-12 shows an external voltage source applied with the other polarity. The **+** terminal connects to the *p*-side and the **−** terminal connects to the *n*-side of the *p-n* junction.

In Fig. 1-12(A), the switch has just closed at this moment. The depletion region has the standard width for an unbiased junction, the same width that was shown in Fig. 1-9. The **−** source terminal attempts to push free electrons through the left side of the depletion region, toward the junction. It is working against the repulsion force exerted by the concentrated **−** charge on the right side of the junction. The **+** source terminal attempts to push holes through the right side of the depletion region, toward the junction. It is working against the repulsion force exerted by the concentrated **+** charge on the left side.

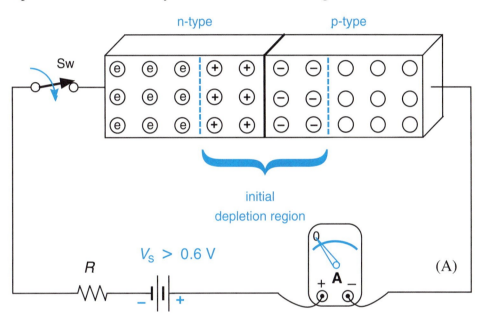

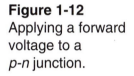

Figure 1-12
Applying a forward voltage to a *p-n* junction.

(A) Initially, a standard silicon depletion region exists.

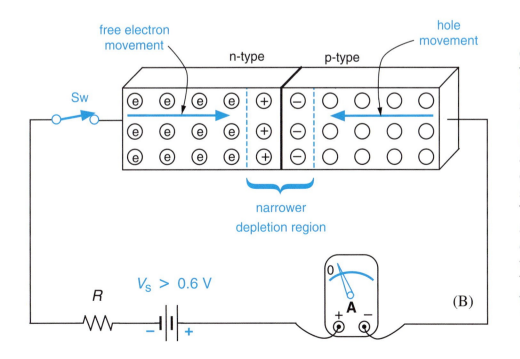

(B) A short time later, the depletion region has shrunk. There is a small value of current in the circuit. (Actually, no real ammeter could react quickly enough to show this current. This is an imaginary ammeter. It is drawn only to help us visualize events. The same is true for the ammeter in Fig. 1-10[A], on page 12.)

Here is a fact that will keep reoccurring in our later explanations of electronic circuits.

In order to overcome the repulsion forces produced by the depletion region of a silicon *p-n* junction, the external applied voltage must be at least 0.6 V.

Thus, if V_S in Fig. 1-12(A) were less than 0.6 V, say 0.4 V, it would not be able to overcome the opposition of the depletion region. Then no current would flow; the ammeter would remain at zero.

But if the applied voltage is 0.6 V or greater, free electrons from the *n*-side and holes from the *p*-side will move into the depletion region. This starts to undo the depletion region, as shown in Fig. 1-12(B). After sufficient time has passed (measured in nano- or microseconds), the depletion region disappears. Then a voltage slightly larger than 0.6 V exists across the junction. This is shown in Fig. 1-12(C).

Figure 1-12 (continued)

(C) A little later yet, the depletion region has disappeared. Free electrons can now move through the *n*-region, cross the junction, and continue on through the *p*-region. Holes can now move through the *p*-region, cross the junction, and continue on through the *n*-region. The *p-n* junction has become a short circuit to that part of the source voltage that is greater than 0.6 V.

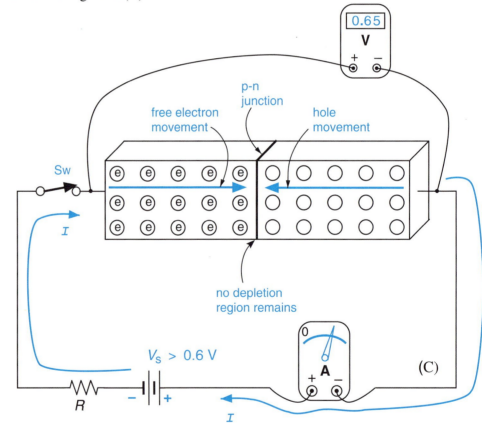

Summarizing the circuit action in Fig. 1-12, we can say:

A *p-n* junction will **conduct** current when an external voltage of at least 0.6 V is applied, with the **+** terminal to the *p*-side and the **−** terminal to the *n*-side.

Applying an external voltage as shown in Fig. 1-12 is called **forward-biasing** the *p-n* junction. The part of the source voltage that is greater than about 0.6 V appears across external resistance *R*. That value of voltage and the resistance value *R* set the overall circuit current, by Ohm's law. Overall circuit

Some resistance *R* must be present in the circuit to limit current *I*.

current *I* is produced by the combined effects of the two flow mechanisms shown in Fig. 1-12(C).

Figure 1-13 shows several circuits with forward bias.

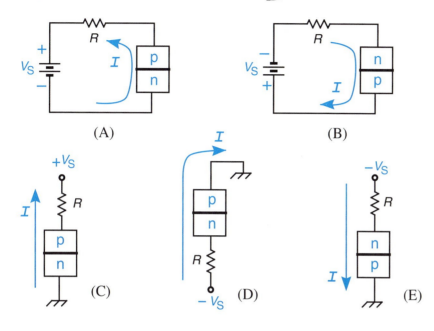

(A)

(B)

(C)

(D)

(E)

**Figure 1-13
Examples of
forward-biasing a
p-n junction.**

In this book, as in *Electricity: Fundamental Concepts and Applications*, Delmar Publishers, we will show all circuit current directions as electron-flow (from the source's negative terminal to its positive terminal). If you prefer the conventional-flow view of current, you can reverse the directional arrows.

   This advice holds true in all schematic diagrams throughout the book, including transistor schematic diagrams.

EXAMPLE 1-1

For the potentiometer bridge circuit in Fig. 1-14:
a) What pot adjustment range will cause the *p-n* junction to block current?
b) What pot range will cause the *p-n* junction to conduct current?

SOLUTION

a) To get started, assume for the moment that the *p-n* junction is not conducting. This causes R_1 to be in series with R_2. By voltage division,

$$\frac{V_2}{-10\ \text{V}} = \frac{R_2}{R_1 + R_2}$$

$$V_2 = -10\ \text{V}\left(\frac{20\ \Omega}{30\ \Omega + 20\ \Omega}\right) = -4\ \text{V}$$

The voltage on the *n*-side of the junction is −4 V, relative to ground. To make the junction start conducting, the voltage applied to the *p*-side must be 0.6 V more positive than −4 V. Therefore the pot's voltage division must produce a voltage on the pot wiper of

$$-4\ \text{V} + 0.6\ \text{V} = -3.4\ \text{V}$$

relative to ground.
   Applying the voltage division rule to the pot,

$$\frac{R_{\text{bot}}}{R_{\text{pot}}} = \frac{V_{\text{bot}}}{V_S} = \frac{-3.4\ \text{V}}{-10\ \text{V}}$$

$$R_{\text{bot}} = (1000\ \Omega)\left(\frac{3.4}{10}\right) = 340\ \Omega$$

The 340-Ω position is the critical pot position. If R_{bot} is greater than 340 Ω (the wiper is closer to the top than the 340-Ω point), the two voltage-division actions produce a voltage difference across the junction that is either:

1) Outright reverse bias, with the p-side more negative than the n-side, or
2) Insufficient forward bias, with the n-side more negative than the p-side, but by an amount less than 0.6 V.

The p-n junction will **block for R_{bot} between 1000 Ω and 340 Ω.**

b) Now let us consider R_{bot} less than 340 Ω – say 310 Ω, for example. The voltage divisions predict

$$V_{n\text{-side}} = -4.0 \text{ V} \quad \text{(unchanged, neglecting any loading effect of the } p\text{-}n \text{ junction)}$$

$$V_{p\text{-side}} = -10 \text{ V} \left(\frac{310 \text{ Ω}}{1000 \text{ Ω}} \right)$$

$$= -3.1 \text{ V (neglecting loading)}$$

The difference between these two voltage values, 0.9 V, might be expected to exist across the p-n junction. This would forward-bias the junction, making it conduct. We can conclude that the p-n junction will **conduct for R_{bot} between 340 Ω and 0 Ω.**

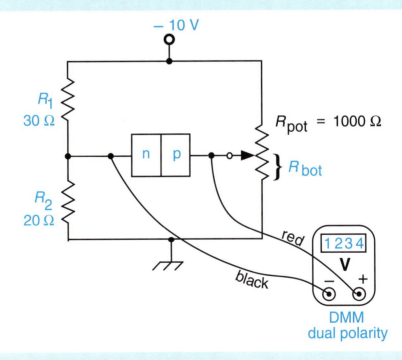

Figure 1-14
For Examples 1-1 and 1-2.

EXAMPLE 1-2

For the circuit in Fig. 1-14, describe the voltmeter response as the pot wiper is adjusted starting from the top end-terminal, then going down to the bottom end-terminal.

SOLUTION

The unloaded voltage difference of 0.9 V, calculated in part **b** of Example 1-1, will not really exist across the p-n junction. This is because a silicon *p-n* junction never permits a voltage magnitude very much larger than 0.6 V to appear. As the junction is made to conduct more and more current, the voltage across the junction increases only slightly. For example, even with the pot in Fig. 1-14 adjusted all the way down to the bottom, the *p-n* junction will allow its terminal voltage to increase by only a few hundredths of a volt. The voltage might increase to 0.64 V or 0.67 V perhaps. It might even rise to 0.7 V.

⑨ ✔

> We use the value 0.7 V as a convenient voltage to assume for a silicon *p-n* junction that is conducting fully, carrying substantial current. It is called forward voltage, symbolized V_F.

Get This

With this fact in mind, here is a description of the voltmeter response in Fig. 1-14.

1 With the pot wiper at the top, the voltmeter reads –6 V, since
$$-10 \text{ V} - (-4 \text{ V}) = -6 \text{ V}$$
2 As the wiper is moved down, the voltmeter measurement decreases in magnitude to –5 V, then –4 V, then eventually to –1 V, then finally to 0 V when the pot wiper is at $R_{bot} = 400 \ \Omega$.
3 As the pot wiper passes through 400 Ω, the voltmeter polarity changes over to + on red, – on black. This causes the digital display to switch from a negative number to a positive number. As the wiper moves from $R_{bot} = 400 \ \Omega$ to $R_{bot} = 340 \ \Omega$, the voltmeter measurement increases in magnitude from 0 V to +0.6 V.
4 As the pot wiper passes through 340 Ω, the *p-n* junction begins conducting. With further downward adjustment, the external driving voltage forward-biases the junction more and more. This causes the junction voltage to rise only slightly. As the wiper moves from $R_{bot} = 340 \ \Omega$ down to $R_{bot} = 0 \ \Omega$, the voltmeter measurement may increase from +0.6 V to about +0.7 V.

SELF CHECK FOR SECTION 1-4

22. When a *p-n* junction is connected to an external voltage source so that the *n*-side connects to the source's + terminal and the *p*-side connects to the source's – terminal, the junction is _____ biased.
23. Repeat Question 22 with the terminal polarities swapped.
24. (T-F) When a *p-n* junction is reverse-biased, its depletion region becomes even wider than it was under zero-bias conditions.
25. (T-F) When a *p-n* junction is forward-biased, it has essentially no depletion region.
26. To cause a silicon *p-n* junction to begin conducting, the external applied voltage must be equal to or greater than _____ V.
27. If the applied voltage is greater than the critical value in Question 26, what additional circuit component is needed to limit the current?
28. Going further with Question 27, if the current is allowed to rise to a substantial amount, say 200 mA, give an estimate of the voltage across the *p-n* junction.
29. Answer Question 28 for a lesser amount of circuit current, say 100 mA.

FORMULA

$V_F \approx 0.7$ V for a silicon *p-n* junction

SUMMARY OF IDEAS

- When a silicon crystal is doped with impurity atoms, it becomes a semiconductor. It is intermediate between a conductor and an insulator.
- If the impurity element has five valence electrons (like arsenic), we call the material an n-doped semiconductor.
- If the impurity element has three valence electrons (like boron), we call the material a *p*-doped semiconductor.
- In an *n*-doped semiconductor, the current flow mechanism is by *free* electrons moving from atom to atom (electron flow).
- In a *p*-doped semiconductor, the current flow mechanism is by *bound* electrons moving into an adjacent hole (hole flow).
- When a piece of *n*-doped material is brought into contact with a piece of *p*-doped material, a *p-n* junction is formed.
- In an unbiased (no applied voltage) *p-n* junction, a depletion region is established. No current can pass through the depletion region.
- If the *p-n* junction is reverse-biased by an external applied voltage, the depletion region becomes even wider. No current can pass through the *p-n* junction.
- If the *p-n* junction is forward-biased by an external applied voltage (at least 0.6 V, – on the *n*-side and + on the *p*-side), the depletion region is eliminated. Then current can pass through the *p-n* junction.
- Once a *p-n* junction begins carrying a substantial amount of current, its forward voltage becomes slightly larger than 0.6 V. We usually approximate it as $V_F \approx 0.7$ V.

CHAPTER QUESTIONS AND PROBLEMS

1. Pure silicon is a(n) _____ . (answer insulator or conductor)
2. (T-F) In a covalent bond, there is a clear-cut transfer of valence electrons away from one atom and into a different atom.
3. (T-F) Atoms with 8 electrons in their outermost shell tend to be chemically and electrically inactive, or stable.
4. The two kinds of impurity atoms that are used to dope silicon are those that have _____ valence electrons, and those that have _____ valence electrons.
5. Referring to your answer to Question 4, give the names of three elements that are of the first kind; repeat for the second kind.
6. Which kind of impurity makes an *n*-type semiconductor? Which makes a *p*-type?
7. Which kind of impurity enables current to flow by movement of negative charge-carriers, or free electrons?

8. Which kind of impurity enables current to flow by movement of positive charge-carriers, holes?

9. (T-F) An isolated block of *n*-type silicon semiconductor has an overall net negative charge.

10. (T-F) An isolated block of *p*-type silicon semiconductor has an overall net positive charge.

11. (T-F) *N*-type silicon semiconductor material is quite widely used, by itself.

12. When a block of *p*-type and a block of *n*-type semiconductor are physically brought together to touch each other, we say that a *p-n* _____ has been created.

13. (T-F) Most practical uses for semiconductors require the touching of *p*-type material with *n*-type material.

14. A *p-n* junction with no external applied voltage forms a _____ region, by random motion of free electrons.

15. In doped silicon, about 10% of all atoms are impurity atoms.

16. (T-F) With a reverse bias across a *p-n* junction, the depletion region becomes wider.

17. (T-F) With a forward bias across a *p-n* junction, the depletion region becomes narrower, or disappears entirely.

18. The minimum applied voltage needed to eliminate the depletion region and begin conducting through a silicon *p-n* junction is approximately _____ V.

19. In Fig. 1-12(C), if the doping density of the *n*-type material is greater than the doping density of the *p*-type material, the free-electron current through the junction will be _____ than the hole current through the junction.

20. In Fig. 1-12(C), what portion of the free electrons that cross the junction moving to the right eventually fall into a hole, roughly speaking?

21. In Fig. 1-12(C), what portion of the holes that cross the junction moving to the left eventually meet up with a free electron, roughly speaking?

22. Give a careful reason for your answers to Questions 20 and 21.

23. (T-F) A silicon *p-n* junction carrying a current of 0.1 A might reasonably be expected to have a forward voltage of 0.68 V.

24. At time t_1, a certain *p-n* junction has a measured $V_F = 0.66$ V. At later time t_2, the same junction has measured $V_F = 0.63$ V. The current at time t_2 is _____ than the current at t_1.

CHAPTER 2

JUNCTION DIODES

As the United States, the European Space Agency, Japan and Canada work toward a permanently manned space station, named *Freedom*, there is need for experimental knowledge about how absence of gravity affects human life processes. These photos show such experiments being carried out aboard the Space Shuttle *Columbia*. At top, electrodes are measuring the subject's head and eye movements. In the bottom photo, the subject is "weighing" herself to keep track of body mass changes. "Weighing" yourself in weightless space can be accomplished only with precise electronic sensors. These photos are also shown on page 2 of the color section.

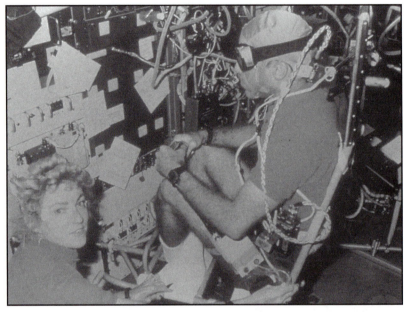

Courtesy of NASA.

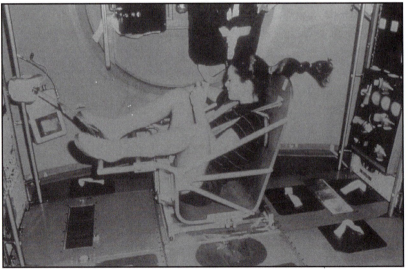

OUTLINE

2-1 Ideal Diode Behavior
2-2 Nonideal (Real) Diodes
2-3 Complete Diode Characteristic Graph
2-4 Troubleshooting Dc Diode Circuits

NEW TERMS TO WATCH FOR

diode   conducts
anode   blocks
cathode   characteristic graph
               peak reverse voltage

The junction diode is the most basic solid-state electronic device. In this chapter we will first look at ideal diode performance, and then at some of the nonideal operating characteristics.

After studying this chapter, you should be able to: ✔

1. Draw the schematic symbol of a diode and label its terminals.
2. Explain the fundamental usefulness of a diode.
3. Given a circuit schematic diagram, tell whether a diode is conducting or blocking.
4. Use the ideal model of a diode to give voltage and current values in a diode circuit.
5. State the difference between the ideal model and the practical model of a diode.
6. Use the practical model of a diode to give voltage and current values in a diode circuit.
7. State the differences between the practical model of a diode and an actual real diode.
8. Interpret and use the characteristic graph of a real diode.
9. Use the ohmmeter test to tell whether a diode is good or bad.
10. Use a dc voltmeter to troubleshoot a diode in-circuit.

2-1 IDEAL DIODE BEHAVIOR

A junction **diode** is simply a *p-n* junction with leads embedded into the *p*-region and the *n*-region. The silicon is enclosed in a sealed package. We call the *p*-region and its lead the **anode**. The *n*-region and its lead are called the **cathode**. The schematic symbol for a junction diode is shown in Fig. 2-1.

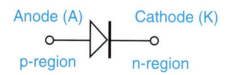

① ✔

Figure 2-1 Schematic symbol and lead identification for a diode. The cathode is given the symbol K rather than C. Think of the cathode as the pointed side of the arrow, and think of the anode as the flat side.

Diodes are useful because they act as one-way conductors of current. If the circuit applies a voltage which is more negative on the cathode (more positive on the anode) then the diode **conducts**. Electron current enters the diode on the cathode lead, and emerges from the diode on the anode lead. Thus, the current direction is against the arrow in the schematic diagram. Figure 2-2 shows several circuit examples with the diode conducting.   ② ✔

Of course, if you are taking the conventional-current view, as mentioned on page 15, then your current direction is with the arrow, not against it. The people who invented the semiconductor diode and gave it its symbol were taking the conventional-current view.

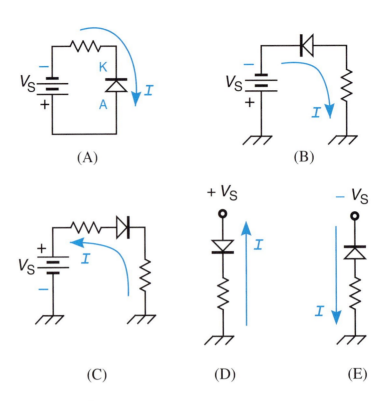

Figure 2-2
Conducting diodes.
Electron current
flows opposite to the
arrowof the diode
schematic symbol.

But if the circuit applies a voltage which is more positive on the cathode (more negative on the anode), then the diode **blocks**. Figure 2-3 shows several circuit examples in which the diode is blocking, or preventing current from flowing.

Figure 2-3
Diodes blocking. In
each case the source
is attempting to force
electron current into
the anode (flat) side.
The diode will not
allow current to flow in
that direction.

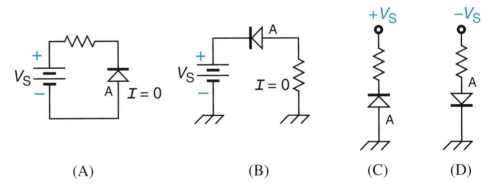

Summarizing the behavior shown in Figs. 2-2 and 2-3, we can say:

A diode conducts when its schematic arrow points
toward the more negative terminal of the source.

When a diode's arrow symbol points toward the more negative terminal, so that it conducts, we say that the diode is forward-biased. When its arrow points toward the more positive terminal, so that it blocks current, we say that the diode is reverse-biased.

An ideal diode can be thought of as a switch. When it is forward-biased it acts like a closed switch. It offers zero resistance and has zero voltage (ideally only) across its terminals, as recorded in Table 2-1. Then we must be sure that there is some resistance present in the circuit to limit the current (drop the supply voltage).

When an ideal diode is reverse-biased, it acts like an open switch. It has infinite resistance and the entire supply voltage appears across its terminals. Any series resistance that is present in the circuit has zero voltage drop. Table 2-1 points out the similarity between an ideal diode and a switch.

Ideal Diode Condition	Is like this switch state	Voltage across diode (or switch) terminals	Voltage across terminals of series resistor R
Forward-biased (conducting)	Closed	0 V	The entire applied voltage, V_S
Reverse-biased (blocking)	Open	The entire applied voltage, V_S	0 V

Table 2-1 Pointing out the identical behavior of an ideal diode and a switch.

EXAMPLE 2-1

For the circuit in Fig. 2-4, find the following:
a) V_{diode}, the voltage across the diode.
b) V_R, the voltage across the resistor.
c) I, the circuit current.
Give voltage polarity and current direction when relevant.

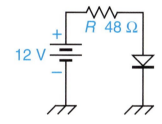

SOLUTION

a) The diode arrow points toward ground, which is the more negative terminal of the source. Therefore, according to Table 2-1, it will act like a closed switch, conducting current. Ideally, $V_{diode} = \mathbf{0\ V}$.
b) With the diode acting like a closed switch, the entire source voltage appears across R. That is, $V_R = \mathbf{12\ V}$, + on the left and − on the right.
c) Applying Ohm's law to R, we get

**Figure 2-4
Analyzing an ideal diode circuit.**

$$I = \frac{V_S}{R} = \frac{12\ V}{48\ \Omega} = \mathbf{0.25\ A}\ ,\ \text{clockwise} \qquad \text{(ideally)}$$

④ ✔

This current passes through both resistor and diode, of course, since they are in series.

EXAMPLE 2-2

Repeat the questions of Example 2-1 for the circuit of Fig. 2-5.

SOLUTION

a) The diode points toward the more positive source terminal. So it acts like an open switch, blocking current, as shown in Table 2-1.

Therefore $V_{diode} = \mathbf{20\ V}$, + on the left and − on the right.

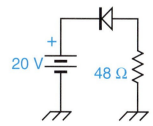

**Figure 2-5
Another ideal diode circuit.**

b) In an open circuit, the voltage across a resistance is **0 V**.

c) In an open circuit, $I = 0$.

2-2 NONIDEAL (REAL) DIODES

In the previous section we took the view that our diodes were ideal, or perfect. In reality, diodes can't quite duplicate the actions of switches. The most important nonideal feature of a real diode is that it doesn't have exactly zero voltage across its terminals when conducting. Instead, it has about 0.6 to 0.8 V, as we know from our discussion of the silicon *p-n* junction in Chapter 1. Assume that the value is 0.7 V, as suggested in Example 1-2. Then we can illustrate this fact about a real diode by drawing the current-versus-voltage **characteristic graph** of the diode. Contrasting a real diode's graph with an ideal diode's graph helps us to visualize their difference.

Figure 2-6(A) shows the I-versus-V graph for an *ideal* diode. The horizontal line going to the left means that when the voltage is reversed (**+** on K, **−** on A) the current is zero. It remains at 0 mA no matter how large the reverse voltage becomes. The vertical line going upward means that when the diode is forward-biased (**+** on A, **−** on K) the voltage drop across the diode is absolutely zero. It remains at 0 V no matter how large the forward current becomes.

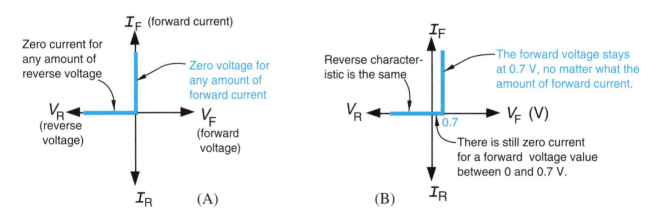

Figure 2-6 (A) Characteristic graph of I versus V for an ideal diode. (B) Approximate characteristic graph for a practical diode. The phrase *characteristic graph* always means a graph of current, on the y axis, versus voltage, on the x axis. It is also called a characteristic *curve*. In this particular example, the graphs have no curves, being just straight lines.

Figure 2-6(B) is an approximate I-versus-V graph for a real diode. Its only difference from the ideal diode is in the forward voltage region between 0 and 0.7 V. In this graph, any value of V_F between 0 and 0.7 V still produces $I_F = 0$. But once V_F reaches 0.7 V, I_F can be as large as it wants, as far as the diode is concerned. Of course, in a circuit, I_F will be limited to some reasonable value by a series resistance.

EXAMPLE 2-3

The circuit of Fig. 2-4 has been redrawn in Fig. 2-7, but with the diode considered to be real, as in Fig. 2-6(B). Find the following:

a) V_{diode}, the voltage across the diode.
b) V_R, the voltage across the resistor.
c) I, the circuit current.

Give voltage polarity and current direction where relevant.

SOLUTION

a) Since the schematic arrow points toward the more negative source terminal, the diode conducts. Forward current flow causes $V_F = \textbf{0.7 V}$, according to Fig. 2-6(B). All V_F values are + on A, − on K.

b) By Kirchhoff's voltage law

$$V_S = V_R + V_F$$
$$V_R = V_S - V_F$$
$$= 12\ V - 0.7\ V = \textbf{11.3 V}$$

The V_R polarity is + on the left, − on the right.

c) Applying Ohm's law to R, we have

$$I_R = \frac{V_R}{R}$$

$$= \frac{11.3\ V}{48\ \Omega} = \textbf{0.235 A}$$

Forward current is entering the diode at K, leaving the diode at A. In Fig. 2-7(B), I_F is counterclockwise.

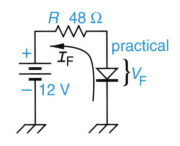

Figure 2-7
For Example 2-3

⑥ ✔

When we visualize a diode's operation in terms of simpler devices, like a switch and battery, we say that we have made a *model* for the diode. Models for new devices help us to understand them.

It is helpful to visualize a practical diode as a switch in series with a 0.7-V source, as illustrated in Table 2-2.

Table 2-2

The two diode models visualized as switch-type circuits.

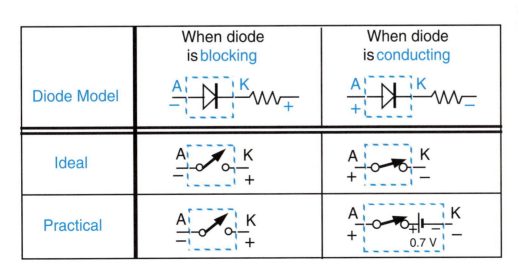

	When diode is blocking	When diode is conducting
Diode Model		
Ideal		
Practical		

Figure 2-8

(A) Circuit for Example 2-4. The diodes are real, not ideal.

(B) Visualizing the real diodes.

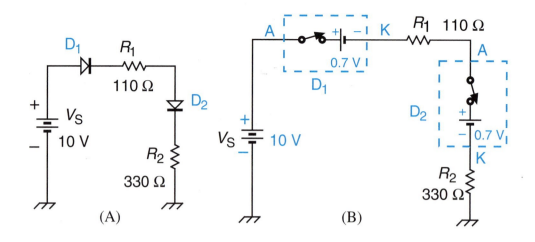

(A)   (B)

EXAMPLE 2-4

The series circuit in Fig. 2-8(A) has 2 real diodes and 2 resistors. Find:
a) The current I.
b) V_{R1}
c) V_{R2}

SOLUTION

a) Both diodes are forward biased. To help visualize the circuit action, redraw the circuit as shown in Fig. 2-8(B). The two 0.7-V forward voltage values combine to equal 1.4 V, opposing the 10-V source. Therefore the net voltage available to drive the resistors is given by

$$V_{Net} = V_S - 2 \cdot (0.7 \text{ V})$$
$$= 10 \text{ V} - 1.4 \text{ V} = 8.6 \text{ V}$$

The total resistance of the series circuit is

$$R_T = R_1 + R_2 = 110 \text{ }\Omega + 330 \text{ }\Omega = 440 \text{ }\Omega$$

Using V_{Net} and R_T in Ohm's law, we get

$$I = \frac{V_{Net}}{R_T}$$

$$= \frac{8.6 \text{ V}}{440 \text{ }\Omega} = \textbf{0.0195 A} \quad \text{counterclockwise}$$

b) Applying Ohm's law to R_1 alone gives

$$V_{R1} = I R_1$$
$$= (0.0195 \text{ A}) (110 \text{ }\Omega)$$
$$= \textbf{2.1 V}, \quad + \text{ on the left}, - \text{ on the right side.}$$

c) Either Ohm's law or Kirchhoff's voltage law (KVL) can be applied to R_2. Using KVL, we get

$$V_{R2} = V_S - 1.4 \text{ V} - V_{R1} = V_{Net} - V_{R1}$$
$$= 8.6 \text{ V} - 2.1 \text{ V} = \textbf{6.5 V}, \quad + \text{ on the bottom}, - \text{ on top.}$$

SELF-CHECK FOR SECTIONS 2-1 AND 2-2

1. Draw the schematic symbol for a diode and label its terminals.
2. In a schematic circuit diagram, if a diode's arrow points toward the more positive terminal of the source, the diode _____ .
3. In a schematic circuit diagram, if a diode's arrow points toward the more negative terminal of the source, the diode _____ .
4. If an ideal diode is forward biased, it acts like a _____ switch.
5. If an ideal diode is reverse biased, it acts like a _____ switch.
6. Six diode circuits are shown in Fig. 2-9. Tell which ones have the diode conducting and which ones have the diode blocking.
7. In Fig. 2-9(B), describe the voltage across the diode, V_{diode}, and the voltage across the resistor, V_R.
8. Repeat Problem 7 for Fig. 2-9(C).
9. Repeat for Fig. 2-9(D).
10. Describe the current in each one of the three circuits in Problems 7, 8 and 9.

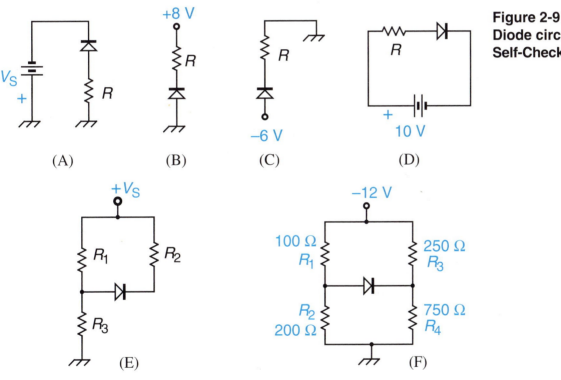

Figure 2-9
Diode circuits for
Self-Check problems.

2-3 COMPLETE DIODE CHARACTERISTIC GRAPH

Even the characteristic graph shown in Fig. 2-6(B) is not a totally accurate description of a real diode. As we learned in Chapter 1, a silicon *p-n* junction *starts* conducting at about 0.6 V of forward voltage. But then, if greater and greater values of current are forced through the diode, its actual value of V_F will rise slightly. This fact is graphed in the true characteristic curve of Fig. 2-10. As this particular diode's forward current is increased from 0 to the 1000-mA $I_{F(\text{max})}$ value, the forward voltage drop increases by a small amount, from 0.6 V to a value greater than 0.8 V.

⑦ ✔

Another fact that is pointed out by Fig. 2-10 is this. A tiny amount of reverse current I_R passes through a real diode when a reverse voltage is applied. This current is only a fraction of a microamp, typically. It is negligible in most circuit applications.

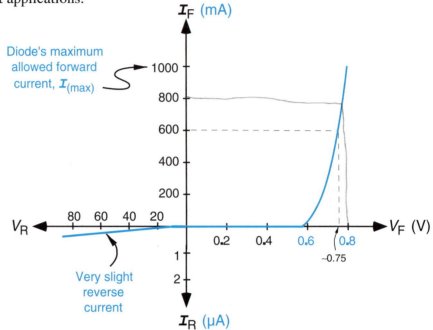

Figure 2-10 Diode characteristic graph. Note that the forward voltage (V_F) scale is small, only 0.2 V per mark, but the reverse voltage (V_R) scale factor is much larger, 20 V per mark. Also the forward current (I_F) scale factor is 200 mA per mark, but the reverse current (I_R) scale factor is tiny, only 1 microamp per mark.

If the characteristic curve were extended further to the left in Fig. 2-10, a V_R voltage would be reached that would cause the graph to turn sharply downward. That value of V_R is called the **peak reverse voltage**. If the diode is subjected to that amount of voltage it is in danger of being destroyed. We normally try to guarantee that such a large reverse voltage is never applied to the diode.

EXAMPLE 2-5

For the diode having the characteristic graph of Fig. 2-10:
a) What will be its actual forward voltage when its forward current is 100 mA?
b) Repeat for 600 mA.

SOLUTION

a) From the 100 mA point on the vertical axis, project a straight horizontal line over to intersect the curve. Then drop a straight vertical line to the x axis. As best we can read the graph, it intersects at a V_F value of approximately **0.65 V**.

b) Repeating the operation of part **a**, we intersect the V_F axis at about **0.75 V**.

The result from this Example helps you to see why the practical model of the diode, described in Sec. 2-2, is a pretty good approximation. In Example 2-5 the current increased by a great deal (600 mA compared to 100 mA is a 500% increase). But the voltage increased by just a small percentage

(0.75 Vcompared to 0.65 V is only a 15% increase). In other words, the current changed drastically, but the forward voltage changed hardly at all.

Figure 2-11 shows the physical appearance of several diodes

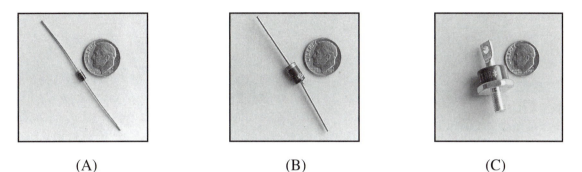

(A)   (B)   (C)

Figure 2-11 Various diodes. (A) Type No. 1N4004 diode, a popular diode for rectifying ac to dc. Diode type numbers often begin with 1N, because a diode contains **1** *p-n* junction. The type Number 1N4004 has an $I_{F(max)}$ rating of 1 A and peak reverse voltage of 400 V.
(B) Larger diode with an $I_{F(max)}$ rating of 3 A. On a cylinder-shaped diode, the striped end is the cathode. In this photo the cathode is on the left. (C) The threaded stud on this diode is for bolting the diode's body to a heat sink. Attached to a large heat sink, this diode has an $I_{F(max)}$ rating of 25 A.

EXERCISING HEART MONITOR

Old-fashioned heart-rate detecting electrodes relied on their conductivity (or resistivity) through the skin. Therefore they were affected by perspiration and the person's physical motion—not an acceptable set of restrictions for people engaged in aerobic exercise.

Today's heart-rate monitoring electrodes are fundamentally capacitive, not conductive. A slight body point-to-body point voltage variation occurs with each heartbeat. These variations must be electronically detected and counted. The stick-on electrode has an insulating dielectric film in contact with the skin, so moisture is irrelevant. The electrode is attached to a small transmitter that is worn under exercise clothing. By that means, every heartbeat is counted by the electronic module on the exercise machine.

The machine shown here is a home model stepping/climbing exerciser. In addition to heart rate, it measures stepping rate, overall distance climbed, elapsed time, an estimate of total calories burned, and other functions. All the information is displayed on the large display module mounted at head level. This photo and a close-up of the electronic display module are shown on page 1 of the color section.

Courtesy of Heart Rate, Inc.

2-4 TROUBLESHOOTING DC DIODE CIRCUITS

Most junction diodes can be tested with a dc ohmmeter. The usual conditions for ohmmeter usage must be met. Namely, the diode must be de-energized, and it must not be paralleled by any other component.

A functionally good diode will indicate a low resistance when forward-biased by a dc ohmmeter. It will indicate a high resistance when reverse-biased by the ohmmeter. If a diode fails either one of these indications, it is bad.

This two-step testing procedure is illustrated in Fig. 2-12. The ohmmeter contains a dc voltage source. If the ohmmeter's negative (–) terminal is connected to the cathode and its **+** terminal to the anode, as shown in Fig. 2-12(A), the diode is forward-biased. It therefore conducts forward current, which the ohmmeter interprets as a low resistance.

But if the leads are switched around, as shown in Fig. 2-12(B), the diode becomes reverse-biased. The reverse current is only a fraction of a microampere, which the ohmmeter interprets as a high resistance.

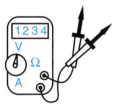

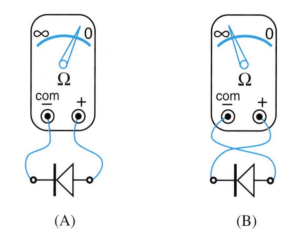

(A)   (B)

Figure 2-12 Testing whether a diode is good. If the diode passes both of these tests, it is good. (A) A forward-biased diode should give a resistance indication less than about 1 kΩ. (B) A reverse-biased diode should give an indication greater than about 100 kΩ.

There are a few things you should be aware of regarding the diode ohmmeter test.

1 You must choose a reasonable resistance scale on the ohmmeter. For a VOM, the best scales are $R \times 10\ \Omega$, $R \times 100\ \Omega$ or $R \times 1\ \text{k}\Omega$. If you were to use the $R \times 10\ \text{k}\Omega$ or $R \times 100\ \text{k}\Omega$ scale, a bad diode's higher-than-normal forward resistance would still allow the meter pointer to move fairly far to the right. This might fool you into thinking that the diode is OK. If you were to use the $R \times 1\ \Omega$ scale, a bad diode's lower-than-normal reverse resistance might still be high enough to keep the pointer fairly far to the left. Again, this might fool you.

2   For a DMM, use the 200 Ω or 2 kΩ scale (or whatever scales are similar on your particular model). If the DMM has both LOW-Ω and HIGH-Ω functions, use HIGH-Ω. The LOW-Ω function causes the test voltage to be about 0.2 V, which is not sufficient to forward-bias a silicon diode.

3   Clearly, you must know which ohmmeter terminal is negative and which is positive. The great majority of ohmmeters are − on the COM terminal, **+** on the Ω terminal. But there are a few units that are the other way around.

It is easy to see that the ohmmeter test can also be used to identify the terminals of an unmarked diode. As Fig. 2-11 illustrates, manufacturers usually mark the cathode and anode leads in some way. But if the identifying mark has somehow been lost, the two-step ohmmeter test will reveal which lead is cathode and which is anode.

If it is difficult to remove a diode from its circuit to isolate it, then the ohmmeter test may not be possible. In that case, a dc voltmeter measurement may be able to prove whether or not a diode is faulty. If your study of the circuit schematic makes you think that a diode is forward-biased, then you expect a dc voltmeter to measure a value between 0.6 and 0.8 V, − on the cathode lead. If an actual voltmeter measurement agrees with your expectation, then you have proved that the diode is functioning properly, at least in the forward direction. This result is shown in Fig. 2-13(A).

On the other hand, if the measured value of forward dc voltage is much greater than 0.8 V, − on the cathode, that suggests that the diode is open-circuited. Figure 2-13(B) shows this condition.

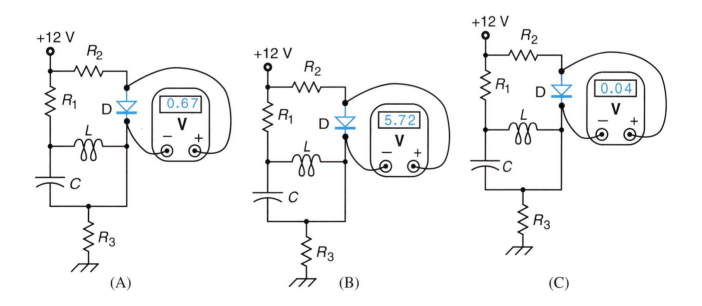

Figure 2-13 Using a voltmeter to test diode operation in-circuit. (A) Normal diode behavior. (B) Voltage measurement too high, suggesting diode is open. (C) Voltage measurement too low, suggesting diode is shorted.

If the measured forward voltage is near zero, as in Fig. 2-13(C), that suggests that the diode is short-circuited. Of course, a short circuit condition may not be the fault of the diode itself. There may be some other parallel path that is shorted between the diode's terminals.

SELF-CHECK FOR SECTIONS 2-3 AND 2-4

11. The forward conduction voltage value for a silicon diode is approximately _____ V.
12. A characteristic graph of an electrical device is a graph of the device's _____ on the *y* axis versus its _____ on the *x* axis.
13. For the diode with the characteristic graph of Fig. 2-10, find the approximate value of V_F when the forward current is 800 mA.
14. (T-F) A real diode, when reverse-biased, exactly duplicates the action of an open switch.
15. A particular diode is tested with a VOM ohmmeter on the $R \times 100\ \Omega$ scale. It measures 240 Ω in one direction and almost infinity in the other direction. Is the diode good, or bad?
16. Another diode is tested as in Question 15. It measures 700 Ω in one direction and about 2 kΩ in the other direction. Is it good, or bad?
17. Another diode is tested as in Question 15. It measures 0 Ω in both directions. Is it good, or bad? Can you make a further statement about the diode?

FORMULAS

For a forward-biased practical-model diode:

$$V_{\text{diode}} = 0.7\ \text{V}$$

$$V_R = V_S - 0.7\ \text{V}$$

For a reverse-biased practical-model diode:

$$V_{\text{diode}} = V_S$$

$$V_R = 0\ \text{V}$$

SUMMARY OF IDEAS

● A junction diode is a one-way conductor. It conducts only when its applied voltage is **−** on the cathode and **+** on the anode.
● You can think of an ideal diode as a switch. It conducts like a closed switch when it is forward-biased (**−** on the cathode). It blocks like an open switch when it is reverse-biased (**−** on the anode).
● A real diode differs from our ideal model in this way: It requires a forward voltage of at least 0.6 V in order to start conducting.
● In a diode-resistor series circuit, we use Ohm's law and Kirchhoff's voltage law to predict the current and the individual component voltages.

● As the forward current gets larger through a real diode, the forward voltage increases slightly above 0.6 V. We usually use the value 0.7 V.

● A diode can be tested with an ohmmeter. A good diode will measure low resistance when forward-biased by the ohmmeter. It will measure high resistance when reverse-biased.

CHAPTER QUESTIONS AND PROBLEMS

1. A diode conducts if its schematic arrow points toward the source's _____ terminal. It blocks if the arrow points toward the _____ terminal.

2. An ideal conducting diode is like a(n) _____ switch. A blocking diode is like a(n) _____ switch.

3. Figure 2-14 contains an ideal diode.
   a) Is the diode conducting, or blocking?
   b) Find V_{diode} (ideally). State the polarity.
   c) Find V_R. State the polarity.
   d) Find the current I. State the direction.

4. Figure 2-15 contains an ideal diode.
   a) Is the diode conducting, or blocking?
   b) Find V_{diode} (ideally). State the polarity.
   c) Find V_R. State the polarity.
   d) Find the current I. State the direction.

5. Explain the main difference between a real practical diode and the ideal model of a diode.

6. Figure 2-16 contains a real diode.
   a) Is the diode conducting, or blocking?
   b) Find V_{diode}. State the polarity.
   c) Find V_R. State the polarity.
   d) Find the current I. State the direction.

7. In Fig. 2-16, describe what would happen if the resistor were replaced with a short circuit.

8. (T-F) A real diode that is reverse-biased carries absolutely zero current.

9. (T-F) A real diode can withstand any amount of reverse voltage, no matter how large.

10. Referring to Question 8, give a reasonable estimate of the amount of reverse current through a reverse-biased diode, like the one in Fig. 2-15.

11. Referring to Question 9, define the meaning of peak reverse voltage.

12. The series circuit of Fig. 2-17 contains two real diodes.
   a) Are both diodes conducting? Explain.
   b) Find V_{D1}. State the polarity.
   c) Find V_{D2}. State the polarity.
   d) Find V_R.
   e) Find the current I.

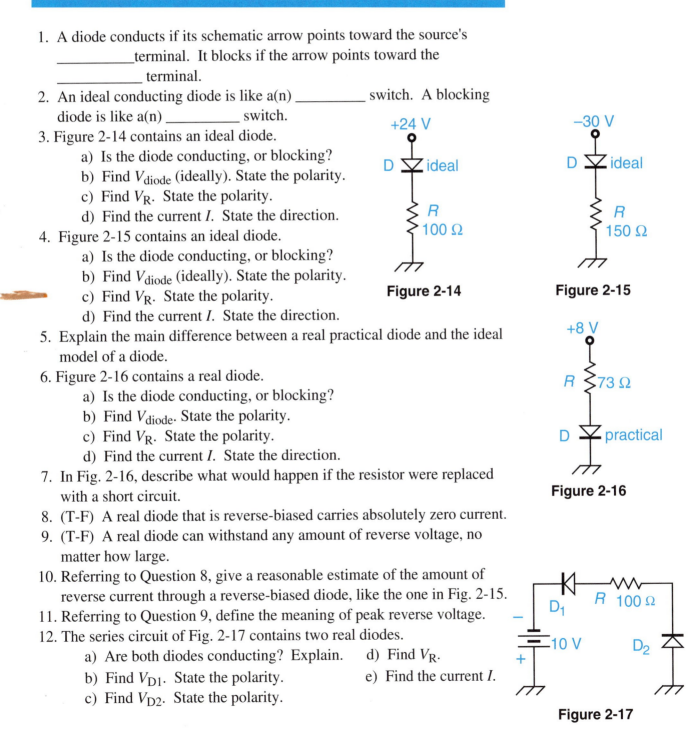

Figure 2-14

Figure 2-15

Figure 2-16

Figure 2-17

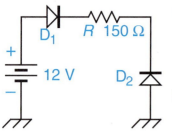

Figure 2-18

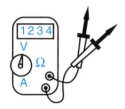

13. Repeat Problem 12 for the series circuit of Fig. 2-18.
    a) Are both diodes conducting? Explain   c) Find V_R.
    b) Find V_{D2}. State the polarity.   d) Find the current I.

14. In a complete diode characteristic graph, the curve starts up from the horizontal voltage axis at about 0.6 V. We call this value the _____ _____ voltage.

15. (T-F) In a complete diode characteristic graph, large forward current causes the forward voltage to rise somewhat higher than 0.6 V.

16. When tested with an ohmmeter, a good diode will measure _____ resistance in one direction and _____ resistance in the other direction.

17. A certain diode is tested with a VOM ohmmeter on the $R \times 100$ scale. It measures about 1 MΩ in one direction and about 2 MΩ in the other direction. Is the diode good or bad?

18. A certain diode is tested with a VOM ohmmeter on the $R \times 100$ scale. It measures about 100 Ω in one direction and about 50 Ω in the other direction. Is the diode good or bad?

19. On a physical diode, the stripe marks the _____ lead. (anode or cathode)

20. A certain diode is tested in-circuit with a DMM dc voltmeter. It is expected that the diode should be conducting. The voltmeter measures 4.8 V, − on the striped end. What is the most likely explanation for this?

CHAPTER 3

DIODE APPLICATIONS

One of the commercial uses of space is in gravity-free growth of protein crystals for the development of new drugs. This photo shows a crystal-related experimental apparatus aboard *Columbia*. Page 3 of the color section shows electron-microscope photographs of virtually perfect space-grown crystals.

Courtesy of NASA

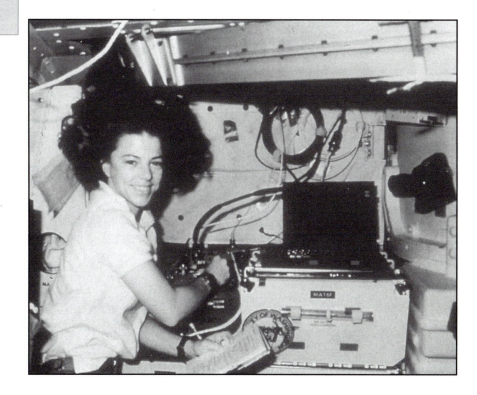

OUTLINE

NEW TERMS TO WATCH FOR

pulsating dc	bridge rectifier	diode clipper
rectify	averaging dc voltmeter	overvoltage spike
half-wave rectifier	filter capacitor	voltage doubler
full-wave rectifier	ripple	dual-polarity supply
center-tapped-transformer rectifier		

Most applications of diodes are based on their ability to conduct in one direction and block in the other direction. The most important use of this ability is for converting ac to dc.

After studying this chapter, you should be able to:

1. Explain the meaning of ac-to-dc rectification
2. Describe the operation of a half-wave rectifier circuit, during each of the half cycles.
3. Sketch the output waveform of a positive half-wave rectifier. Repeat for a negative rectifier.
4. Describe in detail the operation of a center-tapped-transformer full-wave rectifier, during each of the half cycles.
5. Troubleshoot a center-tapped full-wave rectifier to identify a diode fault, a transformer fault, or a shorted load.
6. Describe in detail the operation of a full-wave bridge rectifier.
7. Troubleshoot a full-wave bridge rectifier to identify the exact cause of circuit malfunction (diode(s), transformer, or load).
8. Explain the operation of a filter capacitor in a rectifier circuit.
9. Sketch the output waveform of a capacitor-filtered half-wave or full-wave rectifier circuit.
10. Explain the operation of a voltage-doubling circuit.
11. Explain the operation and application of a diode clipping circuit.

3-1 THE RECTIFIER IDEA

Because a diode can block reverse current, it is capable of converting ac into **pulsating dc**. This process is called **rectifying**. It is illustrated in Fig. 3-1

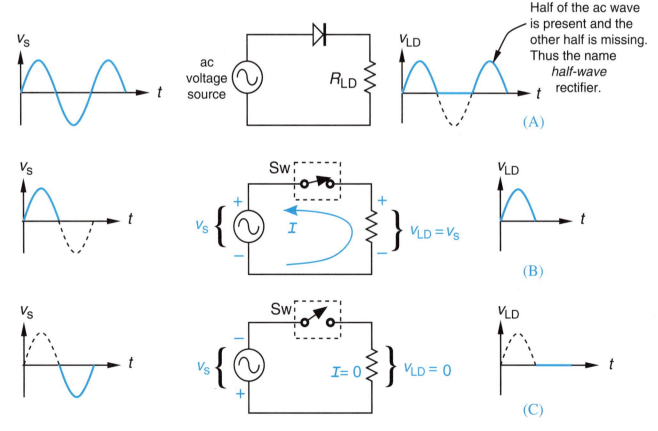

Figure 3-1 An ideal diode half-wave rectifier. (A) Schematic diagram with 1-1/2 cycles of the source (input) and load (output) waveforms. (B) During the positive half cycle the diode acts like a closed switch.
(C) During the negative half cycle the diode acts like an open switch.

Figure 3-1(A) is a schematic diagram for a **half-wave rectifier** circuit. It is simply a diode in series with the load, both driven by an ac source. The source voltage is sine-wave ac. The output load voltage contains the positive half cycles of the sine wave, but it is missing the negative half cycles. This can be understood by studying Figs. 3-1(B) and (C).

Figure 3-1(B) shows the source during its positive half cycle, which is **+** on top, **−** on bottom. Because the diode arrow points toward the negative source terminal, the diode conducts. Using the ideal diode model, we can think of it as a closed switch. Thus, in Fig. 3-1(B), there is current flowing in the circuit and $v_{LD} = v_S$.

The source is shown in its negative half cycle in Fig. 3-1(C). During this time interval the diode arrow is pointing toward the more positive source terminal. Therefore the diode blocks and we can think of it as an open switch. With the switch open, no current flows and $v_{LD} = 0$.

The load voltage waveform shown in Fig. 3-1(A) is called pulsating dc. It qualifies as dc because it never reverses polarity like ac. But since it is not smooth dc, like we get from a chemical battery cell, it cannot be referred to as simply dc.

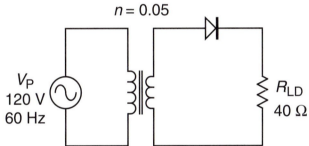

**Figure 3-2
Half-wave rectifying
circuit for Example 3-1.**

Remember the definition of transformer turns ratio n. It is N_S/N_P. Ideally, $n = V_S/V_P$.

EXAMPLE 3-1

Figure 3-2 shows a half-wave rectifier driven by a transformer. The transformer's primary voltage and its turns ratio are given, and so is the load resistance R_{LD}. Assume the practical model of the diode, and:
a) Describe and sketch the load voltage waveform, v_{LD}.
b) Describe and sketch the current waveform, i_{LD}.

SOLUTION

a) From the ideal transformer voltage law,

$$V_{S(rms)} = n\, V_P$$
$$= (0.05)(120\ \text{V}) = 6.0\ \text{V}$$

Therefore the peak value of the secondary voltage waveform is

$$V_{S(pk)} = 1.414\, V_{S(rms)}$$
$$= 1.414\,(6.0\ \text{V}) \approx 8.5\ \text{V}$$

During the positive half cycle, the diode drops 0.7 V, so by Kirchhoff's voltage law, the peak load voltage is

$$V_{ld(pk)} = V_{S(pk)} - V_F$$
$$= 8.5\ \text{V} - 0.7\ \text{V} = 7.8\ \text{V}$$

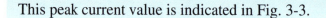

Therefore the V_{LD} waveform is a 7.8-V tall positive pulsation that begins and ends at the instants when $v_S \approx 0.7$ V. This is sketched in Fig. 3-3, showing the time synchronization with V_S.

b) The circuit current i is determined by Ohm's law applied to R_{LD}.

$$i_{(pk)} = \frac{v_{LD(pk)}}{R_{LD}}$$

$$= \frac{7.8 \text{ V}}{40 \text{ }\Omega} = 195 \text{ mA}$$

This peak current value is indicated in Fig. 3-3.

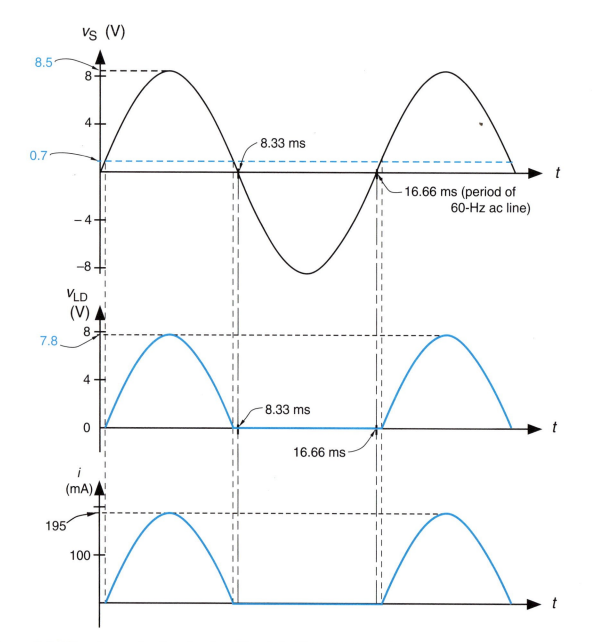

Figure 3-3 Waveforms for the circuit of Fig. 3-2. The v_{LD} pulsations are slightly shorter vertically than the v_S pulsations, due to diode forward voltage drop, V_F. The v_{LD} pulsations are also slightly narrower, in time, than the v_S pulsations, also due to V_F. The i pulsations just follow the v_{LD} pulsations according to Ohm's law.

In this example, the diode drop $V_F \approx 0.7$ V is significant. That is because the transformer secondary voltage is so low, only 8.5 V peak. The value 0.7 V out of 8.5 V is about 8%, which is not negligible. In cases where the ac waveform has a larger peak value, V_F may become negligible. For example, for $V_{S(rms)} = 24$ V, which is a common value, we get

$$V_{S(pk)} = 1.414\,(24\text{ V}) = 33.9\text{ V}.$$

Then V_F represents a much smaller percentage of the overall voltage wave, namely

$$\frac{0.7\text{ V}}{33.9\text{ V}} \approx 2\%$$

which is just about negligible.

Negative Half-Wave Rectifier

By reversing the diode, we can produce negative-pulsating dc. The positive half cycles are eliminated. This assumes that we are regarding voltage that is negative on the top terminal as a negative voltage. A negative half-wave circuit and its waveforms are shown in Fig. 3-4.

Diode points in opposite direction from Figs. 3-1 and 3-2.

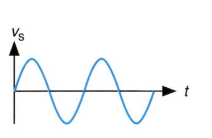 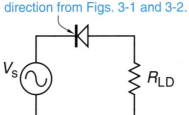 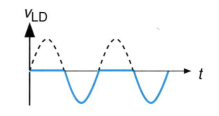

Figure 3-4 Negative half-wave rectifier.

EXAMPLE 3-2

Suppose the circuit of Fig. 3-4 has $V_s = 68$ V rms, 400 Hz, and $R_{LD} = 200\ \Omega$. Sketch the approximate load-voltage and current waveforms.

SOLUTION

The peak value of source voltage is

$$V_{s(pk)} = 1.414\,V_{s\,(rms)}$$
$$= 1.414\,(68\text{ V}) = 96.2\text{ V}$$

The diode drop of 0.7 V is less than 1% of 96.2 V, so it can be neglected.

The waveforms have a period given by

$$T = \frac{1}{f}$$
$$= \frac{1}{400\text{ Hz}} = 2.5\text{ ms}$$

The load voltage waveform, rounded to 96 V, is sketched in Fig. 3-5. For the current waveform, neglect V_F and apply Ohm's law as usual.

$$I_{pk} = \frac{V_{pk}}{R_{LD}} = \frac{96.2\text{ V}}{200\ \Omega} = 481\text{ mA}$$

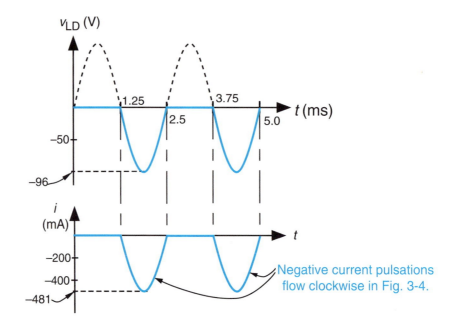

**Figure 3-5
Showing the
solution waveforms
for Example 3-2.**

SELF-CHECK FOR SECTION 3-1

1. How many diodes are there in a half-wave rectifier?
2. Sketch the load voltage waveform for a positive half-wave rectifier driven by a source with $V_{pk} = 45$ V. Assume that the diode is ideal—ignore V_F. Do not scale the time axis.

MEASURING MECHANICAL STRESS

When material is subjected to mechanical stress, it undergoes minute temperature changes. Such temperature changes have an effect on the low-level infrared radiation that is emitted by all objects. By detecting variations in infrared patterns, engineers can find out exactly where mechanical stress is concentrated. This is a great aid to the design effort.

The photo shows an infrared stress scanner and its display unit. Here, it is being set up to measure stress patterns in a truck axle that will be subjected to realistic long-haul driving conditions in a laboratory setting. This photo and a typical color video display are shown on page 7 of the color section.

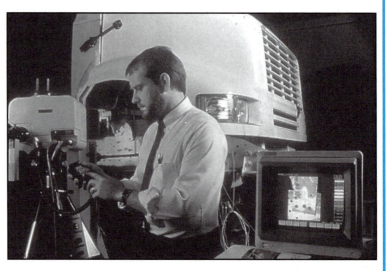

Courtesy of Navistar International Corp.

3. Repeat Problem 2, except that $V_{pk} = 6.2$ V. Assume the practical model of the diode.
4. In Problem 3, if the ac source operates at $f = 60$ Hz, do the load pulsations have a time duration slightly greater than 8.33 msec, or slightly less than 8.33 msec? Explain.
5. Why do we call a pulsating waveform "dc", even though the voltage is not constant?
6. (T-F) The peak value of the load voltage pulsation is always less than the peak value of the ac source that drives a real rectifier.

3-2 FULL-WAVE RECTIFIERS

As you would expect from the name, a **full-wave** rectifier causes both half cycles to be delivered to the load. The tricky part is to make the negative half cycle appear positive. There are two different methods of accomplishing full-wave rectification. The **center-tapped-transformer** method uses two diodes and requires a transformer with a center-tapped secondary winding. The **bridge** rectifier method uses four diodes and doesn't require a transformer at all.

Center-Tapped Transformer Full-Wave Rectifier

The center-tapped method is shown in Fig. 3-6. As the schematic diagram in Fig. 3-6(A) shows, the secondary winding's center-tap connects to the bottom of the load, which is considered circuit ground. The cathodes of the two diodes are tied together, and are connected to the top of the load. This circuit works as follows:

During the positive half cycle of secondary voltage V_S, diode D_1 conducts and applies the top half of V_S to the load. Diode D_2 blocks the bottom half of V_S.

Figure 3-6(B) illustrates this fact.

During the negative half cycle of V_S, diode D_2 conducts and applies the bottom half of V_S to the load, with a positive polarity. Diode D_1 blocks the top half of V_S.

To understand these actions, let us carefully trace out the current paths in Fig. 3-6.

Secondary voltage V_S is in its positive half cycle in Fig. 3-6(B). The top half of V_S, called v_{top}, is **+** at the very top terminal of the winding, **−** at the grounded center tap. This voltage is applied to the series combination of D_1 and R_{LD}. Since the cathode arrow of diode D_1 is pointing toward the more negative terminal of v_{top}, diode D_1 conducts. Therefore D_1 has been replaced with a closed switch in Fig. 3-6(B), taking an ideal view of the diode. It is easy to see

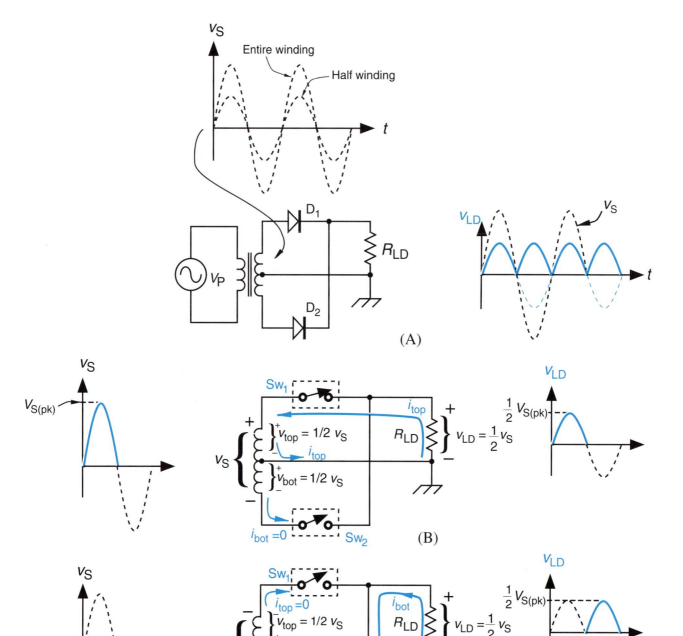

Figure 3-6 (A) Schematic diagram with waveforms. The V_{LD} waveform is only half the height of the V_S waveform, but every pulsation is positive. (B) During the positive half cycle of V_S, diode D_1 turns into closed switch SW_1, while diode D_2 turns into open switch SW_2. Load current and voltage are positive. (C) During the negative half cycle of V_S, diode D_2 turns into closed switch SW_2, while diode D_1 turns into open switch SW_1. Load current and voltage are still positive.

from that drawing that the load is connected directly across the top half of the winding, with voltage equal to v_{top}. Current flows from the − polarity center tap into the bottom of R_{LD}, through R_{LD} and back to the winding's + top terminal. Therefore a positive voltage half cycle is delivered to R_{LD}, which is half as large as the total secondary voltage V_S.

During the positive half cycle of V_S, the voltage on the bottom half of the secondary winding, v_{bot}, has a polarity which is + at the grounded center tap, and − at the very bottom terminal. This idea may be difficult to accept. You may think "Hold it, when we looked at the top half of the secondary winding, the center-tap terminal was −. How can the center tap become + when we look at the bottom half of the secondary winding?"

Here is the answer to that question: When we look at the bottom half of the winding, the center tap is positive *relative to* the very bottom terminal. This does not violate the idea that the center tap is negative *relative to* the very top terminal. In fact, when you think about it carefully, you will agree that the relative polarities must be this way if the two individual voltages, v_{top} and v_{bot}, are to form a series-*aiding* pair. And of course they must form a series-aiding pair in order for the overall secondary voltage to be twice as large as either half by itself.

If you can accept the idea that v_{bot} is instantaneously + on the center tap, − on the very bottom, then look at the circuit formed in Figs. 3-6(A) and (B) by this combination: the bottom half of the winding, diode D_2, and R_{LD}. The cathode of D_2 does not point toward the more negative terminal, namely the very bottom terminal. Instead, it points toward the more positive terminal. It points through R_{LD} toward the center tap. Therefore D_2 is reverse-biased. It is shown as an open switch in Fig. 3-6(B).

All this reasoning is summarized by the first **Get This** on page 41.

Now apply the same reasoning process for the negative half cycle of V_S. The secondary winding polarities are reversed, as shown in Fig. 3-6(C). If you trace this circuit carefully, you will conclude that the bottom half of the secondary winding, with voltage v_{bot}, is connected to R_{LD} through conducting diode D_2. And v_{top} is disconnected from R_{LD} by blocking diode D_1. During this action, the current through the load remains positive, just as it was when D_1 was conducting. This current-flow route is drawn in Fig. 3-6(C). Therefore the load-voltage pulsation also remains positive, as indicated by the waveforms in Figs. 3-6(A) and (C).

All this negative-half cycle activity is summarized by the second **Get This** stated on page 41.

This technician is preparing an aircraft engine for an in-flight test, installed on a flying testbed.

Courtesy of GE Aircraft Engines

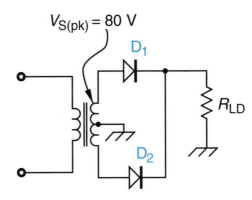

Figure 3-7
Full-wave rectifier for
Example 3-3.

$V_{S(pk)} = 80$ V

D_1

R_{LD}

D_2

EXAMPLE 3-3

The center-tapped full-wave rectifier of Fig. 3-7 has a peak secondary voltage of $V_{S(pk)} = 80$ V, with $f = 60$ Hz.
a) Sketch the waveform of v_{LD}.
b) V_{LD} is not ac, but we can still assign a frequency value to it, because it varies periodically. What is its frequency?

SOLUTION

a) The direct wire connection between the center tap and the bottom of the load is not drawn in this diagram, but the two common ground symbols accomplish the same thing. As before, the load sees one half of the secondary winding at any instant, so

$$V_{ld(pk)} = \frac{1}{2} V_{S(pk)}$$

$$= \frac{1}{2} (80 \text{ V}) = 40 \text{ V}$$

The time duration of a single pulsation is one half the ac cycle period (ignoring the 0.7-V forward voltage drops of the diodes). So with $f = 60$ Hz,

$$t_{pulsation} = \tfrac{1}{2} T = \tfrac{1}{2} (16.66 \text{ ms})$$

$$= 8.33 \text{ ms}$$

The V_{LD} waveform is sketched in Fig. 3-8. Each pulsation takes 8.33 ms.
b) This pulsation waveform repeats its entire sequence of values in just 8.33 ms, so it has period $T = 8.33$ ms. Its frequency is therefore given by

$$f = \frac{1}{T} = \frac{1}{8.33 \text{ ms}} = \mathbf{120 \text{ Hz}}$$

If a frequency counter were connected across R_{LD}, it would measure 120 Hz.

Figure 3-8
A full-wave rectifier
produces a load
waveform with a
frequency twice the
source frequency.

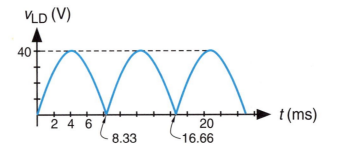

v_{LD} (V)

40

2 4 6

8.33

16.66

20

t (ms)

EXAMPLE 3-4

Suppose the circuit of Fig. 3-7 had both diodes reversed, as shown in Fig. 3-9(A). Sketch the load voltage waveform.

SOLUTION

In this full-wave rectifier the anodes are tied together, not the cathodes. Therefore D_1 conducts during the negative half cycle of V_S, applying a voltage pulsation to R_{LD} that is − on top. Diode D_2 conducts during the positive half cycle of V_S, applying a voltage pulsation to R_{LD} that is also − on top. The load voltage waveform sketched in Fig. 3-9(B) shows negative 40-V pulsations.

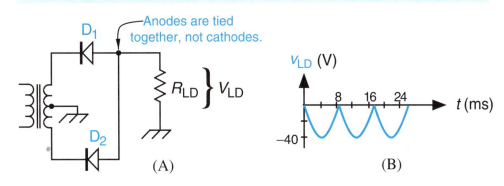

(A)

(B)

Figure 3-9
Negative full-wave rectifier.
(A) Schematic.
(B) Load waveform.

Troubleshooting Center-Tapped Full-Wave Rectifiers

In a transformer-based full-wave or half-wave rectifier, the primary circuit of the transformer almost certainly contains a protective fuse, as shown in Fig. 3-10.

Transformer Faults. If either transformer winding fails in the open condition, V_{LD} becomes 0 V but the primary fuse does not blow. If you encounter $V_{LD} = 0$ V with the fuse intact, run the usual ohmmeter check or high-voltage test on the transformer windings in an isolated state. You are looking for a near-infinite resistance measurement as an indication of a fault.

If either transformer winding fails dead shorted, the primary current will certainly blow the fuse. If either transformer winding fails partially shorted, the primary current will increase to a value that will probably blow the fuse. So if the fuse blows, it is worthwhile to isolate the windings and test them by ohmmeter or high-voltage test setup. You are looking for very low winding resistance with a dead short, or lower-than-normal resistance in the case of a partial short.

Another possible cause of a blown primary fuse is a short between the primary winding and the secondary winding, even though there may not necessarily be any shorted turns in either individual winding itself. If both the primary and secondary circuits are referenced to the earth, such a primary-to-secondary short can cause a large circulating earth-ground current that blows the fuse.

Also, if the transformer brackets or case are connected to the earth through the mounting screws, or if the transformer case is deliberately earth-

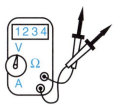

grounded by a bare grounding wire, a short circuit from a ground-referenced primary winding to the case can produce a fuse-blowing ground current.

Therefore, if the fuse blows and neither individual winding tests as shorted, run the ohmmeter or high-voltage test between a primary lead and a secondary lead, and then between a primary lead and the case or mounting screws. Any resistance measurement that is not near infinity is an indication of a fault.

Small transformers are almost never repaired. They are just replaced by an identical unit, or a unit with similar specifications.

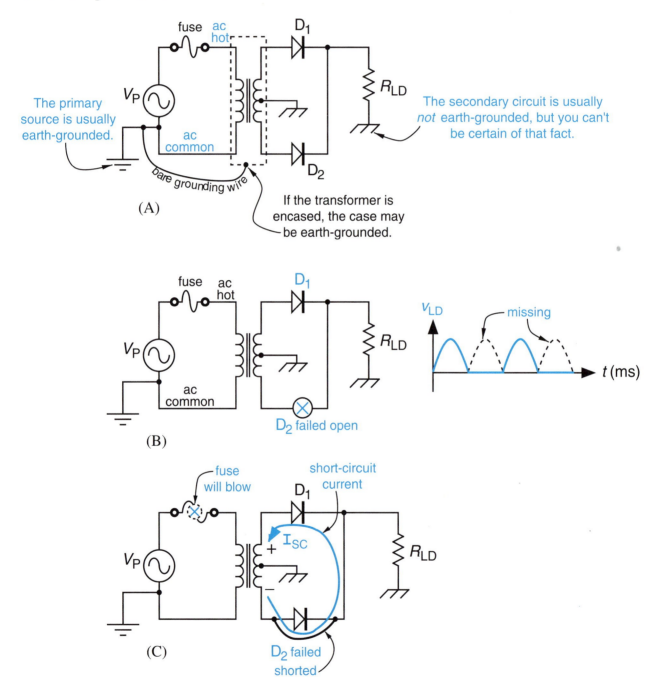

Figure 3-10 (A) Schematic of a standard center-tapped-transformer full-wave rectifier circuit, including grounding details. Use this diagram to understand the effects of winding opens and shorts, and circuit-to-circuit shorts. (B) One open diode causes half the pulsations to be missing. (C) A shorted diode causes an overcurrent in the primary circuit.

Diode Faults. Diodes can fail either open or shorted. A failed-open diode is represented in Fig. 3-10(B). No dramatic malfunction occurs, but the load waveform will be missing half of its pulsations. It will look like a half-wave rectified output. This will be clearly visible on an oscilloscope. With an **averaging dc voltmeter**, like most VOMs and DMMs, the load voltage will measure a value about half as large as it ought to.

A failed-shorted diode will create a short circuit across the entire secondary winding during one of the half cycles. This is illustrated in Fig. 3-10(C) for the case where D_2 is shorted. The excessive secondary current will cause the primary fuse to blow.

You can track down exactly which diode has shorted by isolating and testing each one with an ohmmeter.

Of course, a shorted load in a rectifier circuit will cause the primary fuse to blow by standard transformer current action. This is just like any transformer circuit.

5 ✓

Full-Wave Bridge Rectifiers

A full-wave **bridge rectifier** consists of four diodes arranged as shown in Fig. 3-11. No transformer is required for a bridge rectifier, so we do not show one in Fig. 3-11. Assuming ideal diodes, this circuit works as follows:

However, many real-life bridge rectifiers *do* use a transformer to obtain the desired voltage level and/or to provide electrical isolation.

> During the positive half cycle of source voltage V_s, diodes D_2 and D_1 conduct, applying all of V_s to the load with a positive polarity. Diodes D_3 and D_4 block.

Figure 3-11(B) illustrates this action.

> During the negative half cycle of source voltage V_s, diodes D_4 and D_3 conduct, applying all of V_s to the load, again with a positive polarity. Diodes D_1 and D_2 block.

Figure 3-11(C) illustrates this action.

Get This

Get This

Let us look carefully at the current flow paths in Fig. 3-11 to understand these actions. During the positive half cycle of V_s, current leaves the negative bottom terminal of the source and flows to the bottom corner of the diode bridge, where diode D_2 joins D_3. The current cannot pass through D_3 in the direction of the schematic arrow, so it all passes through diode D_2, against the arrow. This is shown in Fig. 3-11(B). When it arrives at the junction of diodes D_2 and D_4, it cannot pass through D_4, with the schematic arrow. Therefore all the current flows around the loop through the load from bottom to top. This makes v_{LD} − on bottom, + on top, as Fig. 3-11(B) indicates.

When the current arrives at the junction of D_1 and D_3, it takes the D_1 path back to the positive top terminal of the source. This completes the circuit. There is no tendency for current to pass through D_3 against the schematic arrow, because that would be taking it back to the − terminal of the source. Current always flows toward the + terminal of the source.

The bridge rectifier's behavior during the positive half cycle is summarized in the top row of Table 3-1.

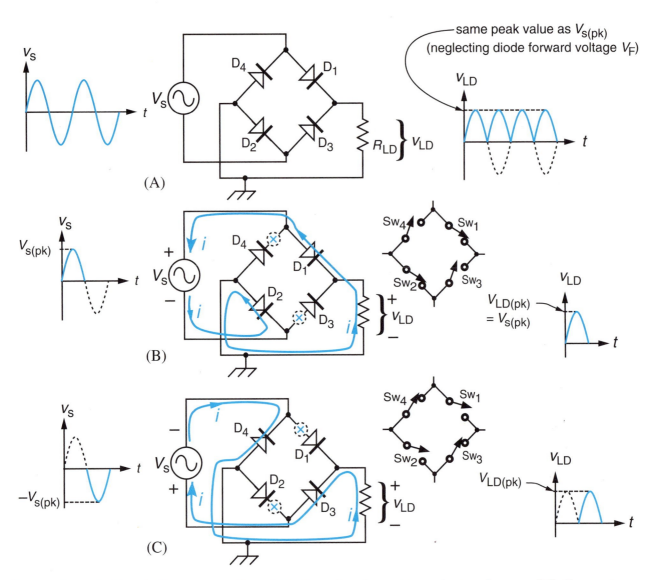

Figure 3-11 Bridge full-wave rectifier. (A) Circuit schematic and waveforms. (B) Current path during positive half cycle of the source. The equivalent switch circuit is also shown. (C) Current path, and equivalent switch circuit, during the source's negative half cycle.

During the source's negative half cycle, the current flow path is the one shown in Fig. 3-11(C). That diagram makes it clear that now diodes D_4 and D_3 conduct, and diodes D_1 and D_2 block. The main feature is that the load's current direction and voltage polarity are the same as they were during the positive half cycle of the source. Trace the circuit carefully to satisfy yourself that these statements are true. The bottom row of Table 3-1 summarizes the bridge rectifier's behavior during the negative half cycle of the source.

⑥ ✔

Table 3-1 Summarizing the performance of a full-wave bridge rectifier.

	Diodes Conducting	Diodes Blocking	Load Polarity
Positive half cycle	D1, D2	D3, D4	+ on top – on bottom
Negative half cycle	D3, D4	D1, D2	+ on top – on bottom

Troubleshooting Full-Wave Bridge Rectifiers

A fused bridge-rectifying circuit is shown in Fig. 3-12. Part (A) shows the situation if a diode fails in the open condition. One of the load's half cycle pulsations disappears. Here the failed diode is D_1. The load result would be exactly the same if D_2 failed open. If either D_3 or D_4 failed open, it would be the other half cycle pulsation that would disappear. An oscilloscope could reveal the missing pulsations clearly. To track down which individual diode failed, you would use the standard technique of isolating and testing by ohmmeter.

If any diode fails shorted, an overcurrent condition will occur during one of the source's half cycles. This will cause the fuse to blow. The situation is pictured in Fig. 3-12(B), for a failure of diode D_1. The ohmmeter test would reveal which particular diode had failed.

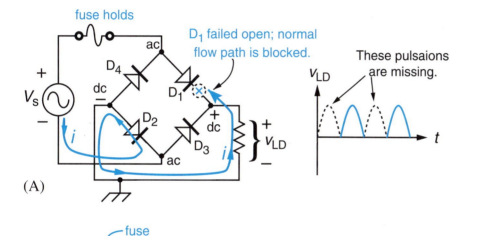

(A)

**Figure 3-12
Trouble in a bridge rectifier.**

(A) Open diode.

(B) Shorted diode.

(B)

SELF-CHECK FOR SECTION 3-2

7. In a center-tapped full-wave rectifier circuit, how many diodes are required?
8. (T-F) In a center-tapped full-wave rectifier circuit, the peak load voltage is equal to twice the peak transformer secondary voltage.
9. (T-F) When the top half of the center-tapped winding is delivering current to the load, the bottom half is carrying virtually zero current.
10. A transformer used in a full-wave rectifier has an overall secondary voltage of $V_{S(p-p)} = 100$ V, at 60 Hz. Draw the load-voltage waveform, assuming ideal diodes.

11. A transformer used in a full-wave rectifier has an overall secondary voltage of $V_{S(p-p)} = 10$ V, at 60 Hz. Using the practical model of the diodes, draw the load waveform.

12. Suppose that a center-tapped full-wave rectifier is blowing its primary fuse. Which of the following circuit faults could cause that problem?
    a) Diode failed open
    b) Diode failed shorted
    c) Secondary winding open
    d) Secondary winding partially shorted
    e) Load failed open
    f) Load shorted

13. For either a center-tapped or bridge full-wave rectifier operated by the 60-Hz ac line, the pulsation frequency is _____ Hz.

14. How many diodes are required in a bridge full-wave rectifier ?

15. At a particular instant in time, diodes D_1 and D_2 are conducting in a bridge rectifier. At that instant, diodes D_3 and D_4 are _____ .

16. Refer to Fig. 3-11(A). Show what circuit changes you would make to change the circuit into a negative rectifier (pulsations that are − on the top load terminal, + on the bottom load terminal).

This technician is conducting a pollutant-absorption test for a particular type of house plant. The plant is located in a sealed chamber. Air with a known amount of typical indoor pollutants (formaldehyde, carbon monoxide, etc.) is forced up through the plant's root system, through a layer of charcoal on top of the soil, and past the leaves. The amount of each pollutant exiting the chamber is then measured electronically, to learn how effective the plant is at removing each one.

Courtesy of NASA Environmental Research Laboratory

3-3 FILTERING (SMOOTHING) A RECTIFIER

The pulsating dc waveforms produced by a plain rectifier are good for simple applications like battery-charging. But many applications require steady dc. Pulsating dc can be smoothed by connecting a capacitor in parallel with the load, as shown in the half-wave rectifier circuit of Fig. 3-13(A).

As the positive half cycle approaches its peak, charge is accumulated on the plates of capacitor C, according to the formula

We are assuming an ideal diode, or that V_F is negligible compared to $V_{s(pk)}$.

$$Q = C \cdot V_{peak}$$ **Eq. (3-1)**

The charge accumulation is − on the bottom plate, + on the top plate, as Fig. 3-13(B) shows.

After the source voltage passes its peak, the capacitor tends to hold its voltage steady near $v_C \approx V_{peak}$. With v_C now larger than the instantaneous source voltage v_s, the diode is reverse-biased. It therefore blocks, as shown in Fig. 3-13(C). The charged capacitor is therefore connected across the load for the remaining (downhill) part of the positive half cycle. During this time, a portion of its charge flows through R_{LD} back to the top plate, as Fig. 3-13(C) shows.

This action continues through the negative half cycle, and then through most of the rising part of the next positive half cycle. The result is the smoothed v_{LD} waveform shown in Fig. 3-13(D). The voltage waveform is not perfectly constant (flat) because the capacitor must give up, or discharge, a portion of its charge in order to keep the load current flowing. The droop, or change, in the v_{LD} waveform can be expressed as

$$\Delta v_{LD} = \frac{Q_{discharged}}{C} \qquad \text{Eq. (3-2)}$$

Equation (3-2) is just a rearranged version of Eq. (3-1), the fundamental relation between charge, voltage, and capacitance.

During the rising part of the next positive half cycle of v_s, a point will be reached where the rising v_s value becomes greater than the slowly falling v_C value. At that point the diode becomes forward-biased again, which allows the capacitor to recharge to the source's peak value. Then the cycle repeats.

8 ✔

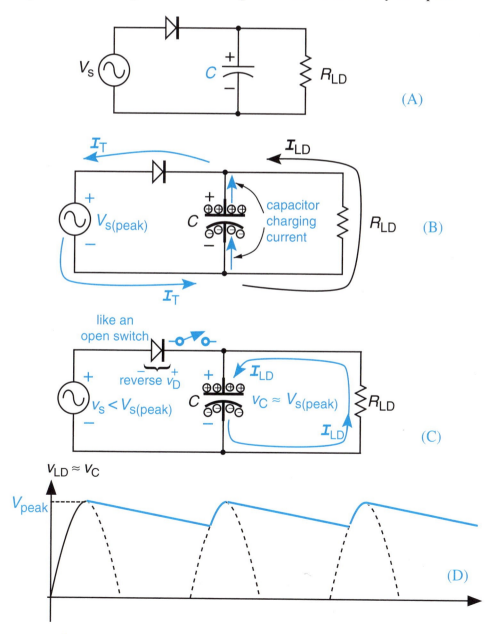

(A)

(B)

(C)

(D)

Figure 3-13

(A) The smoothing capacitor in a rectifier is often called a *filter capacitor*.

(B) Filter capacitor *C* is brought up to full charge at the moment the source voltage reaches its peak.

(C) The situation after the source voltage has passed its peak. The situation is the same when v_s is in its negative half cycle, and also for part of its next uphill climb.

(D) Waveform of v_{LD} showing the slight droop as the filter capacitor discharges.

The time intervals of the rectifier diode conducting and blocking are clearly specified in Fig. 3-14. The voltage droop sketched in Fig. 3-14 and expressed by Eq. (3-2) is called **peak-to-peak ripple**, or simply ripple.

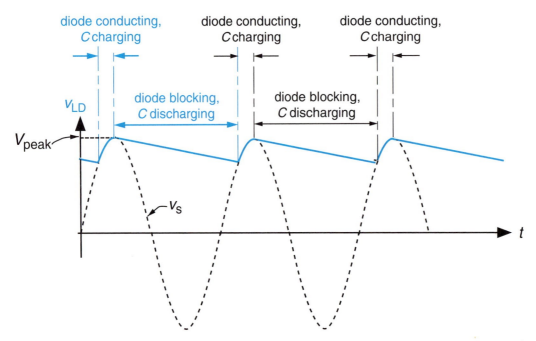

Figure 3-14 The filter capacitor quickly recharges from the moment that rising v_S meets decreasing v_{LD} until the peak instant. It slowly discharges from the peak instant until the next moment that decreasing v_{LD} meets rising v_S.

Figures 3-14 and 3-13(D) show an amount of ripple that is a large portion of the peak voltage. They are drawn this way only for emphasis. Most real-life filtered rectifiers have a much smaller amount of ripple. In fact, many rectified and filtered dc power supplies have ripple that is invisible in comparison to V_{peak}. On an oscilloscope display of the dc output voltage, for example, it may be impossible to even see the ripple. Then, to get a look at ripple, special settings of the scope's controls are required. These settings must eliminate the dc value (essentially V_{peak}) and amplify the ac part (the ripple).

The ripple in Figs. 3-13 and 3-14 could be reduced in two ways:

1 Go to a full-wave rectifier, rather than half-wave. This cuts the time between pulsations in half. Because the filter capacitor discharges for only half as much time, the ripple is also cut in half. This idea is illustrated in Fig. 3-15, for comparison to Fig. 3-13(D).

Figure 3-15 Full-wave ripple has only half the magnitude of half-wave ripple, all other things being equal. The frequency is doubled, typically 120 Hz versus 60 Hz.

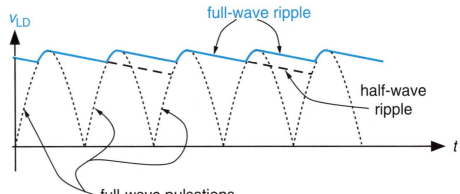

2 Install a larger value filter capacitor. It is clear from Eq. (3-2) that if C is larger, then the peak-to-peak ripple (Δv_{LD}) must be smaller.

As a practical matter, rectifier filter capacitors tend to be high-value electrolytic units. Capacitances of several thousand microfarads (several millifarads) are common.

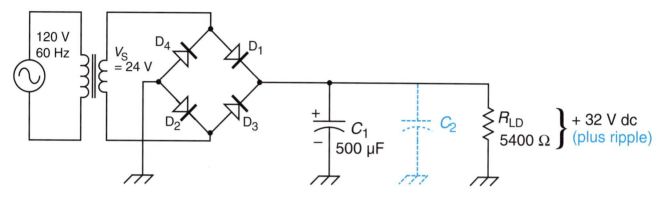

Figure 3-16 Varying the amount of filter capacitance to obtain specified ripple values.

EXAMPLE 3-5

The rectifier circuit in Fig. 3-16 contains a 500-µF filter capacitor. It has a peak-to-peak output ripple of 0.10 V (100 mV).
(a) If a second 500-µF capacitor is connected in parallel with C_1, what will be the new value of ripple?
(b) How much total filter capacitance would be necessary to reduce the output ripple to 25 mV p-p?

SOLUTION

From Eq. (3-2),

$$\text{peak-to-peak ripple} = \frac{Q_{\text{discharged}}}{C_T} \qquad \text{Eq. (3-2)}$$

In this equation, the amount of charge that flows off the plates of the filter capacitor is not affected by the choice of C_T itself. The value of $Q_{\text{discharged}}$ is set by only two things:
1. The load current, which is determined by Ohm's law; and
2. The time of discharge, which is always a little less than 8.3 ms for a full-wave rectifier at 60 Hz.
So $Q_{\text{discharged}}$ can be considered a constant number throughout this example.

(a) For $C_2 = 500$ µF, we get from the parallel capacitance formula
$$C_T = C_1 + C_2$$
$$= 500 \text{ µF} + 500 \text{ µF} = 1000 \text{ µF}$$
which is a doubling of total capacitance. From Eq. (3-2), if C_T is doubled, ripple is halved. Therefore, the new value of ripple is
$$\tfrac{1}{2}(0.10 \text{ V}) = \textbf{0.05 V} \text{ or } \textbf{50 mV (p-p)}$$

(b) Further reduction to 25 mV is another halving of the ripple. Therefore it requires another doubling of total filter capacitance, from 1000 µF to **2000 µF.**

An alternative solution is this: Going back to the initial situation, we can find $Q_{discharged}$ by rearranging Eq. (3-2), getting

$$Q_{discharged} = (V_{ripple}) \cdot C_T$$
$$= (100 \text{ mV}) \cdot (500 \text{ μF}) = 50 \times 10^{-6} \text{ C or } 50 \text{ μC}$$

So, to obtain ripple = 25 mV, we can rearrange the equation again to

$$C_T = \frac{Q_{discharged}}{ripple}$$
$$= \frac{50 \times 10^{-6} \text{ C}}{25 \times 10^{-3} \text{ V}} = 2 \times 10^{-3} \text{ F} = \mathbf{2000 \text{ μF}}$$

Troubleshooting the Filter Capacitor

Like all capacitors, the filter capacitor in a dc power supply can fail shorted or fail open. If it fails shorted, it causes an overcurrent that blows the circuit protective device. This is shown in Fig. 3-17(A) for a half-wave rectifier, and in Fig. 3-17(B) and (C) for the two kinds of full-wave rectifier.

Figure 3-17
When a filter capacitor fails shorted in any rectifier circuit, it causes an overcurrent.

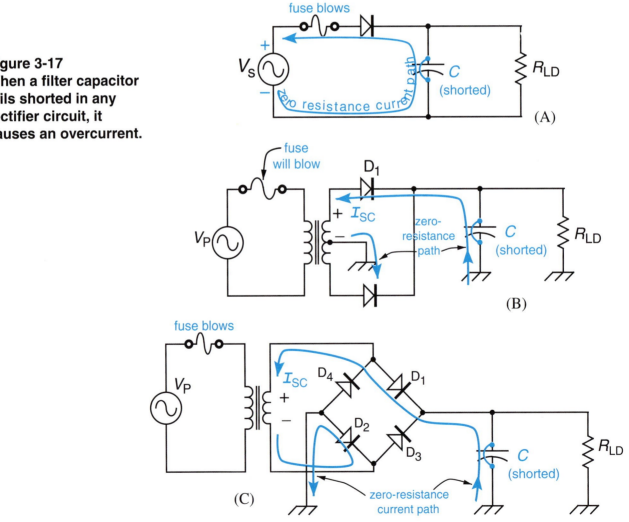

As we have seen, there are several component faults that can cause the fuse to blow in a rectified power supply. To find out if it is the filter capacitor that is causing the trouble, you must isolate the capacitor and test it with an ohmmeter or high-voltage test setup.

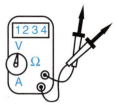

If the filter capacitor fails open, its smoothing effect is lost but the rectifier continues to function in pulsating mode. Another way of saying this is that the ripple increases drastically. An oscilloscope will show this clearly. On an averaging dc voltmeter like a VOM, the measured dc voltage will decrease greatly.

When replacing a failed electrolytic filter capacitor, you must pay especially careful attention to voltage rating. This is because electrolytics, more so than ceramic and plastic capacitors, are marketed with a choice of quite low voltage ratings, even below 10 V.

As always with electrolytic capacitors, you must be careful to make the lead polarity agree with the actual circuit polarity.

Further Filtering

A single filter capacitor by itself can produce a dc output with very low ripple. For certain applications where the ripple must be reduced to an extremely small amount, a further low-pass filter can be inserted between the main filter cap and the load. This is illustrated in Fig. 3-18. An *RC* low-pass filter is shown in Fig. 3-18(A), and an *LC* low-pass filter is used in Fig. 3-18(B).

In both circuits, the combination of the main filter capacitor and the additional low-pass filter forms a circuit schematic that resembles the Greek letter pi (∏). Therefore these combinations are sometimes called ∏-filters.

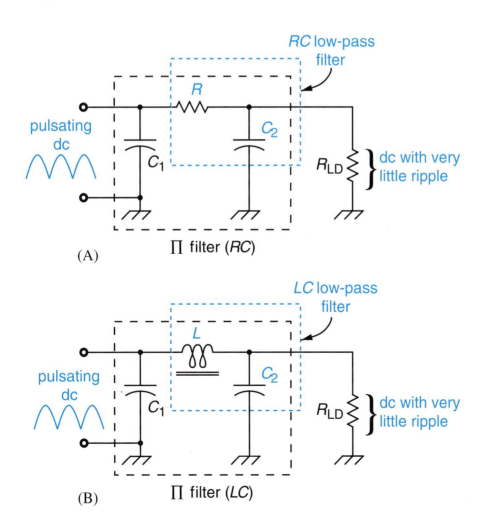

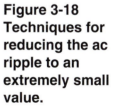

(A) ∏ filter (*RC*)

(B) ∏ filter (*LC*)

**Figure 3-18
Techniques for reducing the ac ripple to an extremely small value.**

SELF-CHECK FOR SECTION 3-3

16. To smooth out the voltage pulsations from a rectifier, we place a _____ in parallel with the load.
17. The larger the filter capacitor, the _____ the ripple tends to be (larger or smaller).
18. For a 60 Hz source, the ripple frequency from a half-wave rectifier is _____ Hz. This results in a capacitor discharge time of approximately _____ ms.
19. Repeat Question 19 for a full-wave rectifier.
20. Draw the schematic diagram of a 36-V CT transformer full-wave rectifier with negative output voltage, filtered by a 200-µF capacitor, driving a 10-kΩ load. Be sure to indicate the polarity of the electrolytic filter cap.
21. Add to the schematic diagram that you have made for problem 20, so that an *RC* low-pass filter is combined with the original filter capacitor. Let $R = 47\ \Omega$ and $C_2 = 300\ \mu F$.

3-4 OTHER DIODE APPLICATIONS — DOUBLERS AND CLIPPERS

Voltage Doublers

A **voltage doubler** is a rectifying circuit containing two diodes and two capacitors. It produces a dc output voltage that is double the value of the peak input voltage, ideally. This is unlike a standard rectifier, which produces a V_{OUT} that is the same value as the peak input voltage. The schematic diagram of a voltage doubler is shown in Fig. 3-19.

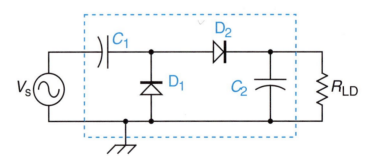

Figure 3-19 Schematic of a voltage doubler. Capacitor C_1 charges to the peak value of V_s. The circuit then puts the C_1 voltage in series with the source voltage itself. This combination voltage $[V_{s(peak)} + V_{C1}]$ is applied to C_2 and the load.

To understand the operation of the voltage doubler in Fig. 3-19, assume the diodes to be ideal. Here is the explanation of the circuit's operation.

During the negative half cycle of V_s, D_1 conducts and D_2 blocks. Study Fig. 3-19 to verify this for yourself. Figure 3-20(A) shows D_1 as a closed switch and D_2 as an open switch. At the source's peak instant, the conducting path through D_1 allows C_1 to charge to a voltage value of $V_{s(peak)}$.

Later, when V_s approaches the peak of its positive half cycle, D_1 blocks and D_2 conducts. Verify this for yourself. Figure 3-20(B) shows D_1 open, or turned off, and D_2 closed, or turned on. Note that capacitor C_2 is driven by the series combination of the source plus the C_1 voltage. At the source's peak instant, this combination charges C_2 to a value equal to $(2) \times V_{s(pk)}$.

After the source passes its positive peak, capacitor C_2 takes over the job of driving load resistance R_{LD}. This is shown in Fig. 3-20(C). As usual, C_2 suffers some amount of voltage droop, which depends on the value of load resistance R_{LD}. Capacitor C_1 likewise does not maintain its maximum value of voltage. It has some droop during the short time interval that C_2 is brought back up to full charge during the next positive half cycle. The overall output ripple depends on the amounts of these two voltage droops. Larger capacitances tend to reduce the ripple, the same as for a standard rectifying circuit.

⑩ ✓

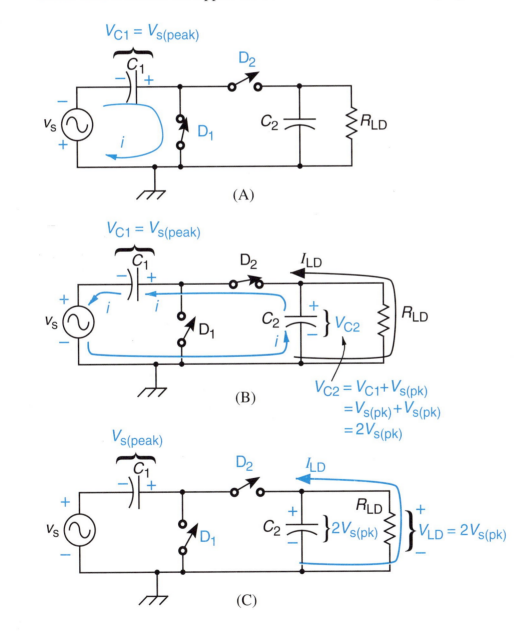

Figure 3-20
Operation of a
voltage doubler.

(A) During the
negative half cycle.

(B) Approaching
the peak of the
positive half cycle.

(C) After the peak
of the positive
half cycle.

Clippers

A diode **clipper** prevents the voltage across the load from exceeding a particular value. That particular value is often 0.7 V, the forward voltage of a practical silicon diode. A clipping example is shown in Fig. 3-21.

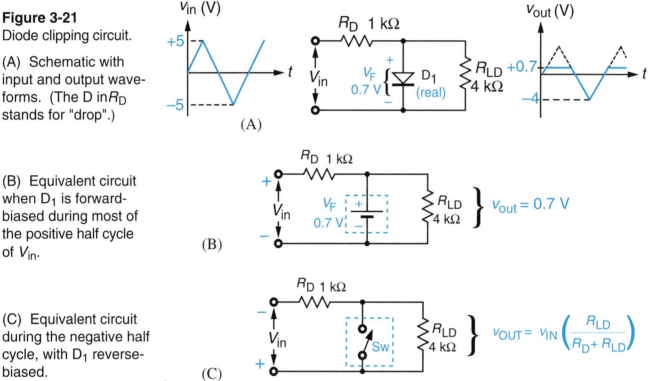

Figure 3-21
Diode clipping circuit.

(A) Schematic with input and output waveforms. (The D in R_D stands for "drop".)

(B) Equivalent circuit when D_1 is forward-biased during most of the positive half cycle of V_{in}.

(C) Equivalent circuit during the negative half cycle, with D_1 reverse-biased.

During the positive half cycle of the input triangle wave in Fig. 3-21, the diode begins conducting when its voltage reaches 0.6 V. Once it begins conducting, it will not allow its forward voltage, V_{diode}, to exceed 0.7 V (using our practical diode model). Normally, if the clipping diode were not there, the rising positive value of V_{in} would voltage-divide between R_D and R_{LD}. For instance, at the moment when v_{in} equals 3 V, the load voltage would normally be given by

$$v_{LD} = v_{in}\left(\frac{R_{LD}}{R_D + R_{LD}}\right) = 3\text{ V}\left(\frac{4\text{ k}\Omega}{1\text{ k}\Omega + 4\text{ k}\Omega}\right)$$

$$= 2.4\text{ V}$$

But with D_1 present in Fig. 3-21, its short-circuiting action in parallel with R_{LD} upsets the series relationship between the two resistors. In effect, v_{in} must voltage divide between R_D and a fixed, unvarying 0.7-V dc source. This is suggested by Fig. 3-21(B).

During the negative half cycle of the V_{in} triangle wave, diode D_1 is reverse-biased. It acts like an open switch in parallel with R_{LD}, as suggested in Fig. 3-21(C). Therefore, it has no effect on the normal voltage division between dropping resistor R_D and load resistor R_{LD}. By voltage division,

$$v_{LD(pk)} = v_{in(pk)} \left(\frac{R_{LD}}{R_D + R_{LD}} \right) = -5 \text{ V} \left(\frac{4 \text{ k}\Omega}{1 \text{ k}\Omega + 4 \text{ k}\Omega} \right)$$

$$= -5 \text{ V} \left(\frac{4}{5} \right) = -4 \text{ V}$$

This value of negative peak voltage is indicated in the output waveform of Fig. 3-21(A).

In effect, the diode has clipped off the part of the output wave that is more positive than +0.7 V.

Understanding the positive clipper of Fig. 3-21, it is easy to see that the diode circuit of Fig. 3-22 is a negative clipper. It prevents the load voltage from ever being more negative than –0.7 V.

[11] ✓

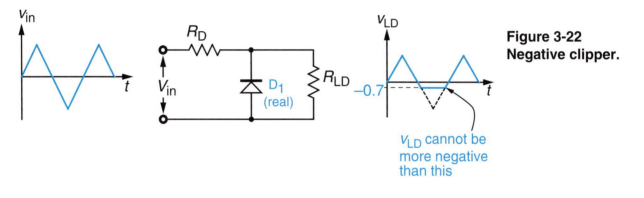

**Figure 3-22
Negative clipper.**

v_{LD} cannot be more negative than this

Clipping at Values Other Than 0.7 V.

By connecting a fixed dc source in series with a clipping diode, it is possible to clip the output waveform at any voltage value we desire.

Get This

For example, Fig. 3-23(A) shows a circuit designed to clip at +3.0 V. It works as follows:

During the negative half cycle of V_{in}, D_1 is reverse-biased. Therefore normal voltage division occurs between R_{LD} and R_D.

During the low-magnitude portions of the positive half cycle, as long as the normal R_D/R_{LD} voltage division produces a load voltage less than about 3.0 V, D_1 remains reverse-biased. The equivalent circuit is shown in Fig. 3-23(B). Normal R_D/R_{LD} voltage division is not interfered with.

When the positive value of V_{in} becomes great enough to produce v_{LD} greater than 3.0 V, the diode becomes forward-biased. This happens because 3.0 V minus an opposing voltage of 2.3 V leaves 0.7 V, enough to force D_1 into the conducting state. Once D_1 becomes forward-biased, it combines with the 2.3-V source to "short out" the load for all voltage above 3.0 V. This effect is represented in Fig. 3-23(C).

Figure 3-23

Biased clipper, designed for clipping at +3.0 V.

(A) Schematic and waveforms.

(B) Equivalent circuit when V_{in} has a small positive value.

(C) Equivalent circuit when V_{in} has a large positive value.

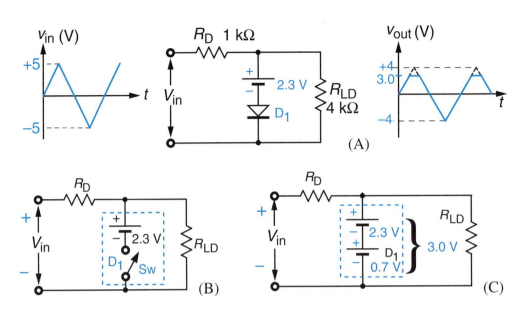

EXAMPLE 3-6

Describe the operation of the circuit in Fig. 3-24. Draw the output waveform.

SOLUTION

This biased clipper will give normal 4/5 voltage division during the positive half cycle, because D_1 will be reverse-biased. During the negative half cycle, a v_{LD} value of −6 V will forward-bias the D_1 − dc source series combination. Therefore load voltages below (more negative than) −6 V are clipped off. The output waveform is drawn in Fig. 3-25.

Figure 3-24
Clipping circuit for Example 3-6.

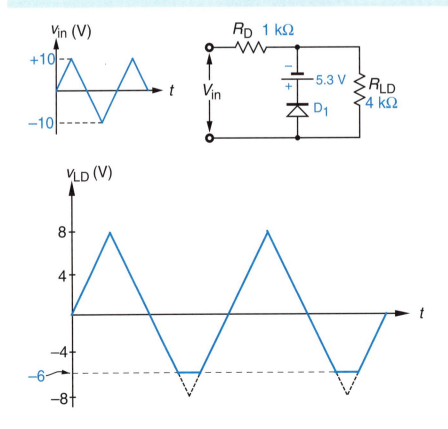

Figure 3-25
Clipped output waveform from circuit of Fig. 3-24, Example 3-6.

Clippers Used to Protect Against Transient Voltage Spikes: One common use of clipper circuits is to protect load devices against **overvoltage spikes**. An overvoltage spike is a large surge of voltage that appears for a very short time, usually as the result of a switch closing or opening in the vicinity. This situation is illustrated in Fig. 3-26.

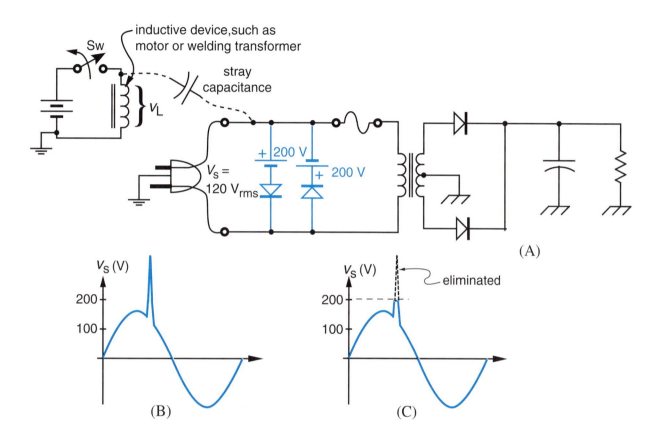

(A)

(B)   (C)

Figure 3-26 Using clippers to protect a circuit from voltage spikes. (A) Equivalent schematic situation. A real high-voltage clipping circuit is equivalent to this, but not actually built like this. (B) Waveform without clipping.   (C) With clipping.

In Fig. 3-26(A), the instantaneous value of the 120-V ac supply should never exceed the V_{pk} value of about 165 V, normally. But if a highly inductive circuit in the vicinity suddenly has its current interrupted by switch SW, the inductive load device may induce a large transient voltage v_L, in an attempt to maintain its current at a steady value. This large voltage spike, shown in Fig. 3-26(B), could be coupled by stray capacitance onto the hot ac line of the rectifying circuit. Or, it could travel directly down the ac power wires. From there, it could be transformed into the secondary circuit, perhaps damaging the rectifier diodes, filter capacitor, or load.

But with the 200-V clippers installed in the primary circuit of Fig. 3-26(A), the voltage spike on the hot line is limited to an instantaneous value of 200 V, as shown in Fig. 3-26(C). This limitation provides protection for the entire rectifier circuitry.

SELF-CHECK FOR SECTION 3-4

22. In Fig. 3-19, suppose $V_{s(rms)} = 24$ V. Find the value of V_{LD}, assuming that the diodes are ideal, and the capacitors are so large that ripple is negligible.
23. In a voltage doubler, what situation tends to make the ripple problem more severe, R_{LD} having a low value or R_{LD} having a high value? Why?
24. Sketch the V_{out} waveform for the clipping circuit of Fig. 3-27.

**Figure 3-27
Sine-wave clipper
for Question 24.**

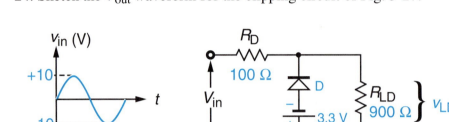

SUMMARY OF IDEAS

● A diode can conduct current in only one direction. If it is reverse-biased, a diode blocks current.

● A diode can be used to rectify ac into pulsating dc.

● A half-wave rectifier uses only one half cycle of the ac sine wave input. It discards the other half cycle.

● A full-wave rectifier uses both half cycles of the ac input.

● There are two kinds of full-wave rectifier: center-tapped transformer and diode bridge.

● The pulsating dc from a rectifier can be smoothed by connecting a filter capacitor in parallel with the load.

● The larger the filter capacitor, the smaller the ac ripple becomes.

● Additional low-pass filtering circuitry can be added to give even smoother dc (even smaller ripple).

● A diode clipping circuit prevents the load voltage from exceeding some particular voltage value.

● A voltage-doubling circuit uses diodes and capacitors to produce a dc output that it twice as great as the peak ac input voltage. There is a current trade-off.

CHAPTER QUESTIONS AND PROBLEMS

1. Draw the schematic diagram of a positive half-wave rectifier driven by a 12-V peak, 60-Hz sine wave source.
2. For the circuit of Problem 1, sketch the load voltage waveform, assuming that the diode is ideal.
3. Repeat Problem 2, assuming that the diode is real. Does the load pulsation last for exactly 8.33 ms? Explain.
4. Repeat Problems 1 and 2 for a negative half-wave rectifier.

5. In Fig. 3-6(A), suppose the transformer secondary has an rms voltage of 48 V, center-tapped.
    (a) Sketch the load waveform.
    (b) Does it matter much whether you regard the diodes as real or ideal? Explain.
6. Sketch the schematic layout of a negative full-wave center-tapped rectifier circuit.
7. In a full-wave center-tapped rectifier circuit, if one diode fails open, describe what happens.
8. In a full-wave center-tapped rectifier circuit, if one diode fails shorted, describe what happens.
9. Figure 3-28 shows a dual-polarity dc power supply It provides both a positive dc voltage relative to ground and a negative dc voltage relative to ground. Explain how it works.
10. Figure 3-29 shows another design for a dual-polarity dc power supply. Explain how it works.

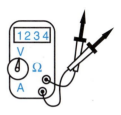

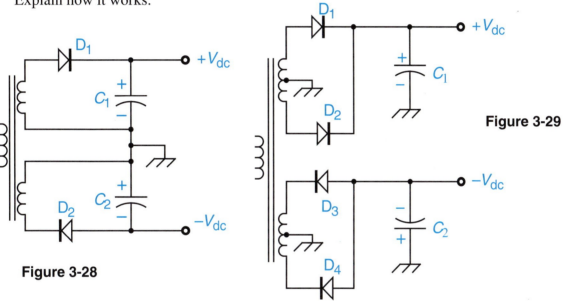

Figure 3-29

Figure 3-28

11. Figure 3-30 uses a transformer with only a single secondary winding (CT). It is the most popular design for a dual-polarity dc supply. Explain how it works.

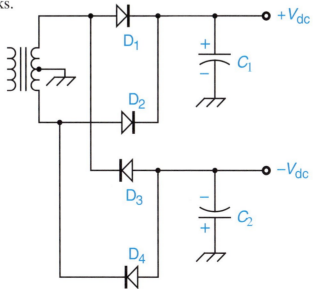

Figure 3-30

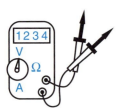

12. Sketch the schematic layout of a full-wave bridge rectifier.

13. For your circuit of Problem 12:
   (a) Trace out the exact path of electron current during the source's positive half cycle.
   (b) Trace out the exact path of electron current during the source's negative half cycle.

14. In a full-wave bridge rectifier circuit, if one diode fails open, describe what happens.

15. In a full-wave bridge rectifier circuit, if one diode fails shorted, describe what happens.

16. If the size of the filter capacitor is doubled, the ripple will be _____ , all other things being equal.

17. A certain rectifier is filtered by 1000-μF capacitor C_1, and it has 0.3 V pk-pk ripple. Another 500-μF capacitor, C_2, is placed in parallel with C_1. What is the new value of ripple?

18. Draw the schematic diagram of a full-wave rectifier with a $\prod$ low-pass filter.

19. In Fig. 3-19, suppose D_1 points down and D_2 points to the left. Describe the dc output voltage across the load.

20. (T-F) A clipping circuit could be given the name "slicing-off" circuit, and that would make sense. Explain.

21. In Fig. 3-24, suppose the dc source has a value of 4.3 V instead of 5.3 V. Draw the v_{LD} waveform.

22. In Fig. 3-27, suppose that diode D points down instead of up, and suppose the dc source is 3.7 V. Sketch the v_{LD} waveform.

CHAPTER 4

BIPOLAR JUNCTION TRANSISTORS

Optical character-readers for blind persons use a small camera-like device containing many light-sensitive transistors. The optical detector is moved slowly across the printed page with one hand. An electronic control unit processes the detector's transistor-generated information and translates it into a vibration code that is then applied to a tactile (sense of touch) pad. In this photo, the user is manipulating the detector with his left hand while the fingers of his right hand are reading the vibration signal.

This tactile technology can be extended to the reading of graphic images on computer screens. See the photo at the start of Chapter 5, page 107.

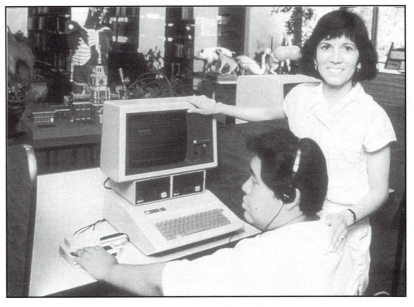

Courtesy of NASA.

OUTLINE

4-1 Transistor Symbols and Terminal Names
4-2 The Basic Transistor Idea
4-3 Internal Construction and Operation of a Transistor
4-4 Applying the Ac Input Signal
4-5 Hitting the Limits—Cutoff and Saturation
4-6 Transistor Characteristic Curves
4-7 Nonideal Transistor Characteristics
4-8 Testing Transistors
4-9 Transistor Physical Appearance

NEW TERMS TO WATCH FOR

base	cutoff
emitter	saturated
collector	active
current gain	linear
collector bias voltage	characteristic curve
current source	breakdown
npn	transistor curve-tracer
pnp	

A transistor is a solid-state device that can amplify (make larger) the power of an electrical signal. Their power-amplifying ability makes transistors extremely useful. The transistor is the basic building-block component of modern electronics.

There are two fundamentally different kinds of transistors. They are the bipolar junction transistor (BJT) and the field-effect transistor (FET). When people use the single word *transistor*, they generally mean the BJT.

We will study the structure and basic operating principles of BJTs in this chapter. FETs will be covered later.

✔ **After studying this chapter, you should be able to:**
1. Draw the schematic diagram and mark the emitter, base, and collector terminals of an *npn* transistor. Do the same for a *pnp* transistor.
2. Use the equation $\beta = I_C \div I_B$ to calculate any one of these three variables, given the other two.
3. Use the proper symbol to stand for a particular aspect of a waveform in which ac and dc are mixed together.
4. Explain the basic operating idea of a transistor, in terms of the B-E junction current having two separate components.
5. Tell which terminal is considered the input terminal, which is the output terminal, and which is the common terminal in the usual transistor circuit arrangement.
6. Given a complete schematic diagram of a transistor amplifier, mathematically analyze the amplifier's performance.
7. While analyzing an amplifier's performance, sketch accurate waveforms of i_B, i_C, and v_C .
8. Define the transistor cutoff condition and sketch the output voltage waveform for a transistor that cuts off.
9. Define the transistor saturation condition and sketch the output voltage waveform for a transistor that saturates.
10. Interpret a family of collector characteristic curves for a transistor.
11. Describe how the characteristic curves of a real transistor are different from an ideal transistor.
12. Test a transistor with an ohmmeter to tell whether it is good or bad.
13. Recognize some of the common transistor packages and identify the leads.

4-1 TRANSISTOR SYMBOLS AND TERMINAL NAMES

The bipolar junction transistor has three terminals. They are called the **base**, the **emitter**, and the **collector**. These terminals are shown in the schematic symbol of Fig. 4-1(A).

In the symbol for a BJT, an arrow identifies the emitter terminal, or lead. The base is the lead in the center, that is perpendicular to the bar inside the circle. The collector is the other slanted lead. Most of the time, but not always, transistors are drawn with this orientation — collector on the top, base on the side, and emitter on the bottom.

The symbol of Fig. 4-1(A) represents an *npn* transistor, as distinguished from a *pnp* transistor This is because its collector is *n*-doped silicon, its base is *p*-doped, and its emitter is *n*-doped. These doping polarities are indicated in the structural view of Fig. 4-1(B).

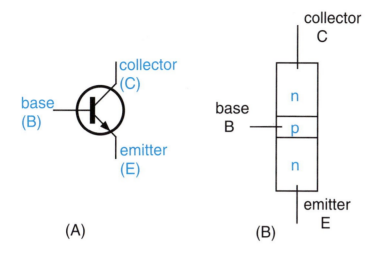

(A)   (B)

Figure 4-1
(A) Schematic symbol of an *npn*-polarity BJT. (B) Structural diagram of *npn* transistor. The comparative thinness of the base region is realistic.

A *pnp*-polarity BJT is symbolized in Fig. 4-2(A). Its only schematic difference is that the arrow on the emitter lead points inward, rather than outward. As the structural diagram in Fig. 4-2(B) shows, the regions of a *pnp* transistor have the opposite type of doping from an *npn* transistor.

①✔

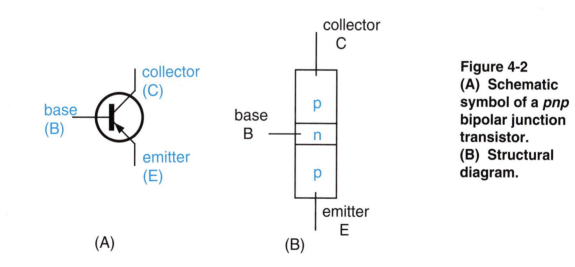

(A)   (B)

Figure 4-2
(A) Schematic symbol of a *pnp* bipolar junction transistor. (B) Structural diagram.

In the transistor schematic symbols the arrow tells the direction of current in the emitter lead. This is like the arrow of a diode. Electron current in a transistor flows against the arrow, the same as in a diode.

Thus, in the *npn* transistor of Fig. 4-1, electron current flows *into* the body of the transistor via the emitter lead. It moves against the arrow. In the *pnp* transistor of Fig. 4-2, electron current flows *out of* the body of the transistor on the emitter lead.

4-2 THE BASIC TRANSISTOR IDEA

A transistor is a current-controlled device. A small amount of base current controls a large amount of collector current. This idea is illustrated in an ac sense in Fig. 4-3.

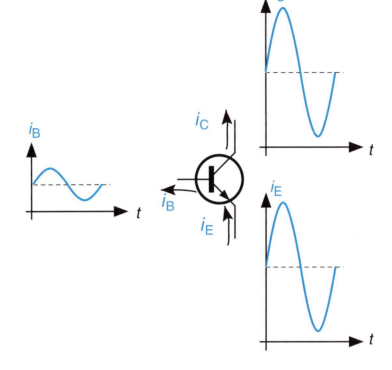

Figure 4-3
The current-amplifying action of a transistor. The large ac currents I_c and I_e are controlled by the small ac current I_b.

As Fig. 4-3 illustrates, a small sine-wave variation in base current i_b causes a much greater sine-wave variation in collector current i_c. Also, for reasons that will be explained when we study the transistor's internal operation, the emitter current i_e goes through an almost identical variation as the collector current. This is indicated in Fig. 4-3.

The current waveforms in Fig. 4-3 show ac sine waves oscillating around a certain dc level. This is correct, since all three currents actually are ac combined with dc. They are not plain ac currents because that would cause one of the half cycles of the emitter current to flow in the same direction as the schematic arrow, which is impossible.

At this time we are going to concentrate on the ac variations only, in order to explain the current-control idea. We will ignore the currents' dc part.

Every transistor has a certain factor by which I_c is greater than I_b. This factor is called the **current gain** of the transistor.

Current gain is symbolized by the Greek letter β (beta). As a formula,

Equations 4-1 and 4-2 refer to the ac parts of the overall currents. The dc parts are not considered here.

$$I_c = \beta\, I_b$$

Eq. 4-1

or

$$\beta = \frac{I_c}{I_b}$$

Eq. 4-2

EXAMPLE 4-1

The transistor in Fig. 4-4 has a current gain $\beta = 150$.

(a) If an external ac source forces 20 μA (rms) to flow in the base lead, how much ac current flows in the collector lead?

(b) We know that the back-and-forth ac collector current I_c enters the transistor via the collector (C) lead. How does I_c *exit* from the transistor?

(c) We know that ac current I_b enters the transistor via the base (B) lead. How does I_b exit from the transistor?

SOLUTION

(a) Applying Eq. 4-1, we have

$$I_c = \beta I_b$$

$$= 150 \, (20 \, \mu A) = 3000 \, \mu A \text{ or } \mathbf{3 \, mA.}$$

(b) I_c exits on the emitter lead.

(c) I_b also exits on the emitter lead.

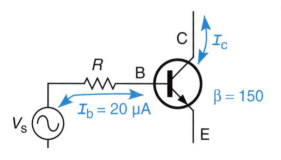

Figure 4-4
Npn transistor with known beta.
I_c stands for the ac collector current, in rms units.
I_b is the rms ac base current.

Since both I_b and I_c leave the transistor via the E lead, we can say

$$I_e = I_b + I_c \qquad\qquad \text{Eq. 4-3}$$

In Example 4-1, Eq. 4-3 gives

$$I_e = I_b + I_c$$

$$= 50 \, \mu A + 3000 \, \mu A = 3050 \, \mu A$$

3050 μA is very close to 3000 μA. In general, we can say

$$\boxed{I_e \approx I_c} \qquad\qquad \text{Eq. 4-4}$$

Equations 4-1 through 4-4 apply just as well to *pnp* transistors as to *npn* transistors. The difference between *npn* and *pnp* transistors is in the direction of the **dc** part of the currents, not ac.

Equation 4-4 gives a good approximation (within 1%) for any transistor with $\beta \geq 100$. Most modern transistors have β values between 100 and 400, so the approximate equation works very well. Some high-power transistors have $\beta < 100$, so Eq. 4-4 should not be used for them.

Figures 4-3 and 4-4 are not complete circuit schematics, obviously. They are just partial schematics, that we are using to emphasize an important idea — the idea of ac current amplification. Be certain that you really accept this transistor idea.

Now that we know what the transistor idea says [Eq. (4-1)], it is worthwhile to point out what the transistor idea does *not* say.

1 It does not say that the amount of collector current depends on the voltage existing in the collector circuit.

2 It does not say that the amount of collector current depends on whatever resistance value is inserted into the collector lead.

We can emphasize these facts by stating the transistor idea as follows:

Get This

For a given transistor with known β, the ac collector current I_c is independent of any collector voltage, and it is independent of any resistance. I_c depends *only* on the ac base current I_b.

EXAMPLE 4-2

The schematic diagram in Fig. 4-5 shows a bit more of the overall construction of an actual transistor amplifier circuit. It goes beyond Fig. 4-4 by showing that:

1 There is a resistor in the collector lead, symbolized R_C.

2 The top of R_C is connected to a dc supply voltage called the collector biasing voltage, symbolized V_{CC}.

3 The emitter is connected directly to the ground reference point of the circuit. V_{CC} and V_s are also both referenced to ground.

(a) From the information given in Fig. 4-5, calculate the value of $I_{c(rms)}$.

(b) Applying Ohm's law to R_C, calculate the value of $V_{Rc(rms)}$.

(c) Suppose V_{CC} is changed from –15 V to –20 V. Calculate the new values of $I_{c(rms)}$ and $V_{Rc(rms)}$.

(d) With V_{CC} remaining at –20 V, let R_C be increased to 1200 Ω. Find the new values of I_C and and V_{RC}.

SOLUTION

(a) This transistor's current gain is 180 and we know that the ac voltage source is delivering 20 μA (rms) into the base. From Eq. 4-1,

$$I_{c(rms)} = \beta\, I_{b(rms)}$$
$$= 180\,(20\,\mu A) = \textbf{3.6 mA}$$

(b) If a known value of ac current is flowing through a known value of resistance, then Ohm's law will give the voltage across that resistance, as always.

$$V_{Rc(rms)} = I_{c(rms)} \times R_C$$
$$= 3.6\text{ mA }(750\,\Omega) = \textbf{2.7 V}$$

(c) The value of V_{CC} is irrelevant to the ac operation of the transistor. The fact that V_{CC} now has a larger magnitude does not affect the ac collector current at all. Therefore I_c and V_{RC} **remain the same** (I_c = **3.6 mA** and V_{Rc} = **2.7 V**).

(d) The amount of resistance in the collector lead, R_C, is irrelevant to the ac current behavior of the transistor. The fact that R_C has been increased to a larger value does not reduce the ac collector current, or affect it in any way. We say that:

Get This

The transistor's collector acts like an ideal ac **current source**, maintaining a constant amount of collector current regardless of the resistance that this current must pass through.

Thus,

I_c = **3.6 mA, remaining the same**.

V_{RC} must be determined by Ohm's law. With I_c constant, increasing R_C will cause V_{Rc} to increase according to

$$V_{Rc} = I_c R_C$$
$$= 3.6 \text{ mA } (1200 \ \Omega) = \textbf{4.32 V (rms)}$$

This higher voltage value is not surprising. All current sources change their output voltage in step with any change in the resistance that is connected to their output terminals.

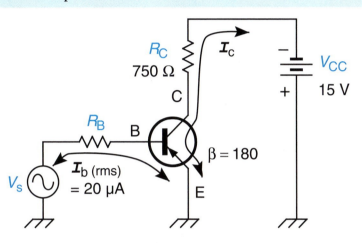

Figure 4-5 Partial schematic of a transistor amplifier circuit.

This circuit happens to be built with a *pnp* transistor, but that is not important to us now. It could just as easily be built with an *npn* transistor, if the polarity of V_{CC} were changed.

Overall Current Directions

Although we have described the basic transistor idea in terms of ac currents, it was mentioned earlier that actual transistor currents are ac combined with dc. That way the dc plus ac combined current in any transistor lead is always in one overall direction. You should know what the overall current direction is in every lead, for both transistor types. This information is summarized in Fig. 4-6. Part (A) of that figure is for an *npn* transistor and part (B) is for a *pnp* transistor.

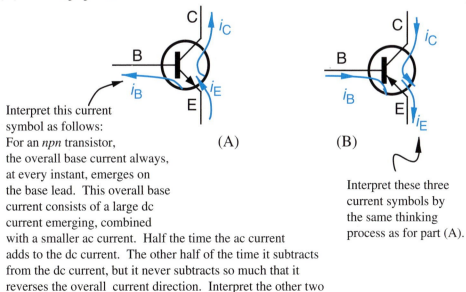

Interpret this current symbol as follows:
For an *npn* transistor, the overall base current always, at every instant, emerges on the base lead. This overall base current consists of a large dc current emerging, combined with a smaller ac current. Half the time the ac current adds to the dc current. The other half of the time it subtracts from the dc current, but it never subtracts so much that it reverses the overall current direction. Interpret the other two current symbols in part (A), i_E and i_C, in the same way.

(A)   (B)

Interpret these three current symbols by the same thinking process as for part (A).

**Figure 4-6
Showing the overall current path and direction for each transistor lead. The currents are electron-flow currents.**
(Reverse all current directional arrows if you are thinking in terms of conventional current.)

(A) *npn* transistor

(B) *pnp* transistor

In Fig. 4-6, base current i_B and collector current i_C are shown entering the transistor's body on one particular lead, and emerging from the transistor's body on another particular lead. However, emitter current i_E is not shown both entering and emerging. This is because i_E is actually the sum of i_B and i_C (Eq. 4-3). In other words, i_E splits apart inside the transistor [Fig. 4-6(A)]. Or, i_E results from a merging together inside the transistor [Fig. 4-6(B)]. That is why i_E cannot be shown appearing at just two particular terminals.

Symbolizing Electric Variables (Voltage and Current) When Ac is Mixed with Dc

When ac and dc mix together, we need to be careful about just what aspect of the combined waveform we are talking about. We must also be careful to use the proper symbol to identify that aspect of the waveform. You may have already noticed a slight change in symbols from Figs 4-4 and 4-5 to Fig. 4-6.

In Figs. 4-4 and 4-5 the letter I is capitalized, but the subscripts b and c are lower case (not capitals). The capital I refers to the magnitude of the waveform's *overall cycle* — not an instantaneous current value. The lower-case b and c subscripts refer to the *ac portion only* — not both ac and dc combined.

In Fig. 4-6, the lower case i refers to the *instantaneous value* of current. The capital B and C and E subscripts mean the total combined effect of ac *and* dc.

The various symbol meanings are identified in Fig. 4-7. They are also summarized in Table 4-1.

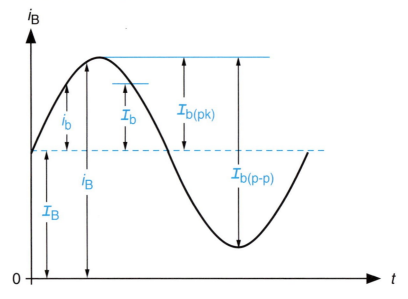

**Figure 4-7
Waveform and symbols for base current i_B.**

I_B — dc value

i_B — instantaneous value of the overall waveform, ac and dc combined

i_b — instantaneous value of the ac part only, ignoring the dc part

I_b [or $I_{b(rms)}$] — rms magnitude of the overall ac cycle, ignoring the dc part

   Also, related to $I_{b(rms)}$, we have:

$I_{b(pk)}$ — peak magnitude of the overall ac cycle, ignoring the dc part

$I_{b(p-p)}$ — peak-to-peak magnitude of the overall ac cycle, ignoring the dc part

Main letter symbol

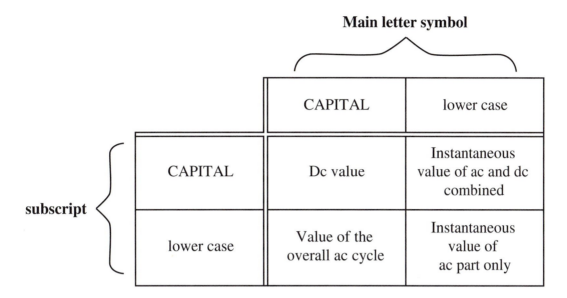

		CAPITAL	lower case
subscript	CAPITAL	Dc value	Instantaneous value of ac and dc combined
	lower case	Value of the overall ac cycle	Instantaneous value of ac part only

Table 4-1 Summary of symbols.

SELF-CHECK FOR SECTIONS 4-1 AND 4-2

1. Draw the schematic symbol for a *pnp* transistor, and label the terminals.
2. For the *pnp* transistor in Question 1, show the flow path and direction of overall base current i_B. (electron flow)
3. Repeat Question 2 for the overall collector current i_C.
4. Show the direction of overall emitter current i_E.
5. Why is it impossible to show the complete flow path of emitter current?
6. Draw the schematic symbol for an *npn* transistor, and label the terminals.
7. For the *npn* transistor in Question 6, show the flow path and direction of overall base current i_B.
8. Repeat Question 7 for the overall collector current i_C.
9. Show the direction of overall emitter current i_E.
10. (T-F) The single word transistor is taken to mean a bipolar junction transistor, not a field-effect transistor.
11. The specific acronym (letters) for a bipolar junction transistor are_____ .
12. What letter is used to symbolize the current gain of a transistor.
13. (T-F) A current gain value of 20 would be typical for a modern transistor.
14. (T-F) In a modern transistor, emitter current i_E is almost the same as collector current i_C.
15. How should we think of the collector circuit of a transistor, as an ac voltage source or as an ac current source?
16. (T-F) The overall current in a transistor lead is ac combined with dc.
17. Should a transistor be described as a current source that is current-controlled, or as a current source that is voltage-controlled?
18. You have no doubt noticed that we haven't shown pictorial ammeters and voltmeters measuring the transistor circuits of Figs. 4-4 through 4-6. A pictorial drawing of Fig. 4-5 is shown on page 74, in Fig. 4-8. Make a comment on how much extra work would be involved in drawing all our schematics in that way. Do you expect to see very many schematics drawn that way?

Figure 4-8

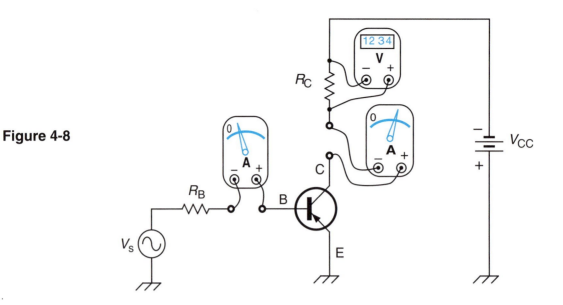

ANIMATED CHARACTERS

Animated robotic characters (animatronics) can reproduce life-like head and face motions, synchronized with electronically recorded speech.

In this photo, the "teaching" robot on the left is guided by a human puppeteer through its motions, while the human speaks the script into a hand-held microphone. Synchronization of mechanical motion and audio speech is therefore natural and correct.

The "learning" robot, which is here a puppy, duplicates these motions exactly, and remembers them electronically. It also remembers the spoken script.

Each robot has five separate motions, called *degrees of freedom* . They are:
1. Shoulder lean: Lean entire assembly forward, or backward.
2. Neck tilt: Tilt head to either side, toward left shoulder or toward right shoulder.
3. Neck turn: Rotate head one way, or rotate head the other way.
4. Neck tip: Tip the face downward toward ground, or upward toward sky.
5. Mouth: From completely closed to fully open.

The teaching robot is shown in detail at the left on the facing page, along with the system's electronic controller module. As that photo shows, the teaching robot contains 5 position-indicating devices, one for each motion axis. Each indicating device, or *resolver*, gives an analog electric signal that represents the present position of its motion axis. This analog electric signal is then converted to a digital binary number by an analog-to-digital converter, or ADC, in the electronic controller module. The combination of a resolver and its ADC can be called a *position-encoder*, or simply an encoder.

For example, the mouth encoder (whose resolver is the one at the top, in the rear, in the left photo below) could be set up to produce the binary equivalent of 0 if the mouth is fully closed, and the binary equivalent of +2047 if the mouth is fully open.

The neck tilt encoder (whose resolver is the one that is second from the bottom) could be set up to produce the binary equivalent of 0 if the neck and head are centered, to produce +2047 if the neck is tilted the maximum distance toward the right shoulder, and to produce –2047 if the neck is tilted the maximum distance toward the left shoulder.

And so on, for the other two neck motion axes and the shoulder tip axis.

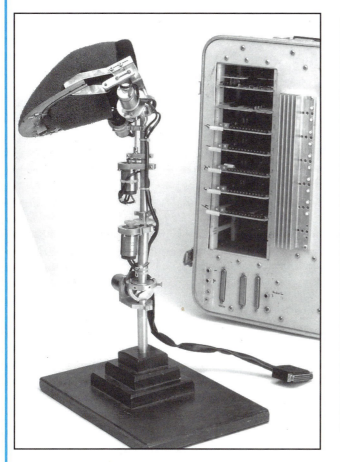

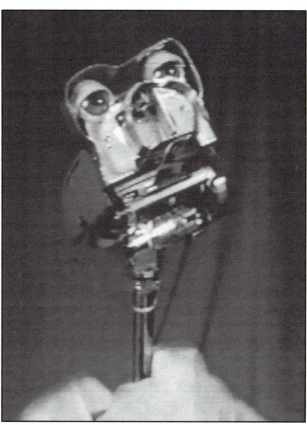

Courtesy of Astrosystems, Inc.

The electronic controller reads the binary-encoded position of each motion axis 50 times per second (250 total reads per second). Each time the controller reads, or samples, it stores the position-encoder's binary number in an electronic memory, and then immediately transfers that information to channel 1 of a stereo audio cassette tape.

At the same time, the controller sends signals to five motors on the learning robot, the puppy. Its internal mechanical structure is shown at the right above. These motors turn the five motion axes on the learning robot so that they duplicate the five axis positions of the teaching robot.

All the while, the puppeteer's voice signal from the microphone is being recorded on channel 2 of the audio tape.

Once the entire motion sequence and script have been recorded, the learning robot can reproduce them again and again by replaying the tape.

4-3 INTERNAL CONSTRUCTION AND OPERATION OF A TRANSISTOR

A transistor has three layers, or regions, of doped silicon. This was shown in Figs. 4-1(B) and 4-2(B). Therefore, a transistor contains two *p-n* junctions, rather than just a single *p-n* junction like a diode. The internal structure of an *npn* transistor is illustrated in Fig. 4-9(A). It is sometimes helpful to think of this as a pair of opposite-pointing diodes, as shown in Fig. 4-9(B).

Figure 4-9
***npn* transistor internal structure.
(A) Solid-state regions.
(B) Equivalent diode circuit.**

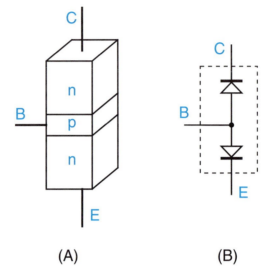

(A)   (B)

The internal operation of a *pnp* transistor is the same, conceptually. But the doping polarities of the three regions in a *pnp* transistor are all opposite from an *npn* transistor. Therefore, all voltage polarities and current directions are opposite too.

We will explain the detailed internal operation of an *npn* transistor. Our explanation is a simplified one. It is not rigorously complete because it ignores certain aspects of transistor construction that also affect its current-gaining ability. But our explanation does capture the essence of the transistor idea.

When a transistor is completely de-energized, with no voltage sources connected, it forms two depletion regions. Such formation was explained in Sec. 1-3. This de-energized state is pictured in Fig. 4-10.

**Figure 4-10
Formation of internal depletion regions before the transistor is biased.**

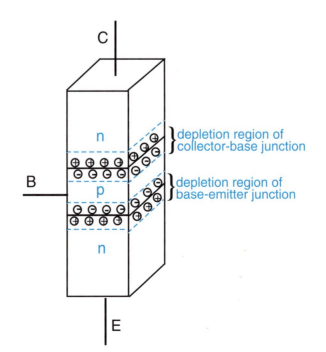

To make the transistor function, external voltages must be applied to the two junctions as follows:

> The base-emitter junction must be forward-biased, and the collector-base junction must be reverse-biased.

Get This

This biasing condition is illustrated in Fig. 4-11. As Fig. 4-11(A) shows, the B-E junction no longer has a depletion region. But the C-B junction's depletion region is wider than before.

In Fig. 4-11(B), there is a V_{CC} voltage of 10 V applied to the top of R_C, and a V_{BB} voltage of 4 V applied to base resistor R_B. With the 10-V value being more positive than the 4-V value, they reverse-bias the C-B "diode".

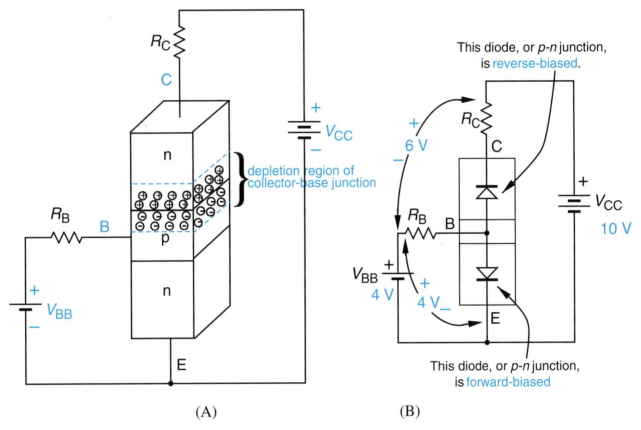

(A)   (B)

Figure 4-11 To prepare the transistor for amplifying ac current, the B-E junction must be forward-biased by voltage source V_{BB}. But the C-B junction must be reverse-biased. This is accomplished by connecting a voltage source V_{CC} to the collector, with the value of V_{CC} greater than V_{BB}. (A) Structural view. (B) Diode equivalent circuit.

The Transistor Idea

With the B-E junction forward-biased, dc current flows through the junction. This dc current consists of two separate flow mechanisms, just like the conducting diode of Fig. 1-12.

[1] Free electrons from the arsenic atoms in the n-doped E region can jump into intact silicon atoms in the B region. This is the free electron-flow component of the current through the junction.

[2] Holes in the boron atoms in the p-doped B region can receive electrons from intact silicon atoms in the E region. This is the hole-flow component of the current through the junction.

The total current through the base-emitter junction is equal to the sum of these two component currents, as indicated in Fig. 4-12(A).

The two current components do not have the same magnitude. The free electron-flow current is much greater than the hole-flow current.

Get This

For the B-E junction, free electron-flow current is β times as great as hole-flow current. This is because the E region has an arsenic doping density that is β times greater than the boron doping density in the B region.

The two regions' doping densities are purposely made much different from one another during the transistor manufacturing process. For example, if a given volume in the base region contained 1000 boron impurity atoms, that same volume in the emitter region might contain 200 000 arsenic impurity atoms. In this example, the ratio of densities is 200:1. Therefore the electron flow will be 200 times as great as the hole flow. We say that β = 200.

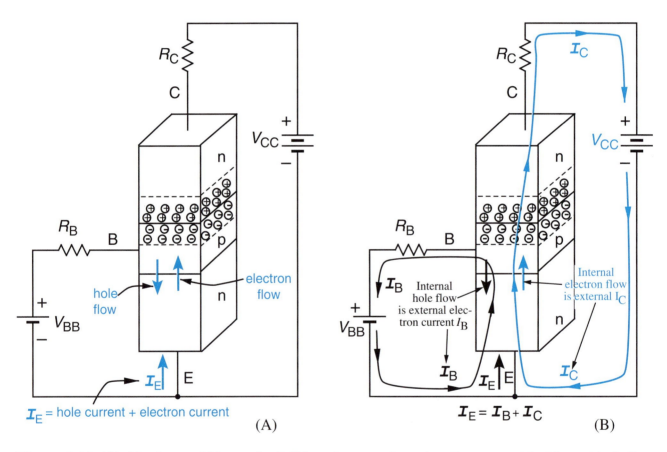

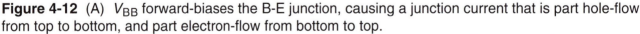

Figure 4-12 (A) V_{BB} forward-biases the B-E junction, causing a junction current that is part hole-flow from top to bottom, and part electron-flow from bottom to top.
(B) The hole-flow from top to bottom through the B-E junction becomes I_B. The electron-flow from bottom to top through the B-E junction becomes I_C.

The hole flow current completes its path by the base circuit loop. This fact is illustrated in Fig. 4-12(B). It has to be this way, because a bonded electron from the emitter region that has recently jumped into a hole in the base layer cannot then move into the depletion region near the C-B junction. Such an electron is repelled by the concentration of negative charge in the depletion region at the top of the base layer. It is true that a new bonded electron must move to enable a new hole to appear in the base layer. So such an electron must move out of the transistor body via the base lead, attracted by the positive voltage V_{BB}. Figure 4-12(B) identifies the resulting current as I_B, the dc bias current in the base lead.

However, the *free* electrons that have entered the very thin base layer from the emitter show different behavior. They *are* able to enter the depletion region because they do not need to break a bond in order to move. In fact, being very mobile, they are more motivated by the large positive voltage V_{CC} to pass through the thin base layer, cross through the depletion region, through the collector region, and emerge from the transistor body on the collector lead. These free electrons are not motivated by the smaller voltage V_{BB} to travel the greater distance to the left in Fig. 4-12(B) to emerge via the base lead.

Figure 4-12(B) makes it clear that the internal electron-flow component of the B-E junction current becomes the external current I_C, the dc bias current in the collector lead.

> The fact that the small hole current flows in the base lead while the large free-electron current flows in the collector lead is the basic operating idea of an *npn* transistor.

Get This

This is the idea that got solid-state electronics started. It is illustrated pictorially in Fig. 4-13.

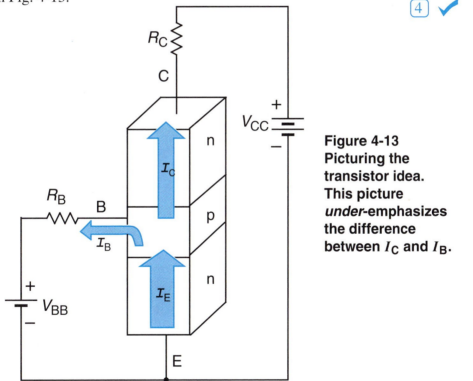

**Figure 4-13
Picturing the transistor idea. This picture *under*-emphasizes the difference between I_C and I_B.**

The two major aspects of transistors that we have ignored are these:

$\boxed{1}$ The *mobility*, or ability to move through the silicon crystal, is not the same for free electrons as it is for holes. To the extent that the free electrons are more mobile as they pass through the base region in Fig. 4-12 than the holes are as they pass through the emitter region, to that extent is β increased.

$\boxed{2}$ The physical width of the base region has an effect on beta. However, if the average distance that a hole moves through the emitter region before it "collides" with a silicon atom (this is called the *hole diffusion distance*) is exactly the same as the width of the base region, then this variable would wash out. We would be completely justified in ignoring it.

We have also ignored the fact that a few of the free electrons that cross the B-E junction in Fig. 4-12 will fall into holes in the base region. This occurrence is called *recombination current*, and it subtracts from the collector current. The same remark applies to a few holes that cross the B-E junction and bump into free electrons in the emitter region, recombining with them. In a modern transistor, recombination current is so small that it is negligible.

We have also ignored the leakage current through the reverse-biased C-B junction in Fig. 4-12. (This is like the diode reverse leakage current shown in Fig. 2-10 on page 28). Again though, in a modern transistor this current is so small that we can rightly ignore it.

EXAMPLE 4-3

In Fig. 4-12(B), suppose $V_{BB} = +4$ V, $R_B = 33$ kΩ, and $V_{CC} = +10$ V. Assume a 0.7-V voltage drop across the conducting B-E junction. The transistor's doping ratio (β) is 200.

(a) Find the magnitude of dc base bias current I_B.

(b) Find the magnitude of dc collector bias current I_C.

(c) Explain why you don't need to know the resistance of R_C in order to calculate I_C.

(d) If V_{CC} were changed to 12 V, would that have any effect on I_C? Explain.

SOLUTION

(a) In the base circuit loop, Kirchhoff's voltage law and Ohm's law must be obeyed, as always. From Kirchhoff's voltage law,

$$V_{BB} = V_{RB} + V_{BE}$$
$$V_{RB} = V_{BB} - V_{BE}$$
$$= 4.0 \text{ V} - 0.7 \text{ V} = 3.3 \text{ V}$$

Then, from Ohm's law,

$$I_B = \frac{V_{RB}}{R_B}$$
$$= \frac{3.3 \text{ V}}{33 \text{ k}\Omega} = \textbf{0.1 mA}$$

(b) Equation 4-1 was introduced in Sec. 4-2 for the ac currents in a transistor. It is the same for the dc currents, as was just explained in this section. That is

$$\boxed{I_C = \beta I_B} \qquad \text{Eq. (4-5)}$$

With β = 200, we get

$$I_C = \beta\, I_B$$
$$= 200\,(0.1 \text{ mA}) = \textbf{20 mA}$$

(c) From an external point of view, the transistor's collector circuit is a 20-mA current-source. It can deliver 20 mA through *any* amount of resistance (within limits). It simply raises or lowers its output voltage if R_C is made larger or smaller.

(d) No. The collector current-source insists on delivering 20 mA, no matter what the value of V_{CC}. The only requirement on V_{CC} is that it must be greater than V_{BB} in order to reverse-bias the C-B junction.

✓ SELF-CHECK FOR SECTION 4-3

19. To make a transistor function as an amplifier, the base-emitter junction must be _____-biased and the collector-base junction must be _____-biased.

20. For an *npn* transistor, the answer to Question 19 means that the dc voltage applied to the base must be _____ with respect to the emitter; the dc voltage applied to the collector must be _____ _____ with respect to the emitter.

21. (T-F) The total current through the B-E *p-n* junction is the same as the total current in the transistor's emitter lead.
22. The total current through the B-E junction consists of a _____-flow component and an _____-flow component.
23. (T-F) The doping density of the emitter region is much greater than that of the base region.
24. For an *npn* transistor, which of the two components (in Question 22) passes in the base lead? Which one passes in the collector lead?
25. (T-F) Part of the explanation for the internal behavior of a transistor has to do with the base region being quite thin, compared to the other two regions.
26. In Fig. 4-12, suppose that $V_{BB} = 8$ V, $R_B = 120$ kΩ, β = 225, $R_C = 680$ Ω and $V_{CC} = 15$ V.
a) Find the value of dc base current I_B. Assume $V_{BE} = 0.7$ V.
b) Calculate the dc collector bias current I_C.
27. In Problem 26, if V_{CC} is increased to 20 V, with everything else remaining the same, the collector current I_C will _____.

4-4 APPLYING THE AC INPUT SIGNAL

After the dc bias conditions have been established, an ac signal voltage is applied between the base and the emitter. This is done through coupling capacitor C_{in}, as shown in Fig. 4-14(A). C_{in} must be present so that none of the dc base bias current flows through the signal source.

The base is considered the input terminal. The emitter is called the common terminal because it is shared in common by the output signal as well as the input signal. We will see this shortly.

⑤ ✔

Ac input current I_b combines with dc bias current I_B to produce the total base current. It is important that the peak value $I_{b(pk)}$ is not as great as dc current I_B. That way, $I_{b(pk)}$ cannot overwhelm I_B and attempt to reverse the direction of the total base current. This must not be allowed to happen, because a reverse bias on the B-E junction would cause the transistor to stop functioning. A typical waveform of i_B is shown in Fig. 4-15(B).

The ac current through the B-E junction, I_e, is produced by the same two flow mechanisms that produced dc bias current. The hole-flow part of I_e becomes I_b. The much larger free electron-flow part becomes I_c, emerging from the transistor body on the collector lead. This is shown in Fig. 4-14(B). A waveform graph of total collector current i_C is given in Fig. 4-15(C). The peak value $I_{c(pk)}$ will automatically be less than dc bias current I_C, since $I_{b(pk)}$ is less than I_B. Thus, we see that:

> The transistor's fundamental operation is as a current amplifier. A small amount of ac input current to the base controls a large amount of ac output current from the collector.

Get This

As the collector ac current source delivers current through collector resistance R_C, it causes an ac voltage to be developed across R_C. The magnitude of this ac output voltage is set by Ohm's law as

$$V_{Rc} = V_{out} = I_c R_C \qquad\qquad \text{Eq. (4-6)}$$

Figure 4-14

Combining the ac signal with the dc bias.

(A) The ac input current in the base circuit.

(B) The transistor becomes a current source, producing an ac output current in the collector circuit. $I_c = \beta\, I_b$.

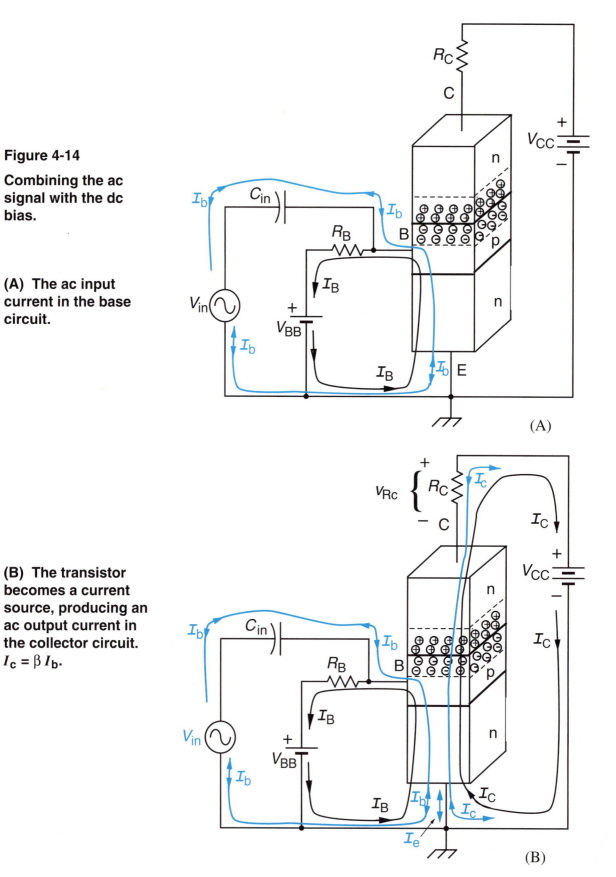

In most circuits, the value of R_C is large enough to develop an output voltage that is larger than the input voltage V_{in}. In that case, the circuit provides voltage gain as well as current gain. The waveform graph of v_{RC} in Fig. 4-15(D) shows this condition.

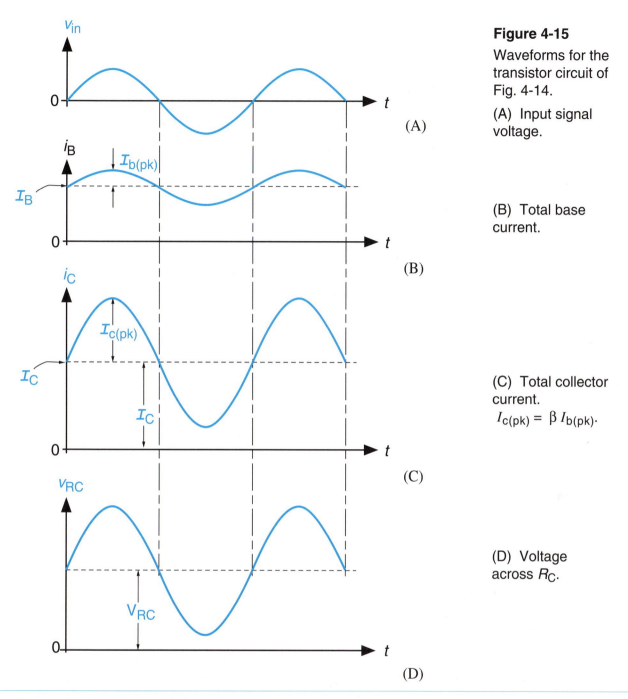

Figure 4-15

Waveforms for the transistor circuit of Fig. 4-14.

(A) Input signal voltage.

(B) Total base current.

(C) Total collector current.
$$I_{c(pk)} = \beta I_{b(pk)}.$$

(D) Voltage across R_C.

Everyone has heard of CAT scan, a medical procedure for making a detailed X-ray examination of the human body. CAT stands for Computer-Aided Tomography, where the word tomography refers to just a single thin plane of body tissue being brought into clear focus. On each successive scan, the CAT control electronics focuses the X-rays on a different body plane.

The same idea can be applied to the X-ray scanning of manufactured parts, as pictured here. This system is called the Advanced Computed Tomography Inspection System, ACTIS.

Also see the photos on page 118, and the ACTIS image on page 6 of the color section.

EXAMPLE 4-4

Figure 4-16 shows a schematic diagram of a transistor amplifier circuit like the one in Fig. 4-14. Resistor $R_{B(ac)}$ is present to limit the base input current I_b.

(a) Find the value of dc base bias current I_B.

(b) Find I_C, the dc bias collector current.

(c) What is the value of V_{RC}, the dc bias voltage across the collector resistor?

(d) Find the value of ac base current I_b. Ignore the ac resistance of the B-E junction.

(e) Find I_c, the ac collector current.

(f) What is the magnitude of V_{out}, the ac output voltage across R_C?

(g) Sketch the waveforms of V_{in}, i_C and v_{RC}.

SOLUTION

(a) The B-E junction will have about 0.7 V across it when it is forward-biased. Therefore the dc voltage available to drive $R_{B(dc)}$ is given by

$$V_{RB(dc)} = V_{BB} - V_{BE}$$
$$= 4\,V - 0.7\,V = 3.3\,V$$

By Ohm's law,

$$I_B = \frac{V_{RB(dc)}}{R_{B(dc)}} = \frac{3.3\,V}{100\,k\Omega} = \mathbf{33\,\mu A}$$

(b) From Eq. (4-5),

$$I_C = \beta I_B$$
$$= 140\,(33\,\mu A) = \mathbf{4.62\,mA}$$

(c) Applying Ohm's law to R_C, we get a dc voltage of

$$V_{RC} = I_C R_C$$
$$= 4.62\,mA\,(820\,\Omega) = \mathbf{3.79\,V}$$

(d) The ac input circuit consists of the series combination of the signal source, resistor $R_{B(ac)}$, C_{in}, and the forward-biased B-E junction. We assume that the reactance X_{Cin} is negligible. The designer must use a capacitance value that makes this assumption true. In general, the conducting B-E junction does present some non-negligible amount of resistance to ac base current, but we are ignoring it here. Therefore, in this example the only limiting resistance is $R_{B(ac)}$, so Ohm's law gives

$$I_B = \frac{V_{in}}{R_{B(ac)}}$$
$$= \frac{0.5\,V}{27\,k\Omega} = \mathbf{18.5\,\mu A\ (p\text{-}p)}$$

(e) From Eq. (4-1),

$$I_c = \beta I_b$$
$$= 140\,(18.5\,\mu A) = \mathbf{2.59\,mA\ (p\text{-}p)}$$

(f) Applying Ohm's law to R_C, we get

$$V_{Rc} = V_{out} = I_c R_C$$
$$= 2.59\,mA\,(820\,\Omega) = \mathbf{2.16\,V\ (p\text{-}p)}$$

(g) These waveforms are sketched in Fig. 4-17, parts (A), (B), and (C).

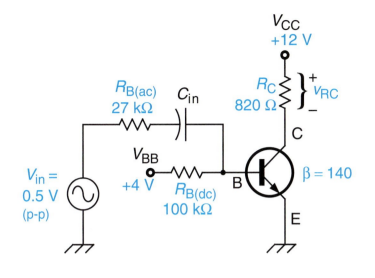

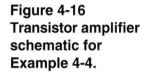

**Figure 4-16
Transistor amplifier
schematic for
Example 4-4.**

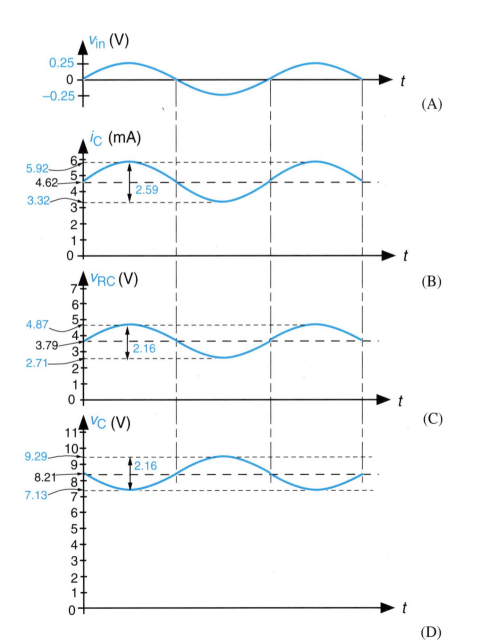

**Figure 4-17
Waveforms for the
amplifier of Fig. 4-16.**

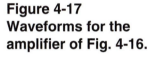

2.59 mA (p-p) from part **e**.

Dc value 3.79 V from part **c**.

Ac value 2.16 V (p-p) from
part **f**.

The same ac value,
2.16 V (p-p), from Eq. (4-7).

In Example 4-4, we regarded the ac voltage across resistor R_C as the output voltage from the amplifier. It is important to realize this fact:

Get This

> The magnitude of the ac voltage waveform across the collector resistor, V_{Rc}, is the same as the magnitude of the ac voltage waveform across the transistor itself, measured from collector to emitter (ground).

As a formula,

$$V_{Rc(p\text{-}p)} = V_{ce(p\text{-}p)} = V_{c(p\text{-}p)} \qquad \text{Eq. (4-7)}$$

This can be understood by referring to Fig. 4-18 and recognizing:

1. When i_C is increasing, for every 1 volt that v_{RC} increases, v_{CE} must decrease by 1 volt, in order that Kirchhoff's voltage law is satisfied. That is,

$$12\text{ V} = v_{RC} + v_{CE}$$

at every instant.

2. When i_C is decreasing, for every 1 volt that v_{RC} decreases, v_{CE} must increase by 1 volt, for the same reason.

Thus, any change in v_{RC} *is* accompanied by an equal-magnitude change in v_{CE}. But the changes are opposite. An increase in one goes with a decrease in the other. This relationship can be seen by comparing the v_C waveform in Fig. 4-17(D) to the v_{RC} waveform in Fig. 4-17(C).

Another way to think of the relation between the ac voltages V_{Rc} and V_c is this: In the collector loop in Fig. 4-18, the sum of the ac voltages must equal zero. This has to be true since there can be no ac voltage existing across dc source V_{CC}. Therefore the ac voltages V_{Rc} and V_c must cancel each other, which can only happen if they have the same magnitude and opposite polarities (are 180° out of phase with one another). Figure 4-17 points out this 180° phase relationship clearly.

Figure 4-18
Comparing v_{RC} and v_{CE} (v_C).
These two voltages must add to 12 V at every instant.

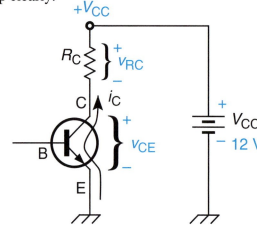

SELF-CHECK FOR SECTION 4-4

30. In Fig. 4-14, if dc base current $I_B = 60$ μA, what is the maximum allowable peak-to-peak ac base current?
31. In Fig. 4-14, if the ac input current has value $I_{b(p\text{-}p)} = 40$ μA, and the transistor has $\beta = 175$, find the peak-to-peak value of the ac output current.
32. The collector circuit of a transistor is an ac _____ source.
33. The ac emitter current I_e is almost _____ to the ac collector current I_c.

34. Should a transistor be thought of as a voltage-controlled current source or a current-controlled current source?

35. Which is slightly larger, I_e or I_c?

36. Suppose we change the circuit of Fig. 4-16 to have the following specifications: $V_{BB} = 10$ V; $R_{B(dc)} = 220$ kΩ; $V_{CC} = 15$ V; $R_C = 1200$ Ω; $\beta = 125$; $V_{in} = 0.1$ V(p-p); $R_{B(ac)} = 10$ kΩ. Find:
    (a) I_B (b) I_C (c) V_{RC} (d) V_C

37. Continuing with problem 36, find the peak-to-peak values of these ac variables:
    (a) I_b (Ignore the ac resistance of the B-E junction.)
    (b) I_c (c) V_{Rc} (d) V_c

38. Combining your information from Problems 36 (dc information) and 37 (ac information), sketch the waveform graphs of:
    (a) v_{in} (b) v_{RC} (c) v_C

39. A *pnp* transistor amplifier is shown in Fig. 4-19. Calculate the following circuit values.
    (a) I_B (c) V_{RC} (e) I_b (Ignore the ac junction resistance.)
    (b) I_C (d) V_C (f) I_c (g) V_{out}

40. The waveforms in Fig. 4-20 on page 88 apply to the *pnp* circuit of Fig. 4-19. Specify the asked-for values in the boxes.

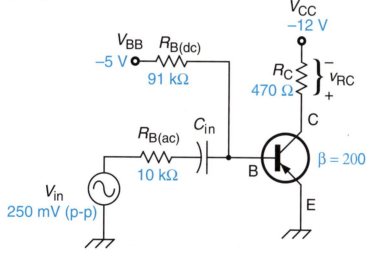

**Figure 4-19
An amplifier built with a *pnp* transistor, used in Problems 39 and 40.**

Crop-duster pilots have to cope with very high temperatures in the cockpit, which can impair their physical performance. The cool-suit shown here circulates chilled liquid through a tube-lined helmet and vest. The electronic pump-controller responds to changes in skin temperature.

Courtesy of Life Support Systems, Inc.

**Figure 4-20
Waveforms for the
pnp amplifier of
Fig. 4-19, showing
the reversal of all
voltage polarities
and current
directions.**

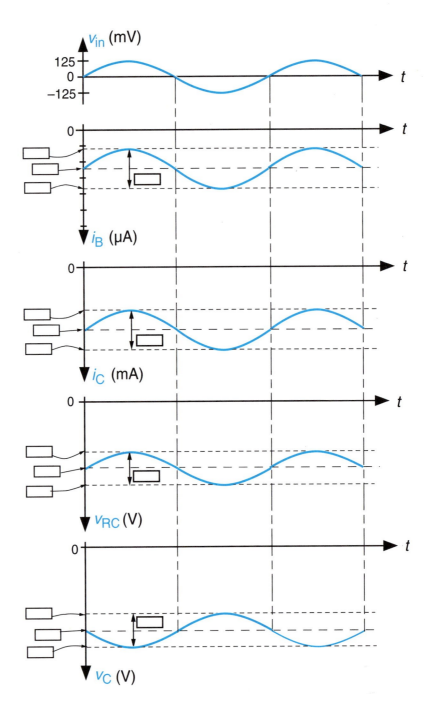

4-5 HITTING THE LIMITS – CUTOFF AND SATURATION

When we described the ac operation of a transistor in the previous section, we made it clear that we could not allow the ac base signal to overcome the dc base bias current. We also made a special point of saying that there were certain limits to the ability of the collector current source to maintain a constant value of current through resistance R_C. In this section we will make a careful study of these two limiting effects.

Cutoff

A transistor is said to be in **cutoff** condition if its base-emitter junction is allowed to become reverse-biased.

Get This

If the B-E junction becomes reverse-biased, a depletion region forms around the junction, as shown in Fig. 4-21. This stops both components of junction current, causing both i_B and i_C to become zero.

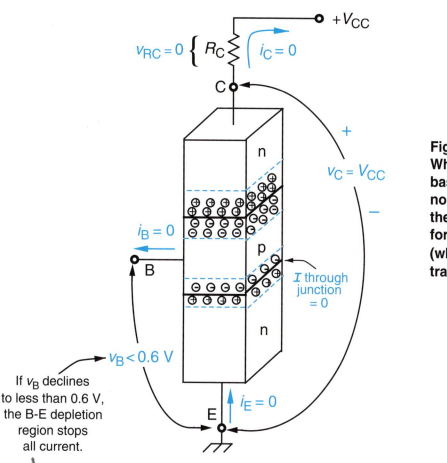

If v_B declines to less than 0.6 V, the B-E depletion region stops all current.

**Figure 4-21
When the external base circuit can no longer keep the B-E junction forward-biased (when $v_B < 0.6$ V), the transistor cuts off.**

As mentioned in Sec. 4-4, transistor cutoff will occur if the ac base current tries to exceed the dc bias base current. The situation shown in Fig. 4-22(A) would allow this to happen.

The calculations shown below the diagram of Fig. 4-22(A) indicate that $I_{b(pk)}$ tries to exceed I_B. This produces transistor cutoff, shown in Fig. 4-22(C), with the i_B waveform being clipped at 0 as it heads down toward -10 μA. The i_C waveform duplicates this action, becoming clipped at 0 instead of reversing direction in Fig. 4-22(E).

v_{RC} cannot become negative when the V_{CC} supply is positive, so it is clipped at 0 V in Fig. 4-22(F). And finally, Fig. 4-22(G) shows the v_C waveform clipped at +12 V, rather than rising to +12.9 V.

You should not get the idea that transistor cutoff is necessarily a bad event that happens only by mistake. In the example of Fig. 4-22, transistor cutoff has certainly caused the output waveform to lose its sine-wave shape, and that *may* be a mistake. But we will study some transistor applications where we *want* to drive the transistor into cutoff.

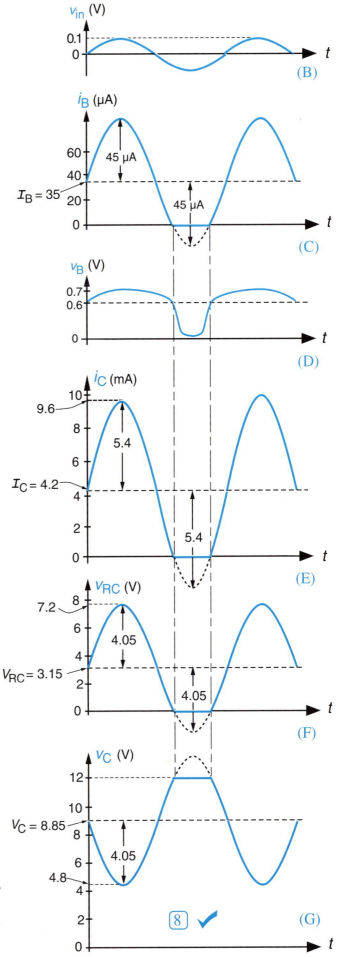

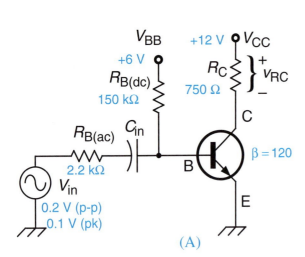

$$I_B = \frac{V_{BB} - 0.7\text{ V}}{R_{B(dc)}} = \frac{(6 - 0.7)\text{ V}}{150\text{ k}\Omega} = 35\text{ }\mu\text{A}$$

$$I_{b(pk)} = \frac{V_{in(pk)}}{R_{B(ac)}} = \frac{0.1\text{ V}}{2.2\text{ k}\Omega} = 45\text{ }\mu\text{A}$$

Figure 4-22 Circuit that causes transistor cutoff, and waveforms that illustrate the cutoff effect.
(A) Circuit schematic and calculations showing that the peak value of I_b is greater than I_B.
(B) through (G) Waveforms with detailed explanations.

(C) If the I_b negative peak could reach 45 μA, it would carry i_B to –10 μA, which is impossible. Cutoff occurs at $i_B = 0$.

(D) Cutoff coincides with v_B dropping below 0.6 V.

(E) $I_C = \beta I_B = 120\ (35\text{ }\mu\text{A}) = 4.2\text{ mA}$
    $I_{c(pk)} = \beta\ [I_{b(pk)}] = 120\ (45\text{ }\mu\text{A}) = 5.4\text{ mA}$
If the I_c negative peak could reach –5.4 mA, it would carry i_C to –1.2 mA, which is impossible. Cutoff occurs at $i_C = 0$.

(F) $V_{RC} = I_C R_C = (4.2\text{ mA})(750\text{ }\Omega) = 3.15\text{ V}$
    $V_{Rc(pk)} = [I_{c(pk)}]\ R_C = (5.4\text{ mA})(750\text{ }\Omega) = 4.05\text{ V}$
If the V_{Rc} negative peak could reach –4.05 V, it would carry v_{RC} to –0.9 V, which is impossible. Cutoff occurs at $v_{RC} = 0$.

(G) $V_C = V_{CC} - V_{RC} = 12\text{ V} - 3.15\text{ V} = 8.85\text{ V}$
    $V_{c(pk)} = V_{Rc(pk)} = 4.05\text{ V}$
If the V_c positive peak could reach 4.05 V, it would carry v_C to +12.9 V, which would be higher than the +12-V V_{CC} supply. This is impossible. Cutoff occurs at v_C =12 V.

Saturation

A transistor is said to be in the **saturated** condition if the collector-base junction loses its reverse-bias.

This happens when the voltage at the collector terminal is no longer more positive than the voltage at the base terminal, but instead v_C becomes more negative than v_B by about 0.6 V. The saturated condition is pictured in Fig. 4-23.

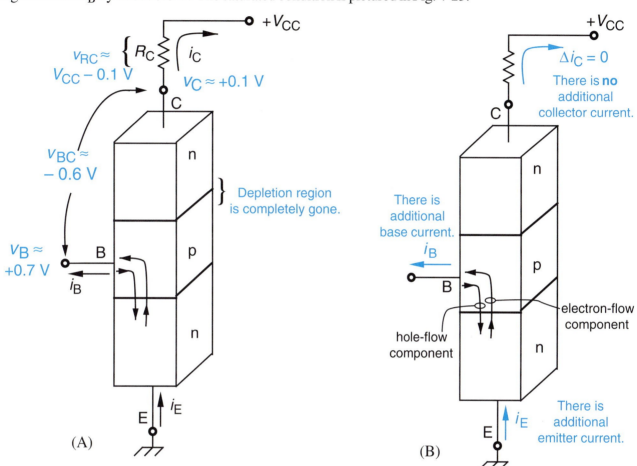

Figure 4-23 Transistor saturation. (A) The C-B depletion region is completely eliminated when the externally applied junction voltage becomes about –0.6 V (v_C more negative than v_B by 0.6 V).

Once a transistor saturates, any additional current that is forced through the B-E junction will all flow in the base lead. This is different from normal *npn* transistor action, in which the electron-flow component of i_E is swept into the collector region. Figure 4-23(B) shows the internal situation. Free electrons from the emitter are shown turning left and passing out the base lead instead of going straight up to pass out the collector lead. This description applies only to *additional* B-E current, the amount over and above the current that caused saturation to begin with.

Here is the event in the external circuit that causes transistor saturation:

Transistor saturation occurs when the collector current becomes so large that the Ohm's-law voltage drop across R_C equals the collector supply voltage V_{CC}. That is,

$$v_{RC} = i_C R_C = V_{CC} \quad \text{(for saturation)}$$

which gives

$$i_{C\,(\text{sat})} = \frac{V_{CC}}{R_C}$$

Eq. (4-8)

In Eq. 4-8, it would be a bit more precise to subtract 0.1 V from V_{CC}, as demonstrated in Fig. 4-23(A).

Combining Eqs. (4-8) and (4-1) gives the critical value of base current.

$$i_{B(sat)} = \frac{i_{C(sat)}}{\beta}$$

$$\boxed{i_{B(sat)} = \frac{V_{CC}}{\beta R_C}} \qquad \text{Eq. (4-9)}$$

EXAMPLE 4-5

Look at the transistor circuit of Fig. 4-24.
(a) Find the critical value of collector current that would cause the transistor to saturate [$i_{C(sat)}$].
(b) Find the critical value of base current that would cause the transistor to saturate [$i_{B(sat)}$].

SOLUTION

(a) To cause saturation, collector current i_C must become so large that it causes the entire supply voltage, V_{CC}, to be dropped across R_C. According to Kirchhoff's voltage law, this leaves zero voltage to appear across the transistor (collector-to-emitter voltage). The internal saturation conditions of Fig. 4-23 take over when $v_C \approx 0$ V. All this happens when i_C reaches the value given by Eq. (4-8):

$$i_{C(sat)} = \frac{V_{CC}}{R_C}$$

$$= \frac{12\text{ V}}{600\text{ }\Omega} = \textbf{20 mA}$$

(b) The ultimate cause of transistor saturation is overdriving the base input circuit. If the base current becomes so large that it makes the collector current-source produce the amount given by Eq. (4-8) (20 mA in this case), then it has driven the transistor into saturation. Applying Eq. (4-9), we get

$$i_{B(sat)} = \frac{V_{CC}}{\beta R_C}$$

$$= \frac{12\text{ V}}{180\,(600\text{ }\Omega)} = \textbf{111 }\boldsymbol{\mu}\textbf{A}$$

In Fig. 4-24 the transistor really will be driven into saturation. Let us demonstrate why it happens, and study the resulting waveforms.

With a 0.7-V drop across the base-emitter junction, we have a voltage across $R_{B(dc)}$ given by

$$V_{RB(dc)} = V_{BB} - V_{BE}$$

$$= 6\text{ V} - 0.7\text{ V} = 5.3\text{ V}$$

Therefore the $R_{B(dc)}$ current, which is dc base current, is given by Ohm's law as

$$I_B = \frac{V_{RB(dc)}}{R_{B(dc)}} = \frac{5.3\text{ V}}{68\text{ k}\Omega} = 77.9\text{ }\mu\text{A}$$

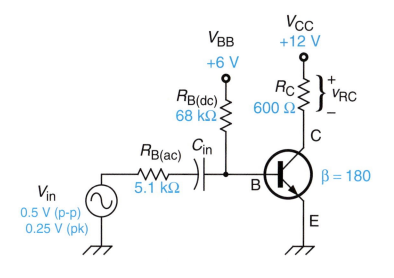

**Figure 4-24
Circuit for
demonstrating
transistor saturation.**

The ac current through $R_{B(ac)}$ flows through the B-E junction on top of (superimposed on) I_B. Its peak value is given by Ohm's law as

$$I_{b(pk)} = \frac{V_{in(pk)}}{R_{B(ac)}}$$

$$= \frac{0.25 \text{ V}}{5.1 \text{ k}\Omega} = 49.0 \text{ }\mu\text{A}$$

This calculation again ignores the resistance of the B-E junction.

The combined base current waveform is graphed in Fig. 4-25(B). Note that i_B exceeds the $i_{B(sat)}$ value of 111 µA from Example 4-5.

Dc collector current is calculated as

$$I_C = \beta I_B$$

$$= 180 (77.9 \text{ }\mu\text{A}) = 14.0 \text{ mA}$$

By Eq. (4-1), peak ac collector current ought to be

$$I_{c(pk)} = \beta [I_{b(pk)}]$$

$$= 180 (49.0 \text{ }\mu\text{A}) = 8.82 \text{ mA}$$

which would cause the i_C waveform to swing down to 5.18 mA and up to 22.8 mA, as suggested in Fig. 4-25(C). But i_C can never actually become larger than 20mA, which is the $i_{C(sat)}$ value from Example 4-5. Figure 4-25(C) shows the i_C waveform clipped at 20 mA.

Once the actual value of i_B reaches the $i_{B(sat)}$ value of 111 µA, i_C cannot follow it higher. As a general statement,

Transistor saturation is the condition in which a further increase in base current i_B produces no further increase in collector current i_C.

Get This

If i_C could reach 22.8 mA, v_{RC} would increase to

$$V_{Rc(pk)} = I_{c(pk)} R_C$$

$$= (22.8 \text{ mA}) (600 \text{ }\Omega) = 13.7 \text{ V,}$$

which is clearly impossible in a circuit with a 12-V supply. The actual v_{RC} waveform is shown in Fig. 4-25(D).

In Fig. 4-24, when the transistor saturates, the collector-to-emitter voltage falls to 0 V, ideally. This is shown in the v_C saturation waveform of Fig. 4-25(E).

Actually, v_{CE} falls to about 0.1 V in small-signal transistors.

Figure 4-25
Waveforms for Fig. 4-24,
showing the saturation effect.

9 ✓

(B) $I_{b(pk)}$ = 49.0 μA superimposed on
I_B = 77.9 μA.

(C) Normal transistor action would drive
i_C up to about 22.8 mA. But the
external collector circuit does not
permit i_C to exceed 20 mA, the
saturation value.

(D) $V_{RC} = I_C R_C = $ (14.0 mA)(600 Ω) =
8.4 V. The negative peak swings down
by (8.82 mA) (600 Ω) = 5.3 V. The
positive half cycle hits saturation at 12V.

(E) $V_C = V_{CC} - V_{RC}$ = 12 V – 8.4 V =
3.6 V. If v_C could swing down 5.3 V,
it would become –1.7 V, which is
impossible in a circuit with a positive
voltage source. At v_C = 0 V, the
transistor has become saturated.

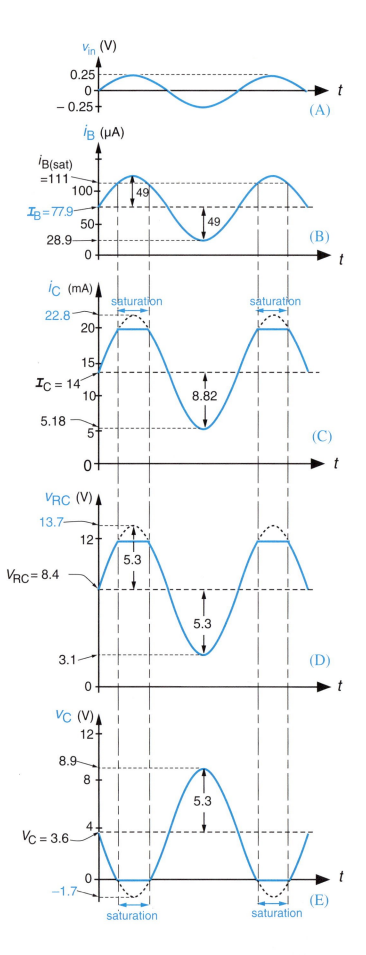

The same remark can be made about transistor saturation that was made about cutoff. It is not necessarily a bad thing. Some circuit applications purposely drive a transistor into saturation.

Comparing Normal Operation, Cutoff, and Saturation

A transistor that is operating normally, with a small change in base current producing a large change in collector current, is said to be in its **active** region. Another word that means the same thing is its **linear** region. It is helpful to summarize the different circuit conditions for each one of these three operating regions. Table 4-2 presents such a summary.

Operating Region	Cutoff	Active (also called Linear or Normal)	Saturated
Schematic of npn transistor	$+V_{CC}$, R_C, $i_C = 0$, $i_B = 0$, R_B, $v_{BE} < 0.6V$, $v_{CE} = V_{CC}$, $i_E = 0$	$+V_{CC}$, R_C, $i_C = \beta i_B$, i_B, R_B, $v_{BE} \approx +0.7V$, $v_{CE} > v_{BE}$, i_E	$+V_{CC}$, R_C, i_C, $i_B > i_{B(sat)}$, R_B, $v_{BE} \approx +0.7V$, $v_{CE} \approx 0V$, i_E
base-emitter junction	reverse-biased	forward-biased	forward-biased
collector-base junction	reverse-biased	reverse-biased	forward-biased
collector current, i_C	0	determined by the current-controlled current source: $i_C = \beta\, i_B$	at its maximum possible value, $i_{C(sat)}$: This is determined by external circuit values: $i_{C(sat)} = \dfrac{V_{CC}}{R_C}$

Table 4-2 Comparing the three transistor operating regions (conditions). For a *pnp* transistor, all the current directions and voltage polarities would be opposite.

SELF-CHECK FOR SECTION 4-5

For Questions 41 through 55, assume that the transistor is *npn*.

41. When a transistor is cut off, describe its base-to-emitter voltage v_B.

42. When a transistor is saturated, describe its v_B.

43. In cutoff, i_B equals _____ .
44. In cutoff, i_C equals _____ .
45. In saturation, i_C is determined by what two circuit values?
46. Write the formula for $i_{C(sat)}$, the i_C value of Question 45.
47. In cutoff, the collector-to-emitter voltage v_C equals _____ .
48. In saturation, v_C equals _____ .
49. In cutoff, the voltage across the collector resistor, v_{RC}, equals _____ .
50. In saturation, v_{RC} equals _____ .
51. In active operation, the base-emitter junction is _____-biased.
52. In active operation, the collector-base junction is _____-biased.
53. In what operating region does a transistor act like a current-controlled current source?
54. In what operating region does the collector current depend on the value of V_{CC}?
55. The formula $i_C = \beta i_B$ is valid in the _____ operating region.
56. For a *pnp* transistor in active operation, the voltage on the collector is _____ with respect to the emitter.
57. For a *pnp* transistor in active operation, describe the voltage at the base with respect to the emitter.

4-6 TRANSISTOR CHARACTERISTIC CURVES

A characteristic graph or characteristic curve of an electrical device is a graph of its current versus its voltage. We saw this in Sec. 2-3 for a junction diode.

A transistor has two kinds of characteristic graphs.

1. The base graph shows dc base current I_B versus dc base-emitter voltage V_{BE}. It is shown in Fig. 4-26.
2. The collector graph, usually called the collector curve, shows dc collector current I_C versus dc collector-emitter voltage V_{CE}. One particular collector curve is shown in Fig. 4-27.

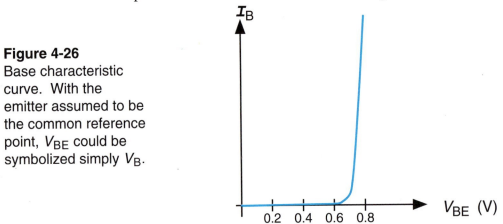

Figure 4-26
Base characteristic curve. With the emitter assumed to be the common reference point, V_{BE} could be symbolized simply V_B.

Base Curve

In Fig. 4-26, the base characteristic graph is the same as that of a silicon junction diode. We should expect this, since the transistor's base-emitter circuit is just that — a silicon *p-n* junction.

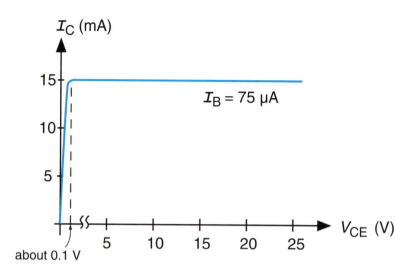

Figure 4-27
Collector characteristic curve for one particular base current value — 75 µA.

The transistor represented by this curve must have β = 200, since 15 mA ÷ 75 µA = 200.

Collector Curve

The collector characteristic curve of Fig. 4-27 shows the collector current I_C versus V_{CE} for a specific value of base current. The meaning of this flat-line graph is that:

> The collector current I_C does not depend on collector voltage V_{CE}, for all V_{CE} values above about 0.1 V.

Get This

This is again what we expect, since the whole transistor idea is that the collector is a constant-current source, regardless of collector voltage.

The small piece of the curve to the left of 0.1 V represents the saturation operating region. This part of the curve tells us that if I_C is unable to rise to 15 mA because the external collector circuit makes the transistor saturate when $I_B = 75$ µA, then V_{CE} will be less than 0.1 V. For example, suppose this transistor's collector circuit contains $V_{CC} = 12$ V and $R_C = 1$ kΩ. Then $I_B = 75$ µA will not produce $I_C = 15$ mA. Instead it will produce only about 12 mA of collector current, since $I_{C(sat)} \approx V_{CC} \div R_C = 12$ V ÷ 1 kΩ = 12 mA.

Family of Collector Curves

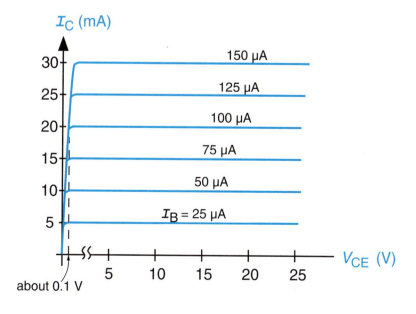

Figure 4-28
Family of collector characteristic curves for a transistor with β = 200.

⑩ ✓

When several collector curves are plotted, each one for a different value of I_B, we say that we have a **family** of curves. Fig. 4-28 presents a family of collector curves for a β = 200 transistor.

The fact that this family shows the saturation region leaning slightly to the right means that if the transistor is carrying larger currents, then the saturation effect will occur a bit sooner — before the V_{CE} value falls to 0.1 V.

Breakdown

The characteristic curves in Figs. 4-27 and 4-28 suggest that V_{CE} could increase to any value, no matter how great, without affecting I_C. Realistically, this is not true. Every transistor has a certain maximum V_{CE} voltage value that it can tolerate. Beyond that value, the collector current surges out of control, as indicated in Fig. 4-29. Exceeding the V_{CE} **breakdown** value can cause the transistor to be destroyed. A particular transistor's breakdown rating is symbolized $V_{CE(max)}$, or BV_{CEO}.

**Figure 4-29
Characteristic curves
showing high-voltage
breakdown. The
breakdown voltage,
symbolized BV_{CEO},
is about 35 V for this
transistor.**

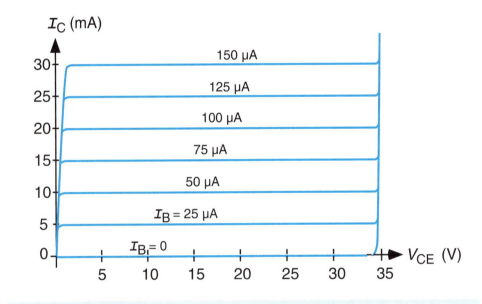

EXAMPLE 4-6

The photograph of Fig. 4-30 was taken from an instrument called a **transistor curve-tracer**. From the curve-tracer's controls, we know:
1. The horizontal axis (V_{CE}) scale factor is 2 V/division.
2. The vertical axis (I_C) scale factor is 2 mA/division.
3. The base current change per step is 10 μA.
    (a) Find the value of collector current I_C, for I_B = 20 μA.
    (b) Repeat for I_B = 60 μA.
    (c) What is the beta of this transistor?

SOLUTION

(a) The I_B = 20 μA curve is basically a flat line at a height of 2.4 divisions. Therefore,

$$I_C = (2.4 \text{ div})\left(\frac{2 \text{ mA}}{\text{div}}\right) = \textbf{4.8 mA}$$

(b) $I_B = 60\ \mu A$ gives an almost-flat horizontal line at a height of about 7.2 divisions.

$$I_C = (7.2\ \text{div})\left(\frac{2\ \text{mA}}{\text{div}}\right) = \textbf{14.4 mA}$$

(c) $\beta = I_C/I_B$. Using the current values from part (a), we get

$$\beta = \frac{I_C}{I_B}$$

$$= \frac{4.8\ \text{mA}}{20\ \mu A} = \frac{4.8 \times 10^{-3}}{20 \times 10^{-6}} = \textbf{240}$$

Using the currents in part (b) gives

$$\beta = \frac{14.4 \times 10^{-3}}{60 \times 10^{-6}} = 240$$

We should expect to get the same β value by using the currents from either part (a) or from part (b). The characteristic curves for an ideal transistor are perfectly evenly spaced, so the current ratio is the same everywhere within the family.

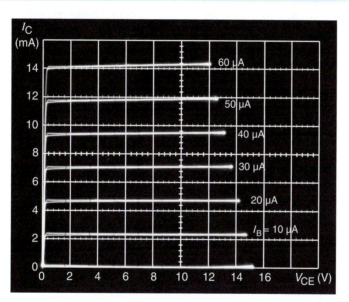

Figure 4-30
Actual family of characteristic curves for a type 2N3643 *npn* transistor.

I_C scale: 2 mA / div
V_{CE} scale: 2 V / div
I_B: 10 μA per step

4-7 NONIDEAL TRANSISTOR CHARACTERISTICS

The characteristic curves in Fig. 4-30 have been taken from a transistor that is nearly ideal. Some transistors, especially high-power types, have characteristic curves that are quite different. For example, Fig. 4-31 is a photograph of the characteristics of a type 2N3055 *npn* high-power transistor.

The first thing we notice about Fig. 4-31 is that the first-from-the bottom (100-μA) curve is very near the zero axis. The higher-value I_B curves that are above it (200 μA, 300 μA, and so on) have greater spacing. This means that the β value obtained from the 100-μA curve is much less than the β value for the higher currents. Thus, this transistor has nonconstant β, which is unlike the ideal.

Furthermore, in Fig. 4-31, the lines are not flat. They rise to the right. This means that the I_C current-source is not independent of V_{CE}, as an ideal current-source would be. Rather, the larger values of V_{CE} are associated with somewhat larger values of I_C. Also, the rising of the curves is not quite straight. It becomes more rapid (steeper) further to the right. This causes the spacing between curves to increase as V_{CE} increases, which translates into higher β at larger values of V_{CE}.

Figure 4-31
Characteristic curve family for a transistor that is very nonideal.

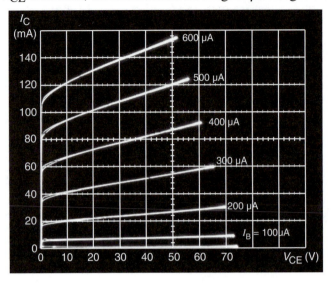

I_C scale: 20 mA/div
V_{CE} scale: 10 V/div
I_B: 100 μA per step

For the transistor shown in Fig. 4-31, it would not be possible to make a simple statement of the β value. Instead, we would have to specify the β value at a particular location within the family (at particular values of V_{CE} and I_C). Of course, if we must do this, then the whole idea of β is not as useful as before.

SELF-CHECK FOR SECTIONS 4-6 AND 4-7

58. (T-F) A flat collector characteristic curve indicates that collector-to-emitter voltage V_{CE} has no effect on collector current I_C.
59. (T-F) The flat curve described in Question 58 would occur with a nearly ideal transistor.
60. Describe how collector-to-emitter breakdown appears on a collector characteristic curve.
61. (T-F) The base characteristic curve for a transistor looks just like the characteristic curve of a silicon junction diode.
62. Sketch a family of characteristic curves for an ideal transistor with β = 150. Show curves for base currents going from 0 to 120 μA in 20-μA steps. Show the vertical and horizontal scale factors clearly. Ignore the saturation effect and the high-voltage breakdown effect.

4-8 TESTING TRANSISTORS

An ohmmeter can be used to test and troubleshoot a bipolar junction transistor. With the transistor isolated, a series of six tests must be made. They are illustrated in Fig. 4-32.

Testing With a VOM Ohmmeter

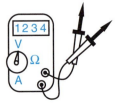

Figure 4-32 shows an *npn* transistor and an analog (VOM) ohmmeter. The VOM should be on a medium resistance multiplier scale. The R × 100 Ω scale is usually best. If the instrument has an R × 10 Ω or an R × 1 kΩ scale, it can probably be used as well. Do not use R × 1 Ω, R × 10 kΩ, or R × 100 kΩ multiplier scales. The reasons for avoiding these scales were given on page 30 in Section 2-4.

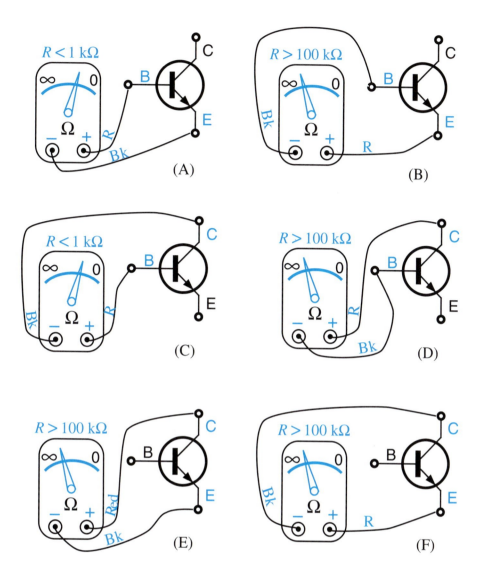

Figure 4-32
Complete testing
procedure for a BJT.

(A) Low resistance
with B-E junction
forward-biased.

(B) High resistance
with B-E junction
reverse-biased.

(C) Low resistance
with C-B junction
forward-biased.

(D) High resistance
with C-B junction
reverse-biased.

(E) and (F)
High resistance be-
tween C and E
for both polarities.

In Figs. 4-32(A) and (B), the *p-n* junction between base B and emitter E is tested for both polarities. This is equivalent to testing a *p-n* junction diode. A good transistor will measure a relatively low resistance value, less than 1 kΩ, in the forward direction. It will measure a relatively high resistance, usually greater than 100 kΩ, in the reverse direction.

In Figs. 4-32(C) and (D), the *p-n* junction between collector C and base B is tested in the same manner.

In Figs. 4-32(E) and (F), the base is disconnected, or open. Therefore the transistor looks to the ohmmeter like a series pair of *p-n* junction diodes, as suggested in Fig. 4-9(B). No matter which polarity of ohmmeter test voltage is applied, one of the junctions will be reverse-biased. So for a good transistor, both measured resistances will be large, greater than 100 kΩ.

> If the transistor passes all six tests, it is probably good.
> If it fails any one test, it is bad and must be replaced.

Get This

Passing the tests means that the transistor is good under the voltage conditions imposed by the ohmmeter. It may not necessarily work properly under higher-voltage conditions.

12 ✔

Another reason we avoid using a VOM's higher resistance multiplier scales is because those scales are usually powered by a higher-voltage internal battery. Internal voltages of 9 V to 30 V are typical. Voltages in this range may exceed a transistor's maximum allowable reverse voltage for its base-emitter *p-n* junction in the Fig. 4-32 (B) and (F) tests.

A transistor's maximum allowable reverse voltage between its emitter and base is called its emitter-base reverse breakdown rating. It is symbolized $V_{EBO(max)}$ or BV_{EBO}. Many transistors have a $V_{EBO(max)}$ less than 10 V.

Testing With a DMM Ohmmeter

The six tests shown in Fig. 4-32 can be performed with a DMM as well. Use the High-Ω function of the DMM, not the Low-Ω function. The equivalent internal test voltage for the Low-Ω function is only about 0.2 V, not enough to forward-bias a silicon *p-n* junction.

When testing for low resistance in Figs. 4-32 (A) and (C), use the 200-Ω or 2-kΩ range. Switch to the 200-kΩ range for the high-resistance tests in Figs. 4-32 (B), (D), (E) and (F).

Testing With a Transistor-Checker

A transistor-checker is an inexpensive instrument that measures a transistor's current gain β. In fact, it is sometimes called a beta-checker. A good transistor will measure a proper value of β — usually greater than 50 for a low-power unit. A bad transistor will measure a very low or zero β value. Figure 4-33 shows a typical transistor-checker.

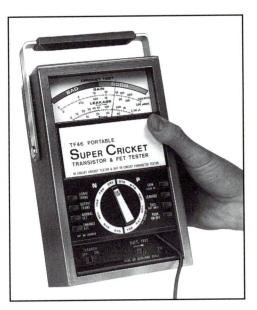

**Figure 4-33
Transistor-checker.**
*Courtesy of Sencore
Instruments, Inc.*

4-9 TRANSISTOR PHYSICAL APPEARANCE

Several of the most popular transistor package styles are shown in Fig. 4-34. The package in Fig. 4-34 (A) is common for small-signal transistors that have relatively low maximum current rating and voltage rating. The unit pictured here is a type number 2N4400 *npn* transistor with a maximum voltage

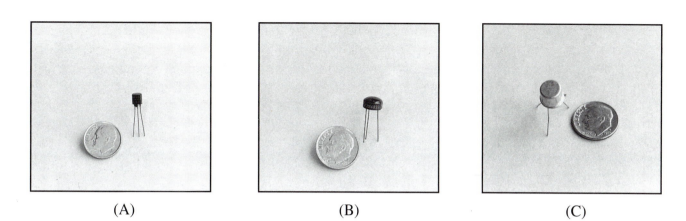

(A)   (B)   (C)

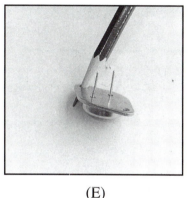

(D)   (E)

Figure 4-34 Commonly seen transistor packages. (A) Package number TO-92. This all-plastic package is the most common one for small-signal transistors. (B) Package number R-124, the bowler-hat-shaped package. (C) Package number TO-5, one of several top-hat-shaped packages. These packages are popular for medium-rated transistors. Starting from the tab, all top hats have a pin configuration of E-B-C. (D) Package number TO-220, having a plastic body with a metal extension that can be screwed onto a heat radiator (heat-sink) to help keep the transistor cool. Such plastic-and-metal packages are popular for transistors with medium-to-high ratings. (E) Package number TO-3, all-metal package. The metal case is the collector terminal. This and similar football-shaped metal packages are preferred for high-power transistors. Most of them have the base on the left and the emitter on the right, from this view.

rating, $V_{CEO(max)}$ of 40 V, and a maximum allowable collector current, $I_{C(max)}$, of 600 mA. The pin configuration for this kind of package is usually E-B-C from left to right.   Usually, there is no specific marking on any transistor package to indicate whether the transistor is *npn* or *pnp* (except the type number, of course).

Figure 4-34 (B) shows a transistor with much higher voltage rating. This type No. 2N4889 *pnp* transistor is rated at $V_{CEO(max)} = 150$ V, $I_{C(max)} = 100$ mA. Such "bowler-hat"-shaped packages generally have a shaved-off flat surface that marks the location of the emitter lead. The collector lead is the one that is farther away from the E lead; the base lead is the one between the C and E leads.

Figure 4-34 (C) shows a "top-hat" package. This particular transistor is a type No. 2N3053, *npn* transistor with $V_{CEO(max)} = 40$ V and $I_{C(max)} = 700$ mA.

A plastic package with metal heat sink extension is shown in Fig. 4-34 (D). This type No. 2N6111 *pnp* transistor has $V_{CEO(max)} = 30$ V and $I_{C(max)} = 8$ A. The pin configuration of this unit is B-C-E from left to right. However, other similar packages may have a different pin configuration.

An all-metal "football" package is shown in Fig. 4-34(E). This unit is a type No. 2N3055 *npn* transistor with $V_{CEO(max)} = 60$ V and $I_{C(max)} = 15$ A. When bolted to a heat sink, this transistor can handle a continuous electrical power input of 100 watts without suffering damage from overheating.

FORMULAS

For ac:

$$I_c = \beta I_b \qquad \text{Eq. (4-1)}$$

$$I_e = I_b + I_c \qquad \text{Eq. (4-3)}$$

$$I_e \approx I_c \qquad \text{Eq. (4-4)}$$

For dc:

$$I_C = \beta I_B \qquad \text{Eq. (4-5)}$$

$$I_E = I_B + I_C$$

$$I_E \approx I_C$$

$$V_{Rc} = V_{out} = I_c R_C \qquad \text{Eq. (4-6)}$$

$$V_{Rc} = V_{ce} = V_c \qquad \text{Eq. (4-7)}$$

$$i_{C(sat)} = \frac{V_{CC}}{R_C} \qquad \text{Eq. (4-8)}$$

$$i_{B(sat)} = \frac{i_{C(sat)}}{\beta} = \frac{V_{CC}}{\beta R_C} \qquad \text{Eq. (4-9)}$$

SUMMARY OF IDEAS

● In the emitter lead of a transistor, electron current flows against the arrow.
● The current gain of a transistor, β, is the factor by which I_c is greater than I_b.
● Since input base current I_b controls the output collector current I_c, a transistor is a current-operated device, not a voltage-operated device.
● *npn* and *pnp* transistors have different dc current directions, but they are no different in their ac behavior.
● For a fixed amount of ac base current I_b, a transistor's collector is a constant ac current-source.
● Collector current I_c does not change if R_C is changed or if V_{CC} is changed.
● To function as an amplifier, a transistor's B-E junction must be forward-biased and its C-B junction must be reverse-biased.
● All transistor operation is based on the idea that the B-E junction current has two components, one of them small in value and the other one larger by a factor of β. The small component flows in the base lead. The larger component is forced to flow in the collector lead.
● The ideas that explain the ac operation of a transistor also apply to the dc operation. They are: 1) $I_C = \beta I_B$ and 2) I_C is independent of R_C and V_{CC}, for a fixed value of I_B.
● The standard transistor configuration has the emitter as the common terminal, the base as the input, and the collector as the output.
● A transistor becomes cut off if the ac input current becomes larger than the dc base bias current, thereby reverse-biasing the B-E junction. All current flow stops. The output waveforms are distorted (clipped) sine waves.
● A transistor becomes saturated if the collector current becomes so large that the Ohm's law voltage drop across R_C becomes equal to V_{CC}. This forward-biases the C-B junction and causes the output waveforms to be distorted (clipped) sine waves.
● A family of collector characteristic curves tells the exact performance details of a transistor.
● A transistor can be tested with an ohmmeter. It can also be tested with a transistor-checker or a curve-tracer.

CHAPTER QUESTIONS AND PROBLEMS

1. In the schematic symbol of an *npn* transistor, which direction does the arrow point, toward the middle or toward the outside? Repeat for a *pnp* transistor.
2. Draw the schematic symbol of an *npn* transistor, label the three terminals, and show the direction of electron current in each lead.
3. Repeat problem 2 for a *pnp* transistor.
4. A certain transistor has $\beta = 200$. If the input source produces a current into the base of $I_b = 26\ \mu A$, what is the value of collector current I_c?
5. In problem 4, you were able to answer the question about I_c even though you do not know the value of the collector supply voltage V_{CC} or the amount of resistance in the collector lead, R_C. Explain fully why you did not have to know V_{CC} and R_C.
6. The most often seen transistor configuration is the common-emitter configuration. In it, the input signal is applied between the _____ and the _____ terminals; the output signal appears between the _____ and the _____ terminals.
7. For a transistor to function as a signal amplifier, the base-emitter junction must be _____-biased and the collector-base junction must be _____-biased. (Answer forward or reverse.)
8. (T-F) A transistor can be described as a voltage-operated current source.
9. Redraw the *npn* transistor amplifier circuit of Fig. 4-16, with the following specifications: $V_{s(p\text{-}p)} = 70\ mV$; $R_{B(ac)} = 3.9\ k\Omega$; $V_{BB} = 8\ V$; $R_{B(dc)} = 160\ k\Omega$; $V_{CC} = 15\ V$; $R_C = 1.2\ k\Omega$; $\beta = 120$.

    (a) Find the value of dc base bias current I_B.
    (b) Find I_C, the dc bias collector current.
    (c) What is the value of V_{RC}, the dc bias voltage across the collector resistor?
    (d) Find the value of ac base current I_b. Ignore the ac resistance of the B-E junction.
    (e) Find I_c, the ac collector current.
    (f) What is the magnitude of V_{out}, the ac output voltage across R_C?

10. For the amplifier of Problem 9, sketch the waveforms of V_s, i_B, i_C and v_{RC}.
11. (T-F) In problems 9 and 10, the magnitude of the ac voltage V_{Rc} is the same as the magnitude of V_{ce} (the ac voltage between collector and emitter).
12. Redraw the *pnp* transistor circuit of Fig. 4-19, with the following specifications: $V_{s(p\text{-}p)} = 100\ mV$; $R_{B(ac)} = 8.2\ k\Omega$; $V_{BB} = -8\ V$; $R_{B(dc)} = 150\ 1\Omega$; $V_{CC} = -20\ V$; $R_C = 680\ \Omega$; $\beta = 270$.
    Calculate the following circuit values.

    (a) I_B (d) V_C (g) V_{out}
    (b) I_C (e) I_b (ignore junction resistance)
    (c) V_{RC} (f) I_c

13. For the amplifier circuit of problem 12, sketch the waveforms of:

    (a) V_s (c) i_C
    (b) i_B (d) v_{RC}

14. (T-F) A transistor will go into cutoff if the peak value of the ac base current is greater than the dc base current.
15. Redraw the *npn* transistor amplifier circuit of Fig. 4-22 with the following specifications:
$V_{s(p\text{-}p)} = 200\ mV$; $R_{B(ac)} = 1.6\ k\Omega$; $V_{BB} = +9\ V$; $R_{B(dc)} = 180\ k\Omega$; $V_{CC} = 20\ V$; $R_C = 1\ k\Omega$; $\beta = 160$.
Calculate: (a) I_B (d) V_C
                    (b) I_C (e) I_b (ignore junction resistance)
                    (c) V_{RC}

16. For problem 15, sketch the waveforms of:
    (a) V_s (c) i_C (e) v_C
    (b) i_B (d) v_{RC}

17. The results of problems 15 and 16 indicate that the transistor has cut off. To what value should $R_{B(ac)}$ be increased, in order to prevent cutoff?

18. Redraw the *npn* transistor amplifier circuit of Fig. 4-24 with the following specifications: $V_{s(p-p)} = 0.7$ V; $R_{B(ac)} = 6.8$ kΩ; $V_{BB} = 8$ V; $R_{B(dc)} = 100$ kΩ; $V_{CC} = 14$ V; $R_C = 820$ Ω; $\beta = 150$. Calculate:
    (a) I_B (c) V_{RC} (e) I_b (ignore junction resistance)
    (b) I_C (d) V_C

19. For problem 18, sketch the waveforms of:
    (a) V_s (c) i_C (e) v_C
    (b) i_B (d) v_{RC}

20. The results of problems 18 and 19 indicate that the transistor has saturated. To what value should $R_{B(ac)}$ be increased, in order to prevent saturation?

21. (T-F) Transistor saturation occurs if the instantaneous total base current (dc bias value plus instantaneous ac value) becomes too large.

22. If a transistor is instantaneously saturated, its collector-to-emitter voltage v_{CE} is equal to _____.

23. If a transistor is instantaneously saturated, the voltage across its collector resistor, v_{RC}, is equal to _____.

    For Questions 24 through 26, refer to the family of transistor collector characteristic curves in Fig. 4-35. Ignore the slight vertical divergence that appears at the left of each new step.

24. What is the current gain, β, for this transistor? Get your information from the center of the curve family.

25. If the transistor had $I_B = 75$ μA, what would be the I_C value?

26. Approximately, what is the breakdown voltage, $V_{CEO(max)}$, for this transistor?

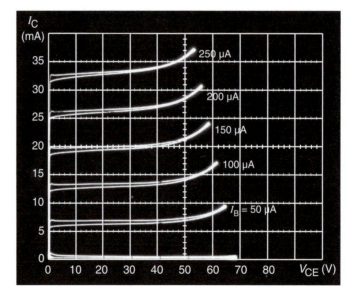

Figure 4-35

I_C scale: **5 mA/div**
V_{CE} scale: **10 V/div**
I_B: **50 μA per step**

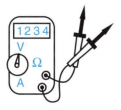

27. The complete ohmmeter test for a transistor requires _____ separate testing steps. (How many?)

28. In the complete ohmmeter test, the C-B junction should measure _____ resistance one way and _____ resistance the other way.

29. In the complete ohmmeter test, with the meter leads connected to collector and emitter, the meter should indicate _____ resistance one way and _____ resistance the other way.

CHAPTER 5

COMMON-EMITTER AMPLIFIERS

With tactile electronic technology, skilled blind persons can read not only printed matter, but complex images on computer screens. This operator is manipulating the computer's mouse with her right hand. She cannot visually see the cursor position, but she can maneuver among screen icons, menus, and printed material, and interpret the screen's graphic image, from the tactile signals detected by her left hand.   *Courtesy of NASA*

OUTLINE

NEW TERMS TO WATCH FOR

batch variation
temperature instability
voltage-divider bias
load-line
voltage gain
current gain

ac ground
power gain
stability
ac collector resistance
cascaded

input resistance
decouple
leakage
driving circuit
driven circuit

As we pointed out in Chapter 4, the common-emitter transistor configuration is only one of three possible configurations, although it is the most widely used. In this chapter we will make a more thorough study of common-emitter amplifier operation. Then we will examine common-collector and common-base operation in Chapter 6.

✓

After studying this chapter, you should be able to:

1. Draw the schematic diagram of an elementary single-supply dc biasing arrangement for a common-emitter transistor amplifier.
2. Explain why the elementary dc biasing circuit does not provide reliable stable bias conditions.
3. Draw the schematic diagram of a common-emitter amplifier with a base voltage-divider and emitter-stabilized bias arrangement.
4. Explain how the base-voltage-divider emitter-stabilized bias circuit is able to produce reliable stable dc bias conditions.
5. Given the component values, analyze a bias-stabilized common-emitter amplifier to predict the dc bias voltages and currents.
6. Show how a load-line can be drawn on top of the transistor's family of collector characteristic curves to visualize changes in the bias conditions.
7. Describe how the ac input signal makes the transistor move up and down the load-line, and show how this movement relates to the amplifier's output signal.
8. Define voltage gain A_v for an electronic amplifier.
9. Explain the meaning of ac resistance of the emitter junction, r_{Ej}, and make an approximate calculation of r_{Ej}.
10. From knowledge of r_{Ej} and the amplifier's component values, predict the voltage gain A_v for a C-E amplifier.
11. Define current gain A_i for an electronic amplifier.
12. Draw the three paths for ac input current in a bias-stabilized amplifier.
13. Explain why a dc power supply terminal (such as V_{CC}) acts like a ground terminal to the ac signals.
14. Define power gain A_P for an electronic amplifier.
15. Draw the schematic and explain the practice of ac-bypassing the emitter resistor R_E. Then draw and explain partial bypassing.
16. Draw the schematic of a capacitor-coupled load on an amplifier and explain how it provides pure ac, rather than ac combined with dc.
17. Explain the meaning of ac resistance in the collector circuit, r_C.
18. Draw a schematic diagram of two (or more) common-emitter amplifier stages cascaded together.
19. Explain the meaning of the ac input resistance seen looking into the base terminal of a transistor, $r_{b\ in}$.
20. Explain the meaning of the ac input resistance of the entire amplifier stage, R_{in}.
21. Explain how the ac input resistance of stage 2 acts like a load in the collector circuit of stage 1.
22. Describe the basic initial steps in troubleshooting an amplifier.
23. Show how to use an isolation transformer to provide safety when troubleshooting a line-operated amplifier.

24. Explain how to check an amplifier's dc power supply integrity and its ground integrity, in the troubleshooting process.
25. Explain how power-supply decoupling circuits work, and how to troubleshoot them.
26. When troubleshooting a common-emitter amplifier stage, relate the dc bias measurements to specific circuit failures, such as resistor shorts or opens, transistor shorts or opens, or capacitor shorts or opens.
27. Describe the systematic way of testing the ac signal to localize the cause of the trouble.
28. Define what is meant by the terms *driving circuit* and *driven circuit*.
29. Explain why disconnecting the driving circuit from the driven circuit sometimes helps the ac test procedure during troubleshooting.

5-1 A REALISTIC DC BIASING METHOD

In all the circuits in Chapter 4, the dc bias condition was set up using two separate supplies, V_{BB} and V_{CC}. In real-life amplifier circuits there usually is no V_{BB} present. Instead, the V_{CC} supply provides both the base bias current I_B and the collector bias current I_C. This is shown in Fig. 5-1(A).

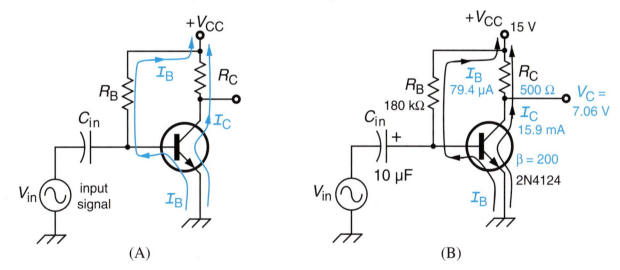

Figure 5-1 (A) Dc bias conditions set up by just a single dc supply rather than two supplies. (B) Specific values of dc bias currents and voltages for a type No. 2N4124 transistor, which has a typical beta value of 200.

For the specific R_B and R_C resistances, V_{CC} value, and β factor shown in Fig. 5-1(B), the dc bias conditions can be found as follows.

$$V_{BE} = 0.7 \text{ V} \text{ for the forward-biased B-E } pn \text{ junction.}$$

$$V_{RB} = V_{CC} - V_{BE} = 15 \text{ V} - 0.7 \text{ V} = 14.3 \text{ V}$$

Applying Ohm's law to R_B gives

$$I_B = \frac{V_{RB}}{R_B} = \frac{14.3 \text{ V}}{180 \text{ k}\Omega} = 79.4 \text{ }\mu\text{A}$$

Collector current I_C is found by

$$I_C = \beta I_B = 200 (79.4 \text{ }\mu\text{A}) = 15.9 \text{ mA}$$

Then, applying Ohm's law to R_C gives

$$V_{RC} = I_C R_C = (15.9 \text{ mA})(500 \text{ }\Omega) = 7.95 \text{ V}$$

Finally applying Kirchhoff's voltage law to the collector-to-emitter loop gives

$$V_C = V_{CC} - V_{RC} = 15 \text{ V} - 7.94 \text{ V} = 7.06 \text{ V}$$

Therefore the transistor's output terminal, its collector, is sitting at 7.05 V dc, waiting for the ac input signal to be applied. Once the V_{in} signal is applied, the instantaneous value of v_C will start to oscillate around the 7.05 V value.

This approach to building a transistor amplifier sounds very simple, but it has a hidden problem, or disadvantage. The problem arises because:

Get This

It is impossible for transistor manufacturers to guarantee that the actual β of a transistor will be close to the target value.

For example, the manufacturer of the 2N4124 transistor in Fig. 5-1(B) had a target value of $\beta = 200$. This is called the typical value on the transistor's data sheet. But because the doping process cannot be precisely controlled during large-volume production, the actual β value of a 2N4124 can vary from a minimum value of 120 to a maximum value of 480. If you intend to mass-produce the circuit of Fig. 5-1(B), then you must buy a large batch of these transistors. In that batch, there are bound to be some individual units that have $\beta \approx 120$ and there are bound to be some that have $\beta \approx 480$.

When a unit with $\beta = 480$ is installed in this amplifier circuit, the results are completely unsatisfactory. Figure 5-2 shows what happens.

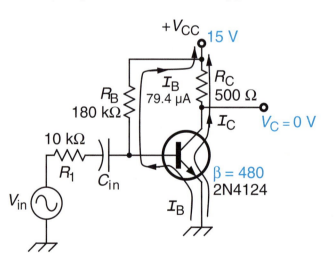

Figure 5-2   Demonstrating the problem of transistor batch variation. The circuit construction is the same as in Fig. 5-1(B), but now the bias conditionsmake the circuit absolutely useless because the transistor's β is too high.

Dc base bias current I_B is the same as before, 79.4 µA. Applying Eq. (4-5) gives

$$I_C = \beta I_B = 480 (79.4 \text{ µA}) = 38.1 \text{ mA}$$

But this amount of collector current cannot flow in Fig. 5-2, because the transistor goes into saturation as soon as I_C becomes large enough to cause the entire V_{CC} supply voltage to be dropped across R_C. That is,

$$I_{C(sat)} = I_{C(max)} = \frac{V_{CC}}{R_C} = \frac{15 \text{ V}}{500 \text{ }\Omega} = 30 \text{ mA}$$

In other words, Eq. (4-5) cannot hold true in this situation because the circuit that the transistor is connected to cannot deliver a current greater than 30 mA.

Thus, we see that the high-β transistor goes into saturation, with a V_C value of nearly 0 V. Since there is no dc voltage to vary around, it is impossible to produce a sine-wave output signal voltage at the collector. The circuit becomes useless as an ac amplifier.

For example, if $V_{in(p-p)} = 500$ mV is applied to a 10-kΩ resistor R_1 in Fig. 5-2, the input and output voltage waveforms will look like those in Fig. 5-3.

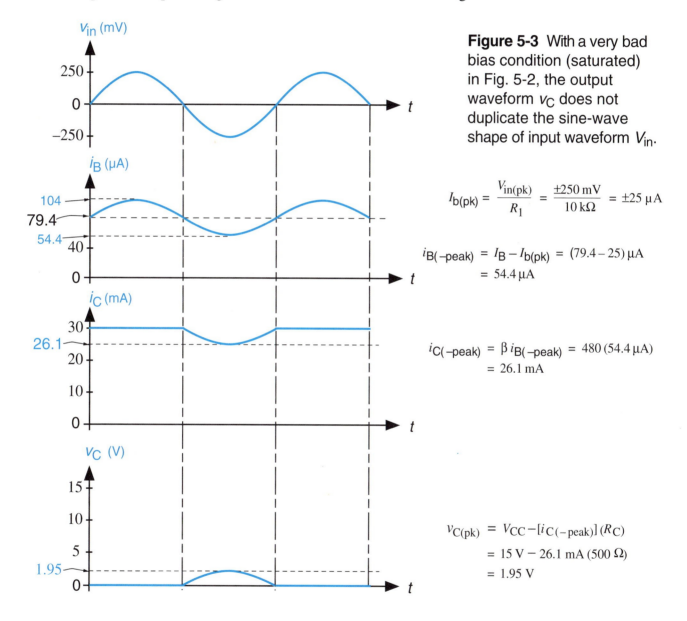

Figure 5-3 With a very bad bias condition (saturated) in Fig. 5-2, the output waveform v_C does not duplicate the sine-wave shape of input waveform V_{in}.

$$I_{b(pk)} = \frac{V_{in(pk)}}{R_1} = \frac{\pm 250 \text{ mV}}{10 \text{ k}\Omega} = \pm 25 \text{ μA}$$

$$i_{B(-peak)} = I_B - I_{b(pk)} = (79.4 - 25) \text{ μA}$$
$$= 54.4 \text{ μA}$$

$$i_{C(-peak)} = \beta\, i_{B(-peak)} = 480\,(54.4 \text{ μA})$$
$$= 26.1 \text{ mA}$$

$$v_{C(pk)} = V_{CC} - [i_{C(-peak)}]\,(R_C)$$
$$= 15 \text{ V} - 26.1 \text{ mA}\,(500 \text{ Ω})$$
$$= 1.95 \text{ V}$$

We must conclude that the bias method shown in Figs. 5-1 and 5-2 is not reliable. It does not work well for transistors with β-values that deviate from the typical value. Actually, the separate V_{BB} supply bias method, shown throughout Chapter 4, has the very same problem.

Large variation in β among transistors of the same type number is called the **batch-variation** problem.

Besides the batch-variation problem, transistors have another problem:

The β of an individual transistor changes as its temperature changes. A transistor will have a higher β when it operates at a higher temperature.

This is called the temperature-variation problem, or the **temperature-instability** problem.

If the amplifier in Figs. 5-1 and 5-2 has to operate over a wide range of temperatures, the dc bias conditions will change as the temperature changes. This is even more unmanageable than the batch-variation problem. At least the batch-variation problem could be dealt with by individually testing every transistor, and installing larger values of R_B for those transistors with higher betas, and lower values of R_B for those transistors with lower β. (Such individual testing would be very expensive for mass-production of the circuit. Therefore it is seldom done.)

Over an operating temperature range of –25°C to +125°C (–13°F to +257°F), an individual transistor's beta may change by more than 50%.

What we need is a better dc bias arrangement, one that can counteract these two problems. Figure 5-4 shows such an improved bias arrangement. It uses three resistors, not just one, to set the dc bias conditions. This arrangement is called **voltage-divider bias**. Here is how it works.

$\boxed{1}$ The two resistors R_{B1} and R_{B2} behave almost like they are in series. This is because the dc base current I_B that flows into their junction point is quite small, compared to their basic dc current, $I_{Divider}$. This idea is pictured in Fig. 5-5(A).

In that figure, suppose $I_{Divider}$ is about 2.5 mA, which is typical of a real amplifier circuit. I_B might be about 100 μA, a fairly typical value. Then,

$$I_{RB2} = I_{Divider} = 2.5 \text{ mA}$$

$$I_{RB1} = I_{Divider} + I_B = 2.5 \text{ mA} + 100 \text{ μA} = 2.5 \text{ mA} + 0.1 \text{ mA}$$

$$= 2.6 \text{ mA}$$

These two resistor current values, 2.5 mA and 2.6 mA, are so close that we consider them approximately equal. Therefore R_{B1} and R_{B2} are approximately in series with each other. Their basic current is given by Ohm's law as

$$I_{Divider} = \frac{V_{CC}}{R_{B1} + R_{B2}}$$

$\boxed{2}$ With R_{B1} and R_{B2} nearly in series, we can apply the voltage-divider formula to find V_{RB2}.

$$\frac{V_{RB2}}{V_{CC}} = \frac{R_{B2}}{R_{B1} + R_{B2}}$$

For example, in Fig. 5-5(B),

$$\frac{V_{RB2}}{15 \text{ V}} = \frac{1.2 \text{ k}\Omega}{4.7 \text{ k}\Omega + 1.2 \text{ k}\Omega}$$

$$V_{RB2} = 15 \text{ V} \left(\frac{1.2 \text{ k}\Omega}{5.9 \text{ k}\Omega}\right) = 3.0 \text{ V}$$

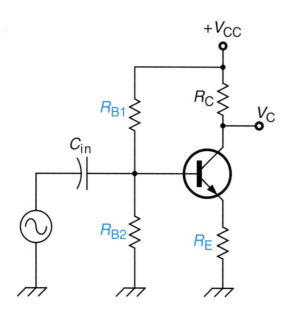

Figure 5-4
A *bias-stabilized* common-emitter amplifier. Bias-stabilized means that variation in beta no longer has a great effect on the dc bias voltage V_C.

3 ✔

3 Now this 3.0 V is V_B, the dc bias voltage at the base with respect to ground. Assuming that the voltage across the forward-biased B-E junction is $V_{BE} = 0.7$ V, we can find the dc bias voltage at the emitter, V_E.

$$V_E = V_B - V_{BE}$$

$$= 3.0 \text{ V} - 0.7 \text{ V} = 2.3 \text{ V}$$

Study Fig. 5-5(B) to see this.

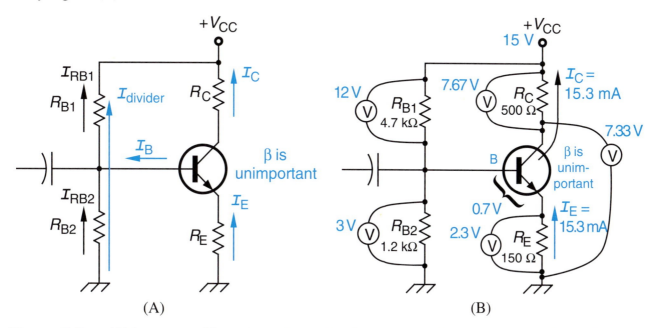

(A)   (B)

Figure 5-5   (A) In an amplifier with voltage-divider bias, we design so that I_B is negligible compared to $I_{Divider}$. This sets the value of V_B.   (B) With V_B set, V_E is bound to be less by 0.7 V. With V_E set, Ohm's law sets I_E, which in turn sets I_C.

4 But this V_E voltage is just the voltage across the emitter resistor R_E, as Fig. 5-5(B) makes clear. Therefore, by Ohm's law,

$$I_E = \frac{V_{RE}}{R_E} = \frac{2.3 \text{ V}}{150 \text{ }\Omega} = 15.3 \text{ mA}$$

⑤ We know that $I_C \approx I_E$ [Eq. (4-4) for the dc case]. Therefore

$$I_C \approx 15.3 \text{ mA}$$

and $\quad V_C = V_{CC} - I_C R_C = 15 \text{ V} - (15.3 \text{ mA})(500 \text{ }\Omega)$

$$= 15 \text{ V} - 7.65 \text{ V} = 7.35 \text{ V}$$

This V_C value, 7.35 V, is about half of V_{CC}. In general, $V_C \approx (1/2)V_{CC}$ is usually a desirable point for a transistor amplifier.

Note carefully what happened in steps 1 through 5. Because the transistor circuit was forced to obey Kirchhoff's voltage law and Ohm's law, the collector current was forced to become a proper value, regardless of the transistor's β. In fact, β never entered into the calculations of steps 1 through 5.

Let us explain the mechanism that caused β to become unimportant to the dc bias. The key is to realize that V_{RE} must become 2.3 V. It cannot become any other value, either larger or smaller.

Figure 5-6
Explaining the self-correcting action of the voltage-divider bias arrangement. The conditions shown in (A) and (B) cannot really occur. We just *imagine* that they occur so we can understand how the circuit would automatically correct itself.

(A) I_E is too small.

(B) I_E is too large.

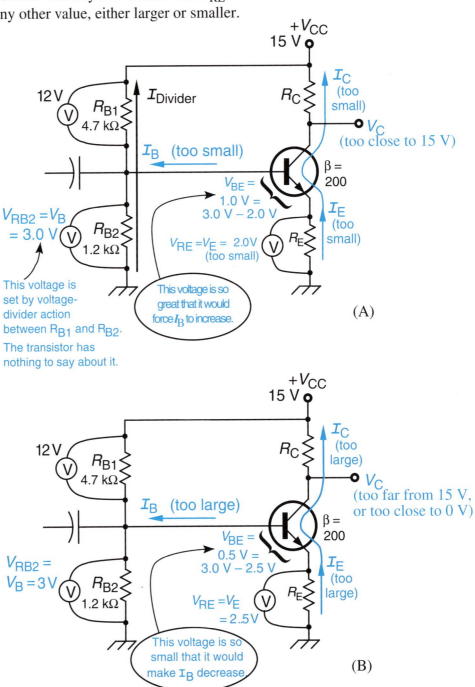

If β is a medium value, say 200, a certain amount of base current I_B must flow in Fig. 5-5. This I_B amount must be the proper amount so that βI_B gives an I_C value that causes 2.3 V to be dropped across R_E. There *must* be 2.3 V dropped across R_E. If there were only, say, 2.0 V across R_E, then the difference between the 3.0-V V_B value and the 2.0-V V_{RE} value would demand that 1.0 V appear across the B-E junction, as pictured in Fig. 5-6(A). This cannot really happen, of course, but the attempt to make it happen would drive the B-E junction harder. The harder drive would make the junction carry more base current, which would increase I_C and I_B, forcing V_{RE} to become larger.

This argument works the opposite way if there were supposedly more than 2.3 V across R_E, let us say 2.5 V, as shown in Fig. 5-6(B). Then the difference between V_B and V_{RE} would be only 0.5 V, which is too small to adequately drive the B-E junction. I_B would therefore decrease, causing a decrease in I_C and I_E. This in turn would force V_{RE} to decrease.

For a β = 200 transistor, the proper amount of I_B that must flow can be found by:

$$I_E = \frac{V_{RE}}{R_E} = \frac{2.3\ V}{150\ \Omega} = 15.3\ mA \approx I_C$$

$$I_B = \frac{I_C}{\beta} = \frac{15.3\ mA}{200} = 76.5\ \mu A$$

The voltage-divider bias setup will automatically produce this amount of base current I_B when β = 200. This is shown in Fig. 5-7(A).

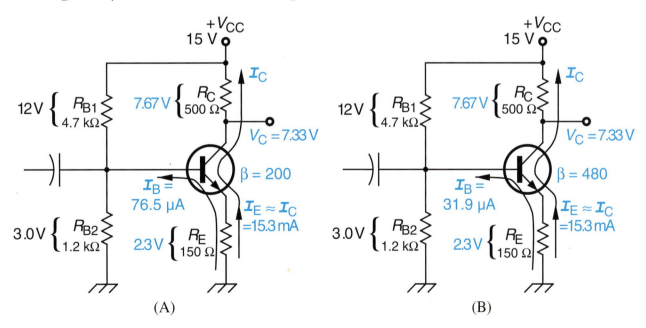

Figure 5-7 The voltage-divider bias arrangement provides a medium amount of base current for a medium-β transistor. (B) It provides a reduced amount of base current for a high-β transistor.

Now suppose that a high-β transistor (β = 480) is placed in the amplifier circuit, as shown in Fig. 5-7(B). There is a *tendency* for the dc collector current I_C to be larger than normal. However, if I_C tries to be a bit greater than 15.3 mA, I_E does the same. This causes V_{RE} to become slightly larger than 2.3 V, say 2.35 V.

Then the difference voltage becomes

$$V_{BE} = V_B - V_{RE}$$
$$= 3.0 \text{ V} - 2.35 \text{ V} = 0.65 \text{ V}.$$

This reduced value of V_{BE} (compared to 0.7 V) reduces the drive on the B-E junction, which cuts back on the base current. Therefore the higher value of β is automatically compensated for by a lower value of I_B. In the end, the transistor winds up carrying the proper value of I_C to give a midpoint bias.

For $\beta = 480$, the proper amount of I_B that must flow is

Still the same key condition:
$V_{RE} = 2.3$ V

$$I_E \approx I_C = \frac{V_{RE}}{R_E} = \frac{2.3 \text{ V}}{150 \, \Omega} = 15.3 \text{ mA}$$

$$I_B = \frac{I_C}{\beta} = \frac{15.3 \text{ mA}}{480} = 31.9 \, \mu\text{A}$$

SELF-CHECK FOR SECTION 5-1

1. (T-F) Modern transistors are manufactured with β tolerance of about $\pm 10\%$.
2. In general, a desirable dc bias voltage V_C is about _____ of V_{CC}.
3. (T-F) Modern transistors are very temperature-stable, showing very little change in β if the ambient temperature changes.
4. In an amplifier that is bias-stabilized by voltage-divider bias, a low-β transistor is automatically given _____ base current; a high-β transistor is automatically given _____ base current (answer greater or less).
5. In the amplifier of Fig. 5-5(B), suppose a particular transistor has $\beta = 140$. Find the value of base current I_B that is provided by the voltage-divider bias setup.
6. For the transistor in Problem 5, suppose the ambient air temperature drops to $-20°$ C, which causes the β to decline to 120. Find the new value of base current.

5-2 OPERATING ALONG THE LOAD-LINE

An amplifier's **load-line** is a line that is drawn on top of the transistor's family of characteristic curves. On graph paper, the position of the load-line is set by the amplifier's circuit values in the collector-emitter flow path, namely V_{CC}, R_C and R_E. Then the bias condition of the amplifier can be identified as a certain specific point along the load-line. This is helpful to us in visualizing transistor bias.

Load-Line for Simple Bias Arrangement

For example, look back to the simple amplifier with no emitter resistor, that was shown in Fig. 5-1(B). The collector-emitter circuit of that amplifier has been redrawn in Fig. 5-8(A), with the same assumed β of 200.

The maximum possible collector current occurs if the transistor goes into saturation, as shown in Fig. 5-8(B). With the transistor saturated,

$$V_{CE} \approx 0 \text{ V}.$$

Therefore, by Kirchhoff's voltage law,

$$V_{RC} = V_{CC} - V_{CE} = 15 \text{ V} - 0 \text{ V} = 15 \text{ V}$$

Applying Ohm's law to R_C gives

$$I_{C(sat)} = \frac{15\text{ V}}{500\ \Omega} = 30\text{ mA}$$

This maximum-current condition identifies one of the end-points of the circuit's load-line. The point is $I_C = 30$ mA, $V_{CE} = 0$ V. This point is marked on the I_C-versus-V_{CE} graph of Fig. 5-8(D).

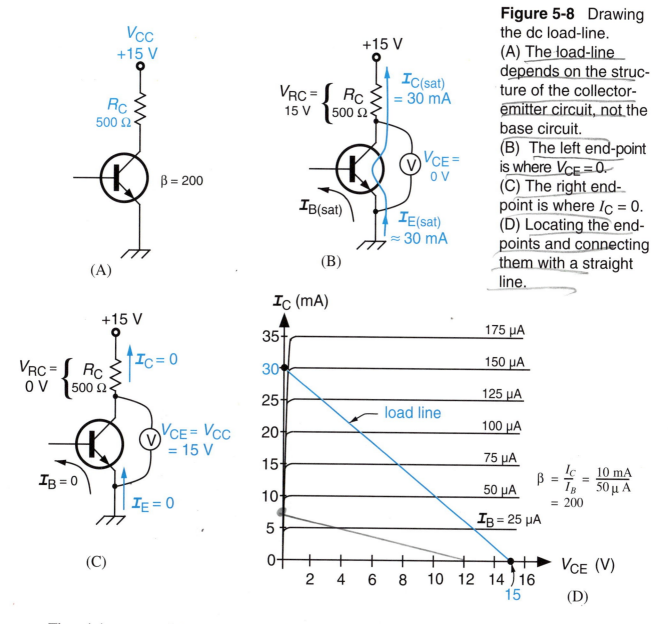

(A)

(B)

(C)

(D)

Figure 5-8 Drawing the dc load-line.
(A) The load-line depends on the structure of the collector-emitter circuit, not the base circuit.
(B) The left end-point is where $V_{CE} = 0$.
(C) The right end-point is where $I_C = 0$.
(D) Locating the end-points and connecting them with a straight line.

The minimum possible current occurs if the transistor cuts off. Then, as shown in Fig. 5-8(C),

$$I_C = 0$$

With zero current, Ohm's law for R_C gives $V_{RC} = 0$ V. So, by Kirchhoff's voltage law,

$$V_{CE} = V_{CC} - V_{RC} = 15\text{ V} - 0\text{ V}$$
$$= 15\text{ V}$$

This minimum-current condition identifies the other end-point of the load-line. The point is $I_C = 0$, $V_{CE} = 15$ V. It is marked on the graph of Fig. 5-8(D).

⑥ ✔

The load-line is formed by joining the two end-points with a straight line, as Fig. 5-8(D) shows. Any pair of values for I_C and V_{CE} must lie on that line.

With the base circuit connected as shown on p. 109 in Fig. 5-1(B), giving $I_B = 79.4$ μA, here is what happens.

[1] The *transistor itself* demands that the circuit operate on the horizontal line for $I_C = 15.9$ mA, corresponding to $I_B = 79.4$ μA.

[2] The *external collector-emitter circuit* demands that the circuit operate on the load-line.

In order for both of these demands to be satisfied, this must happen.

Get This

The actual dc bias point will be where the transistor's horizontal characteristic line intersects the load-line.

This intersection is shown in Fig. 5-9(A). It gives us a clear visual indication of the bias point.

If a transistor with β = 300 is inserted into the circuit, the situation is as shown in Fig. 5-9(B). The higher β is reflected as greater spacing between the individual characteristic curves in the family. It is visually clear that the bias point is not well placed.

For an extreme β of 480, which was shown schematically in Fig. 5-2, we get the indication of Fig. 5-9(C). The relevant collector characteristic curve is $I_C = 38.1$ mA, since

$$I_C = 480 \, (79.4 \text{ μA}) = 38.1 \text{ mA}.$$

But this curve doesn't intersect the load-line. Therefore the bias point is at the load-line's closest point to the curve, namely $I_C = 30$ mA, $V_{CE} = 0$ V (saturation).

This ACTIS system (see p. 83) is more technologically advanced than a CAT scanner in two ways: 1) ACTIS has three separate X-ray sources and three detectors, so it can "slice" the part along three separate axes, without the part having to be moved. A CAT

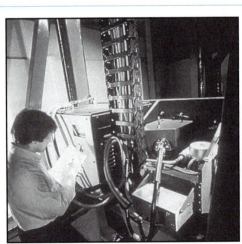

Courtesy of Bio-Imaging Research, Inc.

scanner slices the human body along only one axis. If a different body axis is needed, the human must be rotated. 2) ACTIS has variable geometry, meaning that the distance between an X-radiation source and its detector can be varied. They can be brought closer together or moved farther apart, to adapt to objects of widely differing sizes.

At right, inspecting airplane landing gear with ACTIS.

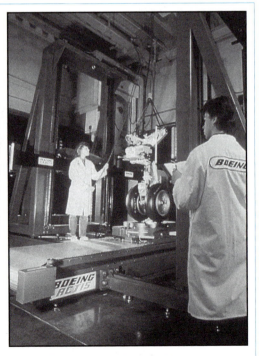

Courtesy of Boeing Aerospace and Electronics Co.

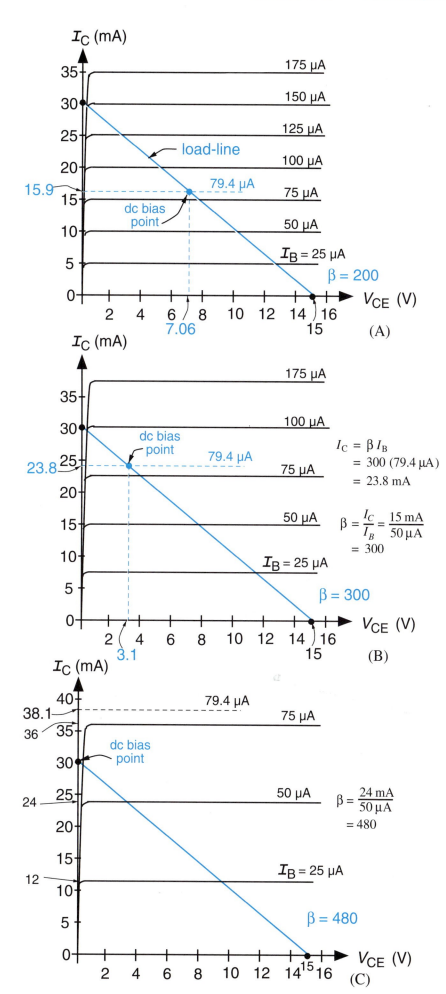

Figure 5-9

The dc bias point is at the intersection of the transistor collector characteristic curve and the load-line.

(A) For the circuit of Fig. 5-1(B), with β = 200, the bias point is at I_C = 15.9 mA, V_{CE} = 7.06 V. This is a desirable bias point because it allows enough room for v_{CE} to move either to the right (increase) or to the left (decrease) once the ac signal is applied.

(B) The same circuit with a higher-β transistor (β = 300). The bias point is at I_C = 23.8 mA, V_{CE} = 3.1 V. This is not as desirable as part (A) because it doesn't allow enough room for v_{CE} to move to the left, once the ac input signal is applied.

(C) The same circuit with a maximum-β transistor (β = 480). The bias point is at V_{CE} = 0 V (saturation). The circuit is clearly unable to respond properly to an ac input signal, because v_{CE} cannot move any farther to the left (cannot make a negative half cycle).

$I_C = \beta I_B$
$= 300\,(79.4\,\mu A)$
$= 23.8$ mA

$\beta = \dfrac{I_C}{I_B} = \dfrac{15\ \text{mA}}{50\ \mu A}$
$= 300$

$\beta = \dfrac{24\ \text{mA}}{50\ \mu A}$
$= 480$

Load-Line for Voltage-Divider Bias Arrangement

Drawing the dc load-line for an amplifier with the voltage-divider bias-stabilized arrangement is almost as easy as for the simple bias circuit. For example, the collector-emitter circuit of Figs. 5-5 through 5-7 is redrawn in Fig. 5-10(A).

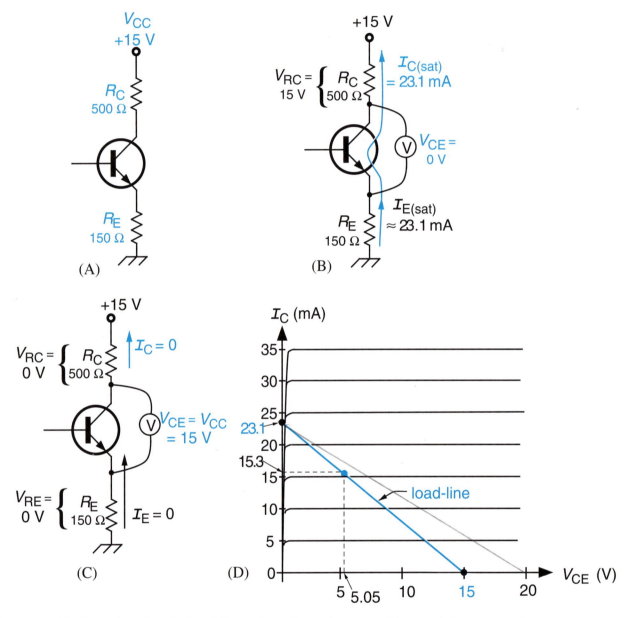

Figure 5-10 Drawing the dc load-line when there is an emitter resistor present.
(A) The collector-emitter path. (B) The left end-point, where $V_{CE} = 0$.
(C) The right end-point, where $I_C = 0$. (D) Connecting the end-points to form the load-line.

The maximum possible collector current occurs if the transistor goes into saturation, as shown in Fig. 5-10(B). With the transistor saturated,

$$V_{CE} \approx 0 \text{ V}.$$

Then, since $I_C \approx I_E$, we can regard the collector resistor R_C to be approximately in series with emitter resistor R_E. Applying Ohm's law to the series circuit, we get

$$I_{C(sat)} = \frac{V_{CC} - V_{CE}}{R_C + R_E}$$

$$= \frac{15\ V - 0\ V}{500\ \Omega + 150\ \Omega} = \frac{15\ V}{650\ \Omega} = 23.1\ mA$$

This maximum-current condition identifies the left end-point of the circuit's load-line, as marked on the I_C-versus-V_{CE} graph of Fig. 5-10(D).

The minimum possible collector current occurs if the transistor cuts off, naturally. Then, as shown in Fig. 5-10(C),

$$I_C = I_E = 0.$$

With zero current,

$$V_{RC} = V_{RE} = 0\ V$$

So by Kirchhoff's voltage law in Fig. 5-10(A),

$$V_{CE} = V_{CC} - V_{RC} - V_{RE} = 15\ V - 0\ V - 0\ V$$
$$= 15\ V$$

Therefore the right end of the load-line is still at 15 V, as marked on Fig. 510(D).

Since this is a bias-stabilized circuit, the particular I_B value will change for different values of transistor β. I_B will become whatever value it has to, in order to locate the bias point at $I_C = 15.3$ mA, $V_{CE} = 5.05$ V. The 5.05-V value for V_{CE} can be understood by applying Kirchhoff's voltage law to the collector-emitter circuit of Fig. 5-10(A).

$$V_{CE} = V_{CC} - V_{RC} - V_{RE}$$
$$= V_{CC} - I_C\, R_C - I_E\, R_E$$
$$= 15\ V - (15.3\ mA)(500\ \Omega) - (15.3\ mA)(150\ \Omega)$$
$$= 15\ V - 7.65\ V - 2.30\ V$$
$$= 5.05\ V$$

SELF-CHECK FOR SECTION 5-2

7. (T-F) The dc load-line represents all the possible dc bias points for a particular amplifier circuit.
8. (T-F) The location of the dc load-line is determined by the collector-emitter circuit values, not the base circuit values.
9. The left end-point of a dc load-line has $I_C = I_{C(sat)}$ and $V_{CE} =$ _____ .
10. The right end-point of a dc load-line has $I_C =$ _____ and $V_{CE} = V_{CC}$.
11. Draw the dc load-line for a simple amplifier circuit like the one in Fig. 5-1, with the following circuit values: $V_{CC} = 12$ V, $R_C = 1.5$ kΩ, $R_B = 270$ kΩ.
12. Suppose a transistor with $\beta = 100$ is installed in the circuit of Problem 11. Draw a few of the collector characteristic curves on the same graph as your load-line. Include the curve for $I_B = 40$ μA. Calculate the actual dc base current I_B and on the load-line mark the amplifier's dc bias point.
13. Draw the dc load-line for an emitter-stabilized amplifier of the type shown in Fig. 5-10(A), with $V_{CC} = 20$ V, $R_C = 750$ Ω and $R_E = 100$ Ω.
14. On the same graph, draw the family of collector characteristic curves for the $\beta = 100$ transistor.
    a) What value of I_B would produce a dc bias point with $V_{CE} = 10$ V?
    b) What value of I_B would produce a dc bias point with $V_{CE} = 6$ V?
    c) What value of I_B would produce a dc bias point with $V_{CE} = 14$ V?

5-3 APPLYING THE AC INPUT SIGNAL
TO AN EMITTER-STABILIZED AMPLIFIER

Once the proper dc bias conditions have been established, the ac input signal is applied through coupling capacitor C_{in}. This is shown in Fig. 5-11(A), using the same amplifier as in Figs. 5-5 through 5-7, and Fig. 5-10.

Figure 5-11
Using a load-line to visualize the total (ac plus dc) operation of an amplifier.

(A) Our familiar amplifier, now with an ac input signal.

(B) If β is 204, the dc base bias current is about 75 μA. If the ac input signal moves i_B back and forth between 100 μA and 50 μA, the transistor moves back and forth between v_{CE} = 1.7 V and v_{CE} = 8.4 V.

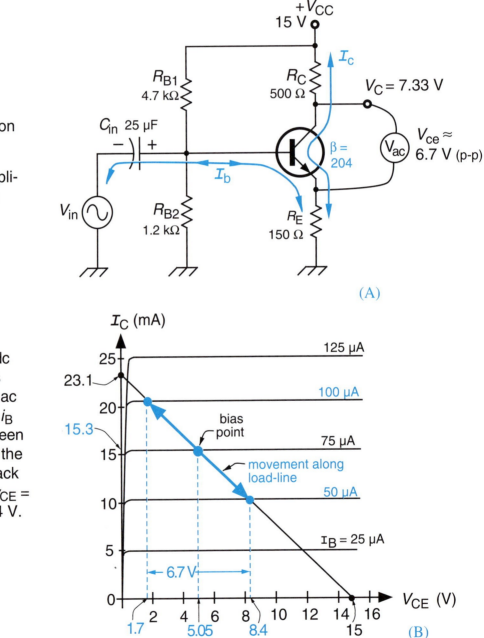

As the total base current i_B varies around its dc value, the instantaneous voltage and current move up and down the load-line. For example, suppose the ac signal causes i_B to swing 25 μA above its 75-μA bias value, up to 100 μA; and to swing 25 μA below its bias value, down to 50 μA. Then the transistor operating point moves up and down the load-line between the two points indicated in Fig. 5-11(B). These two points correspond to V_{CE} = 1.7 V and V_{CE}

7 ✓

= 8.4 V, approximately. In this case the ac voltage between C and E would be given approximately as

$$V_{ce} = \Delta V_{CE} = 8.4 \text{ V} - 1.7 \text{ V}$$
$$= 6.7 \text{ V p-p}$$

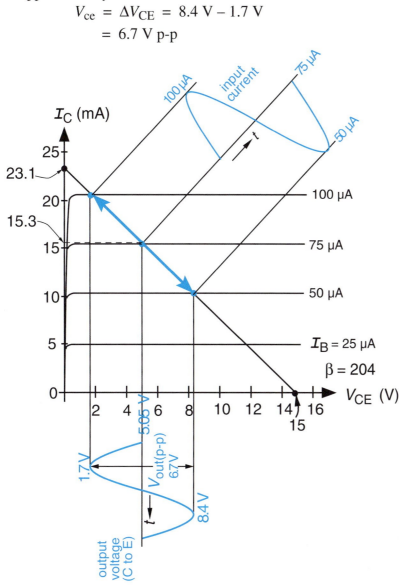

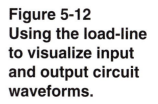

**Figure 5-12
Using the load-line
to visualize input
and output circuit
waveforms.**

This motion along the load-line can be related to sine-wave oscillations as shown in Fig. 5-12. It should be mentioned that this 6.7-V peak-to-peak waveform between C and E is *not* the same as V_c, the ac output voltage taken between the collector and ground. V_c would be less than 6.7 V by an amount equal to V_{Re}. This is because the ac voltage developed across R_E (namely V_{Re}) is always moving in the opposite direction to the motion of V_{CE}. A complete set of waveforms for this circuit is shown in Fig. 5-13.

SELF-CHECK FOR SECTION 5-3

15. Redraw the amplifier of Fig. 5-11(A) with the following changes:
    $R_E = 180 \ \Omega$ and $R_C = 330 \ \Omega$. Everything else stays the same.
    a) Find the maximum possible collector current, $I_{C(sat)}$.
    b) On an I_C-versus-V_{CE} graph, draw the dc load-line for this amplifier.
16. For the amplifier of Problem 15,
    a) What is the value of V_{RE}, the dc bias voltage across R_E?
    b) What is the bias value of I_E and I_C ?
17. Locate the amplifier's dc bias point on the load-line that you drew in
    Problem 15. From the graph, find the bias value of V_{CE}.

18. On the I_C-versus-V_{CE} graph from Problem 17, sketch the collector characteristic curves for a transistor with β = 125. Now suppose that the ac signal causes I_b to be 40 μA (p-p).

    a) From the load-line, what is the peak-to-peak value of I_c ?

    b) From the load-line, what is the peak-to-peak value of V_{ce} ?

19. Repeat Problem 18 for a larger ac input signal, causing $I_b = 120$ μA (p-p).

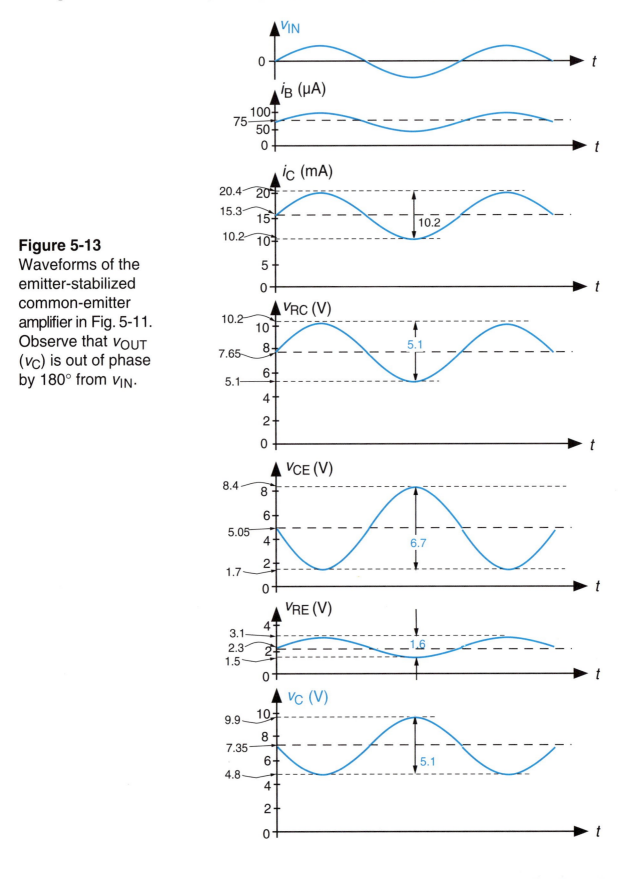

Figure 5-13
Waveforms of the emitter-stabilized common-emitter amplifier in Fig. 5-11. Observe that v_{OUT} (v_C) is out of phase by 180° from v_{IN}.

5-4 VOLTAGE GAIN AND CURRENT GAIN

For any amplifier, here is the definition of **voltage gain**.

Voltage gain, symbolized A_v, is the ratio of ac output voltage, V_{out}, to ac input voltage, V_{in}. As a formula,

$$A_v = \frac{V_{out}}{V_{in}}$$ Eq. (5-1)

Get This

In the CE amplifier of Fig. 5-11, V_{out} is the ac voltage from collector to ground, V_c. This voltage is the same in magnitude as the ac voltage developed across the collector resistor R_C. The equality of these ac voltages was shown in the waveform graphs of Fig. 5-13 and Fig. 4-17. It is pointed out again in Fig. 5-14.

⑧ ✔

V_{in} is the voltage across the terminals of the ac signal source. In Fig. 5-14, this voltage is applied between the outside (left side) of C_{in} and ground.

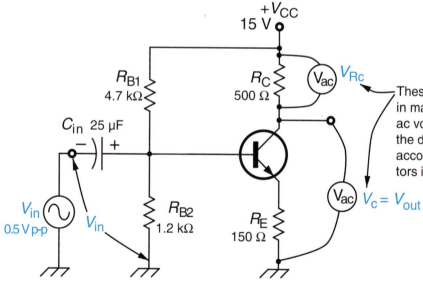

These two ac voltages are equal in magnitude. Imagine that the two ac voltmeters are able to ignore the dc parts. (This could really be accomplished by placing capacitors in series with the voltmeters.)

Figure 5-14

Identifying V_{out} and V_{in}. Output signal V_{out} is considered to be the ac part, V_c, of the voltage between the collector terminal and ground.

In Chapter 4, we mentioned that the base-emitter junction has a certain nonzero amount of resistance to the ac input current. This junction resistance was ignored throughout Chapter 4 because it was negligible compared to the component resistance that was placed in the circuit to limit the base current (for example, the R_{B2} value of 27 kΩ in Fig. 4-16).

It is possible to approximate the ac resistance of the base-emitter junction by the following formula:

⑨ ✔

$$r_{Ej} \approx \frac{30\ mV}{I_E}$$ Eq. (5-2)

Get This

where r_{Ej} is the symbol for the ac resistance of the emitter junction

Equation (5-2) tells us that the ac resistance of the B-E junction is less if the transistor's dc bias current is larger.

This is only a rough approximation. The actual resistance could be anywhere from about 1.6 Ω to about 3.3 Ω.

We generally use lower-case *r* instead of capital *R* whenever we want to emphasize that a resistance is being seen from an ac viewpoint rather than a dc viewpoint.

EXAMPLE 5-1

Estimate the ac resistance between B and E in the amplifier of Fig. 5-14.

SOLUTION

We know from our analysis in Secs. 5-1 and 5-3 that the dc emitter bias current for this amplifier is 15.3 mA. Therefore, from Eq. (5-2),

$$r_{Ej} \approx \frac{30 \text{ mV}}{15.3 \text{ mA}} \approx 2 \text{ } \Omega$$

So r_{Ej} is negligible in this circuit too (Figs. 5-11 and 5-14), since the R_E resistance just beneath it is 150 Ω, far greater than 2 Ω.

In general, r_{Ej} may not always be negligible. For those common-emitter amplifier circuits where it is not negligible, the ac voltage gain is given by

$$A_v = \frac{R_C}{r_{Ej} + R_E}$$

Eq. (5-3)

EXAMPLE 5-2

a) Find the voltage gain of the Fig. 5-14 amplifier, taking ac junction resistance r_{Ej} into account.
b) Recalculate A_v, this time ignoring r_{Ej}. Compare to the result from part (a).
c) For an input voltage $V_{in(p\text{-}p)} = 0.5$ V, find $V_{out(p\text{-}p)}$.

SOLUTION

a) Applying Eq. (5-3) gives

$$A_v = \frac{R_C}{r_{Ej} + R_E} = \frac{500 \text{ } \Omega}{2 \text{ } \Omega + 150 \text{ } \Omega}$$

$$= \frac{500 \text{ } \Omega}{152 \text{ } \Omega} = \textbf{3.29}$$

b) Ignoring r_{Ej} (assuming it is zero), we would get

$$A_v = \frac{R_C}{r_{Ej} + R_E} = \frac{500 \text{ } \Omega}{150 \text{ } \Omega} = \textbf{3.33}$$

⑩ ✔

The two A_v values differ by only about 1%, which is an unimportant difference.

c) Rearranging Eq. (5-1), we have

$$V_{out} = A_v (V_{in})$$

$$= 3.29 (0.5 \text{ V p-p}) = \textbf{1.65 V p-p}$$

1990s LOCOMOTIVE

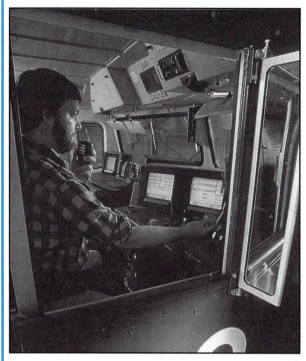

© *1993 Brownie Harris*

Courtesy of GE Transportation Systems.

The *Great Train Robbery* days of shoveling in the coal and watching the boiler pressure gauge creep past the red line are long gone. This modern diesel locomotive cab contains a computerized video display system that gives information on every function of the engines and drive motors, and the status of every wheel brake on the entire train.

The computer menu selections are made by pressing the keypad switches that are located just below the video screens. The system also informs the locomotive engineer about other train traffic on the tracks ahead and behind.

Current Gain

For any amplifier, the **current gain** is defined as:

Current gain, symbolized A_i , is the ratio of the ac output current (the ac current through the load) to the ac input current (the ac current coming into the amplifier from the signal source). As a formula,

$$A_i = \frac{I_\text{out}}{I_\text{in}}$$ Eq. (5-4)

Get This

$\boxed{11}$ ✔

The amplifier's current gain A_i will almost always be less than the transistor's current gain, β. This happens for two reasons:

$\boxed{1}$ Not all of the signal source's ac current actually flows into the base of the transistor. Some of the source's ac current flows to ground through resistors in the base-bias circuit. This definitely will happen in Fig. 5-14, as will be explained in a moment.

We are regarding resistor R_C as the load.

$\boxed{2}$ Not all of the transistor's collector output ac current necessarily flows through the load. This would happen if the load were *not* connected directly into the collector lead, as it is in Fig. 5-14. We will look at this possibility later.

The signal source's input current to the amplifier is illustrated in Fig. 5-15. Ac current I_{in} passes directly through coupling capacitor C_{in}. We assume that capacitive reactance X_{Cin} is so low that it is virtually zero. When I_{in} reaches the tie-point at B, it sees three paths to ground, namely:

$\textcircled{1}$ Through the B-E junction of the transistor, and then through R_E to ground. The current that takes this path is I_b, the ac current that actually operates the transistor. The value of I_b will vary from one transistor to another, being larger for low-β transistors and smaller for high-β transistors.

For an emitter-stabilized (R_E is present) common-emitter amplifier, I_b can be found from

$$I_b = \frac{V_{in}}{\beta\left(r_{Ej} + R_E\right)}$$

Eq. (5-5)

$\textcircled{2}$ The second path is down through R_{B2} to ground. This current is given directly by Ohm's law as

$$I_{Rb2} = \frac{V_{in}}{R_{B2}}$$

Eq. (5-6)

$\textcircled{3}$ The third path is up through R_{B1} to the V_{CC} dc supply.

To an ac signal, a dc voltage terminal is just like a connection to ground. The ac signal "sees" 0 Ω between the V_{CC} point and ground.

Get This

Therefore,

$$I_{Rb1} = \frac{V_{in}}{R_{B1}}$$

Eq. (5-7)

Figure 5-15
At the base node, the source's ac input current I_{in} splits into three paths.

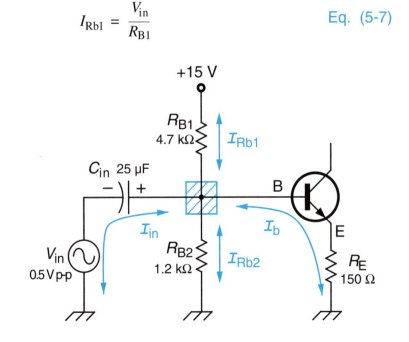

Here is the reason why a dc voltage source terminal is like an **ac ground**: A dc source can never permit any ac voltage to exist across its terminals. If it did, it would no longer be a dc source. The only electrical circuit that never permits voltage to exist is a short circuit to ground. Therefore a dc source is a short circuit (direct connection) to ground, for ac.

⬜13 ✔

An alternative explanation of the ac ground idea is this: A dc power supply usually has a large filter capacitor directly across its output terminals (see Figs. 3-13, 3-16 and 3-18). The ac reactance of this capacitor is so low that it is virtually 0 Ω. It therefore provides a direct path to ground for ac current.

EXAMPLE 5-3

For the amplifier of Figs. 5-14 and 5-15, assume the transistor has $\beta = 200$.
a) Calculate I_b, the ac base current that is amplified by the transistor.
b) Calculate I_{Rb2}.
c) Calculate I_{Rb1}.
d) Find I_{in}, the signal source's total input current to the amplifier.

SOLUTION

a) Applying Eq. (5-5), with $\beta = 200$ and $r_{Ej} = 2\,\Omega$, we get

$$I_b = \frac{V_{in}}{\beta(r_{Ej} + R_E)} = \frac{0.5\text{ V p-p}}{200\,(2\,\Omega + 150\,\Omega)}$$

$$= \frac{0.5\text{ V p-p}}{30\,400\,\Omega} = \textbf{16.4 μA p-p}$$

b) From Eq. 5-6,

$$I_{Rb2} = \frac{V_{in}}{R_{B2}} = \frac{0.5\text{ V p-p}}{1.2\text{ k}\Omega} = \textbf{417 μA p-p}$$

c) From Eq. 5-7,

$$I_{Rb1} = \frac{V_{in}}{R_{B1}} = \frac{0.5\text{ V p-p}}{4.7\text{ k}\Omega} = \textbf{106 μA p-p}$$

d) Applying Kirchhoff's current law to the B node gives

$$I_{in} = I_b + I_{Rb1} + I_{Rb2}$$
$$= (16.4 + 417 + 106)\text{ μA} = \textbf{539 μA p-p}$$

Of this total ac input current of 539 μA p-p, only the 16.4-μA I_b portion gets amplified by the transistor.

EXAMPLE 5-4

For the amplifier of Example 5-3:
a) Calculate the amplifier's current gain, A_i.
b) Calculate the amplifier's **power gain**, A_P.

SOLUTION

a) The ac collector current is given by

$$I_c = \beta I_b$$
$$= 200 \, (16.4 \, \mu A \text{ p-p})$$
$$= 3.28 \text{ mA p-p}$$

In Fig. 5-14, all of I_c passes through the load (R_C). Therefore, $I_{out} = I_c$ = 3.28 mA p-p. Applying Eq. (5-4) gives

$$A_i = \frac{I_{out}}{I_{in}}$$

$$= \frac{3.28 \text{ mA p-p}}{539 \, \mu A \text{ p-p}} = \frac{3.28 \times 10^{-3}}{539 \times 10^{-6}} = \mathbf{6.08}$$

b) For any amplifier that drives a resistive load, this statement is true.

Power gain A_P is equal to the product of voltage gain A_v multiplied by current gain A_i.

$$\boxed{A_P = A_v A_i} \qquad \text{Eq. (5-8)}$$

Applying Eq. (5-8) to this amplifier, we get

$$A_P = A_v A_i = 3.29 \, (6.08) = \mathbf{20.0}$$

In other words, the amount of ac power that the amplifier extracts from the dc power supply and delivers to the load is 20 times as large as the amount of power that the ac signal source delivers to the input of the amplifier. This is the essential idea of an electronic amplifier.

The definition of an amplifier.

Get This

[14] ✓

Ac Bypass Around R_E

Emitter resistor R_E was installed to stabilize the amplifier's dc bias point. Unfortunately, R_E causes voltage gain A_v to be decreased. By connecting a large capacitor in parallel with R_E, we can provide a short-circuit path around R_E for ac current, thereby getting back the lost gain. This is shown in Fig. 5-16.

Figure 5-16
With an emitter-bypass capacitor C_E installed, the amplifier's ac operation is the same as if R_E were not present. This assumes that X_{CE} is near zero ohms.

[15] ✓

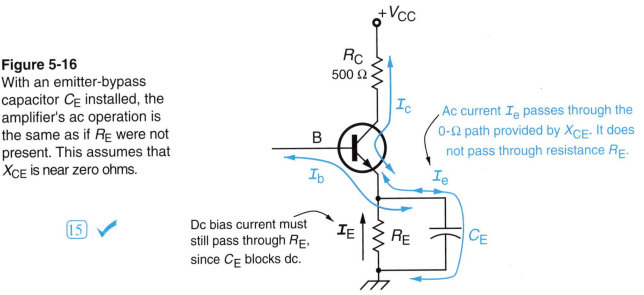

Ac current I_e passes through the 0-Ω path provided by X_{CE}. It does not pass through resistance R_E.

Dc bias current must still pass through R_E, since C_E blocks dc.

As we know, the ac emitter current I_e is the sum of ac base current I_b and ac collector current I_c. Therefore, if I_e passes around R_E, so does base current I_b. Under this condition, the ac base current is found by removing R_E from Eq. 5-5.

$$I_b = \frac{V_{in}}{\beta\, r_{Ej}}$$ Eq. (5-9)

Ac base current when R_E is bypassed by C_E.

Similarly, the amplifier's ac voltage gain is found by removing R_E from Eq. (5-3).

$$A_v = \frac{R_C}{r_{Ej}}$$ Eq. (5-10)

R_E also disappears from the A_v equation.

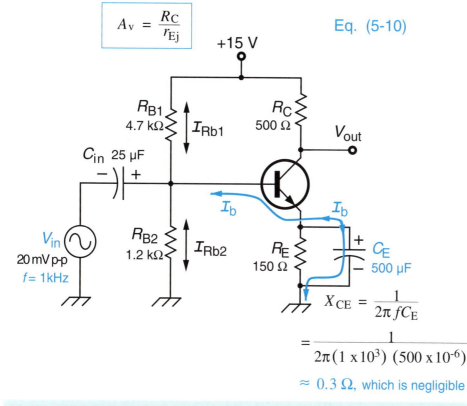

$$X_{CE} = \frac{1}{2\pi f C_E}$$

$$= \frac{1}{2\pi (1 \times 10^3)\, (500 \times 10^{-6})}$$

$$\approx 0.3\ \Omega,\ \text{which is negligible}$$

Figure 5-17 The same bias-stabilized common-emitter amplifier, with R_E bypassed by 500 μF.

EXAMPLE 5-5

Let us install $C_E = 500$ μF in the amplifier of Fig. 5-14. The new schematic diagram is shown in Fig. 5-17, with a much smaller input signal.
a) Give an approximate value for this amplifier's voltage gain, A_v.
b) Give an approximate value for this amplifier's current gain, A_i.

SOLUTION

a) To apply Eq. 5-10, we must recognize this fact: r_{Ej} has the same value that it had in Examples 5-3 through 5-4. This is because the *dc* conditions in the amplifier are not affected by capacitor C_E. Equation (5-2) tells us that r_{Ej} depends only on the dc emitter current I_E. So, with $r_{Ej} \approx 2\ \Omega$, Eq. (5-10) gives

$$A_v = \frac{R_C}{r_{Ej}} \approx \frac{500\ \Omega}{2\ \Omega} = \mathbf{250}$$

b) The new base current is given by Eq. (5-9) as

$$I_b = \frac{V_{in}}{\beta\, r_{Ej}} \approx \frac{V_{in}}{200\,(2\ \Omega)} = \frac{20\ \text{mV p-p}}{400\ \Omega} = 50\ \mu\text{A p-p}$$

Ac collector current I_c is

$$I_c = \beta I_b = 200 \,(50 \, \mu A \text{ p-p}) = 10 \text{ mA p-p} = I_{out}$$

I_{Rb1} and I_{Rb2} are given by Eqs. (5-6) and (5-7) as

$$I_{Rb1} = \frac{V_{in}}{R_{B1}} = \frac{20 \text{ mV p-p}}{4.7 \text{ k}\Omega} = 4.26 \, \mu A \text{ p-p}$$

$$I_{Rb2} = \frac{V_{in}}{R_{B2}} = \frac{20 \text{ mV p-p}}{1.2 \text{ k}\Omega} = 16.7 \, \mu A \text{ p-p}$$

I_{in} from the source is found by

$$I_{in} = I_b + I_{Rb1} + I_{Rb2}$$
$$= (50 + 4.26 + 16.7) \, \mu A \text{ p-p} = 70.9 \, \mu A \text{ p-p}$$

Finally, from Eq. (5-4)

$$A_i = \frac{I_{out}}{I_{in}} = \frac{10 \text{ mA p-p}}{70.9 \, \mu A \text{ p-p}} = \frac{10 \times 10^{-3}}{70.9 \times 10^{-6}} = \mathbf{141}$$

In Example 5-5, both the voltage gain and the current gain are very great. This is the advantage of bypassing R_E. However, there is a disadvantage that goes with this technique. It is that the gains, although large, are quite unpredictable.

The voltage gain equation $A_v = R_C \div r_{Ej}$, Eq. (5-10), depends heavily on r_{Ej}. But our prediction of $r_{Ej} = 30 \text{ mV} \div I_E$ from Eq. (5-2) was only a rough approximation. We do not have a great deal of confidence in the 2-Ω result from that equation. Therefore we cannot have much confidence in our calculated A_v result from Eq. (5-10), namely $A_v = R_C \div r_{Ej} = 250$.

You should realize that lack of confidence was *not* a problem when R_E was unbypassed. Then the voltage gain equation was

$$A_v = \frac{R_C}{r_{Ej} + R_E} = \frac{500 \, \Omega}{2 \, \Omega + 150 \, \Omega} = 3.29$$

in which this 2 Ω doesn't matter very much. r_{Ej} is unimportant when compared to the 150-Ω value of R_E.

Get This

In general, whenever we can build an amplifier so that something that we can't control (such as r_{Ej}) becomes unimportant in comparison to something that we can control (such as R_E), then the amplifier is predictable, also called **stable**.

The left two columns of Table 5-1 (unbypassed and completely bypassed) summarize this situation.

Swamping means that one part dominates another part of the circuit, thereby making it unimportant.

A somewhat similar explanation can be made for the amplifier's current gain A_i. However, even with r_{Ej} made unimportant by the R_E *swamping* effect, A_i is never as stable as A_v. This is because A_i depends to a large extent on the particular transistor β value, as was pointed out in Example 5-4.

	Unbypassed (C_E is not present — Fig. 5-14)	Completely bypassed (C_E in parallel with R_E — Fig. 5-17)	Partially bypassed (C_E in parallel with R_{E2} — Fig. 5-18)
Voltage gain, A_v	Low (3.29, for example)	Very high (250, for example)	Moderate (22.7, for example)
Stability of gain	Very stable. (Could vary only from about 3.26 to 3.30.)	Very unstable. (Could vary from about 150 to about 310.)	Fairly stable. (Could vary from about 21.5 to about 23.1.)

Table 5-1   High voltage gain is paid for by a great degree of instability. But by sacrificing voltage gain, we improve stability.

Partial Bypassing of the Emitter Resistor

With no bypassing of R_E , we get low gain but excellent stability. Complete bypassing of R_E gives very high gain but poor stability. It seems reasonable that we could strike a compromise between these two extreme situations. The compromise would provide moderate gain and fairly good stability.

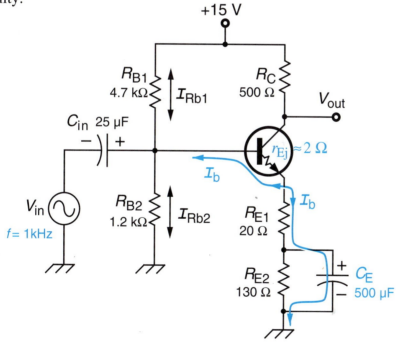

Figure 5-18
Partial bypassing of the emitter resistance. The 20-Ω value of R_{E1} is fairly effective at swamping out variations in junction resistance r_{Ej}. Therefore A_v is fairly stable.

Figure 5-18 shows such a compromise. In it, the 150-Ω emitter resistor has been replaced by two emitter resistors in series, $R_{E1} = 20\,\Omega$ and $R_{E2} = 130\,\Omega$. We bypass R_{E2} with C_E, but leave R_{E1} in the ac current flow path. Then the voltage gain is given by

$$A_v = \frac{R_C}{r_{Ej} + R_{E1}} \qquad \text{Eq. (5-11)}$$

$$\approx \frac{500\ \Omega}{2\,\Omega + 20\,\Omega} = \frac{500\ \Omega}{22\ \Omega} = 22.7$$

Because r_{Ej} cannot be predicted exactly, the actual A_v value could vary from about 21.5 to about 23.1. This is fairly stable behavior, as indicated in Table 5-1.

SELF-CHECK FOR SECTION 5-4

20. Voltage gain A_v is the ratio of _____ to _____ .
21. Write the formula for the voltage gain of a common-emitter amplifier with no bypass capacitor in parallel with R_E.
22. In the emitter-stabilized common-emitter amplifier of Fig. 5-14, suppose that R_C were increased to 600 Ω. For each of the following variables, tell whether it would increase, decrease, or stay the same.
    a) I_C b) V_C c) A_v
23. For the amplifier of Fig. 5-18, the values of R_{E1} and R_{E2} are changed to $R_{E1} = 50$ Ω and $R_{E2} = 100$ Ω. Find the new A_v.
24. (T-F) For the change made in Problem 23, the amplifier's ac gain stability has been improved.
25. How would you alter the values of R_{E1} and R_{E2} to produce a voltage gain of 6.5, with no change in dc bias point?

5-5 CONNECTING THE LOAD TO GROUND

In the amplifiers that we have discussed so far, the load has been the collector resistor R_C. There are many electronic applications where the load must receive a pure ac signal, not a combination of ac with dc. In those applications, the load must be connected to the amplifier as shown in Fig. 5-19(A).

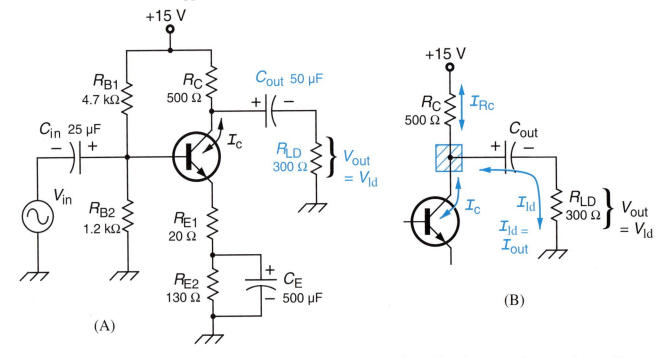

Figure 5-19 (A) Connecting R_{LD} through coupling capacitor C_{out}. This has no effect on the dc bias conditions of the amplifier. (B) The ac collector current I_c splits at the C node into two parts — I_{Rc} going through R_C to ac ground, and I_{ld} going through R_{LD} to actual ground. Now the amplifier's output current is I_{ld}, which is less than I_c.

The ac current I_c that emerges on the transistor's collector lead has a constant value, regardless of whether the R_{LD} connection is present or not. This is because the transistor acts as an ac current source, with value given by $I_c = \beta\, I_b$.

But with R_{LD} connected to ground through C_{out}, it provides I_c with a second path to ground. Now I_c splits into two parts, as shown in Fig. 5-19(B). Both paths allow ac current to flow to ac ground. Therefore the transistor ac current source "sees" the load resistance R_{LD} as if it were in parallel with collector resistance R_C. This is illustrated in Fig. 5-20.

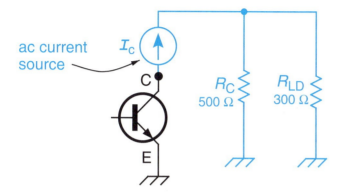

ac current source — I_c

R_C 500 Ω

R_{LD} 300 Ω

C

E

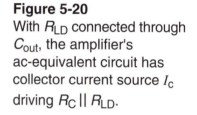

Figure 5-20
With R_{LD} connected through C_{out}, the amplifier's ac-equivalent circuit has collector current source I_c driving $R_C \| R_{LD}$.

The equivalent ac collector-line resistance, symbolized r_C, is the parallel combination of $R_C \| R_{LD}$. As a formula,

$$\frac{1}{r_C} = \frac{1}{R_C} + \frac{1}{R_{LD}} \qquad \text{Eq. (5-12)}$$

or

$$r_C = \frac{R_C\, R_{LD}}{R_C + R_{LD}} \qquad \text{Eq. (5-13)}$$

Get This

[17] ✓

Because the collector's equivalent resistance to an ac signal has been changed (reduced) from R_C to r_C, the new ac voltage gain formula becomes

$$A_v = \frac{r_C}{r_{Ej} + R_{E1}} \qquad \text{Eq. (5-14)}$$

These technicians are measuring the electric power consumption of a drill-press equipped with automatic power-factor-correction circuitry. Such circuitry senses the torque load on the motor shaft and raises or lowers the voltage applied to the motor in order to match the torque demand. This technique conserves electric energy.

Courtesy of Harris Semiconductor Corp.

EXAMPLE 5-6

For the amplifier of Fig. 5-19, with capacitor-coupled output, assume $V_{in} = 0.4$ V p-p, at $f = 1$ kHz.
a) Find the equivalent ac collector resistance, r_C.
b) Calculate the amplifier's voltage gain, A_v.
c) Draw waveform graphs of V_{in} and V_{out} across the load.

SOLUTION

a) From Eq. (5-13)

$$r_C = \frac{R_C R_{LD}}{R_C + R_{LD}} = \frac{500 \,(300)}{500 + 300}$$

$$= \mathbf{187.5 \; \Omega}$$

b) None of the dc bias conditions have changed, since C_{out} is an open circuit to dc. Therefore I_E is still 15.3 mA and r_{Ej} is still about 2 Ω. From Eq. (5-14)

$$A_v = \frac{r_C}{r_{Ej} + R_{E1}} = \frac{187.5 \; \Omega}{2 \; \Omega + 20 \; \Omega}$$

$$= \mathbf{8.52}$$

c) Rearranging Eq. (5-1), we get

$$V_{out} = A_v V_{in}$$
$$= 8.52 \,(0.4 \text{ Vp-p}) = 3.41 \text{ V p-p}$$

V_{out} is 180 degrees out of phase with V_{in}, for the same reason that the ac part of the v_C waveform in Fig. 5-13 was 180 degrees out of phase with V_{in}. The waveforms for this amplifier are drawn in Fig. 5-21.

Figure 5-21

V_{out} is pure ac, with no dc component. It is inverted, or 180° phase-shifted, relative to V_{in}.

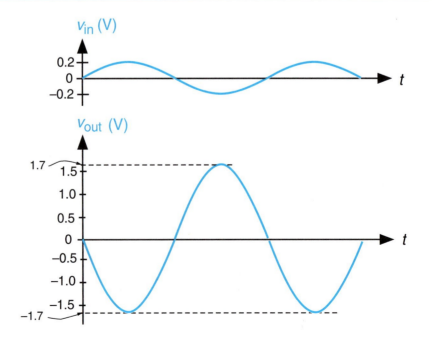

SELF-CHECK FOR SECTION 5-5

26. In Fig. 5-19, suppose that R_{LD} is changed to 200 Ω. The input stays the same as in Example 5-6. Calculate the new voltage gain A_v.
    Problems 27 through 31 refer to the amplifier of Problem 26.
27. Find the value of V_{out}.
28. By applying Ohm's law to R_{LD}, find the output current $I_{out} = I_{ld}$.
29. Suppose the transistor has $\beta = 170$.
   a) Calculate the ac base current I_b by revising Eq. (5-5) to
       $I_b = V_{in} / [\beta (r_{Ej} + R_{E1})]$.
   b) Calculate I_{Rb1} and I_{Rb2}.
   c) Calculate the total ac input current to the amplifier, I_{in}.
30. Using the I_{out} value from Problem 28 and the I_{in} value from Problem 29, calculate the amplifier's current gain A_i.
31. Using Eq. (5-8), calculate the amplifier's power gain.

5-6 MORE THAN ONE STAGE OF AMPLIFICATION

To raise overall power gain, two or more amplifiers can be connected together, or **cascaded**. Figure 5-22 shows such an arrangement. Coupling capacitor C_{cplg} serves as the output coupling capacitor from stage 1 and as the input coupling capacitor to stage 2.

The load on stage 1 is the equivalent ac resistance seen looking into stage 2, from the vantage point of the eye in Fig. 5-22(A). This equivalent resistance is the parallel combination of R_{B1}, R_{B2}, and $r_{b\,in}$, as drawn in Fig. 5-22(B). The symbol $r_{b\,in}$ stands for the equivalent ac resistance seen looking into the base of transistor Q_2.

It can be shown that this ac resistance into the base lead alone is given by the formula

$$r_{b\,in} = \beta (r_{Ej} + R_{E1}) \qquad \text{Eq. (5-15)}$$

Ac Input Resistance

Looking into a complete amplifier stage, the equivalent ac resistance seen by the ac signal is called the **input resistance** of the amplifier stage.

Get This

As indicated in Fig. 5-22(B), the input resistance of a common-emitter stage is given by

$$\boxed{R_{in} = R_{B1} \| R_{B2} \| r_{b\,in}} \qquad \text{Eq. (5-16)}$$

In Fig. 5-22(A), the voltage gain of stage 1 is given by

$$A_{v1} = \frac{r_{C\,(stage\,1)}}{r_{Ej} + R_{E1}} = \frac{R_{C1} \| R_{in\,(stage\,2)}}{r_{Ej} + R_{E1}} \qquad \text{Eq. (5-17)}$$

referring to the Q_1 circuit

where $r_{C\,(stage\,1)}$ is equal to R_C in parallel with $R_{in\,(stage\,2)}$, as diagrammed in Fig. 5-23.

Input resistance is often called **input impedance**. However, that is misleading terminology unless the capacitors' reactances are significant.

The equivalent ac resistance seen by the collector of Q_1 is $R_{C1} \| R_{in\,(stage\,2)}$.

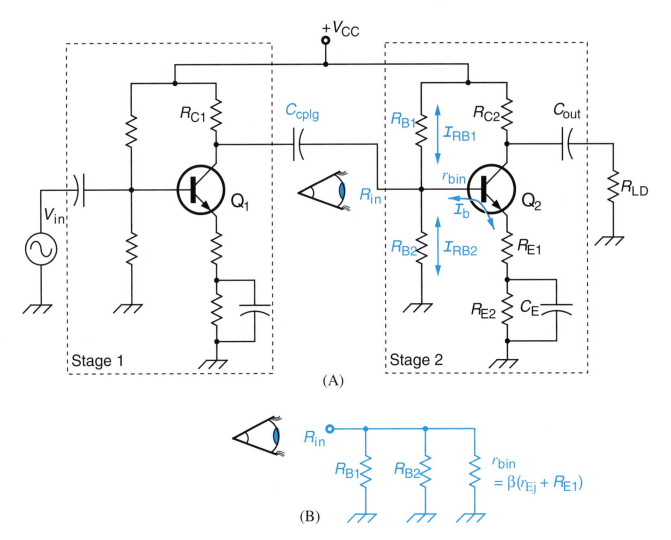

(A)

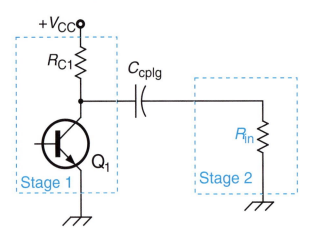

(B)

Figure 5-22 Two-stage common-emitter amplifier.
(A) Overall schematic diagram. The ac load on stage 1 is the equivalent ac resistance seen by an ac signal looking in to stage 2.
(B) The parallel ac resistance seen looking into stage 2 is called the ac input resistance, R_{in}, of stage 2.

**Figure 5-23
Input resistance R_{in}
of stage 2 acts like a
load on stage 1.**

Whatever a stage's output
coupling capacitor leads to,
that is regarded as the load on
the stage.

The overall voltage gain of the entire two-stage amplifier is the product of the two individual voltage gains.

$$A_{v\,(total)} = A_{v1} \cdot A_{v2} \qquad \text{Eq. (5-18)}$$

EXAMPLE 5-7

In Fig. 5-22, suppose the component values in stage 1 are:

$R_C = 1\ \text{k}\Omega$ $r_{Ej} = 2\ \Omega$
$R_{E1} = 51\ \Omega$ $\beta = 150$

and the component values in stage 2 are:

$R_{B1} = 4.7\ \text{k}\Omega$ $\beta = 180$ $R_{LD} = 220\ \Omega$
$R_{B2} = 1.2\ \text{k}\Omega$ $R_C = 620\ \Omega$
$r_{Ej} = 2\ \Omega$ $R_{E1} = 33\ \Omega$

a) Use Eq. (5-15) to find the ac resistance, $r_{b\ in}$, looking into the base lead of transistor Q_2.
b) Find the input resistance of stage 2 by combining $r_{b\ in}$ in parallel with R_{B1} and R_{B2}, as called for in Eq. (5-16).
c) Find the ac equivalent resistance in the collector circuit of transistor Q_1. That is, find $r_{C\ (stage\ 1)}$.
d) Find the voltage gain of stage 1 alone, A_{v1}.
e) Find the voltage gain of stage 2 alone, A_{v2}.
f) Find the overall voltage gain of the entire amplifier in Fig. 5-22, $A_{v\ (total)}$.

SOLUTION

a) With $\beta_2 = 180$, $r_{Ej} = 2\ \Omega$, and $R_{E1} = 51\ \Omega$, Eq. (5-15) gives

$$r_{b\ in} = \beta\ (r_{Ej} + R_{E1})$$
$$= 180\ (2 + 51)\ \Omega = \textbf{9540}\ \Omega\ \text{or}\ \textbf{9.54 k}\Omega.$$

b) From Eq. 5-16, with $R_{B1} = 4.7\ \text{k}\Omega$ and $R_{B2} = 1.2\ \text{k}\Omega$,

$$r_{in} = 4.7\ \text{k}\Omega\ \|\ 1.2\ \text{k}\Omega\ \|\ 9.54\ \text{k}\Omega$$
$$= \textbf{869}\ \Omega$$

c) The equivalent ac collector-line resistance is given by the parallel combination of $R_{C\ (stage\ 1)}$ in parallel with $R_{in\ (stage\ 2)}$.

$$r_{C\ (stage\ 1)} = R_{C\ (stage\ 1)}\ \|\ R_{in\ (stage\ 2)}$$

$$= 1000\ \Omega\ \|\ 869\ \Omega = \textbf{465}\ \Omega$$

d) Applying Eq. (5-14),

$$A_{v1} = \frac{r_{C\ (stage\ 1)}}{(r_{Ej} + R_{E1})_{\ (stage\ 1)}}$$

$$= \frac{465\ \Omega}{(2 + 51)\ \Omega} = \textbf{8.77}$$

e)

$$A_{v2} = \frac{r_{C\ (stage\ 2)}}{(r_{Ej} + R_{E1})_{\ (stage\ 2)}} = \frac{R_{C\ (stage\ 2)}\ \|\ R_{LD}}{(r_{Ej} + R_{E1})_{\ (stage\ 2)}}$$

$$= \frac{620\ \Omega\ \|\ 220\ \Omega}{2\ \Omega + 33\ \Omega} = \frac{162\ \Omega}{35\ \Omega} = \textbf{4.64}$$

f) From Eq. (5-18),

$$A_{v\ (total)} = A_{v1} \cdot A_{v2}$$
$$= 8.77\ (4.64) = \textbf{40.7}$$

SELF-CHECK FOR SECTION 5-6

32. What does it mean to say that two transistor amplifiers are cascaded?
33. For an amplifier stage (or just an amplifier by itself), give an explanation of what we mean when we speak about the input resistance R_{in}.
34. (T-F) When common-emitter amplifier stage 1 drives amplifier stage 2, the equivalent ac resistance in the collector of transistor Q_1 is given by the expression $R_C \| R_{in \, (stage \, 2)}$.
35. For the two-stage arrangement described in Question 34, if stage 2 were disconnected from stage 1, A_{v1} would _____ (increase or decrease).
36. In a two-stage amplifier, we have $A_{v1} = 12.5$ and $A_{v2} = 2.2$. Find $A_{v \, (total)}$.

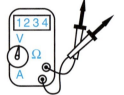

5-7 TROUBLESHOOTING COMMON-EMITTER AMPLIFIERS

To troubleshoot a malfunctioning transistor amplifier, or any other electronic circuit, here is what you should do before you start probing with meters and test instruments.

Make a Careful Visual Inspection

With the power turned off and all high-voltage (above 50 V) capacitors discharged, carefully inspect both sides of the printed circuit board. Use special lighting and/or magnification lenses, if necessary. Be looking for:

a) Burned components (primarily resistors), burned insulation, or burned copper tracks.

b) Broken component leads or broken copper tracks.

c) Transistor leads that are bent out of shape and are touching each other. (The same for other leads that are physically close together.)

d) Transistors that are loose in their sockets (if they are mounted in sockets).

e) Loose or corroded connectors. Both the plug/socket type of electrical connector and the edge-connector that grips a printed circuit board are liable to have these problems.

f) Solder splashes shorting between terminals or between copper tracks.

g) Foreign material (cut-off wire ends, for example) shorting between terminals.

h) Leaking devices, specifically electrolytic capacitors and batteries.

 If any of these conditions are found, make the appropriate repair.

For experienced troubleshooters, there often are specific symptoms that point to certain locations in the circuit or even particular components in the circuit. Naturally, if you can, take advantage of such information that may guide you directly to the problem in a minimum of time.

For example, refer to Fig. 5-24 as our troubleshooting model, and suppose that the load resistor R_{LD} is actually a speaker. If the symptom is that there is no sound whatever, you may know from your experience with this circuit that the speaker coil sometimes fails open-circuited. Then the quickest test to make would be to disconnect one of the speaker leads and apply an ohmmeter, as shown in Fig. 5-25.

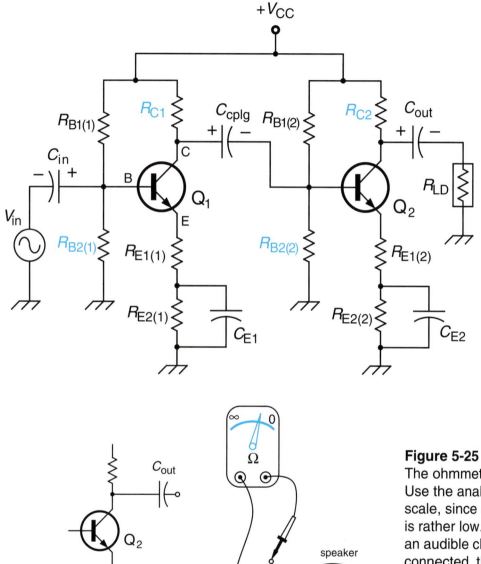

Figure 5-24
Troubleshooting model circuit.
The subscript digits 1 and 2 are used to distinguish between stage 1 components and stage 2 components [for example, R_{C1} and R_{C2}].
But where the component symbol *already* has a 1 or 2 subscript, the stage 1- or stage 2-indicating subscript is put in parentheses. [For example, $R_{B2(1)}$ for stage 1 and $R_{B2(2)}$ for stage 2].

Figure 5-25
The ohmmeter test for a speaker. Use the analog ohmmeter's $R \times 1$ scale, since speaker coil resistance is rather low. If the speaker makes an audible click when the test lead is connected, that provides further proof that the speaker is functioning properly.

If the speaker coil has failed open, the coil resistance will measure $\infty \ \Omega$. If not, the ohmmeter will indicate the expected value of resistance.

But if there are no specific circuit symptoms to guide you, here are some effective methods for finding the trouble. These test procedures must be carried out with the power turned on. For your personal safety, power-on trouble-shooting should be done with the entire circuit and its enclosure isolated from the earth ground, if possible. This can be accomplished with an isolation transformer, as shown in Fig. 5-26.

Figure 5-26   An isolation transformer removes all voltage reference to the earth, reducing the danger of you being shocked by touching only a 120-V hot terminal.

However, not all equipment can be operated this way. Some equipment depends on the earth connection through its power cord for proper functioning. Also, your test instrument may itself be referenced to earth ground, which defeats the purpose of the isolation transformer.

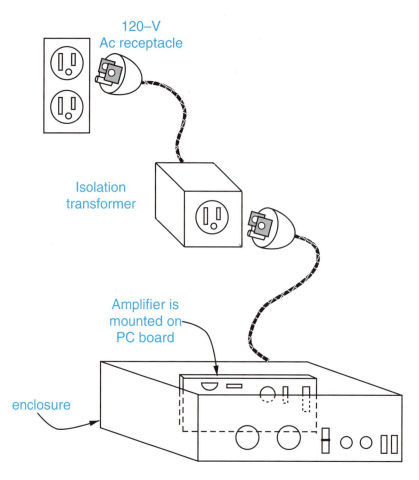

Refer again to Fig. 5-24 as our troubleshooting model.

Check the Dc Power Supply

With a dc voltmeter or oscilloscope, measure the voltage at each point that is supposed to receive V_{CC}. In Fig. 5-24, there are four such points: the tops of resistors $R_{B1(1)}$, R_{C1}, $R_{B1(2)}$, and R_{C2}. Often, a manufacturer's schematic diagram will tell the expected voltage value as well as the maximum allowable deviation from that value.

An oscilloscope is generally a better test instrument for this purpose than an analog or digital voltmeter. This is because a scope can indicate the amount of ripple that is present, as well as the average dc value. For example, a properly operating power supply will have a waveform with very little ripple, as shown in Fig. 5-27(A). Usually, the ripple is so small that it is invisible with the scope set to a 5- or 10-volt / cm vertical sensitivity.

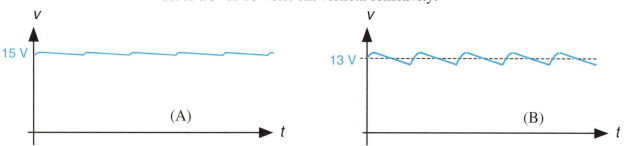

Figure 5-27 Measuring the power supply output V_{CC} using a scope with dc input coupling. (A) Proper smooth dc. Ripple may be invisible.   (B) Excessive ripple.

If the power supply ripple becomes so excessive that it causes amplifier malfunctioning, as shown in Fig. 5-27(B), a dc averaging meter will simply read 13 V. This may not be different enough from your 15-V expectation that it would alert you to a problem with the power supply. Of course, if V_{CC} is not proper, it must be corrected before the amplifier can function.

When checking for the presence of V_{CC}, it is best to touch your test probe to the lead of the resistor itself, not to the copper track or the solder joint on the printed circuit board. This is because there may be an invisible break on the track or a cold solder joint, which could allow your instrument to see and measure V_{CC} on the track even though V_{CC} is not actually present on the resistor lead. See Fig. 5-28.

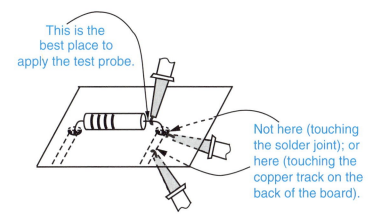

This is the best place to apply the test probe.

Not here (touching the solder joint); or here (touching the copper track on the back of the board).

**Figure 5-28
Best way to apply test probe during troubleshooting. This is true for all troubleshooting tests, not V_{CC} checks only.**

Checking the dc supply voltage, or any voltage for that matter, requires that the ground lead of your test instrument is connected to a chassis ground point. Often, the manufacturer's troubleshooting guide will suggest a specific location to ground your test instrument. There may even be a special terminal provided for grounding purposes.

This brings up the question of whether we have confidence in the ground connections in the amplifier that we are troubleshooting. After all, the grounds are just as important to the functioning of the amplifier as the V_{CC} connections. In Fig. 5-24, signal ground should connect to the following components: signal source, $R_{B2(1)}$, $R_{E2(1)}$, C_{E1}, $R_{B2(2)}$, $R_{E2(2)}$, C_{E2}, and R_{LD}. We will see later how to check for proper ground connections to amplifier components.

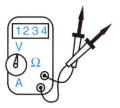

Power Supply Decoupling Circuits

Sometimes, the V_{CC} line does not run directly between amplifier stages as shown in Fig. 5-24. Instead, the stages may be **decoupled** from each other by an RC decoupling circuit, shown in Fig. 5-29.

The idea of an RC decoupling circuit is this: Any high-frequency ac noise that originates in stage 2 must be prevented from travelling down the V_{CC} supply line to stage 1. The RC decoupling circuit does this by acting like a low-pass filter. It passes the 0-Hz dc supply to stage 1. But it stops any high-frequency noise bursts, by voltage-dividing them between the 56-Ω R value and the very low X_C value.

Dc is equivalent to ac with frequency $f = 0$ Hz.

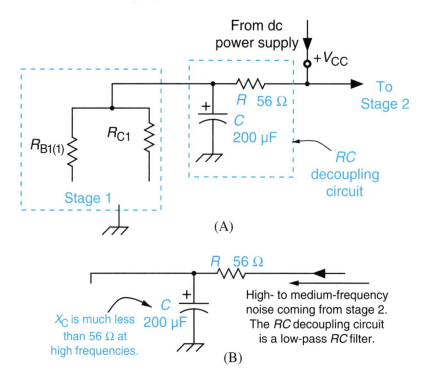

Figure 5-29
(A) An *RC* noise-decoupling circuit between amplifier stages has a low-value resistor in series with the V_{CC} supply line, and a fairly large capacitor to ground, across the stage.
(B) The decoupling circuit can be thought of as a series voltage-divider.

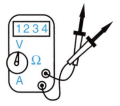

R itself must be rather low so that it does not have a very large voltage drop due to the current in the supply line running to stage 1.

A problem in the *RC* decoupling circuit of Fig. 5-29(A) (an open *R* or a leaking or shorted *C*) can prevent the appearance of V_{CC} at the top of stage 1.

Checking Ground Continuity

As mentioned earlier, the ground connections to the various components in Fig. 5-24 are just as important as the V_{CC} connections. These ground connections are made in the same way as the V_{CC} connections, so they are just as likely to fail as the V_{CC} connections (broken copper tracks, cold solder joints, poor socket seating, etc).

To test ground continuity, put your scope on a very sensitive vertical setting, perhaps 10 mV/cm and put the coupling switch on DC. With the scope ground clipped to the amplifier's power supply ground point, but with the test probe touching nothing, you will probably see a fuzzy 60-Hz noise waveform like the one in Fig. 5-30(B). When you touch the probe hook to the component being tested, say C_{E2}, the trace should settle close to the 0-V position on the screen, with very little fuzziness. This is shown in Fig. 5-30(C). The trace's slight dc offset from zero is due to the small $I \times R$ voltage drop along the ground conductor between C_{E2} and the ground terminal of the dc power supply.

If the ground connection to C_{E2} is not intact, the trace will not settle close to zero as shown in Fig. 5-30(C). Instead, it may retain its appearance as a noise waveform or it may deflect completely off the screen to a large dc level. If it does not settle at zero, look for the same circuit problems that prevent the appearance of V_{CC} — broken copper tracks, cold solder joints, and broken component leads.

Checking Dc Bias Voltages

In Secs. 5-1 through 5-3, we learned about the dc bias values that should be expected at various points in an amplifier. We learned how to predict dc bias voltages by calculation and load-line analysis. Furthermore, manufacturers'

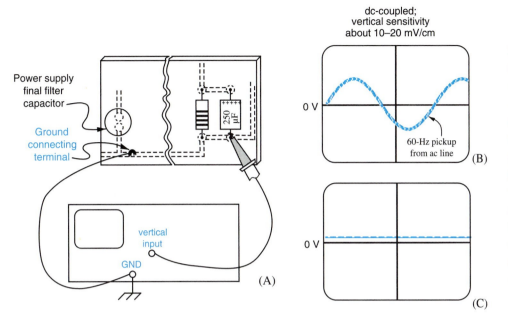

dc-coupled;
vertical sensitivity
about 10–20 mV/cm

Figure 5-30
Checking ground
continuity to C_{E2} in the
amplifier.
(A) Physical place-
ment of oscilloscope
probes.
(B) Scope screen
display *before* touch-
ing probe to ground
test point.
(C) Scope screen
display if ground
connection is intact.

schematic diagrams often tell the dc bias values that should appear at specific points. Using a test instrument to check all the bias voltages is a worthwhile troubleshooting technique.

For example, in stage 2 of Fig. 5-24, suppose our bias expectations are as shown in Fig. 5-31. If every one of those bias voltages is verified within reasonable tolerances, usually about ±10%, it proves that there is no circuit trouble that affects the dc operation of that amplifier stage. (There may still be circuit troubles that affect the ac operation of the amplifier stage though — open capacitors, excess stray capacitance, and other circuit problems may affect ac but not dc.)

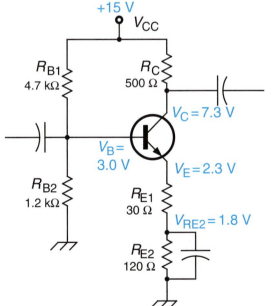

Figure 5-31
**Proper dc bias
voltages.**
Compare the
measurements for
malfunctioning circuits
shown in Figs. 5-32
through 5-44.

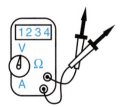

Some of the bias measurements that indicate circuit trouble are shown in Figs. 5-32 through 5-44.

If $V_C = V_{CC} = 15$ V, as in Figs. 5-32 through 5-37, it means that there is no current in R_C and thus no voltage drop across R_C.

Get This

● If the $V_C = V_{CC}$ symptom is accompanied by $V_B = 0$ V and $V_E = 0$ V, it is due to either R_{B1} open-circuited, as shown in Fig. 5-32(A), or R_{B2} shorted, as shown in Fig. 5-32(B).

Figure 5-32 (A) With R_{B1} open, there is no current path to drive the base, so $V_B = 0$ V. With zero base current, I_E and I_C are also zero – the transistor is cut off. Therefore $V_E = 0$ V and $V_C = 15$ V.
(B) With a short circuit around R_{B2}, no current can flow through the B-E junction. All transistor currents are zero, as in part A.

Resistors themselves hardly ever fail shorted; they fail open. However, an external short around a resistor body is quite possible.

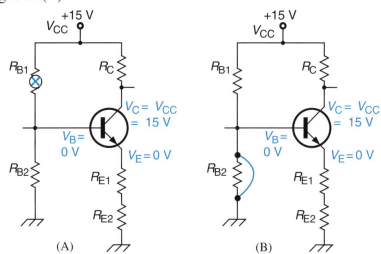

● If the $V_C = V_{CC}$ symptom is accompanied by normal V_B and V_E measurements, the trouble is a shorted collector resistor, as shown in Fig. 5-33.

● If the $V_C = V_{CC}$ symptom is accompanied by very low and equal values of V_E and V_B, the problem is a short circuit in or around the B-E junction, as shown in Fig. 5-34.

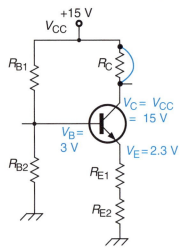

Figure 5-33 With R_C shorted, the transistor delivers its normal value of collector current I_C, but there is no resistance present to develop any voltage. Therefore V_C can't be any different from V_{CC}.

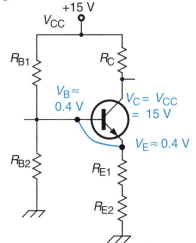

Figure 5-34 With a short around or in the B-E junction, no base current flows and all transistor currents stop. This circuit turns into just a resistive circuit with large R_{B1} in series with the parallel combination of $R_{B2} \| (R_{E1} + R_{E2})$. The low resistance values of R_{E1} and R_{E2} cause the low V_E value.

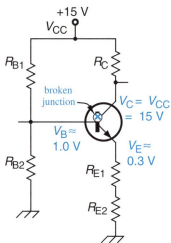

Figure 5-35 With the C-B junction broken open the transistor's current amplification action stops. This circuit becomes just a resistive circuit with large R_{B1} in series with the parallel combination of $R_{B2} \| (B\text{-}E \text{ junction} + R_{E1} + R_{E2})$, similar to Fig. 5-34.

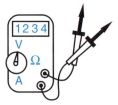

● If the $V_C = V_{CC}$ symptom is accompanied by V_E being very low and V_B greater than V_E by 0.7 V, the problem is due to an internal open circuit in the transistor, between C and B, as shown in Fig. 5-35. Turn the power off, remove the transistor, test it to be certain, and replace it.

● If the $V_C = V_{CC}$ symptom is accompanied by normal V_B, but $V_E = 0$ V, the problem is an internal open in the transistor between B and E, as shown in Fig. 5-36. Remove and test the transistor to be certain, and replace it.

● A similar set of symptoms occurs if a break occurs in the external emitter circuit, as shown in Fig. 5-37. The only difference is that now $V_E = V_B$.

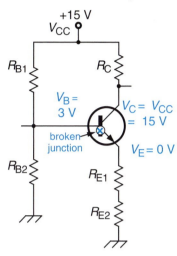

Figure 5-36 With an internal open between B and E, the base voltage-division will be normal, but all transistor currents are zero.

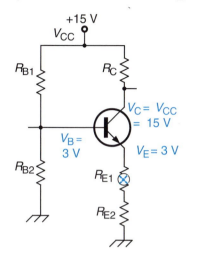

Figure 5-37 With the emitter flow-path broken open, all transistor currents are zero, as in Fig. 5-36. V_B is normal, but with no current through the intact B-E junction, its voltage drop is zero, so $V_E = V_B$.

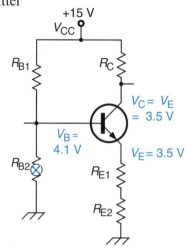

Figure 5-38 With R_{B2} open, the base voltage divider does not work. R_{B1} passes so much current directly to the base that it saturates the transistor.
Of course, it wouldn't have to be R_{B2} itself that failed open. A bad solder joint on the resistor's top lead or a bad ground connection on its bottom lead would produce the same problem.

If $V_C = V_E$, as in Figs. 5-38 and 5-39, it means that the transistor is saturated — V_{CE} is zero.

Get This

● If the saturation condition ($V_C = V_E$) is accompanied by V_B and V_E being just somewhat greater than normal, as shown in Fig 5-38, the trouble is probably an open in the R_{B2} circuit.

● If the saturation symptom is accompanied by V_B and V_E being far too great, near V_{CC}, as shown in Fig. 5-39, the trouble is a shorted R_{B1}. On the other hand, if V_B and V_E are far too low (V_E at 0 V), as shown in Fig. 5-40, the trouble is probably a short circuit around the emitter resistors R_{E1} and R_{E2}.

26 ✔

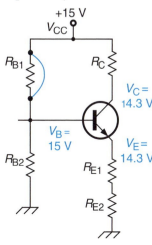

Figure 5-39
A short around R_{B1} pulls the base terminal right up to 15 V. This overdrives the base and saturates the transistor.

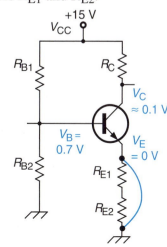

Figure 5-40 With the emitter terminal shorted to ground, R_{E1} and R_{E2} cannot serve to establish a moderate value of I_E, so the transistor is driven into saturation.

Get This

If V_C is too close to V_E (less than the 5-V difference expected in Fig. 5-31) then the transistor has too much main terminal current (I_E and I_C), but not so much to cause saturation.

For example, a short circuit around R_{E1}, shown in Fig. 5-41, causes V_C to be too small.

Sometimes a transistor will **leak**, as described in Fig. 5-42. If the leakage is slight, the transistor will not go into complete saturation. A more severe leak will cause saturation. Leakage is a problem that tends to occur at high temperatures. Any leaking transistor should be replaced.

Another problem that can cause V_C to be too close to V_E is a short circuit between the transistor's collector and base terminals. This is shown in Fig. 5-43.

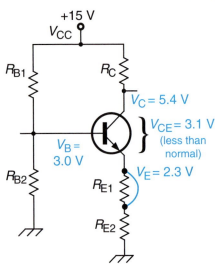

Figure 5-41 With R_{E1} shorted, there is only R_{E2} to limit the main terminal current (I_E and I_C). Therefore that current becomes too large and V_C becomes too close to V_E. If it had been 120-Ω R_{E2} that were shorted, the transistor would have been driven into complete saturation, just like Fig. 5-40.

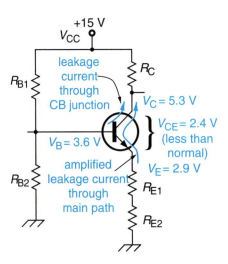

Figure 5-42 A failing transistor may allow a small amount of leakage current to pass through its reverse-biased collector-base junction. This leakage current becomes additional base current, which is then amplified by β to cause additional (larger than normal) collector current. A suspected leaky transistor can be checked on a transistor tester like the one in Fig. 4-33, which has a leakage test function.

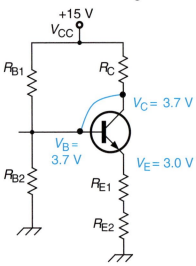

Figure 5-43 A short from C to B causes the transistor's base current to be too large. The transistor does not go all the way into saturation because the short prevents the C-B junction from becoming forward-biased.

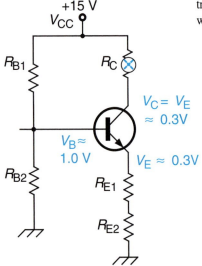

Figure 5-44 With an external open in the collector circuit, transistor current amplification stops. This circuit becomes just a resistive circuit like that in Fig. 5-35. V_C measures the same low value as V_E because the transistor's base-collector diode is internally intact, unlike the one in Fig. 5-35.

Finally, if V_C and V_E are equal to each other and very low, with V_B higher than V_E by 0.7 V, then R_C probably has failed open, as shown in Fig. 5-44.

Shorted Coupling Capacitors

In Fig. 5-24, the amplifier stages depend on their input and output coupling capacitors to provide dc isolation from the rest of the world. For instance, stage 1 depends on C_{in} to dc-isolate its base voltage-divider circuit from the signal source. This allows the transistor's dc bias currents to be completely unaffected by the presence of the signal source, which may have a very low dc resistance tending to short out R_{B2}.

On its other side, stage 1 depends on its output coupling capacitor C_{cplg} to dc-isolate it from the base-bias voltage of stage 2.

If either of these capacitors fails shorted, the dc bias conditions of stage 1 will be completely upset. The same remarks apply to stage 2, regarding C_{cplg} on its input side and C_{out} on its output side.

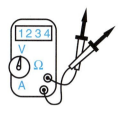

> A shorted coupling capacitor is easy to detect. Measure the dc voltage on both sides. If the voltages are equal, the capacitor is probably shorted.

Get This

For example, if your voltmeter or scope measurement shows an identical dc voltage on both sides of C_{cpl}, as in Fig. 5-45, then C_{cplg} is probably shorted. If no external short is visible, then disconnect and test C_{cplg} for an internal short.

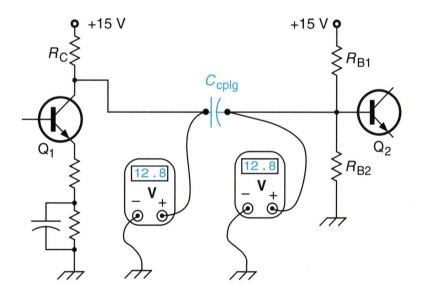

Figure 5-45
The V_C bias voltage for Q_1 and the V_B bias voltage for Q_2 are both completely wrong, and they equal each other. This is the same as saying that C_{cplg} has the same dc voltage on both of its sides, indicating a shorted capacitor.

In the circuit of Fig. 5-24 on page 141, a dc voltage measurement on the left side of C_{in} should show 0 V. Likewise, there should be 0 V dc on the right (load) side of C_{out}.

Testing the Ac Signal

Most amplifier troubles can be uncovered by measuring dc bias voltages, as we have described. But troubleshooting can go quicker if you first find out which stage of the amplifier is malfunctioning. Then you can do the detailed bias measurements on that stage only.

Finding out which stage is malfunctioning can often be done quickly by using an oscilloscope to measure the ac signal at the stage's output and at its input.

Always start testing at the output of the final stage. Verify that the final output signal is no good, then work backward. You are looking for a stage that has a good input signal. When you find a stage that has a bad output but a good input, that is probably the stage that is causing the trouble.

EXAMPLE 5-8

Suppose the two-stage amplifier of Fig. 5-24 (page 141) quits working. An oscilloscope test of the C terminal of Q_2 shows no ac signal present. Moving the scope probe to the B terminal of Q_2 shows a proper ac signal. What is your conclusion?

SOLUTION

There must be trouble in stage 2, since its output is bad even though it has a good input. This assumes that we know the load to be intact—not shorted.

EXAMPLE 5-9

In Example 5-8, suppose that instead of showing a good ac input at the B terminal of Q_2, the scope shows no ac signal present. (The scope must be set to a greater vertical sensitivity [smaller VOLTS/cm factor] to test smaller input voltages.)

What is your next move?

SOLUTION

Move the scope probe to the output of stage 1, the C terminal of Q_1. Then:
  a) If an ac signal is present at the stage 1 output, the trouble must be somewhere in the coupling circuit. Perhaps C_{cplg} is open-circuited.
  b) If there is no ac signal at the stage 1 output, then move to the stage 1 input.

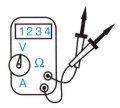

Sometimes it is not possible to identify the trouble location by the simple output/input test above. This can happen because the circuit fault that caused the output to go bad also caused the input to go bad. For example, suppose the E terminal of Q_2 shorts to ground in Fig. 5-24. This will cause a bad output (no ac signal) at the C terminal because the transistor will be saturated. But it will also cause the ac input signal to be bad at the B terminal (the signal will be much smaller than normal). This happens because with R_{E1} and R_{E2} shorted, the ac resistance looking into the base of Q_2 ($r_{b\ in}$) becomes very low, since the B-E junction internal resistance is so low [see Eq. (5-15)]. Therefore R_{in} of stage 2 becomes very low [look at Eq. (5-16)], which destroys the voltage gain of stage 1 [see Eq. (5-17)].

For this example, the output/input test would tend to make you think that the trouble was in stage 1, when really it is in stage 2. When such confusing stage-interaction problems like this occur, they can sometimes be found by temporarily disconnecting the output of stage 1 from the input of stage 2. This is shown in Fig. 5-46.

The disconnection technique is very useful in many areas of electrical troubleshooting. In general, we speak of the **driving** circuit (like stage 1) and the **driven** circuit (like stage 2). If the driving circuit has a bad output, we temporarily disconnect it from the driven circuit, as shown in Fig. 5-47. If the driving circuit's output returns to good, the trouble is probably in the driven circuit.

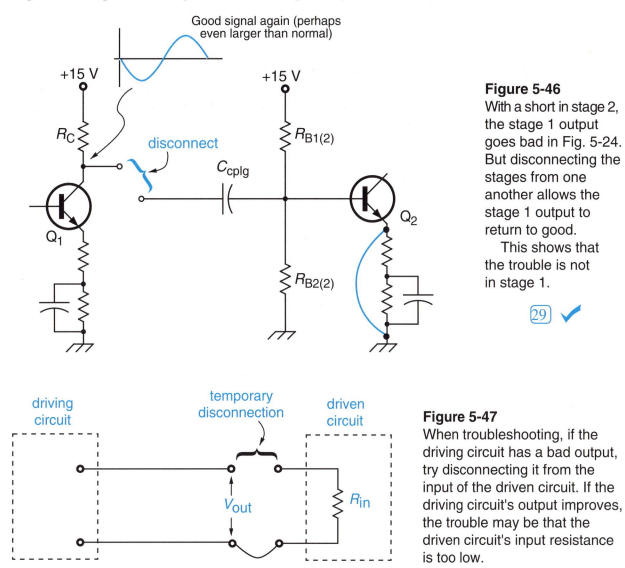

Figure 5-46
With a short in stage 2, the stage 1 output goes bad in Fig. 5-24. But disconnecting the stages from one another allows the stage 1 output to return to good.
 This shows that the trouble is not in stage 1.

29 ✔

Figure 5-47
When troubleshooting, if the driving circuit has a bad output, try disconnecting it from the input of the driven circuit. If the driving circuit's output improves, the trouble may be that the driven circuit's input resistance is too low.

Resistance Measurements During Troubleshooting

When troubleshooting, situations will occur in which you want to measure the resistance of a resistor or other device. If the resistance can be taken in-circuit, then go ahead and take it.

As always with resistance measurements, you must be certain that:
1) The resistor is de-energized — no voltage across it.
2) There is not a second current path in parallel with the resistor.

Get This

For example, suppose you suspected that there was a short or partial short from emitter to ground, as shown in Fig. 5-46. To check your suspicion, you could use an ohmmeter to measure the $R_{E1} - R_{E2}$ combination, as shown in Fig. 5-48.

For the ohmmeter measurement to be useful, the V_{CC} supply must be turned off, and the ohmmeter must not be able to sense the parallel path of R_{B2}

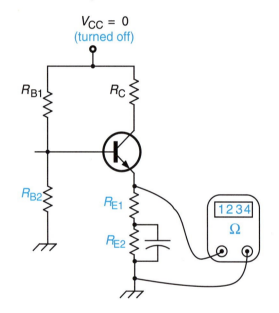

and the B-E junction.   There are several ways to guarantee this second condition:

[1] Temporarily remove the transistor from its socket, if it is mounted in a socket. This is shown in Fig. 5-49(A). [on page 153, facing]

[2] Pay careful attention to the ohmmeter polarity, so you know it reverse-biases, rather than forward-biases, the B-E junction. This is shown in Fig. 5-49(B). When using this approach, you must be careful that the ohmmeter's internal battery voltage does not exceed the maximum allowable reverse voltage across the B-E junction, $V_{EB(max)}$. For many transistors, the $V_{EB(max)}$ rating is quite low, less than 10 V. Therefore you could not use a VOM ohmmeter range with a 9-V or higher internal battery.

[3] Use a DMM ohmmeter in the LOW-ohms function setting. The LOW-ohms setting causes the ohmmeter to apply a test voltage of only about 0.2 V. This is well below the 0.6 V that is required to forward-bias a silicon *pn* junction, so the B-E junction cannot conduct.

Figure 5-48
In-circuit
ohmmeter test.

If an in-circuit measurement is not possible, you can desolder one end of the resistor to isolate it. However, this risks doing damage to the copper track or solder pad, so it should be done only if you have a very strong suspicion that the resistance value is wrong.

Figure 5-50
Clamp-on
current-sensor.

In electronic troubleshooting, current measurements are seldom done, compared to voltage measurements or resistance measurements. This is because standard current measurement is fundamentally more difficult, requiring that the flow-path be broken open. However, there are clamp-on adapters that are used if the conductor can be surrounded by the adapter's jaws. Figure 5-50 shows such a clamp-on device. Of course, they are not usable on printed circuit copper tracks.

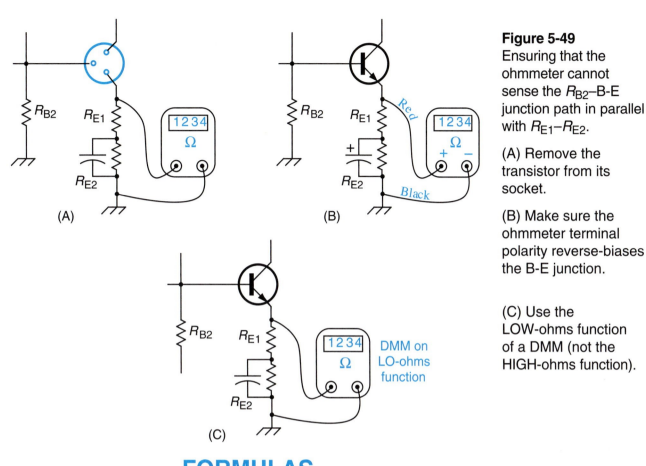

Figure 5-49
Ensuring that the ohmmeter cannot sense the R_{B2}–B-E junction path in parallel with R_{E1}–R_{E2}.

(A) Remove the transistor from its socket.

(B) Make sure the ohmmeter terminal polarity reverse-biases the B-E junction.

(C) Use the LOW-ohms function of a DMM (not the HIGH-ohms function).

FORMULAS

For base voltage-divider bias:

$$V_B = V_{RB2} = V_{CC}\left(\frac{R_{B2}}{R_{B1} + R_{B2}}\right)$$

$$V_E = V_B - 0.7 \text{ V}$$

$$I_E = \frac{V_E}{R_E}$$

$$I_C \approx I_E$$

$$V_{RC} + V_{CE} + V_{RE} = V_{CC}$$

$$A_v = \frac{V_{out}}{V_{in}} \qquad \text{Eq. (5-1)}$$

$$r_{Ej} \approx \frac{30 \text{ mV}}{I_E} = \frac{0.03}{I_E} \qquad \text{Eq. (5-2)}$$

$$A_v = \frac{R_C}{r_{Ej} + R_E} \qquad \text{Eq. (5-3)}$$

$$A_i = \frac{I_{out}}{I_{in}} \qquad \text{Eq. (5-4}$$

$$I_b = \frac{V_{in}}{\beta(r_{Ej} + R_E)} \qquad \text{Eq. (5-5)}$$

$$I_{in} = I_{Rb1} + I_{Rb2} + I_b$$

$$A_P = A_v \cdot A_i \qquad \text{Eq. (5-8)}$$

When R_E is bypassed by capacitance C_E :

$$A_v = \frac{R_C}{r_{Ej}} \qquad \text{Eq. (5-10)}$$

All formulas that contained $(r_{Ej} + R_E)$ now have R_E deleted.

When R_E is partially bypassed by capacitance C_E :

All formulas that contained $(r_{Ej} + R_E)$ now change to $(r_{Ej} + R_{E1})$, where R_{E1} is the unbypassed part of the total emitter resistance.

When R_{LD} is capacitively coupled to the collector output terminal:

$$r_C = R_C \| R_{LD} = \frac{R_C\, R_{LD}}{R_C + R_{LD}} \qquad \text{Eq. (5-13)}$$

$$A_v = \frac{r_C}{r_{Ej} + R_{E1}} \qquad \text{Eq. (5-14)}$$

For a common-emitter amplifier with voltage-divider bias:

$$r_{b\ in} = \beta(r_{Ej} + R_{E1}) \qquad \text{Eq. (5-15)}$$

$$R_{in} = R_{B1} \| R_{B2} \| r_{b\ in} \qquad \text{Eq. (5-16)}$$

For two stages cascaded together:

$$r_{C\,(stage\ 1)} = R_C \| R_{in\,(stage\ 2)}$$

$$A_{v1} = \frac{r_{C\,(stage\ 1)}}{r_{Ej} + R_{E1}} \qquad \text{Eq. (5-17)}$$

$$A_{v\,(total)} = A_{v1} \cdot A_{v2} \qquad \text{Eq. (5-18)}$$

SUMMARY

- The elementary single-base-resistor biasing method is not reliable. Variations in transistor beta cause the dc bias point to be unpredictable (unstable).
- To stabilize the bias of a common-emitter amplifier, a base voltage-divider circuit is installed along with a resistor in the emitter lead.
- An amplifier load-line is a straight line drawn on top of the transistor's I_C -versus-V_{CE} characteristic graphs. The location of the load-line upon the curves is set by the V_{CC} supply voltage and by the resistance values in the amplifier's collector-emitter circuit. The transistor must actually operate at some point along the load-line.
- The factor by which ac output voltage V_{out} is greater than input voltage V_{in} is called the amplifier's voltage gain, symbolized A_v.

● To some extent, A_v depends on the ac resistance of the emitter junction, symbolized r_{Ej} by us.

● When the emitter resistor value R_E is much greater than r_{Ej}, then r_{Ej} has little effect on the amplifier's voltage gain.

● By proper choice of the R_E value, amplifier designers can make A_v almost completely dependent on R_C and R_E; voltage gain is then nearly independent of β and r_{Ej}.

● The factor by which output current I_{out} is greater than input current I_{in} is called the amplifier's current gain, symbolized A_i.

● The V_{CC} supply terminal acts like an ac ground, as far as the ac signal is concerned.

● Because V_{CC} is an ac ground, the source's input current to a common-emitter amplifier splits into three paths: 1) Through R_{B1}; 2) Through R_{B2}; 3) Into the base lead and then through the transistor.

● The power gain of an amplifier, A_P, is the factor by which output power P_{out} is greater than input power P_{in}. Power gain is really the most significant aspect of an electronic amplifier, more so than voltage gain or current gain. However, voltage is the easiest variable to measure, so we concentrate most of our attention on voltage gain.

● By placing a bypass capacitor in parallel with all or part of the emitter resistance R_E, we can effectively remove R_E from the circuit, from the viewpoint of the ac signal. This allows us to trade gain versus gain-stability, without sacrificing dc bias stability.

● When the load is connected through an output coupling capacitor to ground, it has zero dc voltage across it. The output waveform is pure ac.

● When we capacitively couple the load as described above, then the ac collector-line resistance, symbolized r_C, becomes an important idea for understanding the amplifier's voltage gain.

● It is common practice to cascade 2 or more amplifier stages to produce the total overall amplification that is needed. Then $A_{v\,(total)} = A_{v1} \cdot A_{v2}$.

● When an amplifier malfunctions, there is a systematic method for troubleshooting it.

CHAPTER QUESTIONS AND PROBLEMS

1. (T-F) For a specific type number of transistor, the β value can vary greatly from one individual unit to another.

2. (T-F) For an individual transistor, the β value can vary greatly if the transistor's temperature changes.

3. The variation that is described in Question 1 is called _____ variation.

　　Problems 4 through 17 refer to a stabilized common-emitter amplifier like the one in Fig. 5-4 having the following specifications: $V_{CC} = 12$ V, $R_{B1} = 10$ kΩ, $R_{B2} = 2.2$ kΩ, $R_E = 220$ Ω, $R_C = 820$ Ω. Draw a schematic diagram of this amplifier. Include the ac signal source and C_{in}.

4. Assuming that the $R_{B1} - R_{B2}$ combination is virtually a series circuit, calculate $I_{Divider}$ using Ohm's law.
5. Again assuming that the voltage-divider resistors R_{B1} and R_{B2} are virtually in series, calculate the voltage across R_{B2}, which is V_B.
6. Referring to your answer to problem 5, find V_E.
7. Using your V_E result from problem 6, find I_E (and I_C).
8. Using your I_C result from Problem 7, find the voltage drop across R_C.
9. Find V_C, the dc collector voltage relative to ground.
10. In Problems 6 through 9, the transistor's beta value was not used. Explain why the beta value was unimportant (as long as it was not unreasonalby low).
11. For this amplifier, find $I_{C(sat)}$.
12. Draw the amplifier's dc load-line, giving the specific values of the end-points.
13. Locate the position of the dc bias point on the load-line.
14. Assuming that a transistor with $\beta = 150$ is used in this amplifier, what value of dc base current I_B will flow in it?
15. Find this amplifier's approximate ac junction resistance, r_{Ej}.
16. Use Eq. (5-3) to find this amplifier's voltage gain A_v.
17. For $V_{in\ (p-p)} = 0.8$ V, use Eq. (5-1) to find the amplifier's output voltage V_{out}. Where does this output voltage appear (between what specific points in the circuit)?

    Now attach a load resistor, $R_{LD} = 1000\ \Omega$, to the amplifier as shown in Fig. 5-19. Everything else remains the same.

18. Find the equivalent ac resistance of the collector-line circuit, r_C, from Eq. (5-12) or (5-13).
19. What is the new value of voltage gain A_v ?

20. Suppose two amplifier stages are cascaded together. The first stage has $A_{v1} = 2.5$. The second stage has $A_{v2} = 4.8$. Find the overall total voltage gain $A_{v\ (total)}$.
21. When a multistage amplifier malfunctions, what is the first step in the troubleshooting/repair process ?

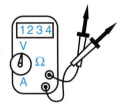

22. (T-F) When testing for the presence of V_{CC} at a circuit location, the best place to touch the test probe is to the copper track that leads up to the circuit component.
23. Besides testing the integrity of all the V_{CC} connections, it is equally important to test all the _____ connections.
24. When trying to isolate the stage where the trouble exists, we look for _____ output signal but _____ input signal.
25. (T-F) If all three dc bias voltages on a transistor are the correct values, that is pretty good evidence that there is no malfunction in that particular stage.
26. When troubleshooting, if the dc collector voltage V_C is the same as the dc emitter voltage V_E, then the transistor is _____ .
27. When troubleshooting, if the dc collector voltage V_C is equal to V_{CC}, then the transistor is _____ .
28. When troubleshooting, if the base-emitter voltage V_{BE} is greater than 1 V, then the transistor is _____ .

CHAPTER 6

COMMON-COLLECTOR AND COMMON-BASE AMPLIFIERS

This turbofan aircraft engine is nearing completion. The large-bladed fan at the front pulls in a great amount of air. Only a small portion of the air enters the core engine's combustion chamber, where it is mixed with fuel for burning. Most of the air flows around the core engine, mixing with the combustion gases as they emerge from the core combustion chamber, at a point slightly forward of the assembly person. The overall mixture then rushes out the engine's exhaust nozzle, producing thrust.

The engine has a sophisticated electronic fuel-injection system that monitors and responds to all aspects of air flow and other engine variables.

Courtesy of CFM International, a 50/50 joint company of GE of the United States and SNECMA of France

OUTLINE

NEW TERMS TO WATCH FOR

common-collector
emitter-follower
resistance match

common-base
high-frequency response

All the transistor amplifiers described so far have been the common-emitter type. There are two other BJT amplifier types, or configurations. As you would expect, they are called the common-collector type and the common-base type.

After studying this chapter, you should be able to:

1. Draw the schematic diagram of a common-collector amplifier and show the locations of the input and output signals.
2. Given the component values, analyze a common-collector amplifier to predict the dc bias currents and voltages. Also analyze its ac performance to predict voltage gain A_v and current gain A_i.
3. Contrast the performance of a common-collector amplifier to a common-emitter amplifier. Explain the advantages and disadvantages of the common-collector configuration.
4. Draw the schematic diagram of a common-base amplifier and show the locations of the input and output signals.
5. Given the component values, analyze a common-base amplifier to predict the dc bias currents and voltages. Also analyze its ac performance to predict voltage gain A_v and current gain A_i.
6. Contrast the performance of a common-base amplifier to common-emitter and common-collector amplifiers. Explain the advantages and disadvantages of the common-base configuration.

6-1 COMMON-COLLECTOR CONFIGURATION

The schematic diagram of a **common-collector** amplifier is shown in Fig. 6-1. It differs from a common-emitter amplifier in two ways:

1 There is no resistor in the collector lead. The C terminal is wired directly to the V_{CC} dc supply voltage. Therefore the C terminal is an ac ground, since V_{CC} is an ac ground.

2 Emitter resistor R_E is now the main resistance in the collector-emitter flow path. It is not a minor low-value resistance placed there only for stability, as it was in the common-emitter configuration.

The input signal is applied between the base terminal and ground, just like a common-emitter amplifier. But the output signal is developed between the emitter terminal and ground, as Fig. 6-1 makes clear. The collector terminal is at ac ground potential (zero volts ac). Therefore the collector is the *common* reference point for both the input and output signals. This is why the circuit is called common-collector.

The phrase **emitter-follower** is used by many people to mean the same thing as common-collector. As we describe the circuit's functioning, it will become clear why this phrase is used.

Actually, emitter-follower is a more popular name than common-collector.

Dc Bias The base voltage-divider bias circuit in Fig. 6-1 works the same way that it did for the common-emitter amplifier in Chapter 5.

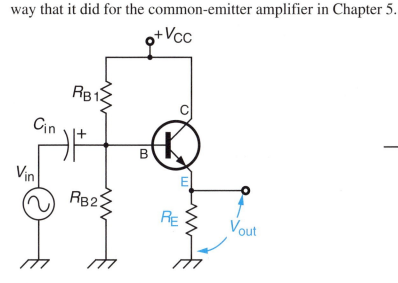

Figure 6-1 In a common-collector circuit, there is no resistor in the collector lead. The output is taken across the emitter resistor, from E to ground.

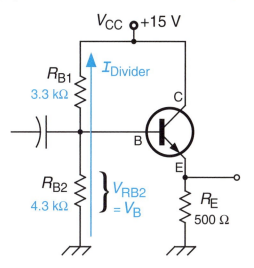

Figure 6-2 We find the dc bias conditions at base and emitter in the same way that we did for common-emitter amplifiers.

Because the main-path resistance is *all* in the emitter lead, the common-collector amplifier automatically has excellent bias stability. Example 6-1 shows why this is true.

EXAMPLE 6-1

For the common-collector amplifier of Fig. 6-2, supose that β can vary from a minimum of 80 to a maximum of 250.
a) Assume for the time being that base-bias current I_B is negligible compared to $I_{Divider}$. Using voltage division, find the dc bias voltages at the base, V_B, and at the emitter, V_E.
b) Using Ohm's law, calculate the dc bias emitter current, I_E.
c) If the transistor has a beta of 250 (maximum) calculate actual I_B. Show that it is negligibly small compared to $I_{Divider}$ and therefore does not disrupt the series-circuit voltage-division between R_{B1} and R_{B2}.
d) Now show the same thing for a beta value of 80 (minimum β).

Like a common-emitter circuit, a common-collector circuit must be designed so that the dc base bias current I_B is negligibly small compared to $I_{Divider}$.

SOLUTION

a) $V_{RB2} = V_{CC} \left(\dfrac{R_{B2}}{R_{B1} + R_{B2}} \right) = 15 \text{ V} \left(\dfrac{4.3}{3.3 + 4.3} \right) = 15 \text{ V} \left(\dfrac{4.3}{7.6} \right)$

$\qquad = \textbf{8.5 V} = V_B.$

With the forward-biased B-E junction dropping 0.7 V as usual,

$\qquad V_E = V_B - V_{BE} = 8.5 \text{ V} - 0.7 \text{ V} = \textbf{7.8 V}$

b) $\quad I_E = \dfrac{V_E}{R_E} = \dfrac{7.8 \text{ V}}{500 \ \Omega} = \textbf{15.6 mA}$

② ✔

Since R_E is the only resistor in the collector-emitter path, V_{RE} is the only resistive voltage drop in that path. This is unlike the common-emitter amplifier, where V_{RC} was the main resistive voltage drop.

c) First we calcualte $I_{Divider}$ as

$\quad I_{Divider} = \dfrac{V_{CC}}{R_{B1} + R_{B2}} = \dfrac{15 \text{ V}}{(3.3 + 4.3) \text{ k}\Omega} = 1.97 \text{ mA}$

With maximum β, I_B is minimum.

$$I_{B(min)} = \frac{I_E}{\beta_{(max)}} = \frac{15.6 \text{ mA}}{250} = 62.4 \text{ μA}$$

This value of I_B is a rather small percentage of $I_{Divider}$, given by

$$\frac{I_{B(min)}}{I_{Divider}} = \frac{62.4 \times 10^{-6} \text{ A}}{1.97 \times 10^{-3} \text{ A}} = 0.032 \text{ or only } \mathbf{3.2\%}, \text{ which is negligible}$$

d) With minimum β, I_B is maximum.

$$I_{B(max)} = \frac{I_E}{\beta_{(min)}} = \frac{15.6 \text{ mA}}{80} = 195 \text{ μA}$$

$$\frac{I_{B(max)}}{I_{Divider}} = \frac{195 \times 10^{-6} \text{ A}}{1.97 \times 10^{-3} \text{ A}} = 0.099 \text{ or } \mathbf{9.9\%} \text{ which is still small} \\ \text{enough to be considered negligible}$$

However, do not think that $V_{CE} = 1/2 V_{CC}$ is always the best bias condition for all amplifiers. In many situations we prefer $V_{CE} < 1/2 V_{CC}$. This will be explained later, in Problems 30-35 on page 208.

The dc bias voltage V_{CE} across the transistor's main terminals is given by KVL as:

$$V_{CE} + V_E = V_{CC} = 15 \text{ V}$$
$$V_{CE} = 15 \text{ V} - V_E = 15 \text{ V} - 7.8 \text{ V}$$
$$= \mathbf{7.2 \text{ V}}$$

As usual, the dc bias conditions have been arranged so that the V_{CC} supply voltage splits roughly half-and-half across the resistance and the transistor itself.

6-2 APPLYING THE AC INPUT TO THE COMMON-COLLECTOR AMPLIFIER

As Fig. 6-1 shows, the ac signal V_{in} is applied to the base through input coupling capacitor C_{in}, just like a common-emitter amplifier. When V_{in} appears at base terminal B, it causes a certain ac base current I_b to flow. The transistor boosts this current by a factor of β to produce ac collector current I_c. Both I_b and I_c flow out the emitter lead, producing I_e. Ac emitter current I_e flows back and forth through R_E, developing output voltage V_e. Resistor R_E is therefore considered to be the load, with $V_{out} = V_{RE} = V_e$.

But there is a somewhat strange relationship between $V_{out}(V_e)$ and $V_{in}(V_b)$. It is this: V_{out} cannot be greater than V_{in}, because if it were, there would be no net difference voltage available to drive ac current through the internal junction resistance r_{Ej}. This idea is illustrated in Fig. 6-3. We can summarize the above-described ac current and ac voltage behavior of a common-collector amplifier as:

| Get This |

A common-collector amplifier does not provide any voltage gain [$A_v < 1$]. But it does provide current gain [$A_i \gg 1$].

When the load is wired directly in the emitter lead, as in Figs. 6-1 through 6-3, it can be shown mathematically that the formulas for A_v and A_i are

You can see that this formula will give an A_v value that is slightly less than 1.0 if r_{Ej} is small compared to R_E.

$$A_v = \frac{R_E}{r_{Ej} + R_E} \qquad \text{Eq. (6-1)}$$

and

$$A_i = \frac{R_{B1} \| R_{B2} \| [\beta(r_{Ej} + R_E)]}{r_{Ej} + R_E} \qquad \text{Eq. (6-2)}$$

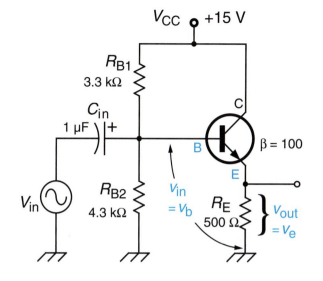

Figure 6-3
It must always be that the instantaneous value of ac voltage at E is less in magnitude than the instantaneous value of ac voltage at B. This must be true in order to have some instantaneous ac voltage left over to drive the internal junction resistance. Only in this way can the circuit cause an instantaneous ac current i_b to flow.

Let us investigate how a common-collector amplifier responds to an ac input signal in the next two Examples.

EXAMPLE 6-2

For the common-collector amplifier of Fig. 6-3, we know from Example 6-1 that I_E = 15.6 mA.
a) Find the ac junction resistance r_{Ej}.
b) Use Eq. (6-1) to find A_v.
c) Use Eq. (6-2) to find A_i.
d) Calculate power gain A_P, from Eq. (5-8).

SOLUTION

a) From Eq. (5-2),

$$r_{Ej} = \frac{30 \text{ mV}}{I_E} = \frac{30 \text{ mV}}{15.6 \text{ mA}} = \textbf{4 } \Omega$$

b) $\quad A_v = \dfrac{R_E}{r_{Ej} + R_E} = \dfrac{500 \text{ } \Omega}{2 \text{ } \Omega + 500 \text{ } \Omega} = \textbf{0.996}$

V_{out} will be slightly less than V_{in} (99.6% as large as V_{in}).

c)
$$A_i = \frac{R_{B1} \| R_{B2} \| [\beta (r_{Ej} + R_E)]}{r_{Ej} + R_E} \qquad \text{Eq. (6-2)}$$

$$= \frac{3.3 \text{ k}\Omega \| 4.3 \text{ k}\Omega \| [100 (502 \text{ } \Omega)]}{502 \text{ } \Omega}$$

$$= \frac{3.3 \text{ k}\Omega \| 4.3 \text{ k}\Omega \| 50.2 \text{ k}\Omega}{502 \text{ } \Omega} = \frac{1800 \text{ } \Omega}{502 \text{ } \Omega} = \textbf{3.59}$$

② ✔

The ac current through the load (R_E in this case) will be 3.59 times as large as the ac current flowing out of the source into the amplifier.
d) From Eq. (5-8),

$\quad A_P = A_v A_i$

$\quad = 0.996 (3.59) = \textbf{3.57}$

The overall power gain is greater than 1.0, which is the mark of a true amplifier.

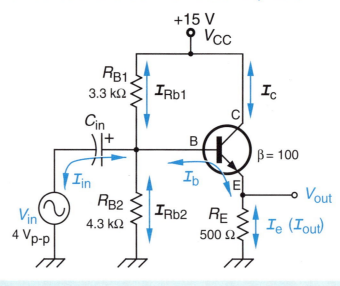

**Figure 6-4
Common-collector
amplifier whose
waveforms are
graphed in Fig. 6-5.**

EXAMPLE 6-3

Suppose $V_{\text{in (p-p)}} = 4$ V to our common-collector amplifier, as shown in Fig. 6-4.

a) Find $V_{\text{out (p-p)}}$.   b) Find the peak-to-peak values of I_{Rb2} and I_{Rb1}.

c) Find the p-p value of I_b, the ac base current that is actually amplified by the transistor.

d) Find $I_{\text{in (p-p)}}$.

e) Find the amplifier's input resistance R_{in} by using Ohm's law with V_{in} and I_{in}.

f) Make waveform sketches of the following amplifier variables: v_{in}, i_{in}, i_B, i_E (i_{out}), v_{out}.

SOLUTION

a) We know from Example 6-2 that $A_v = 0.996$, so

$$V_{\text{out (p-p)}} = A_v [V_{\text{in (p-p)}}]$$
$$= 0.996 (4 \text{ V}) = \textbf{3.98 V}$$

b) The bottom of R_{B2} is connected directly to ground in Fig. 6-4, so V_{in} appears across R_{B2}. From Ohm's law,

$$I_{\text{Rb2}} = \frac{V_{\text{in}}}{R_{\text{B2}}} = \frac{4 \text{ V (p-p)}}{4.3 \text{ k}\Omega} = \textbf{0.930 mA (p-p)}$$

The top of R_{B1} is connected to V_{CC}, which is an ac ground. Therefore the ac voltage V_{in} appears across R_{B1}. From Ohm's law,

$$I_{\text{Rb1}} = \frac{V_{\text{in}}}{R_{\text{B1}}} = \frac{4 \text{ V (p-p)}}{3.3 \text{ k}\Omega} = \textbf{1.21 mA (p-p)}$$

c) The ac resistance seen looking in to the transistor's base terminal is given by the same formula as for a common-emitter amplifier, Eq. (5-15).

$$r_{\text{b in}} = \beta (r_{\text{Ej}} + R_{\text{E}})$$
$$= 100 (2 \text{ }\Omega + 500 \text{ }\Omega) = 50.2 \text{ k}\Omega$$

V_{in} appears across this base input resistance, so Ohm's law gives

$$I_b = \frac{V_{\text{in}}}{r_{\text{b in}}} = \frac{4 \text{ V (p-p)}}{50.2 \text{ k}\Omega} = \textbf{79.7 }\boldsymbol{\mu}\textbf{A (p-p)}$$

d) Apply Kirchhoff's current law to the B node, just as we did in Example 5-3.

$$I_{\text{in}} = I_{\text{Rb1}} + I_{\text{Rb2}} + I_b$$
$$= 0.930 \text{ mA} + 1.212 \text{ mA} + 79.7 \text{ }\mu\text{A} = \textbf{2.22 mA (p-p)}$$

e)

$$R_{in} = \frac{V_{in}}{I_{in}} = \frac{4\ V}{2.18\ mA} = \mathbf{1.83\ k\Omega}$$

Notice that the amplifier's input resistance R_{in} is considerably higher than the 500-Ω load resistance (R_E).

f) The waveforms are sketched in Fig. 6-5. Observe that the common-collector's output waveform v_{out} (= v_e) is in-phase with the input waveform v_{in}. The ac part of v_E "follows" v_{in}, being time-synchronized (in-phase) and almost equal in value to v_{in}. This is why this configuration is often called an emitter-follower.

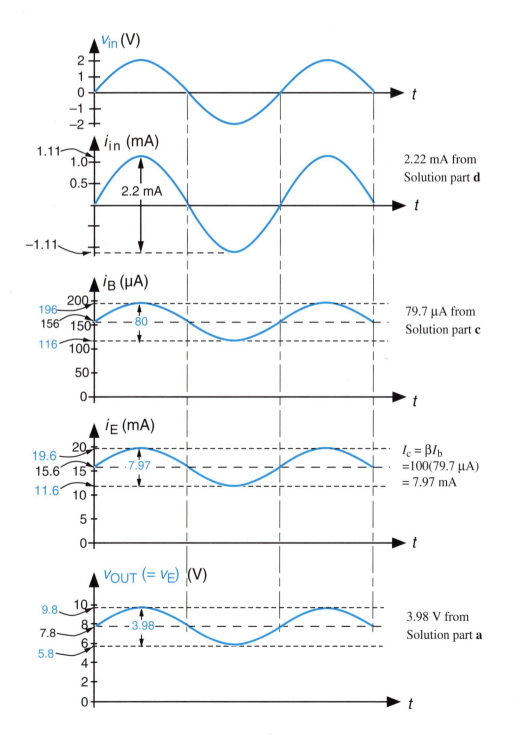

2.22 mA from
Solution part **d**

Figure 6-5
Detailed waveform graphs of the common-collector amplifier of Fig. 6-4 (Example 6-3).

Notice that V_{out} is in-phase with V_{in}, unlike a common-emitter amplifier.

79.7 μA from
Solution part **c**

$I_c = \beta I_b$
=100(79.7 μA)
= 7.97 mA

3.98 V from
Solution part **a**

Connecting an External Load

To remove the dc component of load voltage, we connect the load to the emitter terminal through output coupling capacitor C_{out}, as shown in Fig. 6-6(A). This is the same basic method that was used for common-emitter amplifiers in Ch. 5.

Connecting the load in this manner changes the ac characteristics of the amplifier, just as it did in Ch. 5. Now:

Get This

The equivalent ac resistance looking out of the emitter lead, which we will call r_E, is the parallel combination of R_E and R_{LD}. As a formula,

$$r_E = R_E \| R_{LD} = \frac{R_E R_{LD}}{R_E + R_{LD}} \qquad \text{Eq. (6-3)}$$

Then the voltage gain formula becomes

$$A_v = \frac{r_E}{r_{Ej} + r_E} = \frac{R_E \| R_{LD}}{r_{Ej} + R_E \| R_{LD}} \qquad \text{Eq. (6-4)}$$

The equivalent ac resistance seen looking in to the base terminal becomes

$$r_{b\,in} = \beta(r_{Ej} + r_E) = \beta\,(r_{Ej} + R_E \| R_{LD}) \qquad \text{Eq. (6-5)}$$

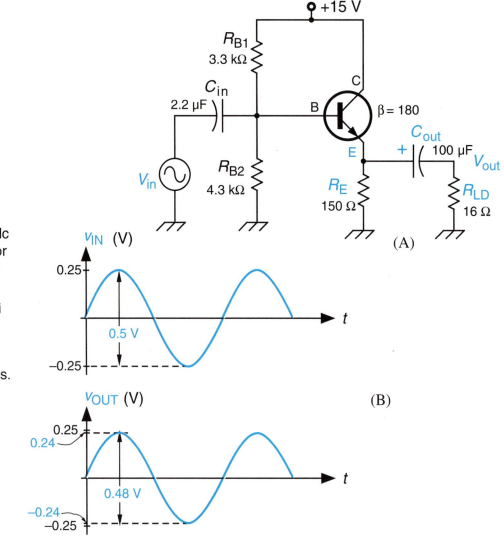

Figure 6-6
Common-collector amplifier with capacitively coupled load.

(A) Schematic. The dc bias emitter current for this amplifier is about 52 mA, so the internal junction resistance r_{Ej} is about 0.6 Ω.

(B) Voltage waveforms. v_{OUT} is almost identical to v_{IN}.

EXAMPLE 6-4

For the common-collector amplifier with capacitively coupled output in Fig. 6-6(A), suppose $V_{in\,(p-p)} = 0.5$ V.
a) Find the voltage gain A_v.
b) Calculate V_{out}.
c) Make waveform sketches of v_{IN} and v_{OUT}.

SOLUTION

a) Applying Eq. (6-3) to get the resistance in the emitter, we have

$$r_E = \frac{R_E R_{LD}}{R_E + R_{LD}} = \frac{150\,(16)}{150 + 16} = 14.5\ \Omega$$

Then from Eq. (6-4),

$$A_v = \frac{r_E}{r_{Ej} + r_E} = \frac{14.5\ \Omega}{0.6\ \Omega + 14.5\ \Omega} = 0.96$$

② ✔

This is a bit lower than the A_v result in Examples 6-2 and 6-3, since the transistor's internal resistance r_{Ej} is no longer completely swamped by the external ac emitter-line resistance r_E.
b) $V_{out} = A_v \cdot V_{in} = 0.96\,(0.5\text{ V p-p}) = \textbf{0.48V p-p}$
c) The v_{OUT} waveform is sketched in Fig. 6-6(B). It is pure ac, with no dc component.

The main usefulness of a common-collector amplifier is that its ac input resistance, R_{in}, tends to be high, much higher than any low-resistance external load that it is driving.

Get This

Because a common-collector amplifier's R_{in} is large, the input current from the source tends to be small. This is a big advantage over a common-emitter amplifier when the signal source has a large internal resistance (output resistance) that limits its current capability.

③ ✔

Figure 6-7
A common-collector amplifier is most often used when the signal source has high internal resistance $R_{s(int)}$. (A microphone is a good example).

The amplifier's high value of R_{in} means that the source voltage V_s divides between $R_{s(int)}$ and R_{in} so that a good share of V_s appears across R_{in}, as the amplifier's V_{in}.

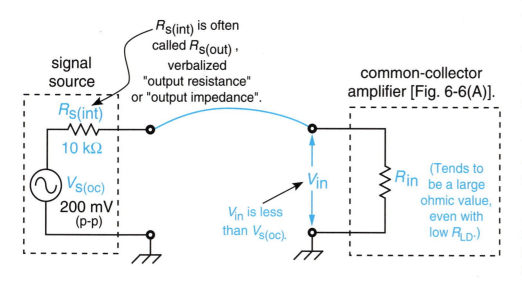

$R_{s(int)}$ is often called $R_{s(out)}$, verbalized "output resistance" or "output impedance".

signal source

common-collector amplifier [Fig. 6-6(A)].

$R_{s(int)}$

10 kΩ

$V_{s(oc)}$
200 mV
(p-p)

V_{in}

V_{in} is less than $V_{s(oc)}$.

R_{in} (Tends to be a large ohmic value, even with low R_{LD}.)

EXAMPLE 6-5

Suppose that the common-collector amplifier of Fig. 6-6(A) is driven by a microphone source, with open-circuit output voltage of 200 mV p-p and $R_{s(int)} = 10$ kΩ. The situation is pictured in Fig. 6-7.

a) Use Eq. (6-5) to find the equivalent ac resistance, $r_{b\,in}$, looking in to the B terminal of Fig. 6-6(A).

b) Equation (5-16), $R_{in} = R_{B1} \| R_{B2} \| r_{b\,in}$, was used to predict the input resistance R_{in} for common-emitter amplifiers. It also works for common-collector amplifiers. The difference is that R_{B1} and R_{B2} now can be made larger, and $r_{b\,in}$ also tends to be greater, even with a low-resistance load in the emitter line. Use Eq. (5-16) to calculate the input resistance of the Fig. 6-6(A) amplifier.

c) Referring to Fig. 6-7, use voltage division to find the value of V_{in}, the actual input voltage to the common-collector amplifier.

SOLUTION

a) We know from Example 6-4 that emitter-line resistance $r_E = 14.5$ Ω.

$$r_{b\,in} = \beta\,(r_{Ej} + r_E)$$
$$= 180\,(0.6\,\Omega + 14.5\,\Omega) = \textbf{2.72 k}\Omega$$

b) $R_{in} = R_{B1} \| R_{B2} \| r_{b\,in}$

$$= 3.3\text{ k}\Omega \| 4.3\text{ k}\Omega \| 2.72\text{ k}\Omega \approx \textbf{1.1 k}\Omega$$

c) By voltage division, we get

$$\frac{V_{in}}{V_{s\,(oc)}} = \frac{R_{in}}{R_{s\,(int)} + R_{in}}$$

$$V_{in} = 200\,\text{mV}\left(\frac{1.1\text{ k}\Omega}{10\text{ k}\Omega + 1.1\text{ k}\Omega}\right) = \textbf{19.8 mV p-p}$$

which is a usable amount.

Compare the V_{in} result in this example to what would have happened if the signal source had been connected directly to the 16-Ω speaker load. The voltage division would be

$$V_{in} = 200\,\text{mV}\left(\frac{16\,\Omega}{16\,\Omega + 10\,016\,\Omega}\right) = \textbf{0.3 mV p-p}$$

which is not a usable amount.

The common-collector amplifier does a good job of obtaining usable input voltage from the signal source. We say that its high input resistance is a good *match* to the high output resistance of the signal source.

SELF-CHECK FOR SECTIONS 6-1 AND 6-2

1. In a common-collector amplifier, the input signal is applied at the transistor's _____ terminal and the output signal is taken at the transistor's _____ terminal.

2. In a common-collector amplifier, which transistor terminal is at ac ground potential?

3. (T-F) In a common-collector amplifier the main resistor in the collector-emitter flow-path is placed in the collector lead.
4. (T-F) A common-collector amplifier provides current gain.
5. (T-F) A common-collector amplifier provides voltage gain.
6. (T-F) A common-collector amplifier provides power gain.
7. For the basic common-collector amplifier of Fig. 6-3, suppose we have the following circuit specifications: $V_{CC} = 20$ V, $R_{B1} = 10$ kΩ, $R_{B2} = 22$ kΩ, $R_E = 1.2$ kΩ, $\beta = 140$, $V_{in} = 3$ V p-p. Find:
    a) V_E b) I_E c) A_v d) V_{out}
8. For the common-collector amplifier of Fig. 6-6(A), suppose we have the following circuit specifications: $V_{CC} = 18$ V, $R_{B1} = 6.8$ kΩ, $R_{B2} = 11$ kΩ, $R_E = 680$ Ω, $R_{LD} = 250$ Ω, $\beta = 200$, $V_{in} = 5$ V p-p. Find:
    a) V_E d) V_{out} g) I_{in} j) A_P
    b) I_E e) $r_{b\,in}$ h) I_{out}
    c) A_v f) R_{in} i) A_i

6-3 COMMON-BASE CONFIGURATION

The schematic diagram of a **common-base** amplifier is shown in Fig. 6-8(A). Figure 6-8(B) is a sideways view of the same circuit, which is preferred by some people. The C-B amp has its main resistance in the collector lead, like a common-emitter amplifier. The emitter lead also contains a substantial resistance R_E, which is used for two purposes.

1 It sets the dc bias current I_E, just like a common-emitter amplifier.
2 It separates the emitter terminal E from ground, so that the input voltage V_{in} can be applied to the E terminal.

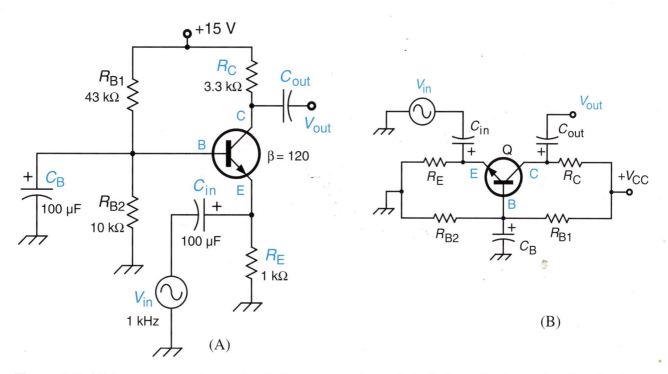

Figure 6-8 (A) In a common-base circuit, there are resistors in both the collector and emitter leads. The input is applied to the emitter and the output is taken from the collector. (B) Alternative schematic view, which is historically popular.

Base terminal B is an ac ground. The ac ground is established by base capacitor C_B, connected between B and power supply ground in Fig. 6-8. Therefore, with V_{in} applied between E and ground, that connection is ac-equivalent to being applied between E and B.

The ac output is taken at the collector terminal C. With V_{out} appearing between C and ground, it is ac-equivalent to appearing between C and B. Thus, we can see that the base terminal B is common to both the ac input and the output — it's a common-base amplifier.

④ ✓

Dc Bias: The voltage-divider bias circuit works in the usual way to establish a stable bias point.

EXAMPLE 6-6

For the common-base amplifier of Fig. 6-8, find:

    a) V_B d) V_{RC}
    b) V_E e) V_{CE}
    c) I_E

SOLUTION

a) Applying the voltage-divider formula, we get

$$\frac{V_{RB2}}{V_{CC}} = \frac{R_{B2}}{R_{B1} + R_{B2}}$$

$$V_{RB2} = 15 \text{ V} \left(\frac{10 \text{ k}\Omega}{43 \text{ k}\Omega + 10 \text{ k}\Omega}\right)$$

$$= \textbf{2.8 V} = V_B$$

b) The transistor junctions are biased as they always are: forward on the B-E junction and reverse on the C-B junction. The forward voltage drop across the B-E junction is approximately 0.7 V, so

$$V_E = V_B - 0.7 \text{ V} = 2.8 - 0.7$$

$$= \textbf{2.1 V}$$

c) From Ohm's law,

$$I_E = \frac{V_E}{R_E} = \frac{2.1 \text{ V}}{1 \text{ k}\Omega}$$

$$= \textbf{2.1 mA}$$

d) Since $I_C \approx I_E$, we can apply Ohm's law to R_C as

$$V_{RC} = I_C R_C \approx 2.1 \text{ mA} (3.3 \text{ k}\Omega) = \textbf{6.9 V}$$

e) From Kirchhoff's voltage law, we have

$$V_{CC} = V_{RC} + V_{CE} + V_E, \text{ or, rearranged,}$$

$$V_{CE} = V_{CC} - V_{RC} - V_E$$

$$= 15 \text{ V} - 6.9 \text{ V} - 2.1 \text{ V} = \textbf{6.0 V}$$

⑤ ✓

6-4 AC OPERATION OF A COMMON-BASE AMPLIFIER

As the ac signal source looks into the common-base amplifier of Fig. 6-8, it sees the circuit sketched in Fig. 6-9.

The reactance of C_{in} is considered to be 0 Ω, so V_{in} appears full strength at E. When the input current I_{in} arrives at the E terminal, it sees two paths to ground. One is the high-resistance path down through 1000-Ω R_E. The other is the low-resistance path up through r_{Ej}, since the top of r_{Ej} is an ac ground (remember that base terminal B is held at ac ground potential by 0-Ω reactance X_{CB}. Virtually all the input current takes the low-resistance r_{Ej} path, as pointed out in Fig. 6-9.

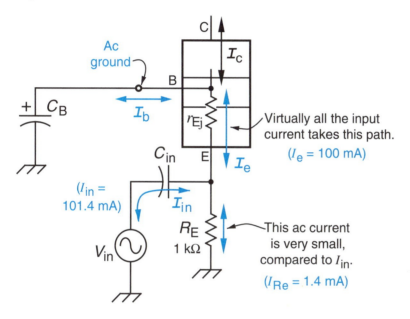

Figure 6-9 Understanding the ac input circuit of a common-base amplifier.

Say $r_{Ej} \approx 14$ Ω and $V_{in} = 1.4$ V. Then the values of I_{in}, I_e, and I_{Re} are as shown.

$(I_e = V_{in}/r_{Ej} = 1.4$ V $/14$ Ω $= 100$ mA).

$(I_{Re} = V_{in}/R_E = 1.4$ V $/1$ kΩ $= 1.4$ mA).

$(I_{in} = I_e + I_{Re} = 100 + 1.4 = 101.4$ mA).

But the ac current passing back and forth through r_{Ej} is the ac emitter current I_e. Therefore, the signal source's input current, I_{in}, is virtually the same as I_e. In other words, the signal source supplies *all* the ac current for the transistor. The transistor's current-gain capability (β) is not being used.

A common-base amplifier does not provide any current gain [$A_i < 1$]. But it does provide voltage gain [$A_v \gg 1$].

Get This

You can think of a common-base amplifier this way. It presents a very low input resistance R_{in} to the source. ($R_{in} \approx r_{Ej}$) If the source itself has good current capability, but operates at a low voltage, then the amplifier's low R_{in} is a good match for it. Here is what happens:

1. The source delivers its current through the low input resistance r_{Ej}.
2. The high-voltage collector circuit grabs almost all that current ($I_c \approx I_e$) and forces it to flow through high-value resistance R_C.
3. The transistor collector, working in conjunction with the V_{CC} power supply, produces the proper amount of ac voltage across R_C to satisfy Ohm's law. That is, $V_{Rc} = I_c R_C$. This voltage then becomes the output voltage V_{out}.

It can be shown mathematically that the formulas for A_v and A_i are

The value $\beta/(\beta+1)$ is defined as a transistor's alpha (α). That is, $\alpha = \beta/(\beta+1)$.

In current terms, $\alpha = I_C / I_E$.

$$A_v = \frac{R_C}{r_{Ej}} \qquad \text{Eq. (6-6)}$$

and

$$A_i = \left(\frac{\beta}{\beta+1}\right)\left(\frac{R_E}{r_{Ej} + R_E}\right) \qquad \text{Eq. (6-7)}$$

Under normal circumstances, Eq. (6-7) gives a result which is slightly less than 1.0.

EXAMPLE 6-7

For the common-base amp of Fig. 6-8, we know from Example 6-6 that $I_E = 2.1$ mA.
a) Find the junction resistance r_{Ej}.
b) Use Eq. (6-6) to find A_v.
c) Use Eq. (6-7) to find A_i.
d) Find the power gain, A_P.

SOLUTION

a) $$r_{Ej} \approx \frac{30 \text{ mV}}{2.1 \text{ mA}} \approx \mathbf{14\ \Omega}$$

A_v is very unstable because r_{Ej} is very unstable. This is really only an estimate of A_v.

b) $$A_v = \frac{R_C}{r_{Ej}} \approx \frac{3300\ \Omega}{14\ \Omega} \approx \mathbf{235}$$

c) $$A_i = \left(\frac{\beta}{\beta+1}\right)\left(\frac{R_E}{r_{Ej} + R_E}\right)$$

5 ✔

$$= \left(\frac{120}{121}\right)\left(\frac{1000\ \Omega}{14\ \Omega + 1000\ \Omega}\right) \approx \mathbf{0.98}$$

which is quite close to 1.0, as expected.

d) From Eq. (5-8),

$$A_P = A_v\, A_i$$

This A_P is only an estimate, since A_v is only an estimate.

$$\approx 235\,(0.98) = \mathbf{230}$$

Courtesy of Pressure Systems, Inc. and NASA

The small device on the left is used to gather air-pressure data from hundreds or even thousands of measurement points, which is the situation encountered in wind-tunnel testing. In the photo on the right, it is going airborne, to gather in-flight pressure data along wings and other airfoil surfaces.

As usual, we can remove the dc component from v_{OUT} by connecting the load through C_{out} to ground. Then the voltage gain formula becomes

$$A_v = \frac{r_C}{r_{Ej}} = \frac{R_C \| R_{LD}}{r_{Ej}}$$ Eq. (6-8)

EXAMPLE 6-8

Suppose that an 800-Ω load resistance is connected from C_{out} to ground in the common-base amplifier of Fig. 6-8, with $V_{in\,(p\text{-}p)} = 50$ mV.
a) Estimate A_v.
b) Calculate V_{out} based on your estimate of A_v.
c) Sketch the waveforms of v_{IN} and v_{OUT}.

SOLUTION

a) From Eq. (6-8),

$$A_v = \frac{R_C \| R_{LD}}{r_{Ej}}$$

$$R_C \| R_{LD} = \frac{3300\,(800)}{3300 + 800} = 644\ \Omega$$

so $A_v \approx \dfrac{644\ \Omega}{14\ \Omega} = \mathbf{46}$

b) $V_{out} = A_v\,V_{in}$

$$= 46\,(50\ \text{mV}) = \mathbf{2.3\ V\ (p\text{-}p)}$$

c) The waveforms are sketched in Fig. 6-10. Notice that V_{out} is in phase with V_{in}, not inverted like a common-emitter amplifier.

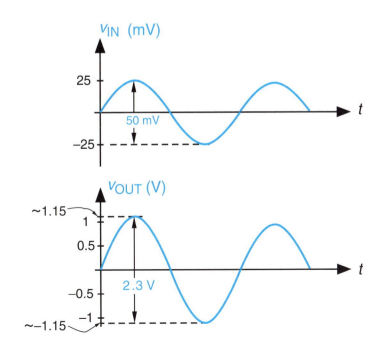

Figure 6-10
Input and output waveforms for the common-base amplifier of Example 6-8.

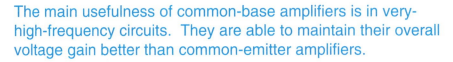

Get This

The main usefulness of common-base amplifiers is in very-high-frequency circuits. They are able to maintain their overall voltage gain better than common-emitter amplifiers.

At frequencies above a few MHz, most bipolar transistor types suffer a reduction in β. That is, they have an effective β that is less than their normal (low-frequency) β. This happens because not all of the free electrons that are emitted by the emitter (*npn* transistor) have enough time to make their way through the base to the collector region.

In a common-emitter amplifier, as the transistor's β decreases further and further at higher and higher frequencies, the $r_{b\,in}$ resistance also decreases. Refer back to Eq. (5-15) to see this. As $r_{b\,in}$ decreases, so does the input resistance R_{in} of the entire amplifier. Therefore less of the source's open-circuit voltage appears across the input terminals as V_{in}. More voltage is lost across the source's output resistance, R_{int}. Of course, this means a reduction in V_{out} — a loss of overall voltage gain.

By overall voltage gain, we mean the ratio of V_{out} to the source's open-circuit voltage, $V_{s(oc)}$.

In summary, a signal source/common-emitter amplifier combination is very much dependent on the transistor's current gain β in order to provide overall voltage gain at high frequencies.

But this effect does not occur in a common-base amplifier. The C-B amp does not depend on β for its input resistance. In fact, its voltage gain doesn't depend on the BJT's current gain β very much at all, since the signal source has to supply the entire current anyway. Therefore a signal source/common-base amplifier combination is able to maintain its overall voltage gain further up the frequency scale. Typically, a common-base circuit may not start to lose voltage gain until frequencies of several tens (or hundreds) of megahertz.

SELF-CHECK FOR SECTIONS 6-3 AND 6-4

9. In a common-base amplifier, the input signal is applied at the transistor's _____ terminal and the output signal is taken at the transistor's_____ terminal.
10. In a common-base amplifier, which transistor terminal is at ac ground potential?
11. (T-F) In a common-base amplifier, the main resistor that has V_{out} developed across it is located in the collector lead.
12. (T-F) A common-base amplifier provides current gain.
13. (T-F) A common-base amplifier provides voltage gain.
14. (T-F) A common-base amplifier provides power gain.
15. For the basic common-base amplifier of Fig. 6-8, assume that the circuit specifications are: $V_{CC} = 18$ V, $R_{B1} = 27$ kΩ, $R_{B2} = 6.8$ kΩ, $R_C = 2.2$ kΩ, $R_E = 750$ Ω, $\beta = 100$, $V_{in(p-p)} = 20$ mV. Draw the amplifier's schematic diagram, and then find:

    a) V_B d) r_{Ej}
    b) V_E e) A_v
    c) I_E f) V_{out}

16. In the common-base amplifier of Fig. 6-8, suppose that a capacitively coupled load is added. The circuit specifications are: $V_{CC} = 25$ V, $R_{B1} = 220$ kΩ, $R_{B2} = 68$ kΩ, $R_C = 13$ kΩ, $R_{LD} = 10$ kΩ, $R_E = 6.8$ kΩ, $\beta = 120$,

$V_{in(p-p)} = 40$ mV. Draw the circuit schematic diagram and then find:

a) V_B g) R_{in} (Use $R_{in} = r_{Ej} \| R_E$, or $R_{in} \approx r_{Ej}$)

b) V_E h) I_{in} (Use Ohm's law with V_{in} and R_{in})

c) I_E i) I_{ld} (Which is I_{out}. Use Ohm's law with V_{out} and R_{LD}.)

d) r_{Ej} j) A_i Use I_{out} / I_{in}. (Eq. [6-7] does not work when the

e) A_v load is capacitively coupled to the C terminal.)

f) V_{out} k) A_P (Use $A_P = A_v A_i$)

17. Suppose that we have two amplifiers side by side. We test both of them at higher and higher frequencies. The first one maintains a steady overall voltage gain $A_v = 25$ until $f = 3$ MHz; beyond that its voltage gain starts decreasing. The second one maintains its voltage gain until $f = 200$ MHz. What configuration did the first amplifier probably have? What configuration did the second amplifier probably have?

EDDY-CURRENT TESTING

Many modern systems have parts that are subject to extreme thermal and mechanical stress. Such parts must be absolutely free from metallurgical flaws.

One method of testing metal integrity is by applying an ac magnetic field to the part to induce an internal ac current called *eddy-current* . The magnitude of the eddy-current can then be measured with an external contact probe. If there is any internal flaw in the metal crystal, it causes a deviation from normal eddy-current magnitude when the magnetic pole passes over the flaw.

Courtesy of NASA

In this test setup, a turbine disc is being rotated slowly past a stationary ac electromagnet, visible at the disc's left edge. The current probe is the hook-shaped object near the center of the disc. Throughout the test procedure, eddy-current data is being recorded by the computer equipment in the background, under the eye of the test technician.

6-5 COMPARING THE THREE TRANSISTOR CONFIGURATIONS

The three bipolar transistor configurations have differing characteristics. These differences make each type useful in particular kinds of applications.

Common-Emitter: This configuration is most useful when the signal source has moderate output resistance R_{int}, and the application calls for both voltage gain and current gain.

Common-Collector (Emitter-Follower): This configuration is most useful when the signal source has a high output resistance R_{int}. (The amplifier must present to the source an R_{in} value that is higher than R_{LD}.) There must be no immediate need for voltage gain, only current gain.

Common-Base: This configuration is most useful for high-frequency operation, where the signal source has a low output resistance. It so happens that high-frequency radio signal sources, like antennas, tend to have a low internal output resistance, often 50 Ω to 300 Ω. There must be no immediate need for current gain, only voltage gain.

Table 6-1 summarizes the characteristics of the three amplifier types.

**Table 6–1
Comparing the features of the three BJT amplifier types.**

	Common-Emitter	Common-Collector (Emitter-Follower)	Common-Base
Voltage Gain	Yes	No	Yes
Current Gain	Yes	Yes	No
Power Gain	Yes (Highest of the 3)	Yes	Yes
Load is in this lead	Collector	Emitter	Collector
Input resistance	Medium	High	Low
Vout – Vin phase	Out of phase	In phase	In phase

Troubleshooting: The methods of troubleshooting common-collector and common-base amplifiers are not basically different from the methods for common-emitter amplifiers. Identify the amplifier stage that has the bad output but a good input. When testing with an oscilloscope, remember that V_{out} from a common-collector stage is not supposed to be larger than V_{in}.

When you find the bad amplifier stage, measure its dc bias voltages and compare to the expected values. Any discrepancy that you find will point toward the exact trouble, as it did in Sec. 5-7.

FORMULAS

For common-collector:

$$A_v = \frac{R_E}{r_{Ej} + R_E} \qquad \text{Eq. (6-1)}$$

$$A_v < 1.0$$

$$A_i = \frac{R_{B1} \parallel R_{B2} \parallel [\beta(r_{Ej} + R_E)]}{r_{Ej} + R_E} \qquad \text{Eq. (6-2)}$$

$$r_E = R_E \parallel R_{LD} \qquad \text{Eq. (6-3)}$$

When R_E is paralleled by R_{LD} to produce r_E in Eq. (6-3), then use r_E in place of R_E in Equations (6-1) and (6-2).

$$r_{b\,in} = \beta\,(r_{Ej} + r_E) \qquad \text{Eq. (6-5)}$$

$$R_{in} = R_{B1} \parallel R_{B2} \parallel r_{b\,in}$$

For common-base:

$$A_v = \frac{R_C}{r_{Ej}} \qquad \text{Eq. (6-6)}$$

$$A_i < 1.0$$

$$r_C = R_C \parallel R_{LD}$$

When R_C is paralleled by R_{LD} to produce r_C, then use r_C in place of R_C in Equation (6-6).

$$R_{in} \approx r_{Ej}$$

SUMMARY

● There are two other transistor amplifier configurations besides C-E. They are called common-collector and common-base. Common-collector is often referred to as emitter-follower.

● The common-collector configuration provides no voltage gain, but it does provide current gain.

● The main advantage of the common-collector configuration is that it has a high value of ac input resistance. It therefore provides a good resistance match to ac signal sources that contain high internal resistance (also called output resistance).

● The common-base configuration provides no current gain, but it provides a large voltage gain.

● The main advantage of the common-base configuration (besides large A_v) is that it can operate at very high frequencies, compared to the other two configurations.

QUESTIONS AND PROBLEMS

1. Draw the schematic diagram of a common-collector amplifier with base voltage-divider bias.
2. In a common-collector amplifier, the input signal is applied between the _____ terminal and ground, and the output signal is taken between the _____ terminal and ground.
3. What is the popular alternative name for common-collector amplifier?
4. In a common-collector amplifier, V_{out} is slightly _____ than V_{in}.
5. In a common-collector amplifier, V_{out} is _____ phase with V_{in}.

For Problems 6 through 12, draw the schematic diagram of a common-collector amplifier with the following specifications: $V_{CC} = 12$ V, $R_{B1} = 5.6$ kΩ, $R_{B2} = 6.8$ kΩ, $R_E = 1200$ Ω, $V_{in(p-p)} = 5$ V, $\beta = 200$.

6. Find the dc base bias voltage V_B, and emitter bias voltage V_E.
7. Find the dc bias current I_E.
8. Using your result from Problem 7, use Eq. (5-2) to approximate the ac resistance of the base-emitter junction, r_{Ej}.
9. Find the amplifier's voltage gain A_v, using Eq. (6-1).
10. Find the amplifier's current gain A_i, using Eq. (6-2).
11. Find the amplifier's power gain, A_P.
12. In this common-collector amplifier, there is actually a slight voltage loss. But the circuit is still a true amplifier. Explain why this is so.

For Problems 13 through 21, connect a load resistance $R_{LD} = 500$ Ω through an output coupling capacitor to ground in the schematic diagram that you drew for Problems 6 through 12.

13. With R_{LD} connected, do any of the dc bias conditions change? Why?
14. Does ac junction resistance r_{Ej} change? Explain.
15. The equivalent ac resistance seen looking out from the emitter lead _____ (increases or decreases). Explain why this happens.
16. Calculate the equivalent ac resistance r_E, seen from the emitter.
17. Find the new value of A_v.
18. Because r_E is less than R_E, the ac resistance seen looking into the base, $r_{b\,in}$, will _____ . (increase or decrease).
19. Because of what happened to the transistor's ac base input resistance $r_{b\,in}$ in Question 18, the overall input resistance of the common-collector amplifier, R_{in}, will _____ (increase or decrease).
20. Because of what happened to amplifier input resistance R_{in} in Question 19, the actual amplifier input voltage V_{in} will _____ (increase or decrease). Assume that the signal source has a high internal resistance R_{int}.
21. Considering your answer to Question 20, even though A_v itself hardly changed when R_{LD} was connected, the final output voltage V_{out} can be substantially _____ than it was previously.

22. Draw the schematic diagram of a single-supply common-base amplifier.
23. In the common-base configuration, the input signal is applied between the_____ and ground. The output signal is taken between the_____ and ground.

For Problems 24 through 27, apply the following specifications to your C-B schematic from Problem 22: $V_{CC} = 20$ V, $R_{B1} = 47$ kΩ, $R_{B2} = 15$ kΩ, $R_E = 1.6$ kΩ, $R_C = 2.7$ kΩ, $V_{in(p-p)} = 20$ mV, $\beta = 175$.

24. Find all the dc bias conditions, V_B, V_E, I_E (and I_C), V_{RC}, and V_{CE}.
25. Make an approximate calculation of r_{Ej}, the emitter junction ac resistance.
26. Use. Eq. (6-6) to make an approximate calculation of voltage gain A_v.
27. Use. Eq. (6-7) to estimate current gain A_i.
28. Based on your answers to Problems 26 and 27, you can say this about the common-base amplifier configuration: it provides a _____ voltage gain, and _____ current gain.
29. The ac input resistance R_{in} of a common-base amplifier is very _____.
30. A common-base amplifier cannot be used with a high-resistance signal source. Why is this?
31. (T-F) A common-base amplifier is fundamentally less dependent on transistor β than either of the other two amplifier configurations.
32. (T-F) Common-base amplifiers are better adapted to handling very high-frequency signals than the other two amplifier configurations.

CHAPTER 7

COUPLING METHODS, FEEDBACK, AND OTHER AMPLIFIER IDEAS

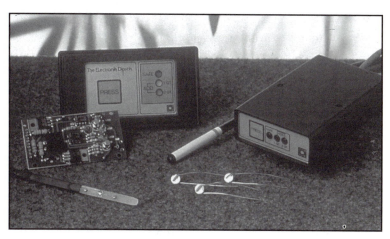

Courtesy of Creative Designs and Inventions, Inc.

This photo shows the component parts of an electronic dipstick, a device that provides visual indication of the level of engine oil or transmission fluid. It uses temperature-sensitive resistors (thermistors) mounted on a plastic stick inside a fiberglass tube. Three isolated thermistors are shown at the center bottom. The plastic stick with the thermistors mounted on the opposite side is at bottom left. The fiberglass tube is poking in from the upper right.

The fiberglass tube is inserted as usual into the engine's steel dipstick tube. Oil is able to enter the hollow fiberglass and contact the thermistors. A thermistor that is bathed in oil stays relatively cool. One that has no oil on it runs hotter. The electronic circuitry on the left senses these temperatures and turns on the proper display light. The readout module at the top center is for trucks, recreational vehicles, and boats. The one on the right is for passenger cars. Also shown on page 1 of the color section.

OUTLINE

NEW TERMS TO WATCH FOR

dual-polarity emitter bias
transformer coupling
primary voltage gain
audio transformer
RF transformer

tuned
transformer-coupled input
dc amplification
direct coupling
negative feedback

quiescent
Q-point
Darlington pair
ac load-line

We studied the operation of the three transistor amplifier configurations in Chapters 5 and 6. There are additional topics about linear transistor amplifiers that you should be aware of. Such things as alternative biasing methods, alternative methods of coupling signals into and out of amplifiers, and several other ideas are important to us.

After studying this chapter, you should be able to:

1. Draw the schematic diagram of a dual-polarity emitter-biased *npn* transistor amplifier. Do the same for a *pnp* transistor.
2. Given the component values in a dual-polarity emitter-bias circuit, analyze its operation to specify the bias current and voltages.
3. Draw the schematic diagram of a base voltage-divider bias circuit with the collector circuit, not the emitter, wired to dc ground. Do this for both *npn* and *pnp* transistors.
4. Explain the working of ideal transformer-coupled output.
5. Use the resistance-reflecting formula, $r = (1/n^2)R_{LD}$, to find the equivalent ac resistance in the primary path of an amplifier with transformer-coupled output.
6. Working with the result of Objective 5, find the overall voltage gain of a transformer-coupled-output amplifier.
7. State the disadvantages of transformer coupling.
8. Explain the operation of a frequency-tuned amplifier with transformer-coupled output. Given the component values, calculate the tuned frequency.
9. Explain the working of ideal transformer-coupled input to an amplifier.
10. Use the resistance-reflecting formula to find the input resistance of an amplifier with transformer-coupled input.
11. Given the specifications of the signal source, and working with the result of Objective 10, find the amplifier's actual input voltage, V_{in}.
12. State the advantages and disadvantages of direct coupling in an amplifier.
13. Describe the idea of negative feedback. List the advantages of negative feedback in an amplifier.
14. Draw the schematic diagram of a Darlington pair. State its advantages.
15. Given the component specifications in an amplifier schematic diagram, draw the amplifier's ac load-line graph, and interpret it.
16. Given a disorganized schematic diagram of a multistage amplifier, redraw it so that the dc bias situation and ac configuration of each stage is made clear.

7-1 DC BIAS METHODS

The base voltage-divider bias method that we have been using since Sec. 5-1 is not the only method available for establishing stable dc bias conditions. The **dual-polarity emitter bias** method is shown in Fig. 7-1.

In that figure, we show two dc power supply connections, plus ground. The positive dc supply, labeled $+V_{CC}$, is connected to R_C in the usual way. The negative dc supply, labeled $-V_{EE}$, is connected to the bottom of R_E, where ground would usually go. The common ground point is connected to base resistor R_B.

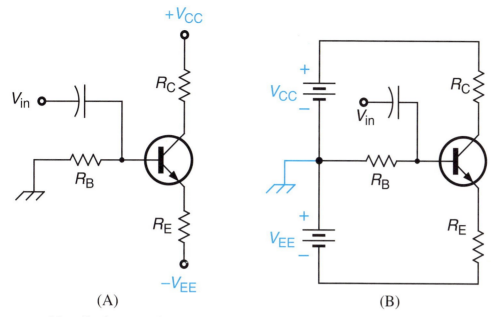

Figure 7-1
Dual-polarity emitter-bias arrangement.

(A) Normal schematic representation.
(B) Deliberately showing the two dc power supplies.

The two opposite-polarity dc supplies could be obtained by any one of the methods shown in Figures 3-28, 29, or 30.

Usually, but not always, V_{CC} and V_{EE} are equal in magnitude. Thus, for example, if $V_{CC} = +10$ V, then $V_{EE} = -10$ V. This equal-magnitude situation is shown in Fig. 7-2, with realistic resistance values. Here is how it works.

$\boxed{1}$ Dc base current I_B is very small, as we know. For most linear transistor amplifiers, I_B is less than 100 μA. R_B is also rather small, in this case only 470 Ω. Therefore the voltage across R_B is very small. For an I_B value of 100 μA, Ohm's law gives

$$V_{RB} = I_B R_B = (100 \times 10^{-6} \text{ A}) (470 \text{ }\Omega) \approx 0.05 \text{ V}$$

With V_{RB} near 0 V, the dc bias voltage at the base terminal is $V_B \approx 0$ V. Figure 7-2(B) shows this.

$\boxed{2}$ The negative dc voltage on the emitter circuit causes the B-E junction to be forward-biased. The voltage across the conducting junction is $V_{BE} = 0.7$ V. Therefore the dc bias voltage at the emitter terminal is given by

$$V_E = V_B - V_{BE} = 0 \text{ V} - 0.7 \text{ V}$$
$$= -0.7 \text{ V}$$

$\boxed{3}$ The dc voltage across the emitter resistor must then be
$$V_{RE} = V_E - V_{EE}$$
$$= -0.7 \text{ V} - (-10 \text{ V})$$
$$= 9.3 \text{ V}$$

as indicated in Fig. 7-2(B).

$\boxed{4}$ The fixed value of V_{RE}, combined with the known value of $R_E = 1$ kΩ, fixes the value of emitter current I_E. By Ohm's law,

$$I_E = \frac{V_{RE}}{R_E} = \frac{9.3 \text{ V}}{1 \text{ k}\Omega} = 9.3 \text{ mA}$$

$\boxed{5}$ With the value of main-terminal current fixed at $I_C \approx I_E = 9.3$ mA, V_{RC} is fixed by Ohm's law as

$$V_{RC} = I_C R_C = (9.3 \text{ mA}) (500 \text{ }\Omega) = 4.7 \text{ V}$$

$\boxed{6}$ By Kirchhoff's voltage law, the total 20-V difference between the dc supply terminals must be dropped by the combination of V_{RC}, V_{CE}, and V_{RE}. That is

$$V_{CC} - V_{EE} = 10 \text{ V} - (-10 \text{ V}) = 20 \text{ V} = V_{RC} + V_{CE} + V_{RE}$$

so
$$V_{CE} = 20 \text{ V} - V_{RC} - V_{RE} = 20 \text{ V} - 4.7 \text{ V} - 9.3 \text{ V}$$
$$= 6.0 \text{ V}$$

$\boxed{2}$ ✔

**Figure 7-2
Deriving the dc bias
conditions in a
circuit with dual-
polarity emitter-bias.**

(A) Schematic diagram.

**(B) Specifying V_B,
V_E, and V_{RE}.**

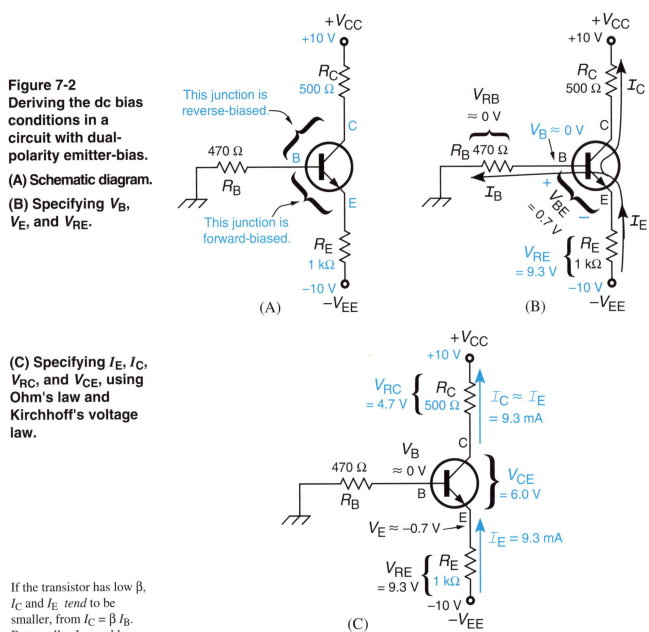

(A)

(B)

**(C) Specifying I_E, I_C,
V_{RC}, and V_{CE}, using
Ohm's law and
Kirchhoff's voltage
law.**

(C)

If the transistor has low β,
I_C and I_E *tend* to be
smaller, from $I_C = \beta I_B$.
But smaller I_E would mean
less voltage across R_E,
which would mean more
voltage available to drive
the B-E junction. There-
fore base current rises to
compensate for low β.
This argument works in the
reverse way for high β.

The only reason R_B was
required in Figs. 7-1 and 7-2
was to separate the B
terminal from ground when
we wish to apply V_{in} at the
B terminal. This happens
in common-emitter and
common-collector amplifiers,
but not in common-base.

The entire derivation of the bias conditions in Fig. 7-2 does not make any mention of β. Instead, the bias conditions are fixed by the circuit having to obey Ohm's law and Kirchhoff's voltage law. This means that the bias conditions are largely β-independent, and therefore are stable.

The dual-polarity emitter-bias arrangement can be used for any one of the three configurations. But it works especially well for the common-base configuration, where we want the base to be ac-grounded anyway. Then we can just eliminate R_B, as shown in Fig. 7-3.

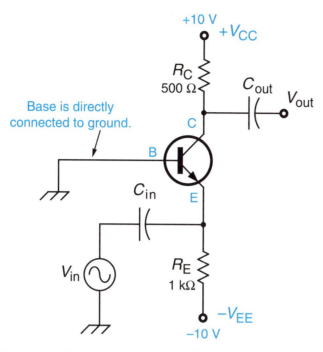

Figure 7-3
Using the dual-polarity emitter-bias method for a common-base amplifier. Now the ac-grounding capacitor C_B in Fig. 6-8 is not needed.

Collector Circuit Connected to Dc Ground

In the base voltage-divider bias arrangement, nothing requires that the emitter circuit be connected to dc ground. It is possible to connect the collector circuit to dc ground and connect the emitter circuit to a $-V_{EE}$ negative dc supply (*npn* transistor). This is shown in Fig. 7-4.

③ ✓

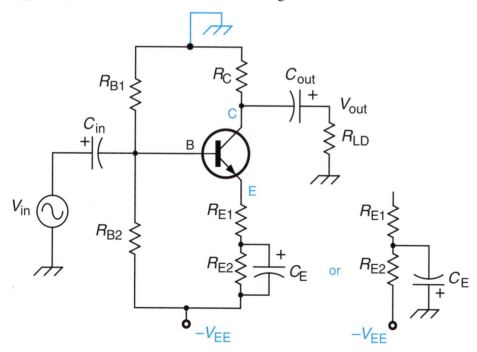

Figure 7-4
For an *npn* transistor the usual bias requirements are that the base be more positive than the emitter and the collector be even more positive than the base. These requirements can be satisfied just as well with the top of the circuit at 0-V potential and the bottom at negative ($-V_{EE}$) potential.

With the dc-grounded collector circuit, the polarities of electrolytic coupling capacitors must be reversed. Now they are − on the inside (toward the transistor), + on the outside (toward the outside world). This polarity reversal is clearly marked in Fig. 7-4. Compare to all schematic diagrams in Chapters 5 and 6. The polarity of the emitter-bypass capacitor is the same as before, though, if C_E is connected directly across R_{E2}.

In Fig. 7-4, it would be proper to connect the bottom side of C_E directly to power-supply ground, as shown at the lower-right. This allows the ac emitter current to reach ground without having to pass down the $-V_{EE}$ supply-line.

Pnp Biasing: As explained in Chapter 4, *pnp* transistors require the opposite bias polarities from *npn* transistors. That is, the base must be negative relative to the emitter, and the collector is even more negative than the base. Examples of correct *pnp* bias arrangements are shown in Fig. 7-5.

Figure 7-5
Proper *pnp* bias and capacitor polarities.

(A) Base voltage-divider bias, dc-grounded emitter circuit, common-emitter configuration.

(B) Base voltage-divider bias, dc-grounded collector circuit, common-emitter configuration.

(C) Dual-polarity emitter bias, common-base configuration.

(D) Base voltage-divider bias, dc-grounded collector circuit, common-collector configuration.

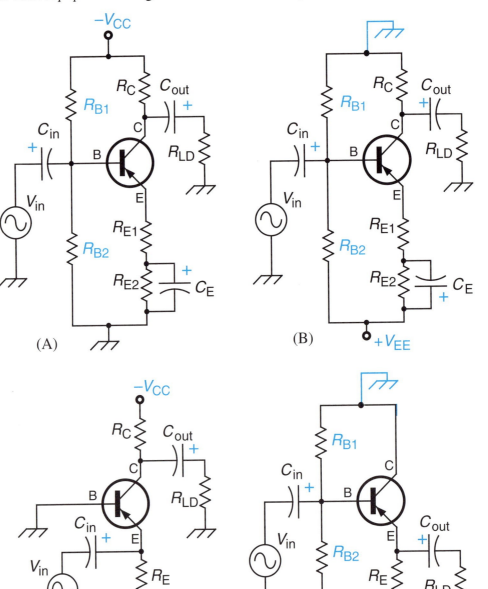

SELF-CHECK FOR SECTION 7-1

1. In dual-polarity emitter biasing, the B-E junction is _____ biased and the C-B junction is _____ biased. (forward or reverse)
2. In dual-polarity emitter biasing, the dc bias voltage at the base, V_B, is approximately equal to _____ .
3. For all linear transistor amplifiers, no matter what the specific biasing arrangement, and no matter what the amplifier configuration, and no matter whether the transistor is *npn* or *pnp*, the B-E junction is _____ biased and the C-B junction is _____ biased. (forward or reverse)

4. In the circuit of Fig. 7-3, suppose the specifications are changed to:
$V_{CC} = +12$ V, $-V_{EE} = -12$ V, $R_C = 1.5$ kΩ, $R_E = 3.3$ kΩ, $\beta = 160$. Find:
   a) V_E d) I_C
   b) V_{RE} e) V_{RC}
   c) I_E f) V_{CE}

5. In the *pnp* circuit of Fig. 7-5(C), suppose we have: $-V_{CC} = -10$ V, $+V_{EE}$
$= +6$ V, $R_C = 3.0$ kΩ, $R_E = 2.7$ kΩ, $\beta = 160$. Find the following. For all
voltages, be careful to express their polarities correctly.
   a) V_E d) I_C
   b) V_{RE} e) V_{RC}
   c) I_E f) V_{CE}

7-2 TRANSFORMER-COUPLED OUTPUT

Up till now, we have used the capacitive coupling method to produce a pure ac output waveform at the load. There is another method of accomplishing this. It is the **transformer-coupling** method, shown in Fig. 7-6.

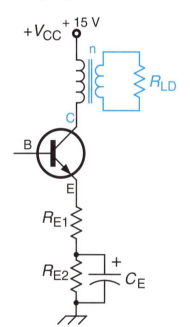

**Figure 7-6
Transformer-coupled output is obtained by placing the transformer primary winding in series in the collector lead. The secondary drives the load.**

Let us make our description of transformer-coupled output as simple as possible by considering the transformer to be ideal. That is, its winding resistances are negligible, and it obeys the transformer voltage law and current law perfectly.

The bias condition in Fig. 7-6 will have a certain dc collector current I_C flowing through the primary winding. But there will be zero dc voltage drop across the winding, since its resistance is zero. Therefore $V_C = V_{CC} = 15$ V. Of course, the transformer does not respond to the steady dc drive, so there is no dc voltage delivered to the load.

When the ac signal is applied, the transistor causes ac collector current I_C to flow through the primary winding. This time-varying ac current does cause the transformer to induce a secondary ac voltage, which is applied to the load.

Transformer-coupled output from the collector lead works for either the common-emitter or common-base configuration.

The output voltage from the transformer secondary can be symbolized V_{out}, V_{ld}, or V_S.

We say that the load resistance R_{LD} is "reflected" into the primary circuit as $(1/n^2)R_{LD}$.

Remember the definition of turns ratio n as the number of secondary turns divided by the nomber of primary turns. $(n = N_S/N_P)$

Figure 7-7 Common-emitter amplifier with the same bias conditions seen throughout Chapter 5, but with transformer-coupled output.

Referring to the complete amplifier schematic of Fig. 7-7, here is the way to find the value of output voltage V_{out}.

The collector's ac current source "sees" an equivalent ac resistance, r_C. The r_C value depends on the load resistance R_{LD} and the transformer's turns ratio, n.

Transformer operation causes the primary circuit to look like an ac resistance given by

$$r_C = \left(\frac{1}{n^2}\right) R_{LD} \qquad \text{Eq. (7-1)}$$

You are probably familiar with Eq. (7-1) from your study of the resistance-matching property of transformers.

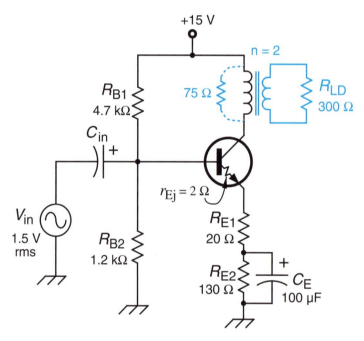

In Fig. 7-7,

$$r_C = \left(\frac{1}{n^2}\right) R_{LD}$$

$$= \left(\frac{1}{2^2}\right) 300 \ \Omega = 75 \ \Omega$$

This means that the ac current source sees the transformer-load combination as an equivalent resistance of 75 Ω. From an ac viewpoint, it is just as if a 75-Ω resistor were wired into the collector lead.

Once we know the value of r_C seen by the collector ac source, we can use our previous equation for A_v. This will give the **primary voltage gain**, which is the gain factor as we go from the input to the primary winding. We can symbolize this voltage gain as $A_{v(primary)}$. Thus, for the common-emitter situation in Fig. 7-7, we get

$$A_{v\,(primary)} = \frac{r_C}{r_{Ej} + R_{E1}} = \frac{75 \ \Omega}{22 \ \Omega} = 3.41$$

Then the voltage across the primary winding in Fig. 7-7 is given by

$$V_P = [A_{v(primary)}]\, V_{in}$$
$$= 3.41\,(1.5\text{ V})$$
$$= 5.1\text{ V}$$

Finally, applying the transformer voltage law gives

$$V_{out} = V_S = n\, V_P$$
$$= (2)\,(5.1\text{ V}) = 10.2\text{ V rms}$$

The amplifier's overall voltage gain is therefore

$$A_v = \frac{V_{out}}{V_{in}} = \frac{10.2\text{ V}}{1.5\text{ V}} = 6.8 \qquad \boxed{6}\ ✔$$

The V_{out} waveform can be made either in phase with V_{in} or out of phase with V_{in}. A change is made simply by swapping the secondary leads.

Putting a transformer into a transistor amplifier may not be desirable. Some transformers have iron cores that are bulky and heavy, which defeats two of the main advantages of the transistor — it is small and lightweight. This problem is more serious with transformers that are designed to work well for low-frequency ac — **audio transformers**. $\qquad \boxed{7}\ ✔$

Transformers that are designed to work well for high-frequency ac, called radio-frequency transformers or **RF transformers**, do not have this problem. They are built with ferrite or air cores, and therefore they are not heavy. Nor are they as expensive as audio transformers, since their windings contain far fewer turns of copper wire.

Tuned Amplifiers

Transformer-coupled output is most useful when we want the amplifier to be frequency-selective. Many radio-frequency (RF) amplifiers are built to be frequency-selective, or **tuned**. Figure 7-8 shows how the tuning is accomplished.

The transformer's primary winding has a certain inductance, L_P. Capacitor C is placed in parallel with the winding, forming a parallel LC circuit, as Fig. 7-8(A) makes clear. This LC combination has a certain resonant frequency, given by the usual formula

$$f_r = \frac{1}{2\pi\sqrt{L_P C}} \qquad \text{Eq. (7-2)}$$

This technician is checking the operation of circuit boards that acquire great amounts of test data in "real time" (just as soon as it happens). The boards then store that data for a while until it is called for by a processing computer.

The rate at which the computer brings in the data will probably not be the same rate at which it occurred in real time.

Courtesy of Macrodyne, Inc.

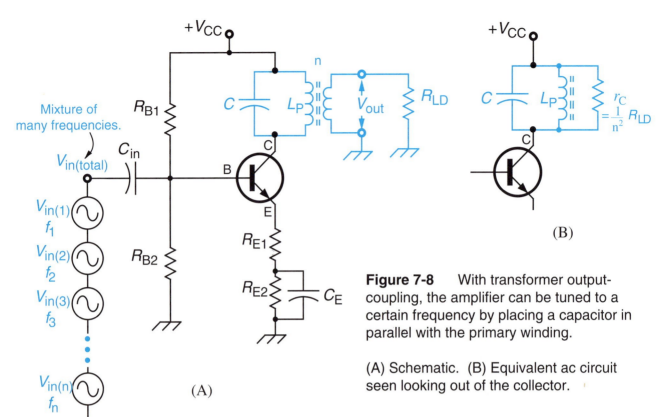

Mixture of many frequencies.

(A)

Figure 7-8 With transformer output-coupling, the amplifier can be tuned to a certain frequency by placing a capacitor in parallel with the primary winding.

(A) Schematic. (B) Equivalent ac circuit seen looking out of the collector.

At frequencies far from resonance, you can think of the *LC* combination as shorting out r_C, reducing the voltage gain.

A signal frequency at this resonant value sees the *LC* combination as an infinite impedance, ideally. Therefore the collector's overall equivalent ac circuit, shown in Fig. 7-8(B), looks like the normal r_C value in parallel with $\infty \, \Omega$. In other words, the collector current source sees just r_C. Then the amplifier behaves as it was explained previously.

But at signal frequencies different from the tuned frequency f_r, the *LC* combination presents a lower value of impedance. Placed in parallel with r_C, this causes the overall impedance in the collector lead to be lowered. Therefore the primary voltage gain, $A_{v(primary)}$, is also lowered. The amplifier is discriminating against frequencies that are different from the tuned frequency.

By making the capacitor variable in Fig. 7-8, we can adjust the tuned frequency. This technique is commonly used for the RF amplifiers in a radio or television receiver.

SELF-CHECK FOR SECTION 7-2

6. (T-F) With transformer-coupled output, there is no dc component in the V_{out} waveform.
7. (T-F) Depending on the frequency range of the signal, different types of transformers must be used for transformer-coupled output.
8. For the tuned common-emitter amplifier of Fig. 7-8, suppose we have the same bias conditions as in Fig. 7-7: $I_E = 15.3$ mA, $r_{Ej} = 2 \, \Omega$. In the collector circuit, the circuit specifications are: $L_P = 8$ mH, $C = 0.01 \, \mu$F, $n = 0.25$, $R_{LD} = 150 \, \Omega$. $V_{in} = 0.1$ V p-p. Find the following:
    a) The tuned (resonant) frequency, from Eq. (7-2). b) r_C
    c) $A_{v(primary)}$, for a signal at the tuned frequency. d) V_P
    e) V_{out} f) Overall voltage gain for a signal at $f = f_r$.

Beginning in part **b**, follow the same basic sequence of calculations shown for the Fig. 7-7 amplifier to handle the transformer-coupled load.

9. For the common-base amplifier shown in Fig. 7-9, find:

   a) I_E d) r_C g) V_{out}
   b) V_{CE} e) $A_{v(primary)}$ h) Overall voltage gain
   c) r_{Ej} f) V_P

   Beginning in part **d**, follow the same basic sequence of calculations to handle the transformer-coupled load.

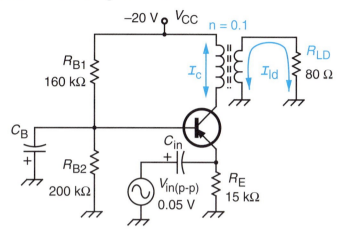

Figure 7-9
Common-base amplifier with transformer-coupled output, for Self-Check Problem 9.

7-3 TRANSFORMER-COUPLED INPUT

A transformer can also be used to couple the input signal to an amplifier. Figure 7-10 shows **transformer-coupled input** to a common-emitter amplifier.

In Fig. 7-10(A), the transformer secondary winding is connected directly to ground and its voltage is applied through dc-blocking capacitor C_{in}. Figure 7-10(B) shows the transformer winding placed directly in the base lead. C_B is needed here to connect the left terminal of the winding to ac ground. This method has the advantage of increasing input resistance R_{in} of the amplifier. R_{in} is larger than in part A because now the transformer's secondary load does not include R_{B1} and R_{B2} in parallel with $r_{b\,in}$. The issue of input resistance will be examined more carefully in the next Example.

Figure 7-10
(A) Using transformer-coupling to apply the input signal to a common-emitter amplifier. Input capacitor C_{in} is still necessary in order to block the dc path to ground.
(B) Placing the transformer secondary winding directly in the base lead.

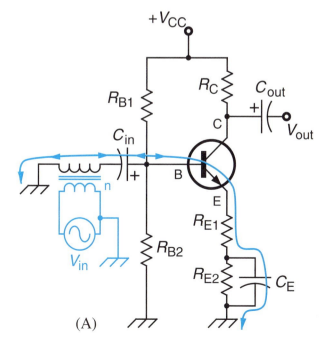

(A)

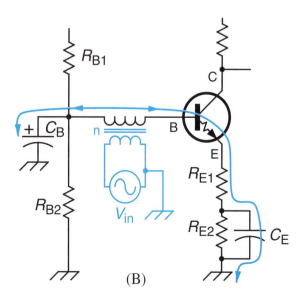

(B)

Figure 7-11 shows two methods of transformer-coupling the input signal into a common-base amplifier.

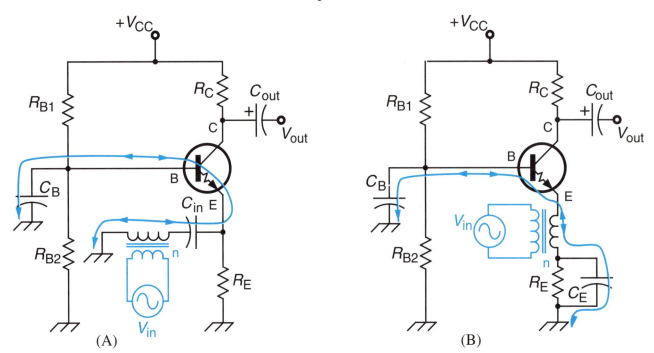

Figure 7-11 Transformer-coupled input to a common-base amplifier. (A) Through C_{in} in the standard way. (B) Directly in series in the emitter lead. This requires that R_E be ac-bypassed by C_E.

For any transformer-coupled input, the input resistance seen by the signal source is greatly affected by the transformer turns ratio n. If we symbolize the ac input resistance seen by the secondary winding as $R_{in(secondary)}$, the input resistance seen by the signal source is given by the usual transformer reflecting formula

$$R_{in\,(primary)} = \left(\frac{1}{n^2}\right) R_{in\,(secondary)} \qquad \text{Eq. (7-3)}$$

In Eq. (7-3), $R_{in(primary)}$ is the overall input resistance seen by the signal source, which is the thing that's important.

EXAMPLE 7-1

In Fig. 7-11(B), suppose that the bias conditions are the same as in our original common-base amplifier, Fig. 6-8. In that circuit, we had $r_{Ej} \approx 14\ \Omega$.

For $n = 0.1$, find the overall input resistance of the amplifier. Assume that the reactances X_{CB} and X_{CE} are virtually zero at the signal frequency.

SOLUTION

The resistance seen by the secondary winding is simply r_{Ej}, since the capacitive reactances are negligible.

$$R_{in(secondary)} = r_{Ej} = 14\ \Omega$$

From Eq. (7-3)

$$R_{in\,(primary)} = \left(\frac{1}{n^2}\right) R_{in\,(secondary)}$$

$$= \frac{1}{(0.1)^2}\ 14\ \Omega = \mathbf{1400\ \Omega}$$

which is the overall input resistance seen by the signal source. Refer to Fig. 7-12.

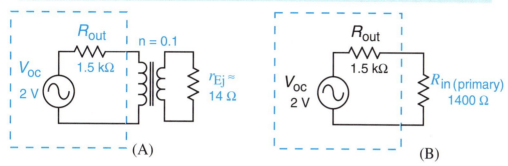

**Figure 7-12
Visualizing the effect of an input-coupling transformer.**

In this example, the common-base amplifier's very low input resistance ($14\,\Omega$) has been raised to a more reasonable value ($1400\,\Omega$) by transformer action.

> One advantage of transformer input coupling is for matching the low input resistance of some transistor amplifiers to the high output resistance of a signal source.

Get This

EXAMPLE 7-2

For the common-base amplifier of Fig. 7-11(B), we know the input resistance to be $1400\,\Omega$ from Example 7-1. If the signal source has $V_{oc} = 2$ V and $R_{out} = 1.5$ kΩ, find the actual transistor input voltage that appears in the emitter circuit.

SOLUTION

The signal source's situation is shown in Fig. 7-12(A). The transformer reflects a resistance of $1400\,\Omega$ into the primary circuit, as drawn in Fig. 7-12(B). By voltage division

$$V_P = V_{Rin} = V_{oc}\left(\frac{R_{in\,(primary)}}{R_{out} + R_{in\,(primary)}}\right)$$

$$= 2\text{ V}\left(\frac{1400\,\Omega}{1500\,\Omega + 1400\,\Omega}\right) = 0.97\text{ V}$$

The transformer steps the voltage down to

$$V_S = n\,V_P = 0.1\,(0.97\text{ V}) = 0.097\text{ V or } \mathbf{97\ mV}$$

$R_{in\,(primary)}$ has been raised by a factor of 100 greater than r_{Ej}, so that a decent amount of the source's open-circuit voltage appears at its output terminals. Then, we must lose most of this voltage in the 10-to-1 step-down action. On balance, we still wind up delivering more signal to the emitter than without the transformer coupling. ⑪ ✓

SELF-CHECK FOR SECTION 7-3

10. (T-F) In Fig. 7-10(A), the input resistance seen by the secondary winding is equal to $R_{B1} \parallel R_{B2} \parallel r_{b\,in}$.
11. In Fig. 7-11(B), describe the input resistance seen by the secondary winding.
12. Explain why C_B is necessary in Fig. 7-10(B).
13. Explain why C_E is necessary in Fig. 7-11(B).
14. In Fig. 7-10(B), suppose that $r_{Ej} = 2\,\Omega$, $R_{E1} = 20\,\Omega$, $\beta = 120$, and $n = 0.4$. Assume that X_{CB} and X_{CE} are zero.
    a) Find $R_{in\,(secondary)}$.
    b) Find $R_{in\,(primary)}$, the overall input resistance of the amplifier.

15. In Problem 14, suppose $V_{in(oc)} = 1$ V, with output resistance $R_{out} = 10$ kΩ.
a) Find the primary voltage V_P.
b) Find secondary voltage V_S, which is the same as V_b.
c) If the base-to-collector voltage gain is 25, find V_{out}. Then find the overall voltage gain from $V_{in(oc)}$ to V_{out}. [$A_{v(overall)} = V_{out} \div V_{in(oc)}$].

ANGIOPLASTY

Arteriosclerosis, hardening of the arteries, is the buildup of fatty deposits called plaque on the inside surface of arteries. It restricts proper blood flow and can lead to heart attack. For some patients, an alternative to major bypass surgery is angioplasty—the insertion of a thin catheter into the clogged artery to break loose the plaque.

There are two kinds of angioplasty—the balloon method and the laser method. In the balloon method, the tip of the catheter has a tiny balloon that is inflated to dislodge the plaque. In the laser method, the catheter contains multiple optical fibers that transmit short bursts of laser energy to clear out the plaque.

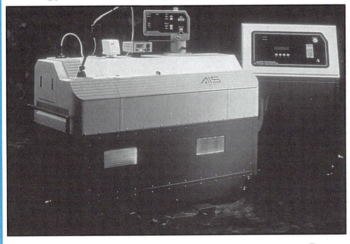
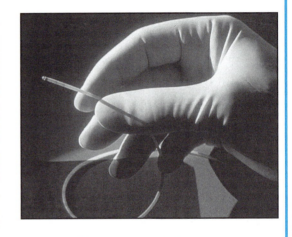

Courtesy of Advanced Interventional Systems, Inc.

The upper left photo shows the electronic system for generating and controlling the 200-nanosecond-long laser pulses. The intensity and duration of the laser pulses must be carefully controlled by the electronic circuitry in order to avoid damaging the tissue of the artery itself. There were 13 of these systems in existence in U.S. hospitals in 1993. A close-up view of the laser-carrying catheter is shown at the top right.

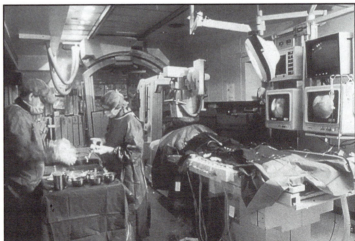

In the bottom photo, a medical team prepares to perform a laser angioplasty. They will work from magnified views of the catheter inside the patient's artery, appearing on the video monitors above the operating table. (See the article on page 224 in the next chapter).

7-4 DIRECT (DC) AMPLIFIER COUPLING

Some amplifiers must be able to handle very low-frequency signals. The extreme of low frequencies is 0 Hz, or dc. Therefore we can lump together low-frequency operation and dc operation. We refer to both of them as **dc amplification**.

A dc amplifier cannot use either capacitive coupling or transformer coupling, on input or on output. Capacitive coupling is impossible because a capacitor has ∞ Ω to dc — it blocks dc. Transformer coupling is impossible because transformers don't work with dc.

A dc amplifier must have direct wire connections at its input, at its output, and between stages. We call this **direct coupling**. A 2-stage direct-coupled amplifier is shown in Fig. 7-13(A).

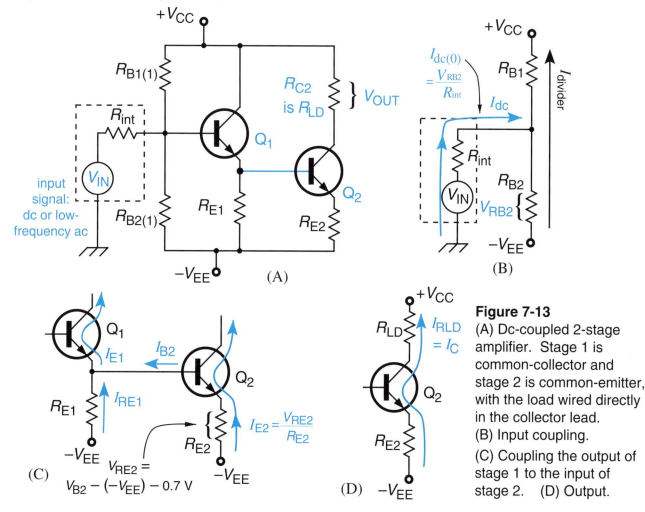

Figure 7-13
(A) Dc-coupled 2-stage amplifier. Stage 1 is common-collector and stage 2 is common-emitter, with the load wired directly in the collector lead.
(B) Input coupling.
(C) Coupling the output of stage 1 to the input of stage 2.   (D) Output.

In the amplifier of Fig. 7-13, the input signal source-R_{int} combination may have some amount of dc voltage applied across it by the base voltage-divider circuit of transistor Q_1. Therefore the source may have some dc current through it, even when its signal voltage is zero. This is shown as $I_{DC(0)}$ in Fig. 7-13(B). The source must be able to tolerate this current. When the source delivers a nonzero dc V_{IN} signal, then the dc *signal* current through it will add to $I_{DC(0)}$ or subtract from $I_{DC(0)}$, depending on the V_{IN} polarity.

The amplifier must be designed so that under the condition of $V_{IN} = 0$, the dc bias voltage V_{E1} is the correct value for establishing a proper bias current in transistor Q_2. In other words, since $V_{E1} = V_{B2}$, the circuit designer must arrange for this voltage to be compatible with both Q_1 and Q_2. The situation is shown in Fig. 7-13(C).

The direct-wired load situation was explained at the beginning of our discussion of transistor amplifiers, back in Sec. 5-4, starting on page 125.

Since it is directly wired into the Q_2 collector, the load must be able to tolerate dc current I_C. Figure 7-13(D) shows this.

To make it easier to get the required dc compatibility, designers sometimes use opposite types of transistors from one stage to the next. Therefore dc amplifiers often look like the circuit of Fig. 7-14.

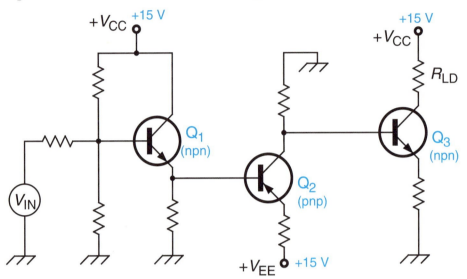

Figure 7-14 Staggered transistor types are often used with direct-coupled amplifiers.

The Dc Drift Problem

Dc amplifiers have a problem that ac-coupled amplifiers do not have. With zero input signal, a slight shift in the dc bias condition of the first stage becomes just like a signal to the second stage. Therefore the second stage amplifies that shift, producing an even larger shift at its output. And that shift is treated like a signal input to the third stage, which amplifies it even further.

In an ac-coupled amplifier, dc drift in one transistor has no effect on the dc bias of the next transistor. So drift does not get amplified.

As this process continues, it can happen that the final output transistor becomes saturated or cutoff, even with no signal present.

Since temperature change causes dc drift, direct-coupled amplifiers are much more temperature-sensitive than amplifiers with capacitive or transformer coupling.

 ✔

Staggering transistor polarities can help control the drift problem. For example, in Fig. 7-14, a temperature increase tends to cause I_E in Q_1 to increase. This would increase V_{E1}, which in turn would tend to decrease the forward bias of the *pnp* Q_2 base-emitter junction. But Q_2 is like Q_1 in that it tends to conduct harder as the temperature increases. Therefore we have opposing tendencies in Q_2, which gives some cancellation of the drift effect.

This hand-held ultrasonic detector is used to monitor the condition of mechanical shaft bearings. When a bearing is approaching wear-out, it starts to produce ultrasonic vibrations. By using this electronic detector to check bearings on a regular schedule, maintenance personnel get a warning well in advance of final failure. This gives them plenty of time to plan an orderly shut-down and replacement, instead of the usual frantic repairs after a surprise breakdown. *Courtesy of UE Systems, Inc.*

Comparing the Coupling Methods

Table 7-1 summarizes the features of the three coupling methods.

COUPLING METHOD ⟹	CAPACITIVE	TRANSFORMER	DIRECT
Dc Amplification?	No. Does not work.	No. Does not work.	Yes. Does work.
Advantages ⟹	Simple to use. Terminals at different dc levels can be ac-coupled.	Can improve the resistance match between signal source and amplifier input or output. Enables the amplifier to be frequency-tuned.	Inexpensive.
Disadvantages ⟹	Cannot handle low-frequency ac unless capacitances are very large. This adds to cost and size.	Cost, size and weight, at audio frequencies. Poor low-frequency response.	Design is more difficult since dc levels must be the same at output of one stage and input of next. Temperature-sensitive due to drift problem.

Table 7-1   Comparing amplifier coupling methods.

7-5 NEGATIVE FEEDBACK

Negative feedback is the most effective way to control drift in a direct-coupled amplifier. In negative feedback, a portion of the output signal is routed back to the input point so that it *subtracts* from the original input signal. An example is shown in Fig. 7-15.

In Fig. 7-15(A), the Q_1 base voltage divider is not driven by V_{CC}. Instead, it is driven by V_{E2}, the emitter voltage from the output stage. With zero input signal, the amplifier is said to be at its **quiescent**, or "quiet", state. The actual quiescent voltages and currents are given in Fig. 7-15(B). Now let us try to understand the basic idea of negative feedback. Refer to Fig. 7-15(C).

[1] Suppose that V_{IN} goes to some positive value. This increases the base drive on transistor Q_1, causing I_B to increase.

[2] The increase in I_B causes I_{E1} and I_{C1} to increase. They become larger than 1.5 mA. Let us imagine that they temporarily *try* to become 1.6 mA.

[3] The increase in I_{C1} causes $V_{RC(1)}$ to increase. $V_{RC(1)}$ temporarily tries to become 10.9 V (1.6 mA times 6.8 kΩ).

Negative feedback is a very powerful idea, useful for many other purposes besides controlling dc drift. It is used for stabilizing all kinds of amplifiers.

The word quiescent is also used for ac amplifiers. We have always called it simply the bias point.

$\boxed{4}$ The increased $V_{RC(1)}$ subtracts from V_{CC}, leaving a reduced voltage at V_{C1}, which is direct-coupled to V_{B2}. Base voltage V_{B2} temporarily tries to decrease to 1.1 V. (12 V – 10.9 V).

$\boxed{5}$ The decreased V_{B2} reduces the base drive on transistor Q_2. This causes I_{B2} to decrease, which makes I_{E2} also decrease. Therefore V_{E2} is also decreased, temporarily trying to decline to 0.4 V. (1.1 V – 0.7 V).

$\boxed{6}$ The reduction in V_{E2} means that the Q_1 base voltage-divider circuit has less voltage, now only 3.4 V [0.4 V – (–3 V) = 3.4 V]. Therefore $I_{Divider}$ decreases, temporarily trying to decline to 85 μA. (3.4 V ÷ 40 kΩ = 85 μA))

$\boxed{7}$ A small part of the "divider" current through R_{B1} was the original quiescent base current I_{B1}. So if the current through R_{B1} tries to decrease, this forces I_{B1} to also decrease. Therefore:

Get This

The initial attempt by the input to *increase* base current I_{B1} has resulted in a change at the output that tends to *decrease* base current I_{B1}. This partially cancels the initial effect of the input. This is the essence of negative feedback.

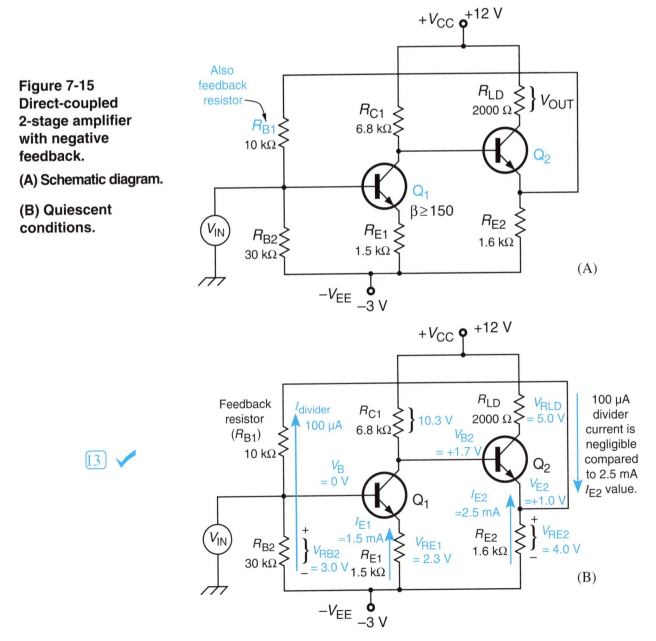

**Figure 7-15
Direct-coupled
2-stage amplifier
with negative
feedback.**

(A) Schematic diagram.

**(B) Quiescent
conditions.**

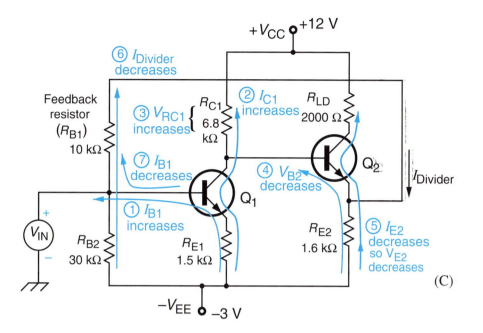

**Figure 7-15(C)
Understanding the
negative feedback
process.**

The cancellation effect of the negative feedback means that the actual changes that do occur in the amplifier are less than they temporarily tried to be. Thus, I_{C1} does increase, but not all the way to 1.6 mA. It may perhaps rise to 1.52 mA. All the other temporarily tried values also fall part-way back toward their quiescent values.

Negative feedback definitely reduces the overall gain of an amplifier. Therefore it may seem to be an undesirable thing. But its purpose is to stabilize the amplifier. In the case of the direct-coupled two-stage amplifier of Fig. 7-15, the negative feedback is effective in controlling dc drift. You can prove this to yourself. Imagine a temperature increase that tends to make Q_1 conduct harder. Then trace through the sequence of events just as we did above, to demonstrate that the temperature-related current increase is cancelled by a reduction in base drive to Q_1.

In most applications we are willing to sacrifice some gain if by doing so we can stabilize our amplifier.

Another major advantage of negative feedback is that it reduces any signal distortion that may occur in the amplifier. If the raw amplifier distorts the input signal for any reason (output waveform not having the same shape as the input waveform), applying negative feedback can nearly correct the problem.

SELF-CHECK FOR SECTIONS 7-4 AND 7-5

16. The only amplifier coupling method that works well at dc is the _____ coupling method.
17. (T-F) A direct-coupled load must be able to tolerate a dc current at all times.
18. What is the problem that direct-coupled multistage amplifiers have that ac-coupled amplifiers do not have?
19. (T-F) Capacitively coupled multistage amplifiers tend to be more sensitive to temperature changes than direct-coupled multistage amplifiers.
20. The practice of allowing an amplifier's output to subtract from or partially cancel the input is called _____ _____ .
21. (T-F) An amplifier's voltage gain and power gain are always increased when negative feedback is used.

22. State the two advantages of negative feedback that we have discussed. (There are others that we haven't discussed.)

7-6 THE DARLINGTON PAIR

A special example of direct coupling between transistors is the **Darlington pair**. It has two transistors with their collectors tied together, and the emitter of the first transistor connected directly to the base of the second transistor, as shown in Fig. 7-16(A).

Figure 7-16
Darlington connection.
(A) Schematic.
(B) I_{B1} is amplified in Q_1 by a factor of β_1, say 200. The resulting current is amplified in Q_2 by a factor of β_2, say 200 again. Therefore the overall current gain from B_1 to C_2 is about $200 \times 200 = 40\ 000$.

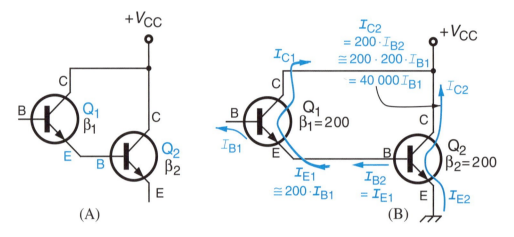

(A)   (B)

The Darlington pair acts like a single transistor with a tremendously high beta. This is because the base current in Q_1 is amplified by a factor of β_1 to create the collector-emitter current I_{E1}. This emitter current I_{E1} is routed directly from the base of Q_2, so it is amplified again, this time by a factor of β_2. This action is shown in Fig. 7-16(B), assuming that both betas are 200.

Often, a Darlington pair is placed into a single transistor package with only three leads brought out of the package. This is shown in Fig. 7-17.

Figure 7-17
A single package containing two transistors in the Darlington connection. Overall β equals β_1 times β_2. This is sometimes referred to as a Darlington transistor, as if it were just one transistor.

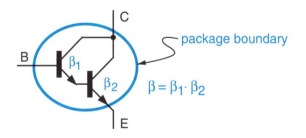

package boundary

$\beta = \beta_1 \cdot \beta_2$

If a Darlington pair is used as an ac amplifier, the equivalent ac resistance seen looking into the base is rather high. This tends to give the amplifier stage a high overall R_{in}. For example, Fig. 7-18 shows a Darlington common-collector amplifier with capacitive coupling at the input and output. An analysis of this circuit would show the following:

$$r_{Ej(2)} \approx 6\ \Omega$$
$$r_{b\,in(2)} = 34.5\ k\Omega \qquad [\text{looking into B(2) in Fig. 7-18}]$$
$$r_{Ej(1)} \approx 1200\ \Omega$$
$$r_{b\,in(1)} = 6.3\ M\Omega \qquad [\text{looking into B(1) in Fig. 7-18}]$$
$$R_{in} = R_{B1} \| R_{B2} \| r_{b\,in(1)} \approx 150\ k\Omega \qquad [\text{looking into the overall amplifier}]$$

Of course, if the signal source has weak current capability (large R_{int}), an amplifier R_{in} value as high as 150 kΩ is a big advantage.

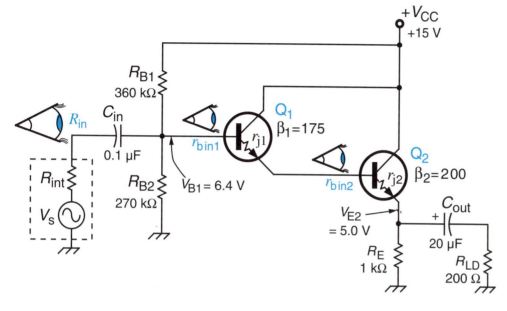

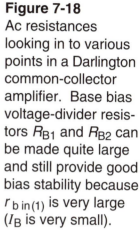

Figure 7-18
Ac resistances looking in to various points in a Darlington common-collector amplifier. Base bias voltage-divider resistors R_{B1} and R_{B2} can be made quite large and still provide good bias stability because $r_{b\,in(1)}$ is very large (I_B is very small).

SELF-CHECK FOR SECTION 7-6

23. Draw the schematic diagram of a Darlington pair of *npn* transistors.
24. Draw the complete schematic diagram of a common-emitter amplifier using a Darlington pair of *npn* transistors. Use standard voltage-divider bias and partially bypassed emitter. Show dc supply polarity.
25. Repeat Problem 23 using *pnp* transistors.
26. (T-F) The current gain of a Darlington pair is very high.
27. In a Darlington pair, if Q_1 has $\beta_1 = 150$ and Q_2 has $\beta_2 = 125$, what is the overall beta of the pair?
28. A Darlington pair amplifier tends to have a _____ input resistance.

7-7 AC LOAD-LINES

In Section 5-2 we described the process of constructing a circuit's dc load-line on top of the transistor's characteristic curves. We then visualized moving back and forth around the bias point along the load-line, in response to the ac input signal. We did this for the situation where the load was placed directly in the collector lead, as in Figs. 5-8 and 5-10. Now we will look at load-line operation when the load is capacitor-coupled or transformer-coupled.

Figure 7-19 shows our usual common-emitter amplifier with capacitively-coupled output. The dc load-line for this amplifier goes from

$$V_{CE} = V_{CC} = 15 \text{ V} \quad (\text{when } I_C = 0)$$

to $\quad I_{C(sat)} = \dfrac{V_{CC}}{R_C + R_{E1} + R_{E2}} = \dfrac{15 \text{ V}}{650 \ \Omega} = 23.1 \text{ mA} \quad (\text{when } V_{CE} = 0).$

This dc load-line is drawn in Fig. 7-20(A).

Figure 7-19
Standard common-emitter amplifier, for studying the idea of ac load-line.

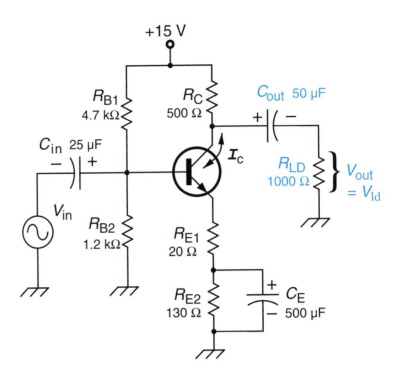

We have seen the bias conditions of this amplifier previously. We know them to be:

$R_{E(T)}$ and $V_{RE(T)}$ refer to the total emitter resistance, R_{E1} plus R_{E2}.

$$V_E = V_B - 0.7\ \text{V} = 2.3\ \text{V} = V_{RE(T)}$$

$$I_E = \frac{V_E}{R_{E(T)}} = \frac{2.3\ \text{V}}{150\ \Omega} = 15.3\ \text{mA} \approx I_C$$

$$V_{RC} = I_C R_C \approx (15.3\ \text{mA})(500\ \Omega) = 7.65\ \text{V}$$

$$V_{CE} = V_{CC} - V_{RC} - V_E = 15\ \text{V} - 7.65\ \text{V} - 2.3\ \text{V} = 5.05\ \text{V}$$

We say that the amplifier's quiescent point, or Q-point, is at $I_C = 15.3\ \text{mA}$, $V_{CE} = 5.05\ \text{V}$. This Q-point is marked on the dc load-line of Fig. 7-20(A).

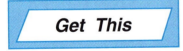

The ac current in the collector-emitter path does not see the same total resistance as the dc current. The ac current sees less resistance, namely

$$r_T = r_C + r_E \qquad \text{Eq. (7-4)}$$

In Fig. 7-19,

$$r_C = R_C \| R_{LD} = 500\ \Omega \| 1000\ \Omega = 333\ \Omega$$

$$r_E = R_{E1} = 20\ \Omega \quad \text{(since } R_{E2} \text{ is ac-bypassed),}$$

so from Eq. (7-4),

$$r_T = r_C + r_E = 333\ \Omega + 20\ \Omega$$
$$= 353\ \Omega$$

Since the ac equivalent resistance is less than the dc resistance in the main current path, it would take a larger amount of peak ac current to saturate the transistor, than it would dc current.

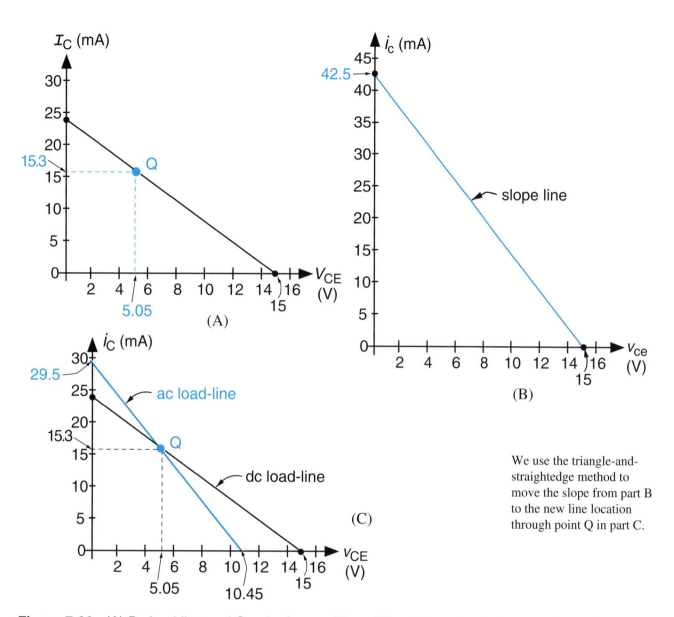

Figure 7-20   (A) Dc load-line and Q-point for amplifier of Fig. 7-19.   (B) Finding the slope of the ac load-line.   (C) Producing the actual ac load-line.   Through point Q, we draw a line with the same slope as the line in part B.

We use the triangle-and-straightedge method to move the slope from part B to the new line location through point Q in part C.

For Fig. 7-19, the peak ac current that it would take to saturate the transistor is given by

$$i_{c\,(sat)} = \frac{V_{CC}}{r_T} \qquad \text{Eq. (7-5)}$$

$$= \frac{15\ \text{V}}{353\ \Omega} = 42.5\ \text{mA}$$

Therefore the *slope* of the **ac load-line** is found by joining the two end points $i_c = 42.5$ mA and $v_{ce} = 15$ V, as shown in Fig. 7-20(B).

But the transistor's actual ac variations must take place around the actual dc quiescent point shown in Fig. 7-20(A). Therefore we must transfer the ac load-line's slope from Fig. 7-20(B) over to Fig. 7-20(A), so that it passes through the Q point. This has been done in Fig. 7-20(C), which creates the actual ac load-line for the amplifier. When an ac input signal is applied, the transistor's instantaneous current and voltage move back and forth (up and down) along the actual ac load-line.

The actual ac load-line is also called the **dynamic load-line or** the **signal load-line**.

For example, Fig. 7-21 shows the amplifier swinging between points P and R on the ac load-line. We would interpret this to mean that the instantaneous collector current swings between 21.0 mA and 9.7 mA, while the instantaneous collector-to-emitter voltage swings between 3 V and 7 V. The waveforms of i_C and v_{CE} are shown in Fig. 7-21(B).

⑮ ✔

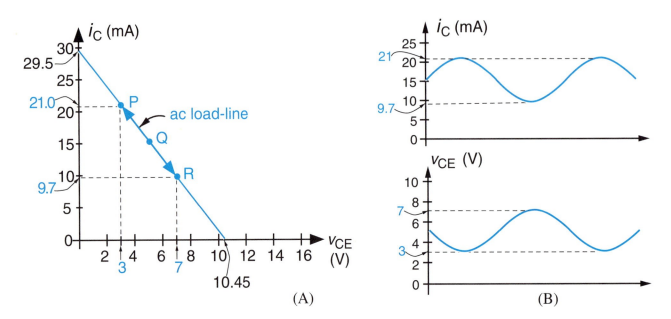

Figure 7-21   (A) The transistor in Fig. 7-19 moves on the ac load-line between points P and R. (B) Waveforms of i_C and v_{CE}.

When you look at any ac load-line operation, you must remember that the voltage being indicated is the collector-to-emitter voltage v_{CE}. It is not the output voltage v_C, which appears between the collector and ground in Fig. 7-19. (In a different amplifier that had no resistors in the emitter lead, the load-line voltage *would* be the output voltage.)

The most useful way to view ac load-line voltage is to look at the overall peak-to-peak travel, $V_{ce(p\text{-}p)}$. This ac voltage is equal to the sum of the ac voltage developed across r_C plus the ac voltage developed across R_{E1}. As a formula,

$$V_{ce} = V_{Rc} + V_{Re1} \qquad \text{Eq. (7-6)}$$

Of course, V_{Rc} is the same as V_{out}. So Eq. (7-6) can be rewritten as

$$V_{out} = V_{ce} - V_{Re1} \qquad \text{Eq. (7-7)}$$

EXAMPLE 7-3

The amplifier of Fig. 7-19 is operating along the ac load-line as shown in Fig. 7-21. Find the amplifier's output voltage V_{out}.

SOLUTION

The peak-to-peak voltage from point P to point R is 4 V. So $V_{ce(p\text{-}p)} =$ 4 V. The peak-to-peak current from point P to point R is given by

$$I_c = 21.0 \text{ mA} - 9.7 \text{ mA} = 11.3 \text{ mA (p-p)}$$

This ac collector current develops an ac voltage across R_{E1} given by Ohm's law as

$$V_{Re1} = I_c R_{E1} = (11.3 \text{ mA}) (20 \text{ } \Omega) = 0.23 \text{ V (p-p)}$$

Applying Eq. (7-7), we get

$$V_{out} = V_{ce} - V_{Re1}$$
$$= 4 \text{ V} - 0.23 \text{ V} = \textbf{3.77 V (p-p)}$$

Of course, it would have been equally correct to just apply Ohm's law to the equivalent ac resistance seen by the collector, r_C. Then,

$$V_{out} = I_c r_C$$
$$= (11.3 \text{ mA}) (333 \text{ } \Omega) = 3.77 \text{ V (p-p)}$$

Remember that
$r_C = R_C \| R_{LD}$
$= 500 \text{ } \Omega \| 1000 \text{ } \Omega$
$= 333 \text{ } \Omega$ from Fig. 7-19.

Ac Load-Line With Transformer-Coupled Output

When an amplifier has transformer-coupled output, the ac load-line is formed by using the equivalent ac resistance that the transformer reflects into its primary.

Get This

This resistance is given by

$$r = \frac{1}{n^2} R_{LD} \qquad \text{Eq. (7-1)}$$

For example, the common-base amplifier in Fig. 7-9 on page 187 has a Q-point of $V_{CE} = 9.6$ and $I_C = 0.69$ mA. You already know this from **Self-Check** Problem 9. This Q-point is indicated in Fig. 7-22.

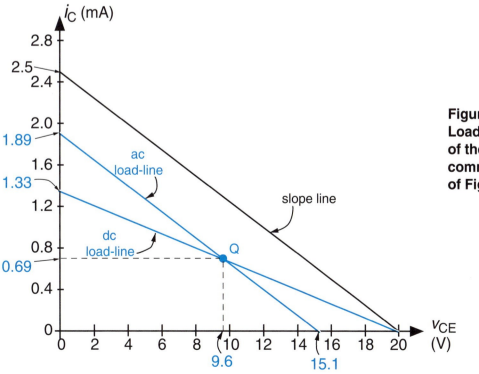

Figure 7-22
Load-line analysis
of the transformer-coupled
common-base amplifier
of Fig. 7-9.

The equivalent ac resistance seen in the collector lead is given by Eq. (7-1) as

$$r_C = \frac{1}{n^2} R_{LD} = \frac{1}{(0.1)^2} \, 80 \, \Omega$$

$$= 8 \text{ k}\Omega$$

Another way to think about R_E being out of the main ac current path is to consider it shorted out by C_{in} and the source.

This is essentially the only resistance that is seen by the ac current in the transistor's main flow path. R_E is not in the main ac flow path, since the ac emitter current comes directly from the signal source.

So in Eq. (7-4), r_E is zero and r_T is equal to just r_C. Then Eq. (7-5) gives us

$$i_{c\,(sat)} = \frac{V_{CC}}{r_T} = \frac{V_{CC}}{r_C}$$

$$= \frac{20 \text{ V}}{8 \text{ k}\Omega} = 2.5 \text{ mA}$$

The slope of the ac load-line is obtained by running a line from $i_C = 2.5$ mA to $v_{CE} = 20$ V in Fig. 7-22. Then we transfer the slope of the line over to the Q-point to get the ac load-line, as shown.

Ac load-line analysis can be used for common-collector amplifiers too.

7-8 SCHEMATIC APPEARANCE

You will often see schematic diagrams of amplifiers, especially multi-stage amplifiers, in which the transistors and their supporting components are not shown in their standard positions. These diagrams can be difficult to interpret.

To help in understanding such diagrams, you may want to redraw them in a more standard format.

For example the 3-stage amplifier schematic in Fig. 7-23(A) is probably difficult for you to interpret. You can make it easier by redrawing it as follows:

⟨1⟩   The *pnp* Q_1 stage can be seen to be a common-emitter stage with voltage-divider base bias, with the collector circuit wired to dc ground, by redrawing it as shown on the far left of Fig. 7-23(B). This makes it clear that 20 kΩ resistor R_1 is our usual R_{B1}. Also, 4.7-kΩ R_2 is our usual R_{B2}. Capacitor C_2 ac-bypasses emitter resistor R_E by connecting directly to ground rather than the $+V_{EE}$ supply-line. This bypass method was presented earlier on page 181. Transformer T_1 couples the output of stage 1 to the input of stage 2. The 100-kΩ resistor R_5 is a negative feedback resistor; this feedback situation will become clearer as we do further rearrangement of the diagram.

⟨2⟩ Redraw the Q_2 stage with the collector at the top and the emitter at the bottom position. The +15-V dc supply now becomes the $+V_{CC}$ supply for the *npn* transistor. The R_6-R_7 base voltage-divider connects to ground as expected. Capacitor C_4 becomes the ac-grounding capacitor for the B terminal of a common-base amplifier.   3.3-kΩ resistor R_4 serves as the emitter-bias resistor in this common-base stage. The T_1 secondary winding is wired in series with the emitter, so the input current path is through the emitter junction resistance $r_{Ej\,(2)}$, through base-bypass capacitor C_4 and through emitter-bypass capacitor C_3 .

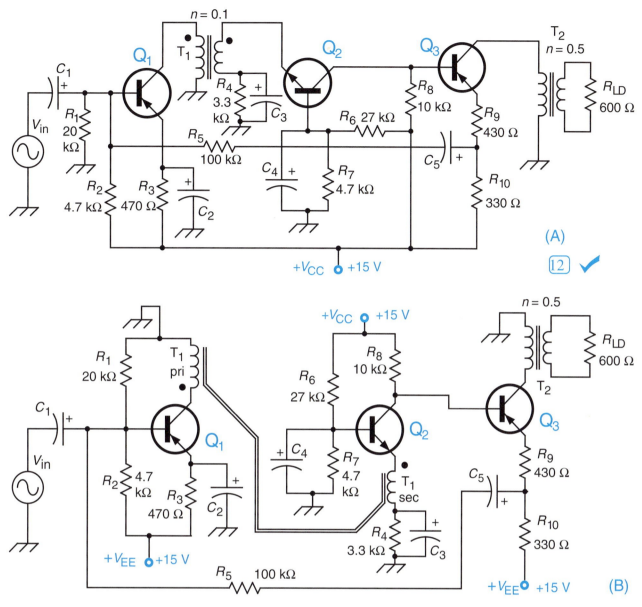

Figure 7-23   Three-stage amplifier.   (A) Schematic diagram drawn for efficient use of space on the paper.   (B) Schematic diagram redrawn for ease of understanding.

[3] Q_3 forms another common-emitter stage, this one with transformer-coupled output. There is direct coupling from the output of Q_2 (the collector) to the input of Q_3 (the base). The circuit will be designed so that the dc collector bias voltage at Q_2 is compatible with the dc base bias voltage at Q_3. ($V_{C2} = V_{B3}$).

[4]   A portion of the final stage-3 ac signal is fed back through 100-kΩ resistor R_5 to stabilize the overall amplifier gain, and to reduce distortion. You can see that this is negative feedback by reasoning carefully as follows:

Referring to Fig. 7-23(B), consider an instant when V_{in} has just entered its negative half cycle. The negative-going value of v_{in} is instantaneously driving the B-E junction of Q_1 harder, producing an increase in all Q_1 currents, including collector current i_{C1}. This produces an instantaneously positive-going ac voltage at the bottom (dot-marked) end of the T_1 primary winding.

The T_1 winding polarization is such that its secondary ac voltage is instantaneously positive-going on its top terminal, in the emitter circuit of Q_2. This fact is conveyed by the phasing dot being drawn at the top end of the T_1 secondary winding.

This instantaneously increasing positive-on-top secondary voltage reduces the drive on the B-E junction of *npn* transistor Q_2. This causes a decrease in all Q_2 currents, including collector current i_{C2}. With i_{C2} instantaneously decreasing, the collector voltage v_{C2} is instantaneously increasing, moving closer to +15 V. This increasing voltage appears at the base of *pnp* transistor Q_3, as v_{B3}. It reduces the drive on the B-E junction of Q_3, by the opposite reasoning process that applied to *pnp* transistor Q_1. Therefore all Q_3 currents are instantaneously decreasing in magnitude.

The decreasing current through resistor R_{10} causes the voltage at its top terminal (the R_9-R_{10} junction) to move closer to +15 V, since the V_{R10} voltage drop is decreasing in value. That is, the R_9-R_{10} pick-off voltage is instantaneously increasing.

Capacitor C_5 passes the ac part of this increasing voltage to 100-kΩ feedback resistor R_5. So we have a *positive-going* ac voltage on the right side of R_5, feeding the amplifier input. This tends to counteract the instantaneously *negative-going* ac voltage from the signal source, which was the beginning point of our discussion. Therefore we have successful negative feedback action.

FORMULAS

For transformer-coupled output:

$$r_C = \frac{1}{n^2} R_{LD} \qquad \text{Eq. (7-1)}$$

$$A_{v(\text{primary})} = \frac{r_C}{r_{Ej} + R_E}$$

$$f_r = \frac{1}{2\pi \sqrt{L_P C}}$$

For transformer-coupled input:

$$R_{\text{in (primary)}} = \frac{1}{n^2} R_{\text{in (secondary)}} \qquad \text{Eq. (7-3)}$$

For a Darlington pair:

$$\beta_T = \beta_1 \beta_2$$

For an ac load-line:

$$r_T = r_C + r_E \qquad \text{Eq. (7-4)}$$

$$V_{\text{out}} = V_{ce} - V_{Re} \qquad \text{Eq. (7-7)}$$

SUMMARY

- A transistor can be dc-biased by the dual-polarity emitter-bias method. This method uses $+V_{CC}$ and $-V_{EE}$ for an *npn* transistor.
- It is possible to bias an *npn* transistor with the collector circuit connected to dc ground and the emitter circuit connected to a negative dc supply, $-V_{EE}$. The same idea holds for a *pnp* transistor, with the emitter circuit connected to positive $+V_{EE}$.
- An amplifier may use a transformer for coupling its ac output to the load.
- When an amplifier uses transformer-coupled output, the equivalent ac resistance that appears in the transistor's main current path is given by $r = (1/n^2) R_{LD}$. We arrange for this to provide an improved resistance match between the load and the amplifier.
- It is possible to frequency-tune an amplifier with transformer-coupled output, by placing a capacitor in parallel with the primary winding.
- An amplifier can use a transformer for coupling its ac input signal to the transistor.
- When an amplifier uses transformer-coupled input, the transformer can help in matching the input resistance of the amplifier to the output resistance of the signal source.
- Amplifiers that must handle dc signals cannot be capacitor- or transformer-coupled. They must be direct-coupled.
- Direct-coupled multistage amplifiers are subject to the dc drift problem. This makes them more temperature-sensitive than ac-coupled amplifiers.
- The dc drift problem can be controlled by using negative feedback. This involves taking a portion of the output signal and routing it back to the amplifier's input so that it subtracts from the original signal input.
- Negative feedback is used for stabilizing amplifiers in all respects, both dc and ac. It also reduces output signal distortion, which is an idea of major importance.
- A Darlington pair is a direct-coupled pair of transistors that acts like a single transistor with extremely high current gain, β.
- The ac load-line technique is a graphical way to analyze an amplifier's ac performance.

This researcher is developing an electronic system that scans a printed page and displays the words one-at-a-time on a television screen. It may be able to expand the reading opportunities for low-vision persons.

Courtesy of Jet Propulsion Laboratory

QUESTIONS AND PROBLEMS

1. Draw the schematic diagram of an *npn* common-emitter amplifier with dual-polarity emitter biasing. Show the polarities of the V_{CC} and V_{EE} supplies.

2. Repeat Problem 1 for a *pnp* transistor.

3. In your schematic diagram from Problem 1, suppose that the component values are: $V_{CC} = +12$ V, $V_{EE} = -12$ V, $R_C = 1.2$ kΩ, $R_E = 2.4$ kΩ, $R_B = 15$ kΩ. Mark these values on your schematic and then find:

   a) V_{RE} c) V_{RC}

   b) I_E d) V_{CE}

4. For the bias conditions calculated in Problem 3, find the values of these voltages, relative to ground.

   a) V_B

   b) V_E

   c) V_C

5. Draw the schematic diagram of a single-supply base voltage divider bias circuit for an *npn* transistor, with the collector circuit wired to dc ground. Specify the polarity of the dc supply.

6. Repeat Problem 5 for a *pnp* transistor.

7. In your schematic diagram from Problem 5, suppose the component values are: $R_C = 680$ Ω, $R_E = 180$ Ω, $R_{B1} = 7.5$ kΩ, $R_{B2} = 1.6$ kΩ, and $V_{EE} = -16$ V. Mark these values on your diagram and then find:

   a) V_{RB2} d) V_{RC}

   b) V_{RE} e) V_{CE}

   c) I_E

8. For the bias conditions calculated in Problem 7, find the values of these voltages, relative to ground.

   a) V_B

   b) V_E

   c) V_C

9. Give one advantage of transformer-coupled output over capacitor-coupled output.

10. Give one disadvantage of transformer-coupled output compared to capacitor-coupled output.

11. In the common-emitter transformer-coupled amplifier of Fig. 7-7, suppose the component values are changed to: $V_{CC} = 20$ V, $R_{B1} = 11$ kΩ, $R_{B2} = 3.0$ kΩ, $R_{E1} = 47$ Ω, $R_{E2} = 270$ Ω, $n = 0.25$, and $R_{LD} = 150$ Ω. Draw the schematic diagram and then find:

   a) $V_{RB2} = V_B$ c) I_E

   b) V_E d) r_{Ej}

12. In the common-emitter amplifier of Problem 11, suppose $V_{in} = 0.1$ V.

   a) Find the equivalent ac resistance in the collector lead, r_C.

   b) Find $A_{v(primary)}$, the primary voltage gain.

   c) Find V_P, the primary voltage.

   d) Find V_S, the secondary voltage.

   e) Find A_v, the amplifier's overall voltage gain.

13. For the amplifier of Problems 11 and 12, suppose $\beta = 175$. Find the amplifier's input resistance R_{in}.

14. Suppose the amplifier of problems 11, 12 and 13 is driven by a signal source with $V_{oc} = 0.5$ V, and $R_{out} = 800\,\Omega$. What will be the value of V_{in}, the actual input voltage to the amplifier?

15. For your amplifier of Problems 11 through 14, suppose that the inductance of the transformer primary is $L_P = 5$ mH. If a 50-nF capacitor is connected in parallel with the primary, as shown in Fig. 7-8, find the tuned frequency, f_r.

16. Refer back to Fig. 7-10(B), which is a common-emitter amplifier with transformer-coupled input. In **Self-Check** Problem 14, we assumed $r_{Ej} = 2\,\Omega$, $R_{E1} = 20\,\Omega$, $\beta = 120$, and $n = 0.4$. You should have found that $R_{in\,(secondary)} = 2640\,\Omega$ and $R_{in\,(primary)} = 16.5$ kΩ. Then, in **Self-Check** Problem 15, you should have found that $V_P = 0.623$ V, $V_S = V_b = 0.249$ V, $V_{out} = 6.23$ V, and $A_{v(overall)} = 6.23$.

Instead of transformer input coupling, suppose the amplifier of Fig. 7-10(B) had used capacitor input coupling. Suppose $R_{B1} = 4.7$ kΩ and $R_{B2} = 1.2$ kΩ, as previously.

   a) Find R_{in} with capacitor input coupling. Label it $R_{in\,(cap)}$.
   b) For the same signal source as in **Self-check** Problem 15, ($V_{in(oc)} = 1$ V, $R_{out} = 10$ kΩ), find V_{in} (equal to V_b) when capacitor input coupling is used. Label it $V_{b\,(cap)}$.
   c) For the same value of base-to-collector voltage gain, 25, find V_{out}. Label it $V_{out\,(cap)}$.
   d) Find the new value of $A_{v(overall)}$, the overall voltage gain from $V_{in(oc)}$ to V_{out}. Label it $A_{v\,(overall-cap)}$.
   e) Which input-coupling method does a better voltage-amplifying job here, capacitor or transformer? Explain.

17. Now compare current gains for the transformer-coupled and capacitor-coupled inputs.

   a) First, assume that the amplifier's load resistance is $R_{LD} = 550\,\Omega$. Calculate the output current I_{out} (equal to I_{ld}) for both coupling methods. Use Ohm's law, $I_{out} = V_{out} / R_{LD}$. Label the two output currents as $I_{out\,(cap)}$ and $I_{out\,(Xformer)}$.

   b) For both coupling methods, draw the input circuits as seen by the signal source. They will have $V_{in(oc)} = 1$ V, in series with $R_{out} = 10$ kΩ, in series with R_{in}. The value of $R_{in\,(Xformer)}$ is 16.5 kΩ, from **Self-Check** Problem 14. The value of $R_{in\,(cap)}$ is known from part (a) of the previous problem. From these diagrams, use Ohm's law to calculate both input currents. $[I_{in} = 1\,V / (10\,k\Omega + R_{in})]$ Label the two input currents as $I_{in\,(cap)}$ and $I_{in\,(Xformer)}$. Use the formula $A_i = I_{out} \div I_{in}$ to calculate the current gain for both coupling methods. Label the two current gains $A_{i\,(cap)}$ and $A_{i\,(Xformer)}$. Which input coupling method, capacitor or transformer, does a better current-amplifying job here?

18. Use the results from the previous 2 problems, Numbers 16 and 17, to calculate the overall power gain for both coupling methods. Use the formula $A_{P(overall)} = A_{v\,(overall)} \cdot A_i$. Label the results $A_{P(overall-cap)}$ and $A_{P(overall-Xformer)}$. In this situation where a high-resistance signal source (10 kΩ) is driving a transistor with lower input resistance, which input

coupling method is superior? This is the essential idea of matching resistances with a transformer.

19. (T-F) Using a transformer to create a better resistance-match between an amplifier with low R_{in} and a signal source with high R_{out} can greatly improve the amplifier's overall power gain.

20. What is the basic advantage of direct coupling in amplifiers?

21. Give some of the disadvantages of direct coupling.

22. The technique of using a portion of the output signal to subtract from (reduce) the effect of the input signal is called _____ .

23. Draw the schematic diagram of two *npn* transistors connected as a Darlington pair.

24. Repeat for two *pnp* transistors.

25. If the two transistors in Problem 24 have $\beta_1 = 250$ and $\beta_2 = 170$, what is the effective beta value of the Darlington pair?

    For Problems 28 through 35, refer to the common-base amplifier with transformer-coupled output in Fig. 7-9. The dc and ac load-lines for this amplifier are drawn in Fig. 7-22. When making these graphs, the negative voltage and current in the *pnp* circuit have been treated as if they were positive. Copy and enlarge Fig. 7-22.

28. If the transistor's ac operation causes v_{CE} to swing 1 V larger than 9.6 V and 1 V smaller than 9.6 V (between 8.6 V and 10.6 V), what values does the collector current swing between. Use the ac load-line.

29. For a different V_{in}, if the transistor's ac operation causes i_C to swing 0.6 mA larger and smaller than 0.69 mA (between 1.29 mA and 0.09 mA), what values does v_{CE} swing between?

30. (T-F) From your answer to problem 29, it is clear that the transistor is in greater danger of cutting off ($i_C = 0$) than of saturating ($v_{CE} = 0$).

31. To avoid cutoff, what is the maximum possible peak-to-peak ac voltage that can be produced across the transistor [$v_{CE(max)}$] ? Use the ac load-line.

32. In Fig. 7-22, the quiescent Q-point is approximately in the center of the dc load-line ($V_{CE} = 9.6$ V, out of $V_{CC} = 20$ V). Has this center bias point hindered the maximum ac response from the amplifier? Explain.

33. To maximize the ac output of the amplifier, should the Q-point be shifted to the left or to the right along the dc load-line? Explain.

34. Let us change the base bias voltage-divider resistors in Fig. 7-9 to $R_{B1} = 110$ kΩ and $R_{B2} = 240$ kΩ.
    a) Calculate the new dc bias conditions.
    b) Locate the new Q-point on the dc load-line of Fig. 7-22.

35. Shift the ac load-line to its new position through the new Q-point. Now what is the value of $V_{ce(max)}$?

    On page 160 in Sec. 6-1, we promised that we would explain why it often is a good idea to have $V_{CE} < 1/2\ V_{CC}$. In answering Problems 30 through 35, you have done that.

CHAPTER 8

OSCILLATORS

This electronic temperature controller enables the bather to set the exact desired water temperature. It works by varying the mixture of hot and cold incoming water. The up-pointing button raises the temperature by one degree Fahrenheit each time it is pressed. The down-pointing button lowers it by 1°F. It is shown in use on page 4 of the color section.

Courtesy of Memry Corporation

OUTLINE

LC Oscillators
Oscillator Stability
Crystal-Controlled Oscillators
RC Oscillators
Troubleshooting Oscillators

NEW TERMS TO WATCH FOR

LC oscillator	variable-frequency oscillator
LC tank circuit	frequency stability
feedback loop	crystal
positive feedback	buffer

An oscillator is a circuit that produces an ac output signal even though no ac input signal is applied. It is different from an amplifier, which produces an ac output by boosting a small ac input. Naturally, an oscillator requires a dc power supply.

There are oscillators for producing various shapes of ac waveforms, including sine waves, square, triangle and sawtooth waves. At this time, we will concentrate on sine-wave oscillators, because they are important in the operation of radio receivers, the topic of the next chapter.

After studying this chapter you should be able to:

1. Describe the natural oscillation process of an *LC* (inductor-capacitor) circuit.
2. Given the *L* and *C* values, calculate the natural oscillation frequency.
3. Explain how an amplifier with positive feedback from the *LC* tank circuit can keep a circuit's oscillations going continuously.
4. State the two specific requirements for successful oscillation in a feedback loop.
5. Recognize a Hartley, Colpitts, or Armstrong oscillator from a schematic diagram.
6. Name the steps that can be taken to improve the frequency stability of an *LC* oscillator.
7. Explain the basic behavior of a quartz crystal.
8. Recognize a Pierce crystal oscillator from its schematic diagram, and explain its method of operation.
9. Recognize a crystal-stabilized Hartley or Colpitts oscillator from its schematic diagram.
10. State the advantages and disadvantages of *RC* oscillators compared to *LC* and crystal oscillators.
11. Troubleshoot a malfunctioning oscillator.

8-1 *LC* OSCILLATORS

The sine-wave oscillators that are easiest to understand are those that use a parallel inductor-capacitor (*LC*) combination. To understand such *LC* oscillators, first let us consider the operation of a plain *LC* circuit.

Figure 8-1

(A) Putting an initial charge on the capacitor, to get the action started.

(B) With SW thrown to the right, *L* is connected to *C*. This creates a parallel *LC* circuit, often called a *tank circuit*.

(C) A tank circuit oscillator naturally producing a sine waveform.

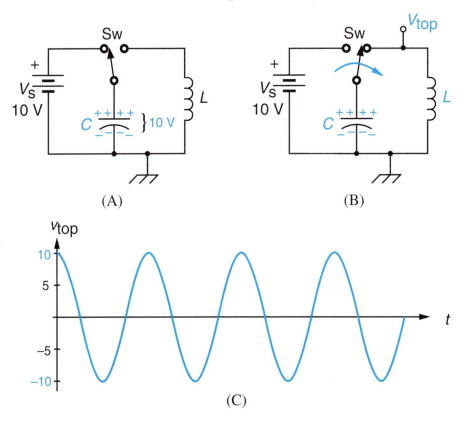

In Fig. 8-1(A), the switch is in its left position. This causes the capacitor to be charged to 10 V, negative on the bottom. The inductor is disconnected. The switch is thrown to the right in Fig. 8-1(B). Now the inductor *L* is connected to the charged capacitor *C* and the voltage source V_S has been disconnected. Ignore V_S from this point forward.

An interesting thing now occurs:

The parallel *LC* circuit begins oscillating in a sine-wave fashion.

Get This

That is, the voltage at the top of the *LC* parallel combination, V_{top} in Fig. 8-1(B), becomes a sine waveform. This is shown in Fig. 8-1(C).

Here is an explanation for why the *LC* tank circuit oscillates: The capacitor, in discharging through the inductor, transfers all of its stored energy to the inductor. When the inductor has the energy, we think of the energy as being contained in its magnetic field. Now there is a current-carrying, energy-containing inductor connected to a discharged zero-energy capacitor. Given this state of affairs, the natural outcome is that the inductor will recharge the capacitor to 10 V, but in the opposite polarity. In this way, the energy that was in the inductor's possession is transferred back to the capacitor.

Then the process repeats itself, but in the opposite direction. At the conclusion of this reverse process, the *LC* circuit is right back to its starting condition shown in Fig. 8-1(B). It then begins the entire process again.

1️⃣ ✔

If we had an ideal inductor and an ideal capacitor, and connecting wires with zero resistance, this oscillating action would go on forever. Such never-ending oscillation is suggested by the sine waveform of Fig. 8-1(C), in which the peaks are not decreasing in magnitude.

Figure 8-2 gives a more detailed, moment-by-moment description of an *LC* **tank** oscillation.

The frequency of these natural oscillations is given by the familiar formula

$$f_{osc} = \frac{1}{2\pi \sqrt{LC}} \approx \frac{1}{6.28 \sqrt{LC}}$$ Eq. (8-1)

2️⃣ ✔

You will recognize Eq. (8-1) as the same formula that predicted the resonant frequency f_r for an *LC* combination.

We often call the oscillation frequency of an *LC* tank circuit its *resonant* frequency.

EXAMPLE 8-1

In Fig. 8-1(B), suppose $C = 20$ nF and $L = 70$ mH. What will be the oscillation frequency of this circuit?

SOLUTION

From Eq. 8-1,

$$f_{osc} = f_r = \frac{1}{2\pi\sqrt{LC}}$$

$$= \frac{1}{6.28 \sqrt{(70 \times 10^{-3}\ \text{H})(20 \times 10^{-9}\ \text{F})}}$$

$$= \textbf{4.25 kHz}$$

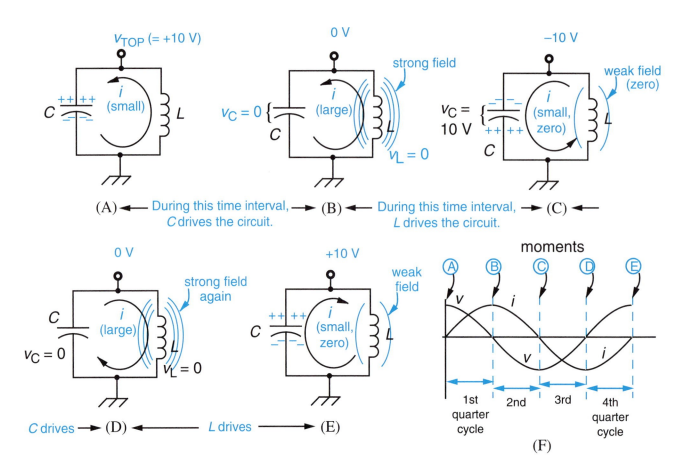

Figure 8-2

(A) When *C* first starts discharging through *L*, current *i* is small (zero) because the inductor *L* induces voltage v_L in opposition to v_C.

(B) Throughout the first quarter cycle, current *i* becomes larger and larger as *L* gives up its opposition voltage v_L. Exactly one quarter cycle after the beginning, v_C and v_L are both zero, but current *i* is at its peak.

(C) With no driving voltage applied to the inductor, it has nothing to operate against. Now, it takes over the duty of driving the circuit. We sometimes say that "the inductor's magnetic field collapses", causing it to create voltage v_L. Throughout the second quarter cycle, *L* causes more and more charge to build up on the capacitor. Therefore v_C gets larger and *i* becomes smaller and smaller. At the exact end of the 2nd quarter cycle, current *i* becomes zero.

(D) During the 3rd quarter cycle, *C* discharges through *L*, but this time in the clockwise direction. As *L* gives up its opposition v_L, current *i* becomes larger, reaching its negative peak at the end of the 3rd quarter cycle.

(E) As the inductor's magnetic field collapses again, it again takes over the job of driving the circuit, but this time in the CW direction. At the exact end of the 4th quarter cycle, *L* will have restored all the original charge on *C*.

Higher-*Q* components (*L* and *C*) allow the natural oscillations to continue longer. With lower *Q*, they decay more quickly.

In the real world, components *L* and *C* are not ideal (they have non-infinite Q-factors) and the wires have some nonzero resistance. This causes a small amount of energy to be lost on every transfer between *L* and *C*. Therefore the oscillations become weaker and weaker and eventually die out, as shown in Fig. 8-3.

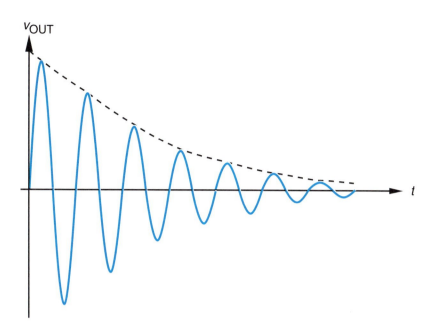

**Figure 8-3
Decaying oscillations
of a real *LC* tank
circuit.**

Keeping the Oscillations Going

To keep the oscillations going, here is what we do.

$\boxed{1}$ We tap the inductor at a point close to one end, and connect the tap to ground as shown in Fig. 8-4(A). Then when the oscillations occur, a small portion of the total oscillation voltage appears between the bottom terminal and ground, as v_{BOT}. This voltage v_{BOT} is inverted, compared to v_{TOP} and total voltage v_C.

For example, in Fig. 8-4(B) the inductor has been tapped at two-tenths of the distance from the bottom. At the positive peak instant, total $v_C = 10$ V, + on top. At that instant, v_{TOP} is +8 V relative to ground, and v_{BOT} is –2 V relative to ground, as that figure shows. When the *LC* tank starts oscillating, the waveforms of v_C, v_{TOP}, and v_{BOT} will be as shown in Fig. 8-4(C).

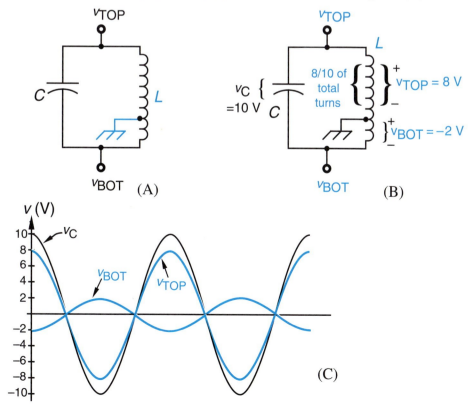

Figure 8-4

(A) Tapping the inductor.

(B) The inductor is tapped at 2/10, or 0.2.

(C) Waveforms of v_C (total tank voltage), v_{TOP} and v_{BOT}. Voltage v_{BOT} is 180° out of phase with the others.

[2] We amplify the small voltage v_{BOT}. For example, we could capacitively couple v_{BOT} to the input of a common-emitter amplifier, as shown in Fig. 8-5(A). Then the voltage waveforms would be as shown in Fig. 8-5(B).

Notice a very important thing about the phase relations in Fig. 8-5(B):

Get This

v_{OUT} is in phase with v_{TOP} and v_C.

This happened because the common-emitter amplifier inverted the v_{BOT} signal, which was itself inverted relative to v_{TOP}. Two inversions puts the resulting voltage back into phase with the reference voltage.

Another way to say this is that two 180° phase-shifts produce a 360° phase-shift, which is the same as 0°, or zero phase-shift.

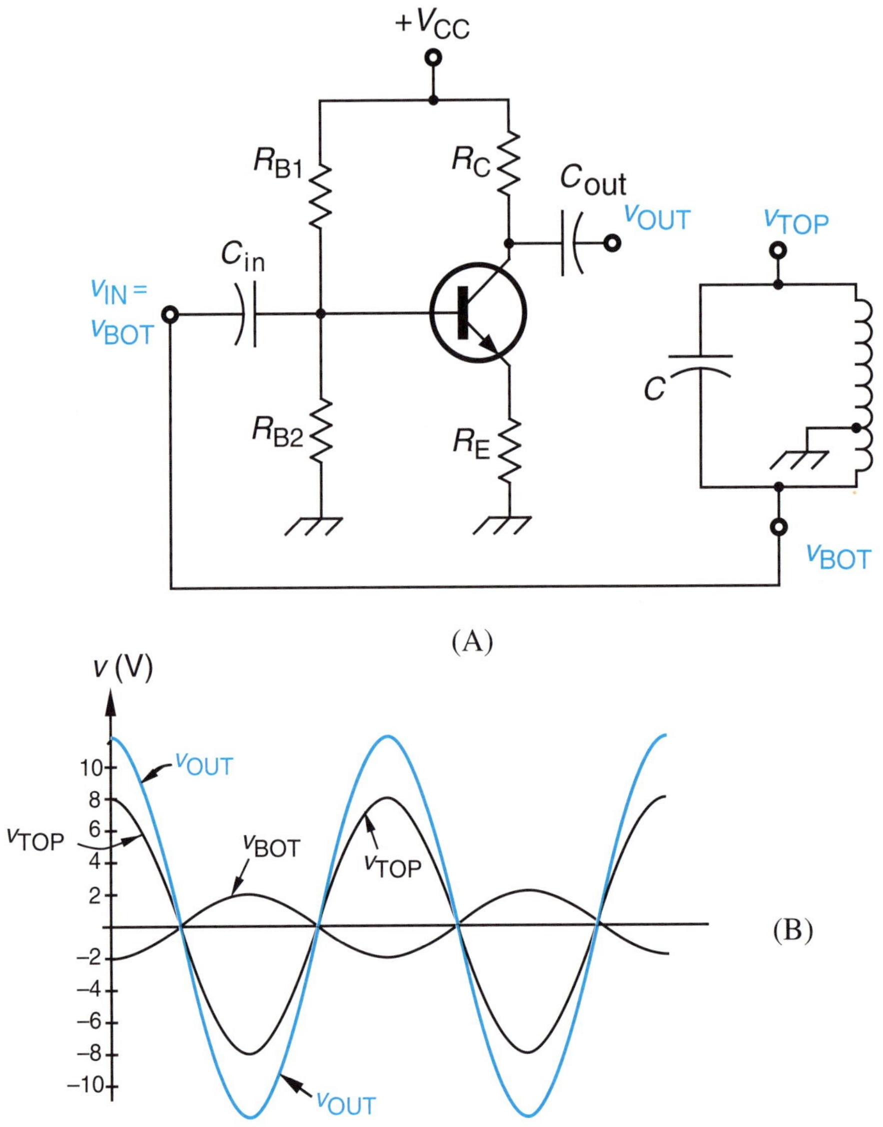

Figure 8-5

(A) Amplifying v_{BOT}.

(A)

(B) The amplifier's output voltage is in-phase with the tank voltage, because of the double inversion.

(B)

v_{OUT} must be greater than v_{TOP}, even after taking into account the loading effect of the LC tank circuit on the amplifier's output. For example, in Fig. 8-5(B), v_{OUT} would shrink if it were connected to a load.

[3] ✔

[3] We connect v_{OUT} to the top of the tank, as shown in Fig. 8-6. As long as v_{OUT} is larger than v_{TOP}, the LC tank circuit will be replenished and the oscillations will continue forever.

You can think of the process this way: When the capacitor voltage oscillates back to its positive peak, it will tend to be slightly smaller than 10 V. This happens because the tank circuit lost a small amount of energy during the cycle. But the amplifier output brings the capacitor back up to its full 10 V, replacing the lost energy. Therefore the next cycle begins just as strong as the previous cycle; and so on.

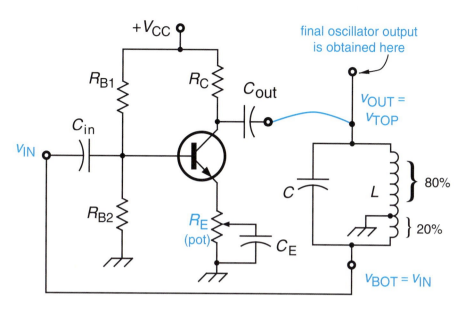

**Figure 8-6
Complete Hartley oscillator. The amplifier's A_v is adjustable by adjusting the R_E pot.**

In Fig. 8-6, the voltage gain of the amplifier might be so great that it tries to make v_{OUT} much larger than the natural value of v_{TOP}. In that case the sine wave will be distorted. The complete oscillator will work, but it will not produce a properly shaped sine wave. This is shown in Fig. 8-7(A).

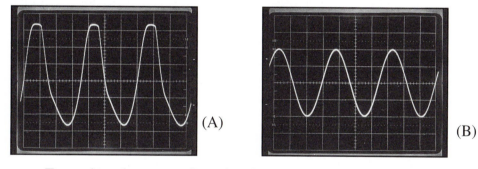

(A)   (B)

**Figure 8-7
(A) Too much amplifier gain.
(B) Correct amount of gain.**

To produce the proper distortion-free sine wave, the amplifier's A_v must be adjusted to produce a v_{OUT} value that is *just barely* larger than v_{TOP}. Voltage gain adjustment is accomplished in Fig. 8-6 by adjusting potentiometer R_E. When the pot wiper is adjusted toward the top of R_E, capacitor C_E will ac-bypass a large portion of the emitter resistance. That leaves less resistance in the emitter's ac path, which increases A_v. When the pot wiper is adjusted toward the bottom of R_E, A_v is decreased.

In Fig. 8-6, the combination of the amplifier and the *LC* tank circuit can be thought of as a **feedback loop**. This is illustrated in Fig. 8-8.

In successful distortion-free oscillation, the amplifier is just barely reaching saturation at one sine-wave peak and just barely reaching cutoff at the other peak.

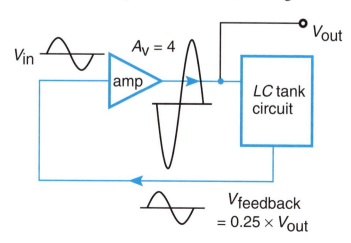

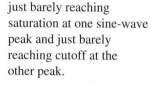

**Figure 8-8
Visualizing an oscillator as a feedback loop.**

Think of it like this: The amplifier provides an output which drives the *LC* tank circuit. A portion of the tank circuit's oscillation is fed back to the amplifier input. That input is amplified to produce the output.

It so happens that many oscillators, not just *LC* types, can be thought of as feedback loops like Fig. 8-8. Then we can state the two requirements for successful oscillation as follows.

Get This

For proper oscillation:

1. The overall gain of the loop must be 1.0, or slightly greater than 1.0.
2. The overall phase-shift of the loop must be zero.

For example, in Fig. 8-6 the tank circuit feeds back a signal, v_{BOT}, that is one-fourth as large as v_{TOP}, which is the same as v_{OUT} (2 V compared to 8 V, say). We can think of the "gain" that the *LC* tank contributes to the loop to be one-fourth, or 0.25. Therefore the amplifier itself must have gain of A_v = 4.0. That way, the overall loop gain is equal to $(0.25) \times (4.0) = 1.0$. These factors are shown in Fig. 8-8.

Regarding requirement No. 2, the phase-shift requirement, think of it by starting at the input of the amplifier, v_{IN} in Fig. 8-6. The common-emitter amplifier shifts v_{IN} by 180°, in producing v_{OUT}. The *LC* tank receives v_{OUT} and feeds back a portion of it. This portion is phase-shifted by an additional 180° (v_{BOT} has the opposite ac polarity from v_{TOP}). Therefore v_{BOT} arrives back at the base input terminal perfectly in phase with the original v_{IN}. The overall phase-shift is zero. These phase relations are portrayed in Fig. 8-8.

Feedback that is in phase with the original v_{IN} is called **positive feedback**. It is the opposite of negative feedback.

Realistic Version of the Hartley Oscillator: An *LC* oscillator with a tapped inductor is called a Hartley oscillator, after the person who invented it. The way that it is shown in Fig. 8-6 makes it easiest to understand. However, real Hartley oscillators are usually built with the *LC* tank circuit directly in the collector lead, as shown in Figs. 8-9(A) and 8-9(B).

Figure 8-9
Realistic Hartley oscillators.

(A) Common-emitter design.

(B) Common-base design.

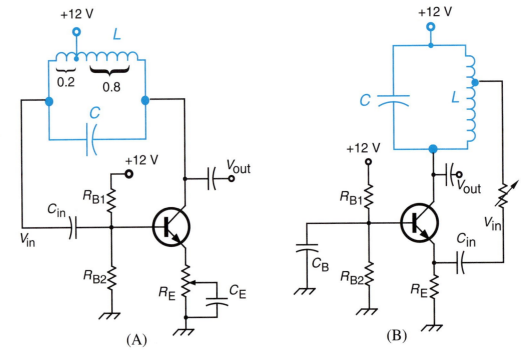

Other *LC* Oscillators

In a Hartley oscillator, the feedback is obtained from the inductor side of the *LC* tank circuit. It is just as workable to get the feedback from the capacitor side, if *C* is replaced by a series combination of C_1 and C_2. This is shown in Fig. 8-10, making a Colpitts oscillator.

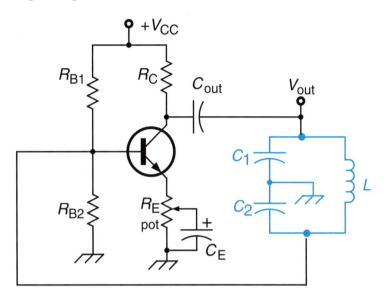

Figure 8-10
Colpitts schematic diagram. In a realistic Colpitts oscillator, the *LC* tank circuit is usually wired directly in the collector lead, like the Hartley oscillators of Fig. 8-9.

In the Colpitts design, oscillation frequency is set by total capacitance C_T, where

$$C_T = \frac{C_1 \, C_2}{C_1 + C_2} \qquad \text{(series capacitors)}$$

Since we want to feed back a small fraction of the tank's total oscillation voltage, the voltage across C_2 should be smaller than the voltage across C_1 in Fig. 8-10. This will happen if X_{C2} is smaller than X_{C1}, which means that C_2 should be *larger* than C_1.

EXAMPLE 8-2

In Fig. 8-10, suppose $L = 20$ mH, $C_1 = 15$ nF and $C_2 = 100$ nF.
a) Find the total capacitance C_T.
b) What is the frequency of oscillation? Use Eq. (8-1).
c) What fraction of the oscillation output voltage is fed back to the amplifier's input?

SOLUTION

a) From the product-over-the-sum formula for series capacitors,

$$C_T = \frac{C_1 \, C_2}{C_1 + C_2} = \frac{(15 \text{ nF}) \, (100 \text{ nF})}{15 \text{ nF} + 100 \text{ nF}}$$

$$= \textbf{13.0 nF}$$

b)
$$f_{osc} = \frac{1}{2\pi \sqrt{LC}} = \frac{1}{6.28 \sqrt{(20 \times 10^{-3} \text{ H}) \, (13.0 \times 10^{-9} \text{ F})}}$$

$$= \textbf{9.88 kHz}$$

c)
$$X_{C1} = \frac{1}{2\pi f C_1} = \frac{1}{2\pi (9.88 \times 10^3)(15 \times 10^{-9})}$$
$$= 1070 \ \Omega$$

$$X_{C2} = \frac{1}{2\pi f C_2} = 161 \ \Omega$$

V_{out} appears across $X_{C1} = 1070 \ \Omega$. Feedback voltage V_{in} appears across $X_{C2} = 161 \ \Omega$. Therefore

The feedback ratio, or factor, is usually symbolized B. In this example, $B = 0.15$.

$$\frac{V_{feedback}}{V_{out}} = \frac{X_{C2}}{X_{C1}} = \frac{161 \ \Omega}{1070 \ \Omega} = \mathbf{0.15}$$

which is the same as the capacitance ratio.

Armstrong oscillator: Another method of obtaining feedback from an oscillating LC tank circuit is by transformer action. The Armstrong oscillator shown in Fig. 8-11 does this.

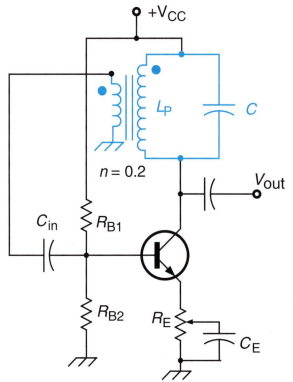

**Figure 8-11
Armstrong oscillator
schematic.**

5 ✔

The oscillation frequency is set by C and L_P, the inductance of the primary winding. The phasing dots in Fig. 8-11 ensure that the transformer secondary winding leads are connected to provide positive feedback. The dotted lead of the secondary winding has the same instantaneous polarity as the dotted lead of the primary winding. (They are both positive together and both negative together.)

An oscillator with adjustable frequency is sometimes called a variable-frequency oscillator, or VFO.

Adjustable-Frequency Oscillators: Any of the three types of LC oscillators can be built to have adjustable frequency. For the Hartley and Armstrong types, it is convenient to use a variable capacitor in the tank circuit. For the Colpitts type, it is more convenient to use a variable inductor in the tank circuit. This avoids the problem of causing a change in the feedback factor whenever the frequency is varied by adjusting a variable capacitor.

SELF-CHECK FOR SECTION 8-1

1. A parallel LC circuit is often called a _____ circuit.
2. A parallel LC circuit has $C = 0.01$ μF and $L = 12$ mH. Find its frequency of oscillation.
3. For the LC circuit in Problem 2, give the reasons why the oscillations do not continue indefinitely.
4. An LC oscillator consists of an LC tank circuit properly connected to an _____ .
5. The feedback from the LC tank circuit to the amplifier must be _____ feedback. (positive or negative)
6. In Fig. 8-6, suppose $C = 0.05$ μF and $L = 25$ mH, tapped at the 10% point.
   a) Find the frequency of oscillation.
   b) For undistorted sine-wave oscillations, what should be the value of the amplifier's voltage gain?
7. In Fig. 8-10, suppose $C_1 = 0.02$ μF, $C_2 = 0.2$ μF, and $L = 50$ mH.
   a) Find the frequency of oscillation.
   b) For undistorted sine-wave oscillations, what should be the value of the amplifier's voltage gain?

8-2 OSCILLATOR STABILITY

Capacitors and inductors are always temperature-sensitive to at least some extent. Therefore Eq. (8-1) predicts that the frequency of an LC oscillator will vary as the temperature changes. An oscillator with good **frequency stability** will keep such frequency variations to a minimum. This is accomplished by using high-quality capacitors with good temperature-stability and high-Q inductors in the LC tank circuit.

A further problem is this: In a real-life LC oscillator, the actual frequency is slightly higher than the ideal resonant frequency given by Eq. (8-1). The reason for this has to do with maintaining an exact $0°$ phase-shift in the loop; it is explained in advanced courses. At this time, we will just mention that the input resistance of the amplifier has an effect on the overall loop phase-shift. Therefore, variations in the amplifier's input resistance can produce variations in frequency. Of course, the amplifier's R_{in} depends on the transistor's β, which is itself temperature-dependent and batch-dependent.

⑥ ✔

To minimize the entire effect, we have a general rule:

The higher the amplifier's input resistance, the better the oscillator's frequency stability.

Get This

For this reason, some LC oscillators use a two-stage amplifier with a common-collector in stage 1. An example is shown in Fig. 8-12.

Figure 8-12 Very stable Colpitts oscillator. For frequency-stability, a common-base amplifier is preferable to common-emitter because of its better output characteristic. To minimize any loading effect on the tank circuit, the overall oscillator output is taken from the emitter of the common-collector stage. The R_{F1}-R_{F2} feedback divider determines the overall feedback ratio and loop gain.

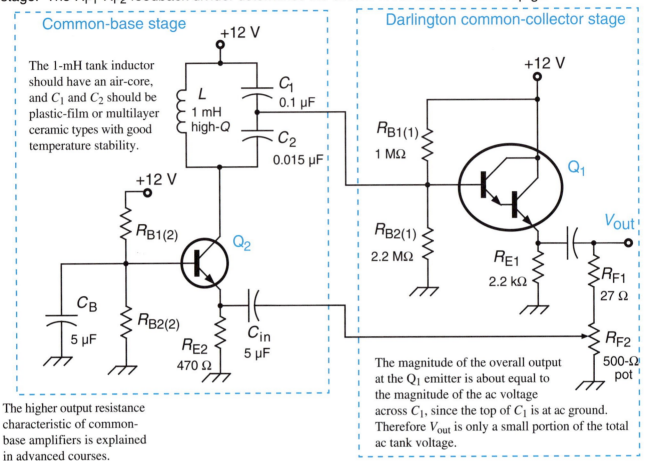

Common-base stage

The 1-mH tank inductor should have an air-core, and C_1 and C_2 should be plastic-film or multilayer ceramic types with good temperature stability.

The higher output resistance characteristic of common-base amplifiers is explained in advanced courses.

Darlington common-collector stage

The magnitude of the overall output at the Q_1 emitter is about equal to the magnitude of the ac voltage across C_1, since the top of C_1 is at ac ground. Therefore V_{out} is only a small portion of the total ac tank voltage.

8-3 CRYSTAL-CONTROLLED OSCILLATORS

To get extremely stable oscillation frequency, we use a mechanical crystal in our oscillator circuits. A crystal is a piece of quartz that has been machined to exact mechanical dimensions. An example is shown in Fig. 8-13(A).

Figure 8-13
(A) Mechanical dimensions of a 2.000-MHz crystal.
(B) The crystal is sandwiched between metal plates to make it electrically functional.
(C) Schematic symbol.
(D) Physical appearance of the packaged component.

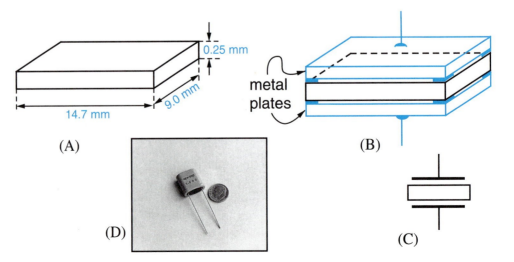

The crystal is placed between two metal plates with attached leads, as shown in Fig. 8-13(B). The structure reminds us of a capacitor, and in fact the entire crystal structure carries with it a significant amount of parallel capacitance. The schematic symbol is shown in Fig. 8-13(C). The entire crystal structure is encased in a metal package like the one shown in Fig. 8-13(D). The package leads may be directly soldered or they may plug into a socket in the oscillator circuit.

Crystals can be rectangular, as shown in Fig. 8-13, or they can be circular.

There are three features of crystals that cause them to function.

1 If a crystal is mechanically compressed (squashed very slightly), it will undergo tiny vibrations when the compression force is removed. We can visualize this mechanical vibration as shown in Fig. 8-14.

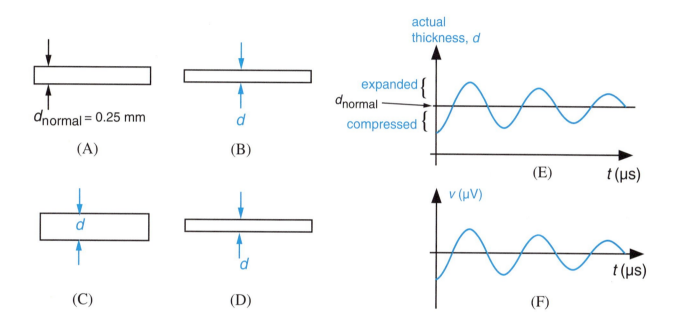

Figure 8-14 A crystal will vibrate (oscillate) mechanically.
(A) The crystal's normal height, or thickness. (B) Initially compressed.
(C) Expanded. (D) Vibrates back to compressed, naturally. (E) The deformations occur in a sine-wave fashion. (F) Waveform of piezoelectric voltage.

Like all such natural oscillations, a crystal's vibrations eventually die out as the initial compression energy is dissipated as heat into the surrounding air.

2 As the mechanical vibrations occur, an atomic effect in the crystal's structure produces a very small voltage between its two sides, or faces. This effect is called the piezoelectric effect (pronounced pea - ay´ - zo). The voltage appears between the plate leads in Fig. 8-13(B). A waveform of the piezoelectric voltage is shown in Fig. 8-14(F).

3 The piezoelectric effect also works in reverse. That is, if an external voltage is applied to the plate leads of Fig. 8-13(B), that voltage will cause the crystal either to compress or to expand, depending on the voltage polarity.

For a fixed length and width in Fig. 8-13(A), a smaller thickness (thinner crystal) produces higher-frequency vibrations. By machining to particular dimensions, we can manufacture crystals to operate in the range of frequencies from about 100 kHz to about 10 MHz.

These three features of crystal behavior bring to mind a feedback loop. If we can get the crystal vibrating, it will produce a small voltage. We can then amplify that small voltage and apply the output back onto the crystal with positive feedback. The applied voltage reinforces the mechanical vibrations and keeps them going forever. This idea is shown in Fig. 8-15.

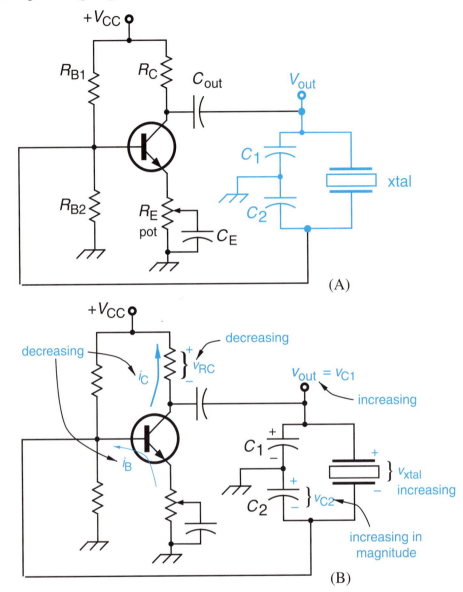

**Figure 8-15
Pierce oscillator.**

(A) Schematic.

**(B) Feedback
conditions at an
instant when the
crystal voltage is
+ on top and
increasing.**

In some cases, C_1 is provided by the circuit's stray capacitance. Then no actual C_1 component is present.

C_2 is larger than C_1 so that the feedback voltage is a small portion of the output, like a Colpitts oscillator.

The circuit of Fig. 8-15 is called a Pierce oscillator. The crystal is connected in parallel with the $C_1 - C_2$ series combination. The purposes of C_1 and C_2 are to set the feedback ratio and to provide the phase inversion necessary for positive feedback. C_1 and C_2 have almost no effect on the oscillation frequency, which is set by the crystal itself.

To understand the positive feedback/zero phase-shift nature of the Pierce oscillator, look at Fig. 8-15(B). This shows the instantaneous behavior of the circuit's voltages and currents at an instant when the crystal voltage, v_{xtal}, is increasing in the positive polarity The voltage across the $C_1 - C_2$ combination is equal to v_{xtal}, so the individual capacitor voltages are also positive on top and increasing in magnitude. Therefore, v_{C2}, the feedback voltage on the un-grounded bottom terminal of C_2, is negative and becoming larger in magnitude. This tends to oppose the forward bias on the B-E junction, reducing the transistor's base drive. Collector current i_C is therefore decreasing in magnitude, which causes v_{RC} to decrease and v_C to increase. The increase in v_C

reinforces, or adds to, the instantaneously increasing crystal voltage. Thus, the feedback is positive and the crystal's oscillations continue indefinitely.

A high-frequency Pierce oscillator may use an inductor in place of collector resistor R_C in Fig. 8-15. This is done to take advantage of the large inductive reactance that occurs at high oscillation frequencies. Large inductive reactance in the collector tends to increase the amplifier's A_v, which might otherwise decrease because of stray capacitance effects. Also, a large X_L value helps isolate the V_{CC} power supply line from the high-frequency output. This may be useful in preventing high-frequency noise from being injected into other circuits that are connected to that same dc power supply.

The inductor design is also used for other types of high-frequency oscillators that do not already have an LC tank circuit in the collector.

Frequency Fine-Adjustment: Although a crystal is essentially a fixed-frequency device, it is possible to make *very slight* frequency adjustments in a crystal oscillator. This is usually accomplished by placing a low-value trimmer capacitor in series with the crystal itself, as shown in Fig. 8-16. The presence of C_3 has a slight effect on the overall phase-shift of the entire feedback loop. It causes the overall phase-shift to be slightly different from $0°$ at the natural crystal frequency. Therefore the crystal is forced to oscillate at a slightly different-from-natural frequency, one that brings the overall phase-shift of the loop back to exactly $0°$.

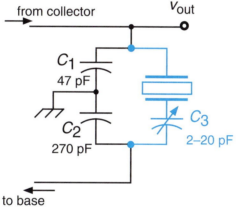

Figure 8-16
A method for getting slight adjustments in the frequency of a crystal oscillator. Frequency adjustments on the order of $\pm0.01\%$ (±100 parts per million) can be achieved in this way.

Crystal − LC Combinations: In a Pierce oscillator the quartz crystal is the basic oscillating element. It is also possible to use a crystal as a support component, not as the basic oscillation element. When used this way, a crystal is placed in series with the LC tank circuit to serve as a very-high-Q filtering element, as shown in Fig. 8-17. Its job is to act like a zero-resistance path at its

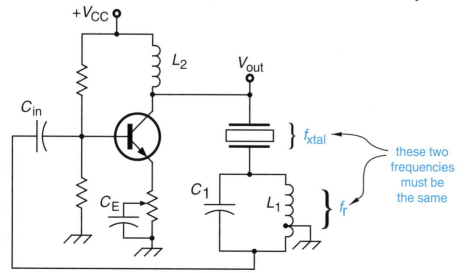

Figure 8-17
Crystal-stabilized Hartley oscillator. The resonant frequency of the L_1C_1 tank circuit must be the same as the natural resonant frequency of the crystal.

natural resonant frequency, thereby completing the feedback loop and allowing oscillation to occur. But at any other frequency the crystal acts like a high-impedance device, thereby interfering with the overall loop gain and preventing oscillation, In this way, the crystal prevents the oscillator frequency from changing due to temperature or batch variations.

VIDEO IMAGING FOR ANGIOPLASTY

To maximize the probability for a successful angioplasty, the medical personnel must have clear magnified views of the progress of the catheter through the patient's artery. This is accomplished by shining X-rays through the arterial area onto a flat plate of X-ray-sensitive semiconducting material. The flat plate is electronically scanned to produce the video image.

By rotating the X-ray apparatus, as shown in this photo, the arterial area can be viewed from any desired vantage angle. Also, using electronic data storage and retrieval, previously recorded images can be maintained on certain monitors while the present "real-time" image is displayed on another monitor. In this way, the cardiologist can observe the progress of the procedure.

The bottom photo is a stored image of a recently treated artery. The arrow shows the location where a severe plaque blockage was cleared away.

Artery-imaging system with remote observation facility for medical education. Courtesy of Philips Medical Systems North America Co.

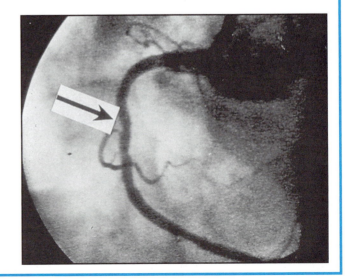

SELF-CHECK FOR SECTIONS 8-2 AND 8-3

8. (T-F) For good frequency stability, an *LC*-type oscillator must use capacitors with low temperature coefficients (not sensitive to temperature changes).

9. For good frequency stability, oscillator designers try to give the amplifier a _____ input resistance.

10. (T-F) When a compressed crystal is released, it immediately springs back to its normal thickness and remains right there.

11. The effect in which compressing a crystal causes a small voltage to appear between the leads is called the _____ effect.

12. Applying an external voltage of one polarity to a crystal causes it to compress; applying an external voltage of the opposite polarity causes it to _____ .

13. A crystal oscillator uses _____ feedback to keep the mechanical / electrical oscillations going indefinitely.

14. (T-F) Crystal oscillators have better frequency stability than *LC* types.

8-4 *RC* OSCILLATORS

It is not really necessary for an electronic oscillator to contain a naturally oscillating element or circuit segment, such as a crystal or an *LC* tank. The only things that are required for continuous oscillation are a loop gain of 1, and a loop phase-shift of $0°$. These requirements can be met with resistor-capacitor-transistor circuits. By eliminating the crystal and the inductor, an oscillator can be miniaturized in an integrated circuit (IC). This is not feasible with crystal-controlled or *LC* oscillators because inductors and crystals cannot be sufficiently miniaturized.

Setting aside the miniaturization issue, it is difficult to build very low-frequency oscillators using inductors. This is because a very large inductance value is required in order to obtain low oscillation frequencies. Equation (8-1) indicates this.

For example, the largest commercially available capacitance in a non-electrolytic capacitor is about 50 µF. Let us say that we wish to oscillate at 7 Hz, a very low frequency. Rearranging Eq. (8-1) to solve for the required inductance value, we get

$$L = \frac{1}{4\pi^2 f^2 C}$$

$$= \frac{1}{(4\pi^2)\,(7^2)\,(50 \times 10^{-6})} = 10 \text{ H}$$

A 10-henry inductor would be large, heavy and expensive.

An *RC* oscillator for operating at 7 Hz is shown in Fig. 8-18. Each *RC* combination shifts the feedback voltage by $60°$. After passing through three successive 60-degree phase-shifts, the base input voltage to Q_1 is $180°$ phase-shifted from V_{out}. There is zero phase-shift through the common-collector

Not just oscillators, but all low-frequency circuits that contain inductors have the same basic problem. To provide the necessary reactance at low *f*, the inductance value tends to be unreasonably large.

stage, followed by a 180° phase-shift through the common-emitter stage. Summing all the phase-shifts in the loop gives the overall phase-shift of 0°. The voltage loss in the *RC* phase-shift network is made up by the voltage gain of the Q_2 stage. Therefore both requirements are met and oscillation is maintained.

Because they have no sharp (narrow-bandwidth) tuning segment, *RC* oscillators are not as frequency-stable as *LC* or crystal types.

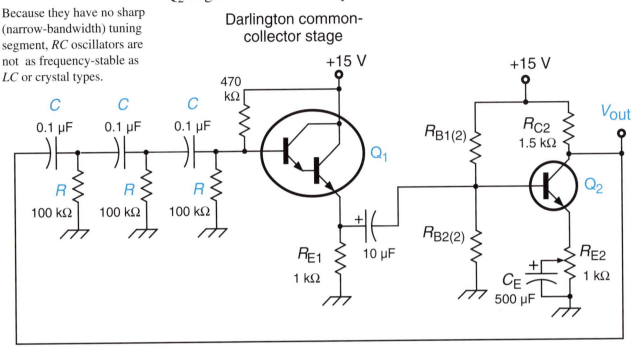

Figure 8-18 *RC* phase-shift oscillator for use at low frequencies. The 7-Hz frequency is set by the values of *R* and *C* in the phase-shift network.

8-5 TROUBLESHOOTING OSCILLATORS

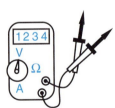

An oscillator can give trouble in two ways. 1) It can stop altogether. 2) It can produce an improper output. Examples of improper output are incorrect frequency, continually drifting frequency, and distorted waveshape.

When troubleshooting an oscillator, keep in mind the model shown in Fig. 8-19(A). As that figure suggests, it is wise to disconnect the oscillator from its driven load (isolate the V_{out} terminal) when troubleshooting.

Stopped Oscillating: If the oscillator stops completely, check its amplifier as you would any amplifier. Check its dc power supply and ground connections as explained in Sec. 5-7. Check bias conditions. An incorrect bias voltage will give you some clue to the cause of trouble, as described in Sec. 5-7. If the dc bias conditions are OK, you must inject an ac signal at the amplifier's input to check its ac operation. This is best accomplished by breaking open the input path as shown in Fig. 8-19(B). In some cases it may not be necessary to disconnect the input path – you may be able to simply connect the signal generator.

Set your signal generator (sig gen) frequency to the oscillator's normal operating frequency. Verify that the amplifier is providing its proper value of voltage gain.

If the amplifier is OK, you may be able to check whether the *LC* tank-circuit oscillating segment is still capable of oscillation. This is accomplished

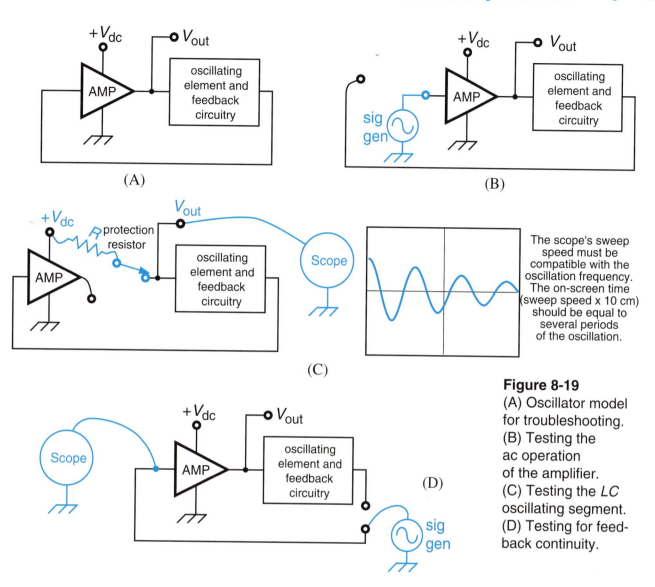

Figure 8-19
(A) Oscillator model for troubleshooting. (B) Testing the ac operation of the amplifier. (C) Testing the *LC* oscillating segment. (D) Testing for feedback continuity.

by breaking open the path that connects the amplifier output to the *LC* oscillating segment, as shown in Fig. 8-19(C). In some cases this disconnection may not be necessary. With an oscilloscope connected to the output terminal, temporarily touch the dc supply to that terminal, as Fig. 8-19(C) suggests. This will inject energy into the *LC* tank. When you remove the dc supply, there should be several cycles of decaying oscillation visible on the scope. If they are present, measure their approximate period, and find their frequency by $f = 1/T$. This will verify the correct functioning of the *LC* oscillating segment. If the decaying oscillations are not visible, test the inductors and capacitors by your usual methods.

If the amplifier and oscillating segment are OK, then the trouble must be in the feedback path connecting the oscillating segment to the amplifier input. It may be convenient to inject a proper low-voltage signal, as shown in Fig. 8-19(D). That signal can them be traced back to the amplifier input.

Improper Output: If the output waveshape is distorted, expect that the amplifier's A_v is too high. Adjust it if possible. If the gain is not adjustable, look for the reason that it has increased.

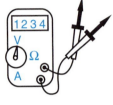

Oscillator circuitry is also subject to loading by an oscilloscope. Be sure that your scope has 10-MΩ, low-capacitance probes.

To accurately measure an oscillator's output frequency, you need a frequency counter. Be careful of the loading effect on the oscillator circuitry of the counter itself. The counter should have an input resistance of at least $10\,\text{M}\Omega$; its input capacitance must be sufficiently low. If possible, connect the counter to a point in the circuit that has plenty of current drive capability, like the Q_1 emitter terminal in Fig. 8-18. Such a point is said to be **buffered** from the rest of the circuit.

If the frequency is incorrect, some capacitance or inductance value has changed, or an excessive amount of stray capacitance has somehow appeared in the circuit. The transistors' stray capacitances might be the trouble, especially in high-frequency oscillators. These problems can be difficult to locate.

FORMULAS

$$f_{\text{osc}} = \frac{1}{2\pi\sqrt{LC}} \qquad \text{Eq. (8-1)}$$

$$C_T = \frac{C_1\,C_2}{C_1 + C_2} \qquad \text{(for a series combination)}$$

SUMMARY

- An inductor-capacitor tank circuit will oscillate naturally in a sine wave fashion, at a frequency $f = 1/2\pi\sqrt{LC}$.
- Natural LC oscillations die out because of energy losses in the nonideal circuit components.
- To keep LC oscillations going, we feed a small portion of the sine wave voltage back to the input of an amplifier, and use the output of the amplifier to drive the LC tank circuit.
- For undistorted sine-wave output, the amplifier's gain must not be too large. The overall loop gain should be 1.
- To produce oscillations, the feedback from the LC tank circuit to the amplifier must be positive feedback, in-phase with the original input.
- The three basic types of LC oscillators are the Hartley, Colpitts, and Armstrong types.
- Any LC oscillator can have its frequency adjusted, by varying either the C value or the L value.
- An oscillator's transistor amplifier can have common-emitter, common-base, or common-collector configuration. It can also be a two-stage amplifier.
- An LC oscillator's frequency stability can be improved by using tempera-ture-stable components, especially capacitors, and by making the amplifier's input resistance as high as possible.

● Quartz crystals are natural mechanical vibrators (oscillators). The oscillation frequency depends on the crystal's mechanical dimensions.
● Quartz crystals exhibit the piezoelectric effect. Mechanical deformation produces a voltage between crystal faces, and applying a voltage between crystal faces causes mechanical deformation.
● Crystal oscillators are much more frequency-stable than *LC* or *RC* oscillators. Of course, they are limited to a fixed frequency.
● A crystal can be used in two distinct ways in an oscillator: 1) It can be the basic oscillating element, as in a Pierce oscillator. 2) It can be placed in series with an *LC* tank circuit of matching resonant frequency. The tank circuit is then the basic oscillating segment and the crystal serves a frequency-filtering function.
● Crystal oscillators are capable of very slight frequency adjustment, usually by a variable capacitor.
● *RC* phase-shift oscillators contain no inductor and no crystal. This gives them two advantages: 1) They can be miniaturized in an IC; 2) They can operate at very low frequencies.
● When troubleshooting an oscillator, the feedback loop model should be kept in mind.

QUESTIONS AND PROBLEMS

1. (T-F) In an *LC* tank circuit, oscillations occur because the inductor and capacitor transfer energy back and forth between them.
2. (T-F) In an *LC* tank circuit, the capacitor contains all the energy when the voltage is at peak, and the inductor contains all the energy when the voltage is at zero (which is when the current is at peak).
3. (T-F) In an *LC* tank circuit, a low-*Q* inductor enables the natural oscillations to continue for a longer time than a high-*Q* inductor would.
4. An *LC* oscillator consists of an *LC* tank circuit combined with a(n) _____.
5. What is the defining feature of a Hartley oscillator?
6. Answer Question 5 for a Colpitts oscillator. Answer again for an Armstrong oscillator.
7. (T-F) The higher the voltage gain of the amplifier, the better the sine wave shape of an oscillator.
8. Ideally, the overall loop gain of an oscillator should be _____ .
9. An oscillator has an overall loop phase-shift of exactly _____ degrees.
10. (T-F) An oscillator utilizes the technique of negative feedback for its amplifier.
11. (T-F) An oscillator's amplifier must always be in the common-emitter configuration, in order to get the overall loop phase-shift to be 0°.
12. In Example 8-2 the feedback ratio was $B = 0.15$. Ideally, what is the proper value of amplifier voltage gain A_v ?
13. For the Armstrong oscillator of Fig. 8-11, what is the ideal value of amplifier A_v ?
14. (T-F) An oscillator always runs its amplifier from saturation to cutoff.

15. In the Colpitts oscillator of Fig. 8-10, suppose the component values are changed to: $L = 100\ \mu H$, $C_1 = 2.7$ nF, $C_2 = 10$ nF.
   a) Find the total capacitance C_T from the product-over-the-sum formula.
   b) Calculate the frequency of oscillation.

16. In the oscillator of Problem 15, suppose $V_{out} = 5$ V.
   a) What is the value of feedback voltage produced across C_2 ?
   b) What is the value of feedback ratio, B ?
   c) What amplifier A_v is proper?

17. In Problem 16, what is the value of total tank voltage, which is the same as the inductor voltage V_L ?

18. In the 2-stage oscillator of Fig. 8-12, the very high input impedance of the Q_1 stage contributes to:
   a) The oscillator frequency being very temperature-stable.
   b) The actual oscillator frequency being very close to the tank resonant frequency, $1/2\pi\sqrt{LC}$.
   c) Good buffering of the load from the oscillator (changes in load resistance having very little effect on oscillator operation).
   d) all of the above.

19. The _____ effect causes one polarity of voltage to appear when a crystal is deformed in one direction and the opposite polarity to appear when the crystal is deformed in the other direction.

20. (T-F) Thinner crystals oscillate at higher frequencies than thicker crystals, other things being equal.

21. (T-F) In a Pierce oscillator like the one in Fig. 8-15, the crystal itself is the oscillating element.

22. (T-F) A crystal-stabilized Hartley oscillator like the one in Fig. 8-17 has better temperature stability than a plain Hartley oscillator.

23. (T-F) In the crystal-stabilized Hartley oscillator of Fig. 8-17, the crystal itself is the main oscillating element.

24. Explain the advantages of an *RC* phase-shift oscillator over an *LC* or crystal oscillator.

Aerobic stepping / climbing exercise machine, described on page 29.   At right, the eye-level display module.   *Courtesy of Heart Rate, Inc.*

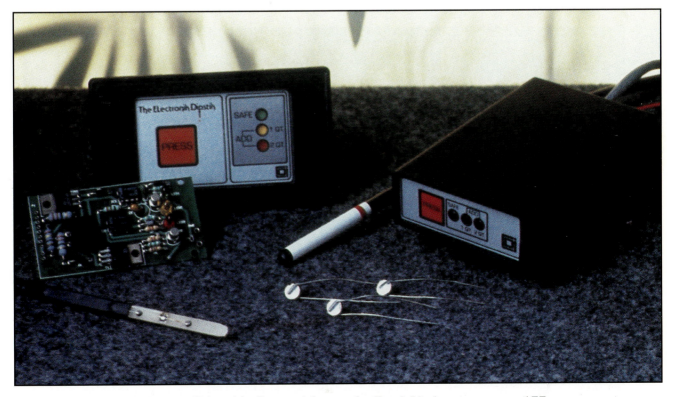

Electronic oil-level indicator (electronic dipstick) shown on page 177.
Courtesy of Creative Designs and Inventions, Inc.

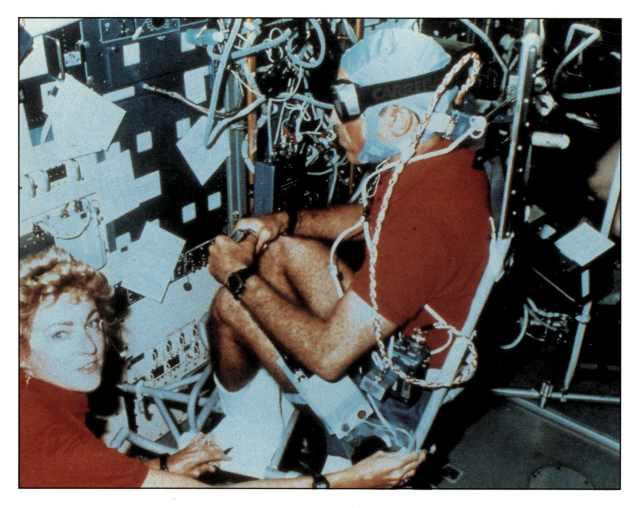

Life-function experiments aboard the *Columbia* space shuttle, shown on page 20.

Courtesy of NASA

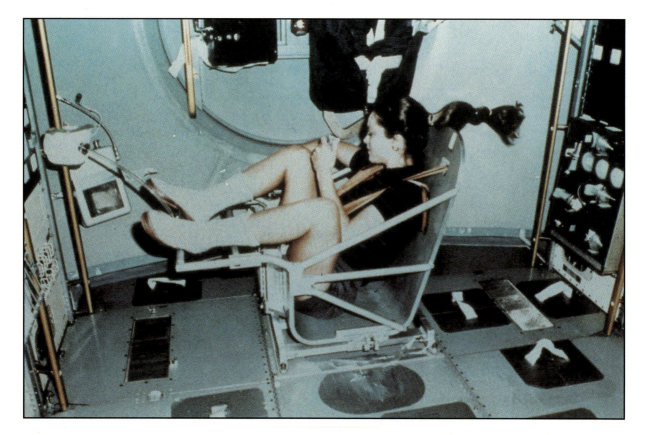

Protein crystals grown in zero-gravity conditions, free of imperfections. See page 35.

Courtesy of NASA

The Hubble space telescope in preparation for its launch. The optical replacements conducted from *Endeavour* in December 1993 will determine whether Hubble can accomplish all of its original mission or only about 70% of it, over the remainder of its 15-year lifetime.

Courtesy of NASA

Measuring plankton activity on a coral reef. See page 347.   *Courtesy of Biospherical Instruments, Inc.*

Electronic precision water-temperature control, described on page 209.   *Courtesy of Memry Corporation*

Ballerina who benefited from spinal-correction device shown on page 407.

Courtesy of Copes Foundation

Soil-less greenhouse from page 461.
Courtesy of NASA

This laser-based apparatus gives precise measurement of the rate of fluid flow by measuring the portion of the initial laser energy that is absorbed as the beam passes through the moving fluid stream.

This apparatus is used for physics mechanics experiments. Refer to laser description on p. 437.

Both courtesy of Dantec Measurement Technology

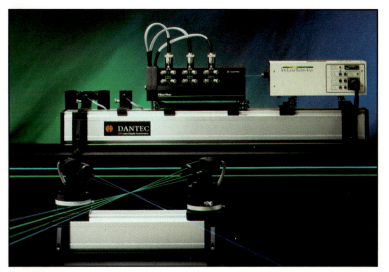

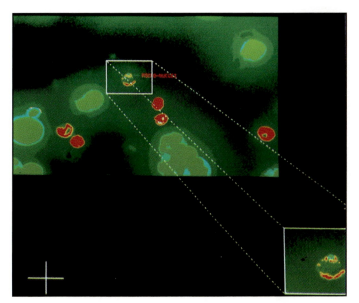

Atmospheric thermal images at multiple heights, from UARS. See page 251.

Courtesy of NASA

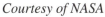

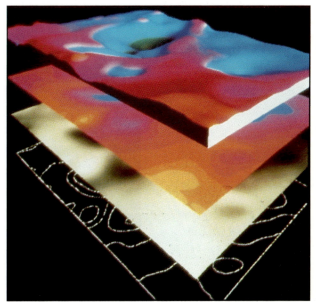

This video display is from the remote medical imaging system on page 247. The larger red objects are normal cells in a tissue sample. The boxed-in cell is abnormal, possibly malignant. The imaging system can magnify it for closer inspection, as seen in the lower right corner.

Courtesy of Medical Image Management Systems

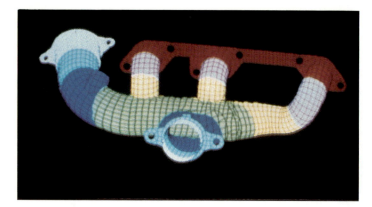

In the design of automobile parts, computer simulation is very useful for predicting a part's performance before a prototype is actually built and tested, or before a design change is implemented. This saves a great deal of time and expense. In this image, computer analysis with color-simulation is predicting temperature distribution in an exhaust manifold.

The computer's prediction of stress distribution in a drive-shaft universal-joint.

Both courtesy of Analytical Design Service Corporation

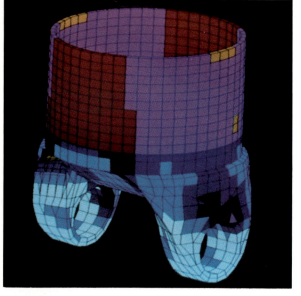

Below: Prediction of relative air turbulence at points on the surface of a moving car.

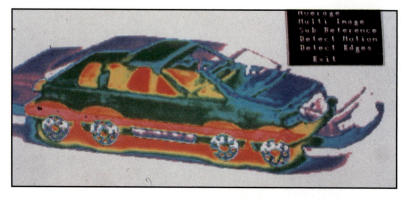

Courtesy of Bio-Imaging Research, Inc.

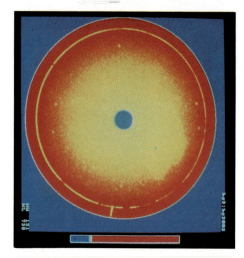

A single ACTIS "slice" of a circular metal part. At the bottom, just to the left of 6 o'clock, notice the radial yellow line between the outer edge and the circular channel (also yellow) that has been machined into the part. The radial line indicates a void in the metal, a manufacturing flaw. Refer to the photos on pages 83 and 118.

Infra-red stress scanning, described on page 40. Below, right: Stress pattern for the axle assembly that is under test. Each color correlates with a specific range of stress. For example, the black area near the center (and the very small dark purple area to its lower left) are the areas of greatest compression (squashing) stress — black means between 288 and 343 pounds per square inch; dark purple between 344 and 389. The light blue areas have the greatest tension (stretching) stress — between 100 and 114 pounds per square inch.

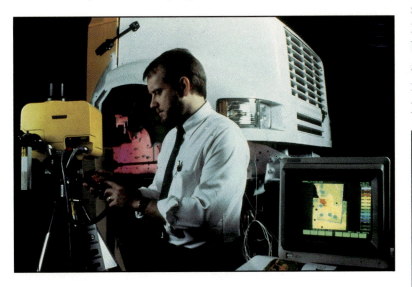

Courtesy of Navistar International Corp.

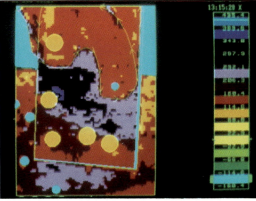

SATCOM communications satellite, under construction, also shown on page 232. It shares some features with the class of earth-orbiting satellites that gather meteorological data, or perform earth-science research measurements. Images from such satellites (named TIROS, NIMBUS, LANDSAT, and UARS) are shown on the next page.

Courtesy of GE Astro Space and NASA

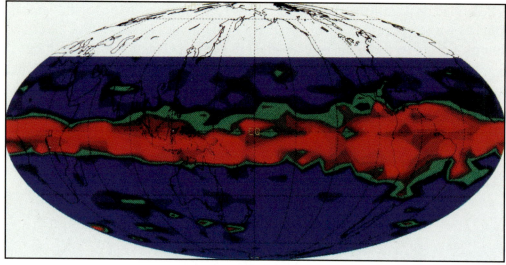

Sulfur dioxide concentrations five months after the 1991 eruption of Mount Pinatubo in the Phillipines, measured by UARS. See page 251.

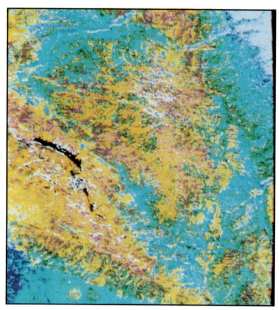

Various types of vegetation in New Mexico's Cibola National Forest, sensed by a LANDSAT satellite.

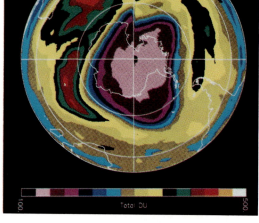

Ozone concentrations in the Southern hemisphere, measured by UARS.

Land vegetation and ocean productivity, detected by NIMBUS and TIROS research satellites.

All courtesy of NASA and GE Astro Space

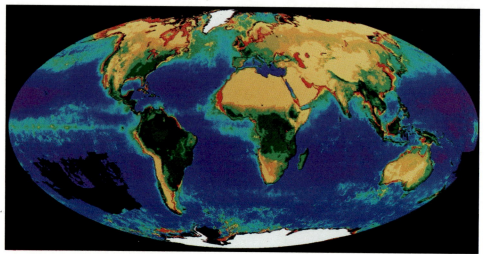

For Questions 25 through 39, refer to the Colpitts oscillator of Example 8-2, Fig. 8-10. The tank circuit values are given in Example 8-2. Suppose the other component values are: R_{B1} = 8.2 kΩ, R_{B2} =1.8 kΩ, R_C = 1.5 kΩ, R_E = 500-Ω pot, C_E = 20 μF, C_{out} = 0.1 μF and V_{CC} = 15 V.

25. If the oscillator stopped oscillating, what is the first thing you would do?
26. If your effort in Question 25 did not get the oscillator running, what would you check next?
27. Suppose that your checks in Question 26 revealed V_C = 4.3 V, V_B = 4.3 V, V_E = 3.6 V. What is the trouble here?
28. Suppose that your checks in Question 26 revealed V_C = 8.6 V, V_B = 2.8 V, V_E = 2.1 V. What can you conclude?
29. Proceeding from your conclusion in Question 28, what would you do next?
30. Suppose that your effort in Question 29 revealed that the amplifier has A_v = 3, and it does not change as the 500-Ω pot is adjusted. What is the trouble here?

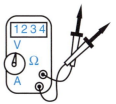

31. Suppose that your effort in Question 29 revealed that the amplifier has A_v = 0 (V_{out} = 0 V) for all pot adjustments. The bias voltage measurements are still the values given in Question 28. Give one possible explanation for this trouble.
32. Suppose that your effort in Question 29 reveals that A_v is adjustable over the range from A_v = 3 to A_v > 20. What can you conclude?
33. Proceeding from your conclusion in Question 32, what would you do next?
34. (T-F) To carry out your efforts in Question 33, a proper value of oscilloscope sweep speed would be 20 ms/cm.
35. What would be a proper value of scope sweep speed for your test in Question 33?
36. Suppose that your efforts in Question 33, at the sweep speed stated in Question 35, revealed that there were absolutely no oscillations present. What is your conclusion?
37. Based on your conclusion in Question 36, what would you do next?
38. Suppose that your efforts in Question 33 revealed that there were several cycles of oscillation at about 10 kHz, dying out. What can you conclude?
39. Based on your conclusion in Question 38, what would you do next?

CHAPTER 9

RADIO RECEIVERS

These technicians are preparing to perform a final antenna alignment on a communications satellite. This satellite, dubbed SATCOM C-1 by its GE Astro-Space builders, is designed for long-distance transmission of telephone and facsimile messages, radio and television programming, and business data. Its receiving (uplink) antenna circuitry is tuned to a radio frequency of 6 GHz. Its sending (downlink) antenna uses an RF of 4 GHz.

Operating on a 35-V solar-cell-powered dc supply, it is expected to remain operational for at least 12 years. It also appears on page 7 of the color section. Page 8 of that section shows several images of our planet and atmosphere, which were sent back by SATCOM's relatives in the GE family of satellites.

Courtesy of GE AstroSpace Division, General Electric Co.

OUTLINE

The Basic Radio System
  Radio Waves
  The Receiving Circuit
Receiver Sensitivity and Selectivity
Amplitude Modulation
  Receiving AM
AM Bandwidth

Music and Speech
  Microphones
The Heterodyne/Intermediate Frequency Idea
Automatic Gain Control
Frequency Modulation
  Producing an FM Waveform
FM Receivers
Receiver Troubleshooting

NEW TERMS TO WATCH FOR

electric field	upper sideband frequency	local oscillator
electromagnetic waves	lower sideband frequency	image frequency
modulation	frequency spectrum	automatic gain control (AGC)
demodulation	frequency domain	frequency modulation (FM)
sensitive	time domain	resting frequency
selective	spectrum analyzer	frequency deviation
tune	fidelity	varicap diode
radio-frequency (RF) amplifier	stagger tuning	guard bands
carrier frequency	microphone	limiter
audio amplifier	heterodyne receiver	double-tuned discriminator
amplifier alignment	frequency converter	automatic frequency control (AFC)
amplitude modulation (AM)	intermediate frequency (IF)	

Radio communication is one of the most important technical developments of all time. In this chapter we will study the principles of radio wave transmission and the operation of radio receivers.

Every radio communication system has two parts: A transmitter and a receiver. We will concentrate on the receiver now, because the receiver contains low-power electronic amplifiers that operate in the linear mode – never cut off. This is the amplifying mode that we have covered in Chapters 5, 6 and 7. The transmitter usually contains high-power electronic amplifiers that operate in a nonlinear mode – cut off part of the time. That mode of amplification will be the topic of Chapter 10.

After studying this chapter, you should be able to: ✔

1. Describe the characteristics of a radio wave.
2. Draw the schematic diagram of a basic radio receiver circuit, and explain how it is able to receive key switch-coded messages.
3. Explain what it means to modulate a radio wave.
4. Explain what it means to demodulate a radio wave.
5. Give a general definition of a radio receiver's sensitivity and its selectivity.
6. Draw the block diagram of a tuned radio frequency receiver, including its filtering circuits. Explain why this receiver has better sensitivity and selectivity than a basic radio receiver.
7. Draw a voltage-versus-time waveform of an amplitude-modulated radio signal. Tell how the modulating frequency and the modulation percentage affect the AM waveform.
8. Given the radio carrier frequency and the audio modulating frequency, draw the frequency spectrum graph of an AM radio signal.
9. From a frequency spectrum graph, give the values of the upper sideband, the lower sideband, and the bandwidth.
10. Given the carrier frequency and the modulating frequency spectrum, make a detailed sketch of the frequency-response curve of the filters and RF amplifiers in a tuned radio frequency (TRF) receiver.
11. Make a drawing showing the relationship among the frequency spectrums of the commercial AM radio stations.

12. Explain how we prevent interference between AM radio stations.
13. Explain the function of microphones.
14. Describe the operation of a dynamic microphone; do the same for a crystal microphone.
15. Explain the basic practical problems with the TRF receiver.
16. Explain the basic practical advantage of a superheterodyne/IF radio receiver, compared to a TRF receiver.
17. Describe the mixing, or heterodyning, process.
18. Given the frequencies that are being mixed, tell the four frequencies that appear in the spectrum of the mixer output.
19. Identify which one of the four mixer output frequencies becomes the intermediate frequency, IF.
20. State the advantages of automatic gain control (AGC).
21. Draw a time-domain waveform of a frequency-modulated carrier. Show the effects of modulating magnitude and modulating frequency.
22. Explain the use of a varicap diode in producing an FM waveform for transmission.
23. Draw the frequency spectrum of an FM waveform. Show the effects of modulating magnitude and modulating frequency.
24. Describe the frequency allocation for commercial FM radio broadcasting, including carrier frequencies, signal bandwidths, and guard bands.
25. Draw the block diagram of a superhet FM receiver, and state its frequency specifications.
26. Explain the operation of a double-tuned discriminator, the simplest FM demodulator.
27. Use an oscilloscope to systematically troubleshoot a radio receiver.

9-1 THE BASIC RADIO SYSTEM

A simple-as-possible radio system is shown in Fig. 9-1. A high-frequency oscillator is connected through a switch to an antenna, which is simply two pieces of very long wire suspended horizontally at some height above the earth. For convenience, we will show short metal rods rather than long pieces of wire.

This particular antenna construction is called a dipole antenna.

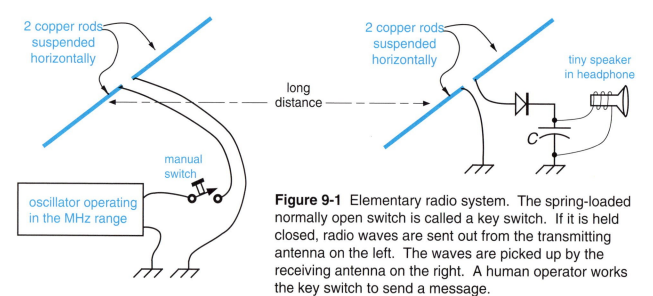

Figure 9-1 Elementary radio system. The spring-loaded normally open switch is called a key switch. If it is held closed, radio waves are sent out from the transmitting antenna on the left. The waves are picked up by the receiving antenna on the right. A human operator works the key switch to send a message.

Although we cannot see a complete current flow-path from one antenna rod to the other, nevertheless there is ac current in the rods when the key switch is closed, as Fig. 9-2(A) shows. This current exists because the high-frequency oscillations cause energy (power, on a time basis) to be radiated out into the surrounding space. It is a fundamental principle that $P = V \times I$. Therefore, in order for high-frequency power P to be radiated, as it must be, high-frequency current I must flow.

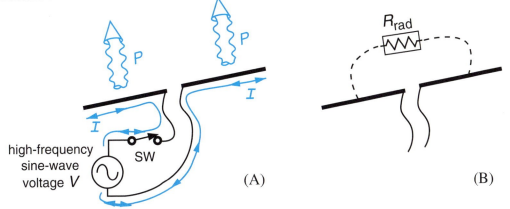

(A)   (B)

Figure 9-2 (A) At high frequencies (especially), ac current flows even though there is no visible complete circuit. (B) Think of the circuit being completed by an imaginary load, radiation resistance R_{rad}, connected between the rods. That load accounts for the consumption of power P, and its dissipation out into the surrounding space.

EXAMPLE 9-1

One popular method for sizing a dipole antenna produces a fictitious radiation resistance $R_{rad} = 72 \ \Omega$. If such an antenna is driven by a 30-V rms high-frequency oscillator, find:
  a) The ac current I in the antenna rods.
  b) The power P that is radiated into the surrounding air.

SOLUTION

a) From Ohm's law,

$$I = \frac{V_{osc}}{R_{rad}} = \frac{30 \ \text{V}}{72 \ \Omega} = \textbf{0.417 A rms}$$

b) $P = VI$
$$= 30 \ \text{V} \times 0.417 \ \text{A} = \textbf{12.5 W}$$

Radio Waves

In an antenna circuit, the resistive load R_{rad} causes current to be in-phase with voltage. At the positive peak instant, the antenna-rod current I creates a peak-strength magnetic field B, that is circular around the rods. This is shown in Fig. 9-3(A). Observe that the B flux lines lie in vertical planes, perpendicular to the rods.

Review your basic electricity text to refresh your understanding of magnetism.

Voltage is also instantaneously maximum in Fig. 9-3. Here is a fact of physics that we haven't yet talked about.

The existence of voltage between two objects in space causes an **electric field** to be created. The electric field is represented by electric flux lines, given the symbol E, that start at the positive object, curve through space, and end on the negative object.

Get This

The electric field for a dipole antenna is three-dimensional, shaped like a football. None of the *E* flux lines are in planes perpendicular to the rods. Figure 9-3(B) shows only the electric flux lines that lie in the horizontal plane.

As the oscillation cycle continues beyond the positive-peak instant shown in Fig. 9-3, both the *B*-field and the *E*-field will decrease in strength in sine-wave fashion. One quarter cycle later, *B* and *E* will be instantaneously zero.

Another quarter cycle later than that, the *i* and *v* sine waves will be at their negative peaks. The *B*- and *E*-fields will be at their maximum strengths, but in the opposite directions from Fig. 9-3. That is, *B* will become clockwise and *E* will become right-to-left.

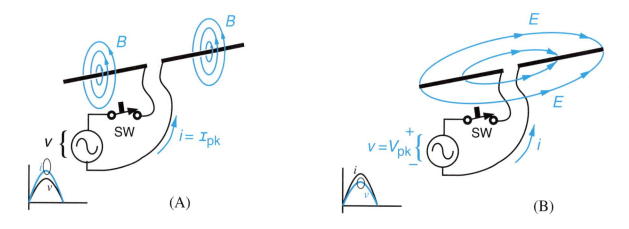

(A)　　　　　　　　　　　　　　　　　　　(B)

Figure 9-3 The positive peak instant. (A) The current I_{pk} produces a peak-strength magnetic field that is circular around the rods, counterclockwise when viewed from the left end. (B) At that same instant, with $v = V_{pk}$, the electric field is at its peak strength, curving through the air from the left rod to the right rod.

The high-frequency oscillations of the electric *E*-field and the magnetic *B*-field near the antenna rods cause this result:

Get This

Electromagnetic waves are radiated into the surrounding space. The waves consist of a sine wave of magnetic flux and a sine wave of electric flux, oriented at a 90-degree angle.

If it helps your understanding, you can picture the breaking-away mechanism as based on the fields' inability to return their energy to the source oscillator quickly enough. With the vary fast sine-wave source having left the fields lagging behind, hanging in the air, they have no alternative but to break loose and fly away.

These electromagnetic waves are created as the *B*- and *E*-fields "break away" from the antenna rods.

Figure 9-4 shows the "main" electromagnetic wave, the one that travels on a flat, straight, horizontal trajectory over the surface of the earth. In that figure, the *B* sine wave is vertical and the *E* sine wave is horizontal. Thus, *B* and *E* are perpendicular to each other, and both of them are perpendicular to the wave's direction of travel — namely the straight horizontal trajectory over the earth's surface.

Look carefully at Fig. 9-4 to see that the electric field *E* and the magnetic field *B* in the waveform are in phase with each other. At an instant in time, a particular location on the ray that is experiencing peak *E* field will also be experiencing peak *B* field. This is to be expected, since the original E and B fields around the antenna were in-phase with each other, in Fig. 9-3.

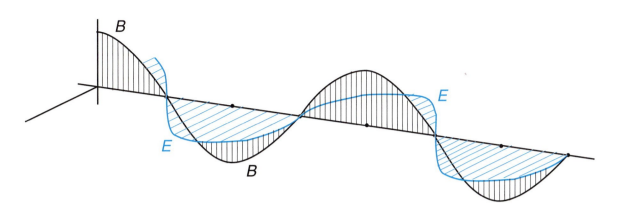

Figure 9-4 An electromagnetic wave. This is a "snapshot" view of the wave, showing its sine variations versus distance, at a fixed instant in time. The electric field (*E*) and the magnetic field (*B*) are in phase with each other, but they are physically rotated by one-quarter turn, or 90°.

There are several points that should be understood about such electromagnetic waves.

1   They are radiated in all directions. This is not revealed in Fig. 9-4, which focuses on just one particular wave, in order to make clear the mechanical orientation of *E*, *B*, and travel direction. Figure 9-5 shows that the waves have the same intensity everywhere within a vertical plane that is perpendicular to the antenna rods.

2   Figure 9-4 shows the orientation of the two fields for horizontally placed antenna rods. However, if the antenna rods had been placed vertically, which is possible, then the *E*- and *B*-fields would exchange position. *E* would then be vertical and *B* would be horizontal.

3   Electromagnetic waves do not need air or any physical medium in order to exist. They are unlike sound waves in this respect. In fact, electromagnetic waves are slowed down by air or water. They move (propagate) fastest in complete emptiness, like that in outer space.

4   These waves move at the speed of light. This is because they are essentially the same thing as light. Light is just a specific range of electromagnetic waves, in the frequency range from about 4×10^{14} Hz to 7.5×10^{14} Hz. Their speed in an empty vacuum is 3×10^8 meters/sec, or 186 450 miles/sec. They are only slightly slower in air. Thus, they can move between any two points on the earth in a small fraction of a second.

Waves are also emitted in planes that are not perpendicular to the antenna rods. But the wave intensity in those other planes is less than in the plane of Fig. 9-5.

① ✔

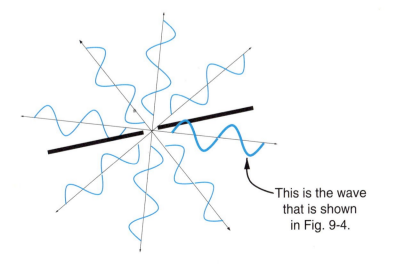

This is the wave that is shown in Fig. 9-4.

Figure 9-5 Electromagnetic waves are radiated with the same intensity in every direction, within a plane that is perpendicular to the antenna rods.

Radio Frequencies: As a general rule, the higher the oscillation frequency, the stronger the electromagnetic waves tend to be, other things being equal. Also, the structure of the antenna has a great effect on the strength of the waves. Real antennas are often much more elaborate than the simple dipole type shown in Figs. 9-1, 9-2, and 9-3.

By using the proper type and the proper physical dimensions for the antenna, radio communication can be accomplished by electromagnetic waves as low as about 10 Hz or as high as about 300×10^9 Hz (300 GHz). Most radio communication is done at frequencies ranging from 20 kHz to 300 GHz. So we can consider **radio waves** to be electromagnetic waves in this range of frequencies.

The Receiving Circuit

A few of the radio waves sent out by the transmitting antenna will strike the receiving antenna in Fig. 9-1. When these waves strike, a small sine wave voltage is induced in the receiving antenna, as shown in Fig. 9-6(A). The E field and the B field both contribute equally to inducing this voltage, which has the same frequency as the oscillator in Fig. 9-1.

The *magnitude* of the induced ac voltage depends on the intensity of the radio waves. The transmitter has some means of varying the wave intensity. At this time, we are not concerning ourselves with how the transmitter accomplishes this.

The induced antenna voltage is rectified by the ideal diode in Fig. 9-6(A), so capacitor C charges. While the incoming radio waves have a greater intensity, capacitor voltage v_C will be larger. While the incoming radio waves have a lesser intensity, v_C will be smaller. The varying v_C is applied to the tiny speaker, making it vibrate and producing audible sound.

Figure 9-6

(A) When a receiving antenna is struck by radio waves, it produces a radio-frequency voltage. This voltage can be rectified and capacitor-filtered to produce a usable output.

(B) When no waves are present, the receiver circuit returns to its de-energized condition.

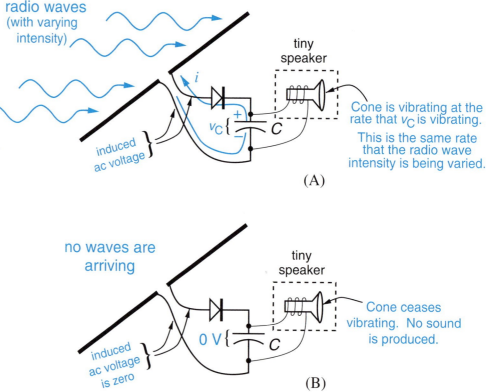

Sound is produced only during a period when radio waves are striking the receiving antenna, which is when the key switch is held closed in Fig. 9-1.

When the key switch is released, the transmitting antenna stops emitting radio waves. Therefore the receiving antenna stops inducing ac voltage, as shown in Fig. 9-6(B). The capacitor quickly discharges through the speaker winding, making $v_C = 0$. With zero voltage applied to the speaker, it stops vibrating, becoming silent.

If the key switch operator uses her finger to rapidly close and open the key switch, there will be a short period of sound followed by a short period of silence, followed by another short period of sound, and so on. Let us call each period of sound a "beep". The operator can make short beeps by holding the switch closed for only a very short time, say 1/4 second. Or, she can send longer beeps by holding the key switch closed for a longer time, say 3/4 second. The very short beeps are called "dots" and the longer beeps are called "dashes". Each letter of the alphabet is coded by a particular combination of dots and dashes.

For example, the letter **A** is represented by a single dot followed by a single dash. This scheme, called the Morse code, is shown in Table 9-1.

A	• —	J	• — — —	S	• • •	1	• — — — —
B	— • • •	K	— • —	T	—	2	• • — — —
C	— • — •	L	• — • •	U	• • —	3	• • • — —
D	— • •	M	— —	V	• • • —	4	• • • • —
E	•	N	— •	W	• — —	5	• • • • •
F	• • — •	O	— — —	X	— • • —	6	— • • • •
G	— — •	P	• — — •	Y	— • — —	7	— — • • •
H	• • • •	Q	— — • —	Z	— — • •	8	— — — • •
I	• •	R	• — •	0	— — — — —	9	— — — — •

Table 9-1 Morse Code for letters and numerals. Each dot (•) is a quick press of the key switch. Each dash (—) is a longer-held press.

The essential idea here is this: In order to send a message, the radio waves must be changed in some manner. The changing process is called **modulation**. Turning the wave on and off with a switch is the simplest form of modulation.

When the radio receiver circuit extracts the message from the transmitted radio waves, it is said to **demodulate** the radio signal. Demodulation is the process of recovering the original pattern of modulation.

Summarizing what we have learned about radio communication, we can represent a radio system in block diagram form as shown in Fig. 9-7.

A continuous uninterrupted radio wave carries no message.

Don't consider the transmitter's variation of the wave intensity to be a *change* in the sense that we mean here. That variation is not under our control, let's assume. If we can't control it, then we can't make it work for us to send a message. Therefore it cannot be regarded as modulation, in the sense that we mean here.

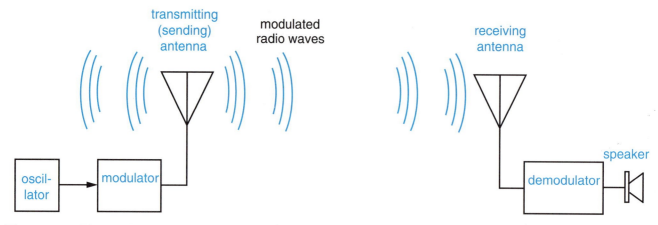

Figure 9-7 Block diagram showing the essential functional blocks of a radio system. This diagram uses the schematic symbol for a general antenna, not necessarily a simple dipole antenna.

SELF-CHECK FOR SECTION 9-1

1. A radio system uses two _____ , one for sending and one for receiving.
2. Radio transmission is done by _____ ; these waves contain a sine-varying _____ field and a sine-varying _____ field.
3. If a dipole transmitting antenna is mounted horizontally, its radio waves that are propagated horizontally contain E fields that are in the _____ plane and B fields that are in the _____ plane.
4. (T-F) If the transmitting antenna is mounted horizontally, then the receiving antenna should also be mounted horizontally. And the same for vertical mounting.
5. (T-F) Electromagnetic waves use air molecules to move from place to place.
6. Speaking approximately, radio waves are electromagnetic waves that have a frequency in the range from _____ Hz to _____ Hz.
7. The process of putting a message pattern onto a radio wave is called _____ .
8. The process of getting a message pattern out of a radio wave is called _____ .

9-2 RECEIVER SENSITIVITY AND SELECTIVITY

The system of Fig. 9-1 is useful for understanding the basic ideas of radio, but it is not practical for two reasons.

1   *Sensitivity.* The receiving antenna voltage is probably too small to even forward-bias the diode, let alone make an audible sound from the speaker. The receiver of Fig. 9-1 is not **sensitive** eno;ugh.

2   *Selectivity.* The receiver will respond equally to two or more transmitters operating at different radio frequencies. For example, if three transmitters are in the vicinity, as shown in Fig. 9-8, the receiver will respond to all three radio signals. This causes the three messages to be garbled together. No single message is understandable. The receiver of Fig. 9-1 is not **selective**.

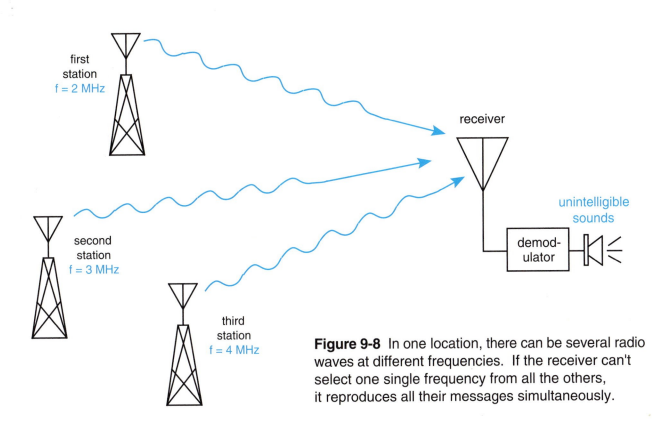

Figure 9-8 In one location, there can be several radio waves at different frequencies. If the receiver can't select one single frequency from all the others, it reproduces all their messages simultaneously.

To solve the two problems of sensitivity and selectivity, a radio receiver can be improved as shown in Fig. 9-9. The LC_1 circuit is a series-resonant band-pass filter. It will select one particular radio frequency, namely its resonant frequency, to be passed to R and to the amplifier. If we wish to change the frequency that is selected, we can **tune** the receiver by adjusting C_1 to give a new resonant frequency.

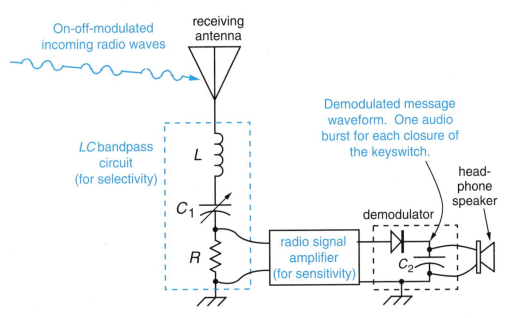

Figure 9-9
Improved radio receiver. The *LC* band-pass filter passes its resonant frequency to the amplifier input. It rejects the other signal frequencies that are picked up by the antenna. The selected frequency has its signal voltage and power amplified in the RF amplifier.

The weak antenna voltage is then boosted to a larger value by the radio signal amplifier. Such amplifiers are called **radio-frequency amplifiers**, or **RF** amplifiers, to distinguish them from amplifiers that are designed to work at lower frequencies. The RF amplifier output drives the diode-capacitor demodulator circuit and the load (the speaker).

The series-resonant *LC* band-pass circuit of Fig. 9-9 is simplest to understand. However, most radio receiving antennas are tuned by the method shown in Fig. 9-10, or by some variation on this method. Figure 9-10 shows three radio signals being picked up by the receiving antenna. The first station, operating at a **carrier** frequency (oscillator frequency) of 2 MHz, is sending the message dot–dot–dot, the code for the letter *S*. The second station, operating at a 3-MHz carrier frequency, is sending the message dot–dot–dash, the code for the letter *U*. The third station, with a 4-MHz carrier, is sending the message dash–dash, code for the letter *M*. The person using the receiver has adjusted C_1 to 281 pF, tuning it to 3 MHz. The *LC* tuning filter rejects the 2-MHz and 4-MHz antenna signals. Those signals are dropped across antenna resistance *R*. Only the 3-MHz signal is passed to the RF amplifier. Therefore the RF amplifier output signal is just a larger version of the second station's 3-MHz radio signal. The D–C_2 demodulator changes this large-magnitude RF waveform to an adequate-magnitude **audio** waveform. The audio signal is applied to the speaker, giving the sound signal dot–dot–dash, which the person recognizes as the letter *U*.

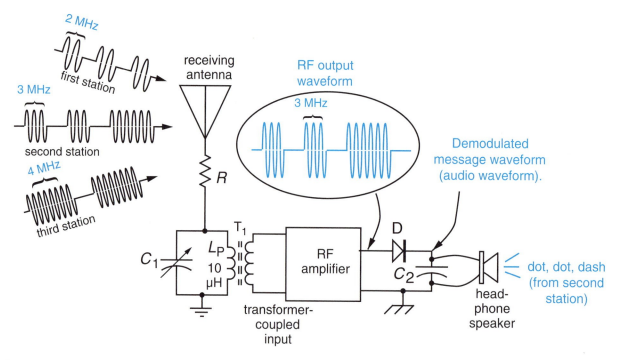

Figure 9-10 **More common tuning circuit. With C_1 adjusted to 281 pF, the $L_P C_1$ filter is tuned to a resonant frequency of 3 MHz. Therefore the audio output is a reproduction of the key switch pattern at the second station (the station with the 3-MHz oscillator).**

Improving the Receiver Further – Better Sensitivity and Selectivity

The radio receiver of Fig. 9-10 contains a single frequency-tuning circuit and a single amplifier stage. By introducing several frequency-tuning circuits and more amplifying stages, the radio receiver will perform better yet. This is shown in the block diagram of Fig. 9-11.

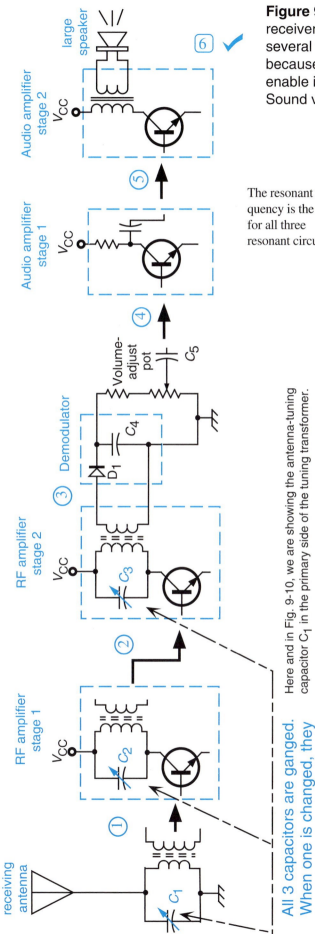

Figure 9-11 Block diagram of a *tuned radio frequency* (TRF) receiver. Such a receiver has good selectivity because it has several tuned circuits, not just one. It has good sensitivity because of its large RF voltage gain. Its audio amplifiers enable it to drive a real loudspeaker, not just a headset. Sound volume is adjusted by the control pot at point ④.

In that figure, there are two RF amplification stages. Each transistor stage uses tuned transformer-coupled output. The amplifiers' tuning capacitors, C_2 and C_3, are ganged with the antenna tuning capacitor, C_1. Therefore all three resonant circuits change together. That way, if a small amount of an unwanted signal frequency does manage to pass through the antenna tuner into the first RF stage, at point ① in Fig. 9-11, it does not get amplified. The same idea repeats at point ② —any small amount of unwanted signal frequencies that are still alive do not get amplified in RF stage 2.

By point ③ the tuned-frequency signal is tremendously larger than the untuned frequencies. The receiver has been made very selective.

In a high-quality radio receiver, the tuned signal voltage at point ③ is at least 1000 times larger than the strongest untuned signal.

At the same time, the receiver's sensitivity has been greatly improved. Even a very tiny signal voltage picked up by the antenna is amplified large enough to drive the $D_1 - C_4$ demodulator at point ③.

A high-quality radio receiver can operate on an antenna voltage as low as 2 µV.

In the receiver diagrams shown through Fig. 9-10, the demodulator was connected directly to a tiny headset speaker. In Fig. 9-11 the demodulator output is voltage-divided by the volume-control pot, then fed to an **audio amplifier** at point ④. The two audio amplifier stages boost this signal to operate a larger, stand-alone speaker.

The two audio amplifier stages are designed to operate at much lower frequencies than RF. In our key switch-modulated system, the audio beep-producing frequency is the rate at which the transmiter varies the wave intensity — perhaps a few hundred hertz, let us say. In a real home or car receiver, the audio frequencies are all the hearable frequencies that make up music and speech – about 20 Hz to 15 kHz.

A schematic diagram of a TRF radio receiver is shown in Fig. 9-12. The ⌐⟍ symbol in that diagram has the following meaning:

The ⌐⟍ symbol indicates that the device (capacitor, inductor, or transformer) is adjustable, but that the adjustment is done during manufacture and is then sealed in position. It should not be changed unless a major repair is made, and then only with the use of specialized instruments.

Get This

Transformers T_2 and T_3 have factory-adjustable cores. The core positions are adjusted to set the primary inductance at the proper value for the desired tuning range. The core adjustments also affect the degree of magnetic flux coupling between the primary and secondary windings. This alters the transformer's effective turns ratio, which affects the resistance-matching that occurs between the output of one transistor amplifier and the input of the next. The process of adjusting transformers T_2 and T_3, and capacitor C_2, is called **amplifier alignment**. When the amplifiers are RF amplifiers, as they are here, we call it RF alignment.

Typically, each RF amplifier stage in Fig. 9-12 would have a voltage gain of perhaps 200. The overall RF gain is then $A_{v(RF)} \approx 200 \cdot (200) = 40\,000$. Thus, an antenna signal voltage of $10\,\mu V$, say, is boosted to a demodulator voltage of

$$V_{demod} = (10\,\mu V)(40\,000) = 0.4\text{ V}$$

which is sufficient to forward-bias demodulator diode D_1 (germanium).

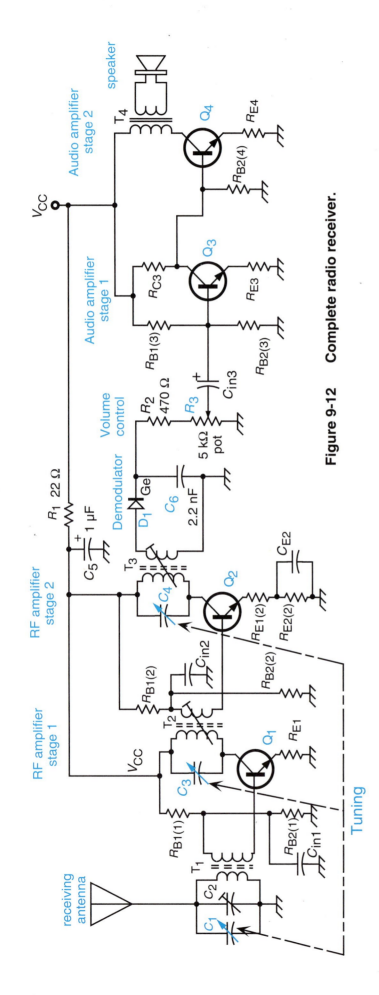

Figure 9-12   Complete radio receiver.

The $R_1 - C_5$ combination is a low-pass power supply filter. It prevents any audio-frequency surges that appear on the V_{CC} line in the audio section from being coupled into the RF section. The radio section V_{CC} line is kept free from noise.

In Fig. 9-12, D_1 is a germanium diode because it requires less forward voltage than a silicon diode.

SELF-CHECK FOR SECTION 9-2

9. Which feature enables a radio receiver to receive weak radio signals and reproduce their message? (Answer either selectivity or sensitivity.)
10. Which feature, selectivity or sensitivity, enables a radio receiver to reproduce just one message, rather than reproducing several different messages garbled together?
11. A radio receiver is given better _____ by including several LC-resonant circuits in its design.
12. A radio receiver is given better _____ by including several RF amplifier stages with high A_V.
13. A radio receiver is given the ability to produce loud, easily heard messages by including _____ amplifiers in its design.

9-3 AMPLITUDE MODULATION

The on-off modulation produced by a key switch is the simplest kind of modulation. But the messages that it sends are primitive – only dots and dashes. To send speech and music messages we need a more sophisticated modulation method.

Amplitude modulation (AM) continuously varies the amplitude, or strength, of the radio wave. The waveform graph of an amplitude-modulated radio signal is shown in Fig. 9-13. This signal must be applied to the transmitting antenna.

7 ✔

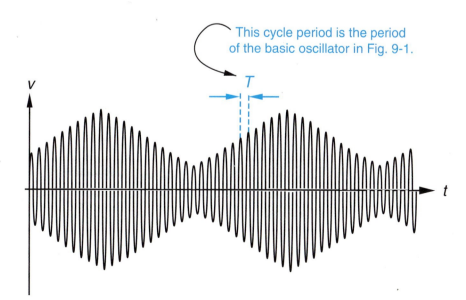

This cycle period is the period of the basic oscillator in Fig. 9-1.

Figure 9-13
Amplitude-modulated radio signal.

Producing AM

To produce an AM signal, we combine a low-frequency audio voltage with the output voltage from a high-frequency oscillator. This is suggested in Fig. 9-14, for a 1-kHz audio signal and a 3-MHz oscillator frequency. The low-frequency audio signal is called the **modulating** signal. It represents the message that is being sent. The high-frequency oscillator output is called the **carrier**.

**Figure 9-14
Using a
low-frequency
audio signal to
amplitude-modulate
a high-frequency
carrier.**

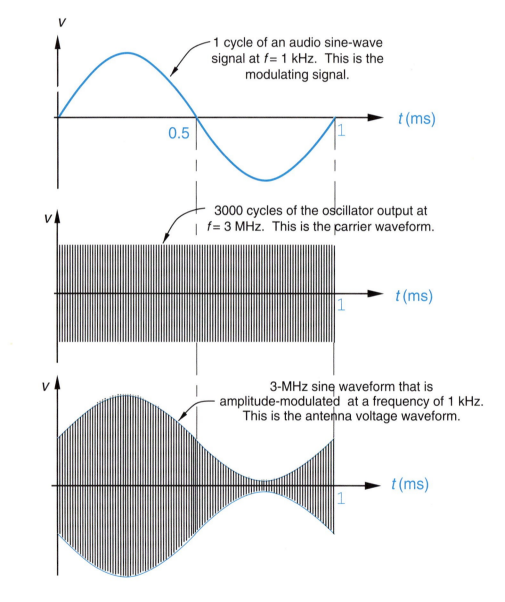

1 cycle of an audio sine-wave signal at f = 1 kHz. This is the modulating signal.

3000 cycles of the oscillator output at f = 3 MHz. This is the carrier waveform.

3-MHz sine waveform that is amplitude-modulated at a frequency of 1 kHz. This is the antenna voltage waveform.

The circuitry that can accomplish amplitude modulation is shown in Fig. 9-15. You can think of the circuit operation like this: Transistor Q does not have forward dc bias on its base-emitter junction. Therefore it is in cutoff condition during the oscillator's negative half cycle. This causes instantaneous collector voltage v_C to be clipped at (to be equal to) the value of voltage that is applied to the top of resistor R_C at that moment.

The oscillator's positive half cycle, on the other hand, is able to drive the transistor all the way to saturation. Therefore, at some point during the oscillator's positive half cycle, v_C will be clipped at 0 V.

For example, suppose the audio secondary voltage v_S in Fig. 9-15 has a peak value of 5 V. During the positive half cycle of v_S, which is defined as **+** on the winding's bottom terminal, v_S adds to V_{CC}, causing the voltage at the top of R_C to be greater than 15 V. Therefore the clipped v_C waveform has a peak-to-peak value that is greater than 15 V.

But during the negative half cycle of v_S, the voltage at the top of R_C is less than 15 V. Therefore the clipped v_C waveform has a peak-to-peak value that is less than 15 V. This circuit operation is pictured in the waveforms of Fig. 9-16. In Fig. 9-16, the audio modulating waveform's time scale has been tremendously exaggerated to demonstrate its effect.

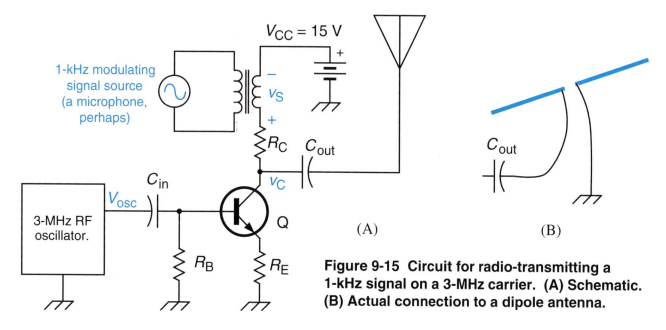

Figure 9-15 Circuit for radio-transmitting a 1-kHz signal on a 3-MHz carrier. (A) Schematic. (B) Actual connection to a dipole antenna.

Do not worry about the distortion of the sine wave in the output waveform. Its undistorted sine-wave shape can be restored by a 3-MHz filter circuit in the antenna feed line, before it gets transmitted. This filter has been left out of Fig. 9-15 for simplicity.

These medical personnel are using a centralized medical imaging system. The system acquires a microscope-based image from a blood- or tissue-sample, digitizes the image, and transmits the digital data by radio and telecommunication lines to a central consulting laboratory. There it is reproduced on video screen for inspection by specialists.

A close-up of a tissue-sample display is shown on page 5 of the color section.

Courtesy of Medial Image Management Systems and NASA

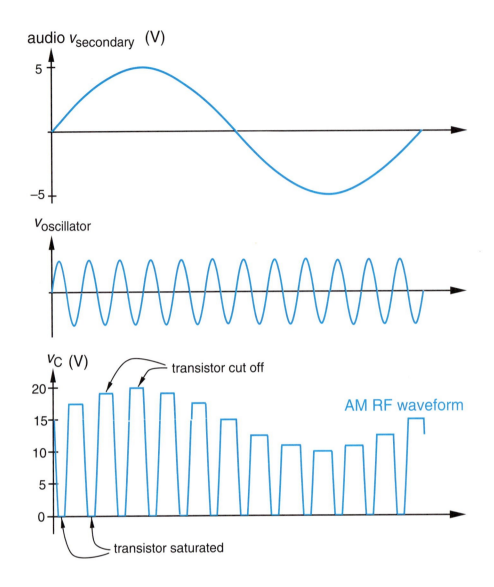

Figure 9-16

The audio waveform slowly (compared to the carrier) changes the effective power supply voltage in the circuit of Fig. 9-15. Therefore the radio frequency waveform of v_C slowly changes its height. It has been amplitude-modulated.

In this figure, we show 1 audio modulating cycle taking the same time as 12 RF carrier cycles. In reality, 1 audio cycle would take as much time as 3000 RF carrier cycles, assuming a 1-kHz audio frequency and 3-MHz carrier.

Receiving AM

In the general TRF receiver of Fig. 9-11, or the more specific receiver of Fig. 9-12, an amplitude-modulated RF waveform like Fig. 9-17(A) appears at the input of the demodulator. We are assuming that the signal has been modulated by a 1-kHz sine wave, as shown in Figs. 9-14 and 9-15. If capacitor C_6 were not present in Fig. 9-12, the waveform at the D_1 cathode would be simply rectified, like Fig. 9-17(B). If C_6 were very small, so that the C_6-R_2-R_3 combination had a short (fast) discharge time constant, comparable to the period of the RF carrier, the waveform would look like Fig. 9-17(C). It would duplicate the original 1-kHz audio modulating signal, but would contain some RF ripple. A properly designed demodulator has a discharge time constant τ that is quite long (slow) compared to the RF period, but short (fast) compared to the period of the audio modulating signal. Then we get the ripple-free audio signal of Fig. 9-17(D).

The dc component of Fig. 9-17(D) is blocked by input coupling capacitor C_{in3} in Fig. 9-12. A 1-kHz ac signal is amplified in the audio amplifiers and applied to the speaker. The sound from the speaker will be a plain 1-kHz audio tone. This duplicates the message from the signal source back in Fig. 9-15.

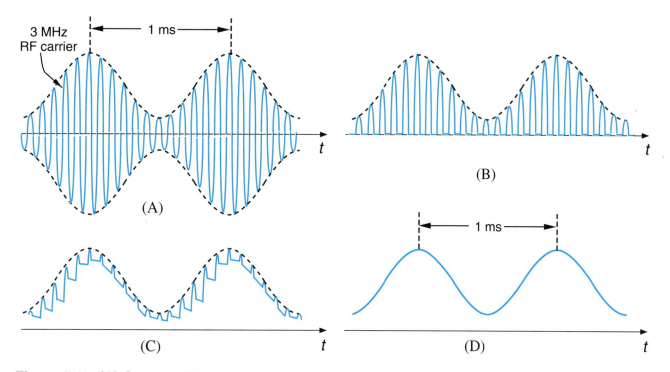

Figure 9-17 (A) Output of RF amplifier, input to demodulator. (B) Rectified only. (C) Demodulator filtering capacitor is too small. τ is too short. (D) Proper demodulator τ.

In the demodulator of Fig. 9-12, $\tau = (R_2 + R_3)C_6 = (5.47\ \text{k}\Omega)\ (2.2\ \text{nF}) = 12\ \mu s$. This τ value is much longer than the RF period, which is

$$\tau_{RF} = \frac{1}{3\ \text{MHz}} = 0.33\ \mu s$$

But the demodulator's τ is short compared to the shortest audio period, given by

$$\tau_{\text{audio (min)}} = \frac{1}{f_{\text{audio (max)}}} = \frac{1}{15\ \text{kHz}} = 67\ \mu s$$

Modulation Percentage: In the transmitter of Fig. 9-15, if the audio modulating signal voltage is small, the amount of amplitude modulation of the carrier will be small. If the audio signal voltage is large, the carrier's amplitude modulation will be proportionally larger. These two situations are compared in Fig. 9-18.

In that figure, the transmitter voltage waveform has a peak amplitude of 10 V when it is unmodulated. The 5-V modulating voltage in Fig. 9-18(A) causes the transmitted waveform to have a maximum change of 5 V. The modulation percentage is defined as

$$\text{modulation percentage} = \frac{\text{maximum voltage change}}{\text{unmodulated voltage}} \times 100\% \qquad \text{Eq. (9-1)}$$

Therefore, Fig. 9-18(A) has a modulation percentage of

$$\frac{5\ \text{V}}{10\ \text{V}} \times 100\% = 50\%$$

7 ✔

Figure 9-18 differs from the $V_{CC} = 15$ V situation shown in Figs. 9-15 and 9-16. In those figures, the transmitter would have peak amplitude = 15 V when un odulated.

In Fig. 9-18(B), we have

$$\frac{10\text{ V}}{10\text{ V}} \times 100\% = 100\%$$

In the radio receiver, a greater percentage of modulation creates a larger-magnitude message signal at the output of the demodulator. It produces a louder sound.

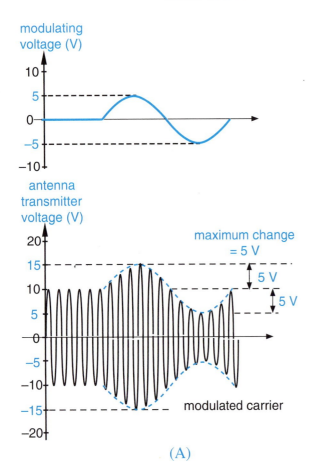

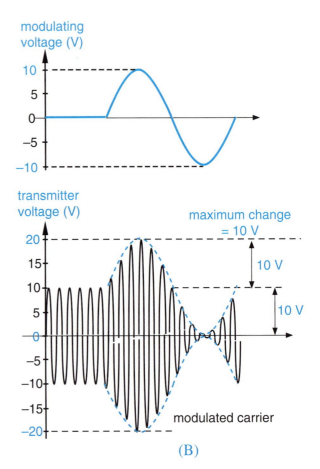

Figure 9-18   (A) 50% modulation.   (B) 100% modulation, the greatest possible. Larger modulation percentage represents a louder sound, all other things being equal.

SELF-CHECK FOR SECTION 9-3

14. (T-F) In amplitude modulation, the peak-to-peak voltage of one RF cycle is exactly the same as the peak-to-peak voltage of the neighboring RF cycle.

15. (T-F) During one cycle of the modulating signal, it is typical for several thousand cycles of the carrier to occur.

16. In an AM demodulator, what would go wrong if the filter capacitance were too small?

17. In an AM demodulator, what would go wrong if the filter capacitance were too large?

18. A certain transmitter has a 20-V-peak carrier, when it is not modulated. Suppose that it is modulated so that the transmitter antenna waveform has a maximum peak of 24 V. What is the modulation percentage?

19. In Question 18, what will be the minimum peak voltage of the transmitter antenna waveform?
20. Repeat Questions 18 and 19 for a situation where the antenna waveform has a maximum peak of 28 V.
21. All other things being equal, which will produce louder sound from the receiver, the situation in Question 18 or the situation in Question 20? Explain.

UPPER ATMOSPHERE RESEARCH

The changing condition of the earth's atmosphere, especially its ozone depletion, has become an issue of vital environmental and political concern. The U.S. National Aeronautic and Space Administration's Upper Atmosphere Research Satellite (UARS) was launched in September 1991 for the purpose of obtaining precise measurements of atmospheric conditions through at least January 1996. The top photo shows the 10-meter-long, 5-meter-diameter craft under construction.

It carries ten electronic instruments/antennas, each of which can be aimed with 1-degree accuracy. The electric power to operate the instruments comes from three 4.8-V nickel-cadmium batteries. These batteries are recharged by the 1.5-m by 20-m array of photovoltaic solar cells, shown deployed in the picture on the right, which is a view of UARS orbiting the earth at a height of 600 km.

Three atmospheric images radioed to us from UARS are shown in the color section, one on page 5 and two on page 8.

Courtesy of GE AstroSpace Division and NASA

9-4 AM BANDWIDTH

Am amplitude-modulated waveform like Figs. 9-14, 9-17(A), and 9-18(A) and (B) can be thought of in quite a different way than we have seen until now. Here is one of the major ideas of radio technology:

A steady-frequency amplitude-modulated wave is exactly equivalent to a series combination of three separate sine-wave sources. These sources have the following frequencies:
1. The carrier frequency, f_c. This is 3 MHz in our examples.
2. The audio modulating frequency *added to* the carrier frequency, $f_c + f_{audio}$. This is 3 001 000 Hz, or 3.001 MHz in our examples.
3. The audio modulating frequency *subtracted from* the carrier frequency, $f_c - f_{audio}$. This is 2 999 000 Hz, or 2.999 MHz in our examples.

The relative voltage magnitudes of the 3 sources depend on the modulation percentage. That is not an essential idea to us right now, so don't concern yourself with it.

This idea is illustrated in Fig. 9-19.

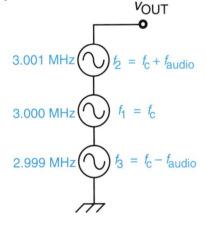

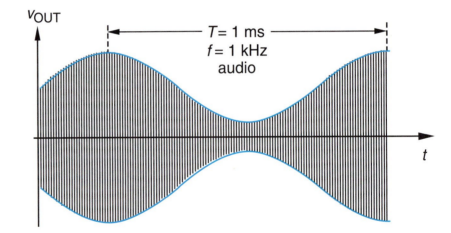

Figure 9-19   An AM wave is equivalent to three sine-wave voltages added together by a series connection.

The frequency that is higher than the carrier is called the **upper sideband frequency**, or just the **upper sideband**. The frequency that is lower than the carrier is called the **lower sideband frequency**, or lower sideband.

The frequencies that make up any AM waveform are referred to as the **frequency spectrum** of the waveform. A frequency spectrum is represented on a bar-graph like Fig. 9-20.

When we think about an AM waveform in this way — in terms of the component frequencies in its frequency spectrum — we say that we are thinking in the **frequency domain**, as opposed to the **time domain**.

voltage
magnitude

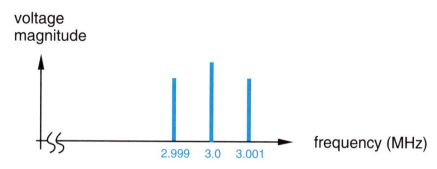

frequency (MHz)

**Figure 9-20
Graph of the frequency
spectrum of our AM
example. This kind of graph
is called a *spectrograph*.**

If you were looking at a real spectrograph, the voltage magnitudes might
be important, but we are not concerning ourselves with that now.

We were thinking in the time domain when we studied the
moment-by-moment production of the transmitter AM waveform in
Fig. 9-16. We were also thinking in the time domain when we looked
at the moment-by-moment demodulating process in Fig. 9-17. Thinking in the time domain comes naturally to us because we live our entire
lives in the time domain. But when we need to think about complex
electronic waveforms like an AM waveform, the time domain is not the
best way to think. That is why you must now accept the idea of thinking
in the frequency domain.

EXAMPLE 9-2

Suppose that our 3 MHz carrier is amplitude-modulated by a
5-kHz audio sine wave, rather than 1 kHz. Graph the spectrum
of the AM waveform.

SOLUTION

We can use the letters USB and LSB to stand for upper side band
and lower side band. Then the USB frequency is given by

$$f_{USB} = f_c + f_{audio}$$
$$= 3\text{ MHz} + 5\text{ kHz} = 3.005\text{ MHz}$$

The lower sideband is

$$f_{LSB} = f_c - f_{audio}$$
$$= 3\text{ MHz} - 5\text{ kHz} = 2.995\text{ MHz}$$

The spectrum is graphed in Fig. 9-21(A).

magnitude

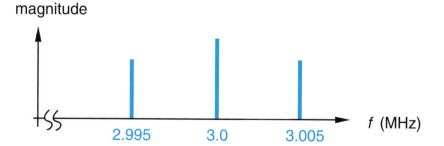

f (MHz)

**Figure 9-21
AM waveform spectrum for
5-kHz modulation frequency.
The sidebands are further
from the carrier than they
were in Fig. 9-20.**

EXAMPLE 9-3

Imagine that our 3 MHz carrier is modulated by an audio signal that is a combination of two equal-magnitude signals, one at 5 kHz and the other at 2 kHz. This modulation signal waveform would be difficult to draw, and so would the AM waveform (in the time domain).

Figure 9-22(A) shows the situation in the modulating circuit.

In the frequency domain, graph the spectrum of the AM waveform.

SOLUTION

With a modulating signal that is a mixture of two frequencies, each AM sideband will have two frequency components. In the USB we have

$$3.0 \text{ MHz} + 5 \text{ kHz} = 3.005 \text{ MHz}$$
$$\text{and} \quad 3.0 \text{ MHz} + 2 \text{ kHz} = 3.002 \text{ MHz}. \left.\right\} \text{USB components}$$

In the LSB we have

$$3.0 \text{ MHz} - 5 \text{ kHz} = 2.995 \text{ MHz}$$
$$\text{and} \quad 3.0 \text{ MHz} - 2 \text{ kHz} = 2.998 \text{ MHz}. \left.\right\} \text{LSB components}$$

The frequency spectrum is shown in Fig. 9-22(B).

Figure 9-22

(A) Modulating with a more complicated audio signal, containing two sine waves at different frequencies.

(B) Resulting frequency spectrum of the AM waveform.

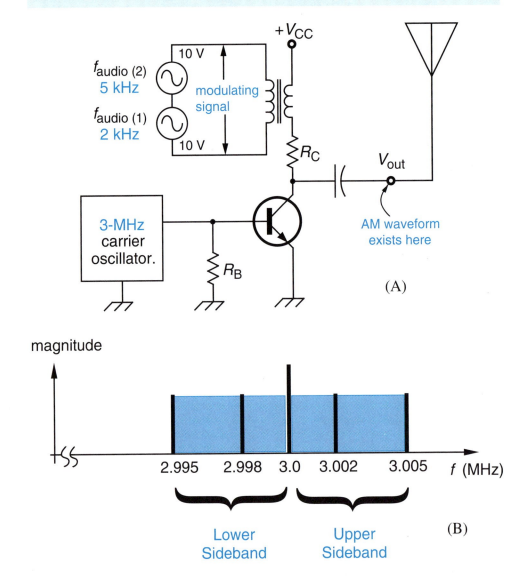

(A)

(B)

Figure 9-22(B) makes an important point. The sidebands are truly *bands* of frequencies, not necessarily just a single frequency. In Fig. 9-22(B) the upper sideband is the entire range of frequencies from 3.0 MHz to 3.005 MHz. The lower sideband is the entire range from 2.995 MHz to 3.0 MHz.

The test instrument that displays a spectrograph on its screen is called a **spectrum analyzer**. Figure 9-23 shows its appearance.

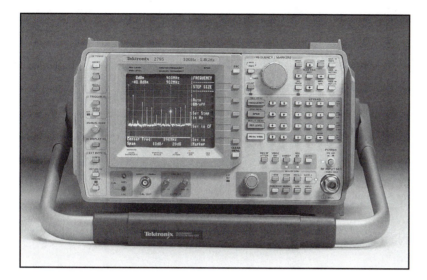

**Figure 9-23
Spectrum analyzer.**

Courtesy of Tektronix.

Bandwidth of the RF Amplifiers in the Radio Receiver

In the TRF radio receiver of Figs. 9-11 and 9-12, we have made it clear that the RF amplifier stages are tuned to a resonant frequency of 3 MHz, the carrier frequency. Now that we understand the frequency spectrum of the AM radio waveform, we can describe the bandwidth, or range of frequencies, that must be successfully amplified by those amplifiers.

> A radio receiver's RF amplifiers must have a bandwidth that covers the range from the lowest frequency of the lower sideband to the highest frequency of the upper sideband.

Get This

For example, to properly receive the AM radio signal described in Fig. 9-22, the 1st- and 2nd-stage RF amplifiers must be able to amplify the range of frequencies from 2.995 to 3.005 MHz. The difference between the high end of this range and the low end of this range is

$$3.005 - 2.995 \ = \ 0.010 \text{ MHz, or 10 kHz.}$$

Therefore the amplifiers' bandwidth (Bw) must be equal to or greater than 10 kHz. This idea is illustrated in Fig. 9-24 for a standard *LC*-resonant bell-shaped frequency response.

This specification of RF bandwidth assumes the double-sideband transmitting system, not single-sideband. Our entire discussion is of the double-sideband system.

Here is another way of specifying the required bandwidth for the RF amplifiers in an AM receiver:

> The RF amplifiers' required bandwidth is 2 times the highest frequency component of the modulating signal. As a formula,

Get This

$$Bw \ \geq \ 2 \left[f_{audio \, (max)} \right] \qquad \text{Eq. (9-2)}$$

**Figure 9-24
Frequency-response
curve of radio amplifier.**

A_v is maximum at the
amplifier's tuned fre-
quency. It must be at
least 70.7% of its
maximum value through
the range of frequencies
in the AM spectrum.

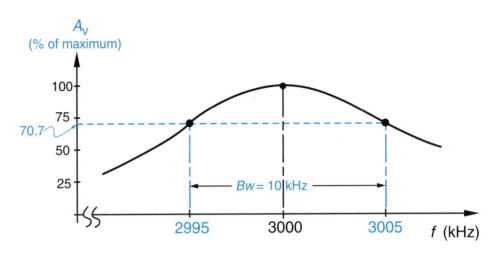

EXAMPLE 9-4

Suppose the modulating signal in Fig. 9-22(A) is made even more
complicated, by adding a third signal frequency at 7 kHz.
a) Draw a spectrograph of the AM waveform.
b) To successfully receive the radio signal, what is the minimum required
bandwidth of the receiver's RF amps?

SOLUTION

a) The frequency spectrum is drawn in Fig. 9-25(A), assuming that the
new frequency component is the same strength (same modulating voltage)
as the other two components.
b) $Bw \geq 2[f_{audio\ (max)}] = 2\ [7kHz] = \mathbf{14\ kHz}$
The amplifier's revised frequency-response curve is sketched in Fig. 9-25(B).

The bell-shaped frequency-response curves shown in Figs. 9-23 and
9-25(B) are not really desirable. Amplifiers with bell-shaped response curves
give poor signal **fidelity**, because they penalize component frequencies that are
further from the carrier. These far frequencies are under-represented in the
final audio waveform because they have not gotten as much amplification as
the center frequencies. In other words, higher audio frequencies (like 7 kHz)
are de-emphasized, compared to lower audio frequencies (like 2 kHz).

To solve this problem, radio designers try to give the RF amplifiers a more
squarish frequency response. This idea is shown in Fig. 9-26(A). One way to
improve the overall amplifier squareness is by **stagger tuning** the amplifier
stages. This is illustrated in Fig. 9-26(B).

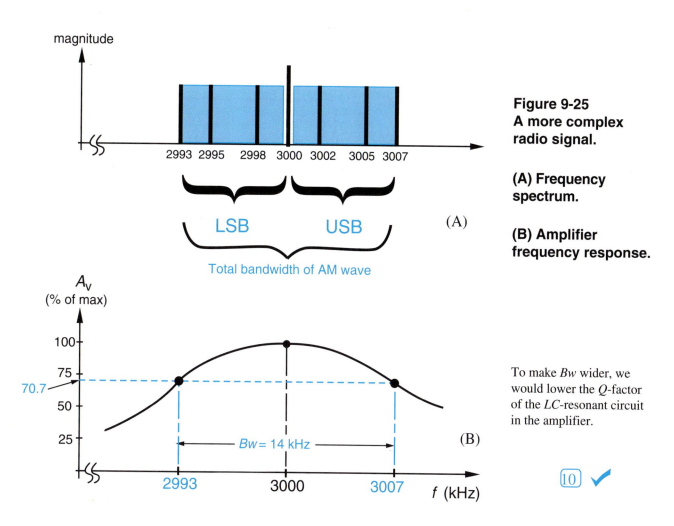

Figure 9-25
A more complex
radio signal.

(A) Frequency
spectrum.

(B) Amplifier
frequency response.

To make *Bw* wider, we
would lower the *Q*-factor
of the *LC*-resonant circuit
in the amplifier.

SELF-CHECK FOR SECTION 9-4

22. Suppose that a 1-MHz carrier wave is amplitude-modulated by a steady
    3-kHz sine wave. The lower sideband frequency is _____ Hz.
    The upper sideband frequency is _____ Hz.
23. Draw the frequency spectrum of the AM wave in Question 22.
24. A waveform graph of instantaneous AM voltage versus time is an
    example of thinking in the _____ domain.
25. A spectrograph like the one you drew for problem 23 is an example of
    thinking in the _____ domain.
26. Suppose a 1-MHz carrier wave is amplitude-modulated by an audio
    signal which is a mixture of several distinct frequencies, namely 500 Hz,
    2.5 kHz and 4 kHz. The bandwidth of the AM wave's frequency spec-
    trum is _____ Hz.
27. Suppose a tuned-frequency radio receiver is receiving the signal in
    Problem 26. To reproduce the original signal with good fidelity, its RF
    amplifiers must have a bandwidth of at least _____ Hz.
28. The trick of tuning the first RF amplifier stage slightly below the carrier
    frequency and the second RF stage slightly above the carrier frequency
    is called _____ _____ .

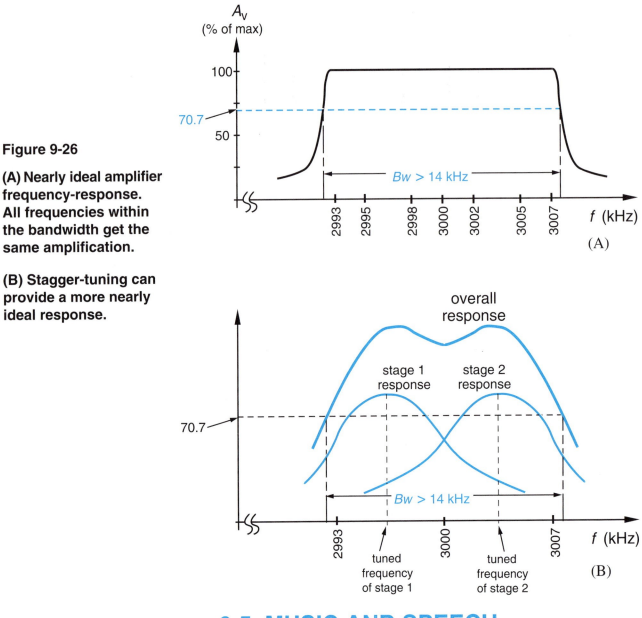

Figure 9-26

(A) Nearly ideal amplifier frequency-response. All frequencies within the bandwidth get the same amplification.

(B) Stagger-tuning can provide a more nearly ideal response.

9-5 MUSIC AND SPEECH

In normal AM radio operation, the message is speech or music. Then the modulating signal is a speech or music waveform. Such waveforms are complex, much more complex than the 2-frequency modulating signal suggested in Fig. 9-22(A), or even the signal of Fig. 9-25(A). A typical time-domain graph of a music waveform is shown in Fig. 9-27(A). In the frequency domain, its spectrum is shown in Fig. 9-27(B). Note that this spectrum is for the audio signal alone, not the AM radio signal.

A real audio signal can be visualized as many sine-wave voltage sources of different frequencies, adding together in a series connection. This is shown in Fig. 9-27(C). This is the same basic idea as shown in Fig. 9-19 for a simple single-frequency AM waveform.

There is some difference of opinion as to what constitutes the total audio range of frequencies. Those humans with the most acute hearing can reportedly

The graph in Fig. 9-27(B) is shown as a bar graph, but actually it would be continuous. Imagine the tops of the bars joined by a continuous curve.

The circuit of Fig. 9-27(C) shows a finite (countable) number of different frequencies. But actually there are an infinite number of frequencies, all blending into one another.

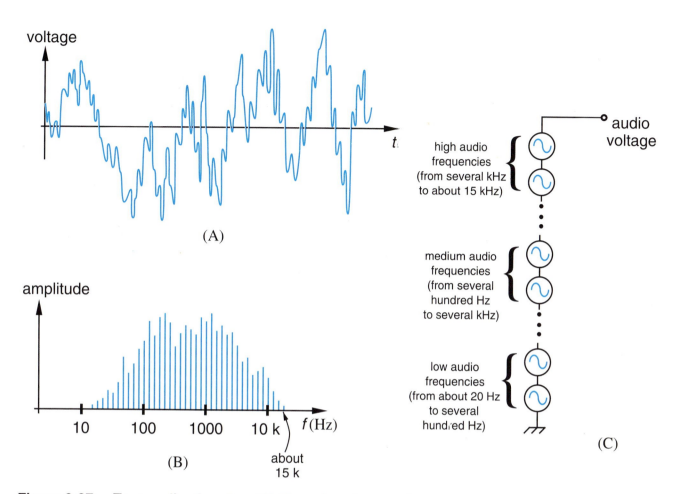

Figure 9-27   True audio signal.   (A) Time-domain waveform.   (B) Frequency spectrum. (C) Visualizing the signal as a mixture of sine waves of many frequencies.

detect sound waves with frequencies as high as 17 kHz. So some people consider 17 kHz to be the upper limit of the audio range.

However, for normal sounds of nature, including music, very little of the sound energy that enters our ears is above 5 kHz in frequency. Thus, by passing only those frequencies below 5 kHz (and ignoring the 5 kHz-to-17 kHz range), music can be reproduced with very good fidelity to the original sound from the musical instrument. To help you appreciate this, consider the following fact: The highest note (tone) on a piano has a frequency of about 4.1 kHz. When we go up to 5 kHz, we are even 20% higher than the frequency of that tone, which is itself subjectively quite high.

So, if we modulate with a real audio signal that is limited to a maximum component of 5 kHz, the AM waveform has sidebands that are 5 kHz wide. Thus, a 3-MHz AM radio waveform would have an USB from 3000 kHz to 3005 kHz. Its LSB is also 5 kHz wide, from 2995 kHz to 3000 kHz. Its overall bandwidth is therefore 10 kHz, as shown in Fig. 9-28.

North American commercial AM radio broadcast stations operate at carrier frequencies from 540 kHz to 1600 kHz (1.6 MHz). Their carrier frequencies are separated by exactly 10 kHz. Thus, the lowest-frequency AM stations (for example, KIEZ in Carmel, California) have a carrier frequency of 540 kHz. The next-lowest stations (for example, WSAU in Wausau, Wisconsin) use $f_{carrier} = 550$ kHz. And so on to $f_{carrier} = 1.6$ MHz (for example, KCRG in Cedar Rapids, Iowa).

At this writing, commercial AM radio is in the process of expanding its band up to 1700 kHz.

**Figure 9-28
Frequency spectrum
of a real AM radio
signal.**

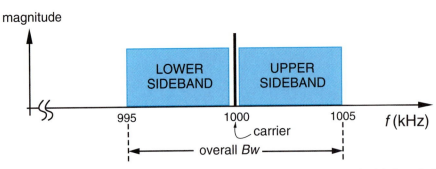

With their carrier frequencies only 10 kHz apart, and each sideband about 10 kHz wide, AM broadcasting stations have the potential to interfere with one another. This is because the upper sideband of one carrier just touches the lower sideband of the next-higher carrier. If a station's transmitter emits any radiation slightly outside its normal sideband limits, which is almost inevitable, that radiation frequency will be within the normal sideband of an adjacent-frequency station. This problem is illustrated in Fig. 9-29.

**Figure 9-29
Showing the side-
band near-overlap
of commercial AM
radio stations.**
The only reason the frequency spectrums are shown in different vertical positions is to separate them visibly. The different heights don't represent any real technical differences.

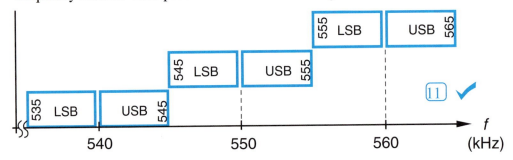

As an example, consider Carmel, California station KIEV at 540 kHz. Its USB reaches up to 545 kHz. If it accidentally emits any signal slightly higher than 545 kHz, it will find itself overlapping the LSB of station WSAU in Wausau, Wisconsin. This means that some of the radio waves broadcast by 540-kHz KIEV would be able to pass through the filters of a radio receiver that was actually trying to receive (was tuned to) the 550-kHz station, WSAU. Thus, the 540-kHz California station has the potential to garble the message of the 550-kHz Wisconsin station.

Of course, the thing that prevents such garbling is the great distance between California and Wisconsin. The U.S. Federal Communications Commission (FCC), in cooperation with Canadian and Mexican authorities, insists that stations operating in the same geographical area use widely separated carrier frequencies. This is pictured in Fig. 9-30.

Figure 9-30
Two stations that are geographically close together must use widely separated carrier frequencies, to prevent interference. But a far-away station doesn't need a widely separated carrier.

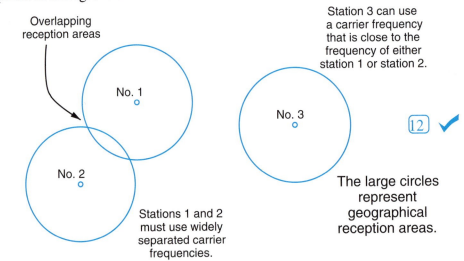

Microphones

To modulate an RF carrier with real music or speech, we need an actual voltage waveform like the one sketched in Fig. 9-27(A). This voltage waveform must be a duplicate of an original sound wave that was produced by musical instruments or by human vocalization.

The device that produces a voltage waveform that duplicates a sound waveform is a **microphone**. There are two basic types of microphones, the dynamic type and the crystal type.

A dynamic microphone is a dynamic speaker operated in reverse. As a sound wave strikes the microphone's flexible diaphragm in Fig. 9-31(A), the diaphragm vibrates in a pattern that duplicates the sound wave. These vibrations are mechanically transferred to the moving coil that is attached to the hollow rear of the diaphragm cone. The coil's rapid back-and-forth motion through the magnetic field causes it to induce voltage, according to Faraday's law. The voltage waveform is an exact duplicate of the original sound waveform.

A sound wave, physically, is a variation in instantaneous air pressure.

Refer to Fig. 9-1 to see the rough structure of a dynamic speaker.

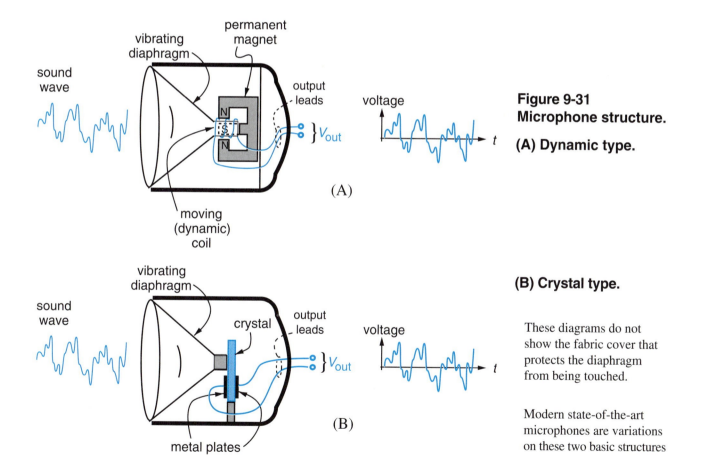

**Figure 9-31
Microphone structure.**

(A) Dynamic type.

(B) Crystal type.

These diagrams do not show the fabric cover that protects the diaphragm from being touched.

Modern state-of-the-art microphones are variations on these two basic structures

The crystal microphone shown in Fig. 9-31(B) has a cone-shaped diaphragm also. But the solid rear of the cone is in contact with a crystal. As the vibrating cone rapidly flexes and releases the crystal, the crystal's piezoelectric effect creates a voltage that is an exact duplicate of the original sound wave.

The microphone's small output voltage is amplified and can then be used as the modulating signal in the radio transmitter, as suggested in Figs. 9-15 and 9-22.

Of course, the microphone's output signal could also be recorded for later playback. The three basic recording methods are: 1) Mechanically, by the depth of a groove in a plastic record; 2) Magnetically, by the strength of the magnetic field on a tape; 3) Optically, by the digitally-coded sequence of pits and flats that are burned by laser light into the surface of a plastic disk.

SELF-CHECK FOR SECTION 9-5

29. A 5-kHz maximum-frequency music or speech AM radio waveform has a sideband width of _____ kHz.
30. A 5-kHz maximum-frequency music or speech AM radio waveform has an overall bandwidth of _____ kHz.
31. A commercial AM radio station operating at a carrier frequency of 580 kHz occupies the frequency band from _____ kHz to _____ kHz.
32. (T-F) The FCC would permit an AM broadcast station with $f_{carrier}$ = 1240 kHz to operate in the same city with another station with $f_{carrier}$ = 1250 kHz. Explain your answer.
33. (T-F) The FCC would permit an AM broadcast station with $f_{carrier}$ = 1240 kHz to operate in the same city with another station with $f_{carrier}$ = 1370 kHz. Explain.
34. A _____ produces a voltage waveform that is an exact duplicate of an original sound waveform.

9-6 ANOTHER MAJOR IMPROVEMENT IN RADIO RECEIVERS – THE HETERODYNE / INTERMEDIATE FREQUENCY IDEA

The tuned radio receivers of Figs. 9-11 and 9-12 have this disadvantage: It is nearly impossible to get all the individual RF stages and the antenna tuning circuit to "track" together. This means that when the receiver's tuning control is adjusted, the new tuned frequency of the antenna circuit will probably not be exactly equal to the new tuned frequency of the 1st RF stage. And we have the same problem with the 2nd RF stage.

The above problem is easy to understand. Here is another problem of TRF receivers, that is a little harder to understand:

Get This

When we change the tuned resonant frequency of an RF amplifier, its bandwidth also tends to change.

This fact can be understood by thinking back to the basic bandwidth formula for an *LC* resonant circuit

$$Bw = \frac{f_r}{Q_T}$$

Eq. (9-3)

Equation (9-3) indicates that as f_r is adjusted, Bw tends to change with it. Basically, a higher tuned resonant frequency tends to produce a wider bandwidth. Q_T, the total circuit Q, will also change as the resonant frequency is adjusted. If we could get the Q_T variation to be exactly proportional to the f_r variation, the two effects would cancel each other. Then Bw would stay constant. In practice, though, it is nearly impossible to get this to happen. Therefore the filter's bandwidth will vary, and it will vary in rather unpredictable ways.

⑮ ✔

EXAMPLE 9-5

In the TRF receiver of Fig. 9-11, suppose we have designed RF stage-1 so that it has a 10-kHz bandwidth when it is tuned to $f_r = 800$ kHz. This is shown in the center of the frequency-response graph of Fig. 9-32. Rearranging Eq. (9-3), we must have a total circuit Q of

$$Q_T = \frac{f_r}{Bw} = \frac{800 \text{ kHz}}{10 \text{ kHz}} = 80$$

Now suppose that we tune the radio receiver to 1200 kHz and experimentally measure the stage-1 amplifier's frequency response. Suppose we get the result shown on the right of Fig. 9-32.
a) Why did Bw get wider? Tell what happened to Q_T as f_r was adjusted to 1200 kHz.
b) What practical problem will this cause for the radio receiver?

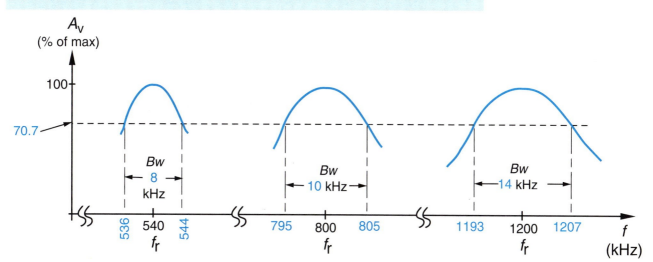

Figure 9-32 In a practical *LC* band-pass filter circuit, adjusting the tuned resonant frequency can cause the bandwidth to change.

SOLUTION

a) The bandwidth got wider because the circuit Q did not increase as much as the frequency increased. From Eq. (9-3),

$$Q_T\big|_{1200 \text{ kHz}} = \frac{f_r}{Bw} = \frac{1200 \text{ kHz}}{14 \text{ kHz}} \approx \mathbf{86}$$

Thus, Q_T increased by a factor of only about 1.1 (86/80 ≈ 1.1) as f_r increased by a factor of 1.5 (1200/800 = 1.5).

b) A not-very-distant radio station at $f_{carrier}$ = 1190 kHz (1 step away from 1200 kHz) will have upper sideband frequencies that reach 1195 kHz. These frequency components can now pass through RF stage-1, whose low cutoff frequency is 1193 kHz. They will not be filtered out. Therefore they may garble the message on the 1200-kHz carrier (interfering sounds mixed into the speech and music).

The same is true for the LSB of a not-greatly-distant station at $f_{carrier}$ = 1210 kHz.

EXAMPLE 9-6

Imagine that we tune this receiver to 540 kHz and again experimentally measure the stage-1 frequency response. Suppose we get the result shown on the left of Fig. 9-32 .

a) Why did *Bw* get narrower? Refer to Eq. (9-3).
b) What practical problem will this cause?

SOLUTION

a) Q_T decreased less drastically than the resonant frequency decreased. From Eq. (9-3),

$$Q_T\big|_{540\,kHz} = \frac{f_r}{Bw} = \frac{540\,kHz}{8\,kHz} \approx \mathbf{68}$$

So, as f_r decreased from 800 to 540 kHz, a start-to-finish ratio of about 1.5, Q_T decreased from 80 to 68, a ratio of only about 1.2.

b) Some of the intended message will be lost because the edges of the sidebands will not be able to pass through RF stage-1. The LSB frequencies between 535 and 536 kHz will be filtered out. So will the USB frequencies between 544 and 545 kHz. This worsens the quality of the sound — poor audio fidelity.

Heterodyning — Converting the Carrier Frequency to an Intermediate Frequency

To solve the tuning problems described above, the **superheterodyne receiver** was invented. It works by **converting** the received radio frequency into an **intermediate frequency**, without affecting the modulation. This way, the receiver's amplifiers can be permanently tuned to the new intermediate frequency. This completely eliminates the amplifier frequency-tracking problem and the variable-bandwidth problem, since the amplifiers' resonant frequency is never changed. The superheterodyne receiver idea is illustrated in Fig. 9-33.

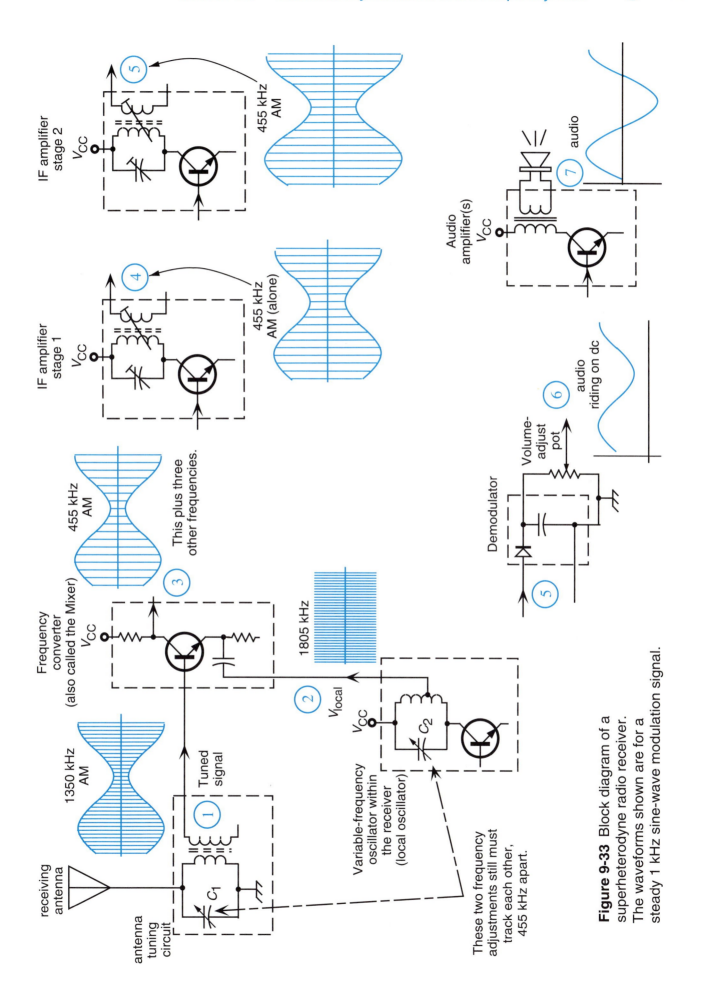

Figure 9-33 Block diagram of a superheterodyne radio receiver. The waveforms shown are for a steady 1 kHz sine-wave modulation signal.

In Figure 9-33, the antenna tuning circuit works as it always does. At point ① the selected frequency appears with greater strength than the other signals picked up by the antenna. The tuned antenna signal is *one of two* input signals fed to the **frequency converter**.

For the moment, think of the frequency converter as a two-input amplifier. It receives one input, the antenna signal, at the transistor's base terminal. It receives a second input at the transistor's emitter terminal.

17 ✓

The second input, at point ②, comes from a variable-frequency oscillator within the radio receiver. It is called the **local oscillator**. Its oscillation frequency can be symbolized f_{local}, and its output voltage can be symbolized V_{local}. The frequency f_{local} is a high (radio) frequency, not a low (audio) frequency.

Get This

The local oscillator is ganged with the antenna tuning circuit, so that f_{local} is always 455 kHz higher than the antenna-tuned frequency. In other words, whatever carrier frequency the antenna is tuned to, the oscillator will be at

$$f_{local} = f_{carrier} + 455 \text{ kHz}$$ Eq. (9-4)

EXAMPLE 9-7

a) In a superheterodyne receiver like Fig. 9-33, suppose the antenna tuning circuit is adjusted to 800 kHz. What will be the frequency of the local oscillator?

b) Repeat part **a**, assuming the radio receiver tuning knob has been adjusted to receive a station at 1350 kHz.

SOLUTION

a) From Eq. (9-4)

$$f_{local} = f_{carrier} + 455 \text{ kHz}$$
$$= 800 \text{ kHz} + 455 \text{ kHz} = \textbf{1255 kHz}$$

b) As the tuning knob adjusts C_1 to tune the antenna to 1350 kHz, C_2 tracks C_1. This is the *only* tracking that must still be maintained in a superhet radio receiver. C_2 changes to the proper value to give

$$f_{local} = 1350 \text{ kHz} + 455 \text{ kHz} = \textbf{1805 kHz}$$

We often abbreviate the word superheterodyne to "superhet".

Get This

When two similar-frequency input waveforms are mixed together, the resulting output waveform has a frequency spectrum containing four frequencies. They are:

The first frequency.
The second frequency.
The sum of the two frequencies.
The difference between the two frequencies.

18 ✓

EXAMPLE 9-8

Suppose that our superhet receiver in Fig. 9-33 is tuned to a station at 1350 kHz, as in Example 9-7.

a) Make a list of the radio frequencies that are in the spectrum of the output of the mixer, at point ③.

b) Sketch a spectrograph of the waveform at point ③ .

SOLUTION

a) From the preceding **Get This** item, we have.

    1. **1350 kHz** ($f_{carrier}$)

    2. **1805 kHz** (f_{local})

    3. **3155 kHz** ($f_{local} + f_{carrier} = 1805 + 1350 = 3155$ kHz)

    3. **455 kHz** ($f_{local} - f_{carrier} = 1805 - 1350 = 455$ kHz)

b) The frequency spectrum is shown in Fig. 9-34.

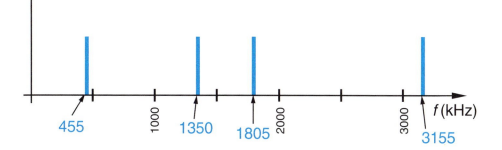

In Fig. 9-34, all magnitudes are shown the same. Relative magnitude is not an important issue to us, so do not concern yourself with it.

Figure 9-34 Frequency spectrum of the mixer output voltage, point ③, for Example 9-8.

The heterodyne idea that we are expressing here is actually the same idea that we expressed in Sec. 9-4, when we explained amplitude modulation. In Sec. 9-4 we said that the spectrum contained 3 frequencies. That wasn't quite true. It contained all 4 frequencies, but only 3 *radio* frequencies. The 4th frequency was the 1-kHz audio modulating frequency. It was so far out of the picture that we ignored it.

Going back to point ③ in Fig. 9-33, it now is a simple matter to filter out the first three frequencies in the list of Example 9-8, namely 1350 kHz, 1805 kHz and 3155 kHz. The amplifier stages are just permanently tuned to 455 kHz, called the **intermediate frequency**, or IF. The 455-kHz IF signal is boosted at point ④ and boosted again at point ⑤. The other three frequencies are blocked by the IF amplifier stages. The IF signal has exactly the same amplitude modulation that the original 1350-kHz carrier had. The AM message is not affected by the mixing/frequency-conversion process.

From point ⑤ onward, the signal is demodulated and audio-amplified just like in a TRF receiver. The waveforms are shown at points ⑥ and ⑦ .

Although Fig. 9-33 doesn't show it, the mixer/frequency-converter stage output can itself be tuned to the IF frequency. Replacing the collector resistor with an *LC* resonant circuit with $f_r = 455$ kHz helps suppress the other three frequencies at point ③.

SELF-CHECK FOR SECTION 9-6

35. (T-F) In a TRF receiver, the bandwidth of an amplifier stage can change when a new station is selected.
36. Explain what it means when we say that the filtering circuits do not track perfectly in a TRF receiver.
37. When two frequencies, an RF carrier and a local oscillator are mixed together in a transistor stage, the process is called _____ .
38. In an AM superheterodyne radio receiver, if the antenna tuned circuit is adjusted to 1270 kHz, the local oscillator frequency will be _____ kHz.
39. For the situation in Problem 38, list the four radio frequencies that are present in the mixer output.
40. For the situation in problems 38 and 39, what happens to the three unwanted frequencies?

9-7 THE FINAL IMPROVEMENT – AUTOMATIC GAIN CONTROL

The superhet receiver shares this problem with the TRF receiver:

> Nearby strong radio stations produce greater voltage in the receiver antenna than far-away weak stations. This causes the final sound volume to be louder for strong stations, and quieter for weak stations.

This is not a major problem, because the listener can adjust the sound volume with the receiver's volume-control pot. A bigger problem is that the antenna reception voltage for a fixed station can vary, as reception conditions change. This makes the listening volume rise and fall, which can be irritating.

To solve this problem, we use **automatic gain control** (AGC). This is negative feedback that decreases the IF amplifiers' voltage gain when the IF input signal voltage is large. It allows IF voltage gain to increase when the IF input voltage is small.

The receiver schematic diagram in Fig. 9-35 shows one way that AGC can work. In the IF amplifier stages, Q_3 and Q_4 are voltage divider-biased, but the R_{B2} resistors are not connected to ground. Instead, they are connected to a negative dc voltage, $-V_{AGC}$. The AGC voltage is developed by averaging the audio signal voltage from the demodulator. This arrangement is drawn in detail in Fig. 9-36 for a single stage.

Demodulator diode D_1 points toward the T_4 secondary winding in Figs. 9-35 and 9-36. This causes the demodulator output voltage to be negative relative to ground, rather than positive as before. This polarity change makes no difference to the audio amplifier, since the average dc component is blocked by input coupling capacitor C_{17} in Fig. 9-35.

The negative audio voltage is applied to the R_3–C_{16} low-pass filter. This *RC* filter removes the audio oscillations and develops a negative dc voltage that

Get This

Automatic gain control is sometimes called automatic volume control.

AGC can also be applied to the RF stages, including the converter / mixer.

 ✔

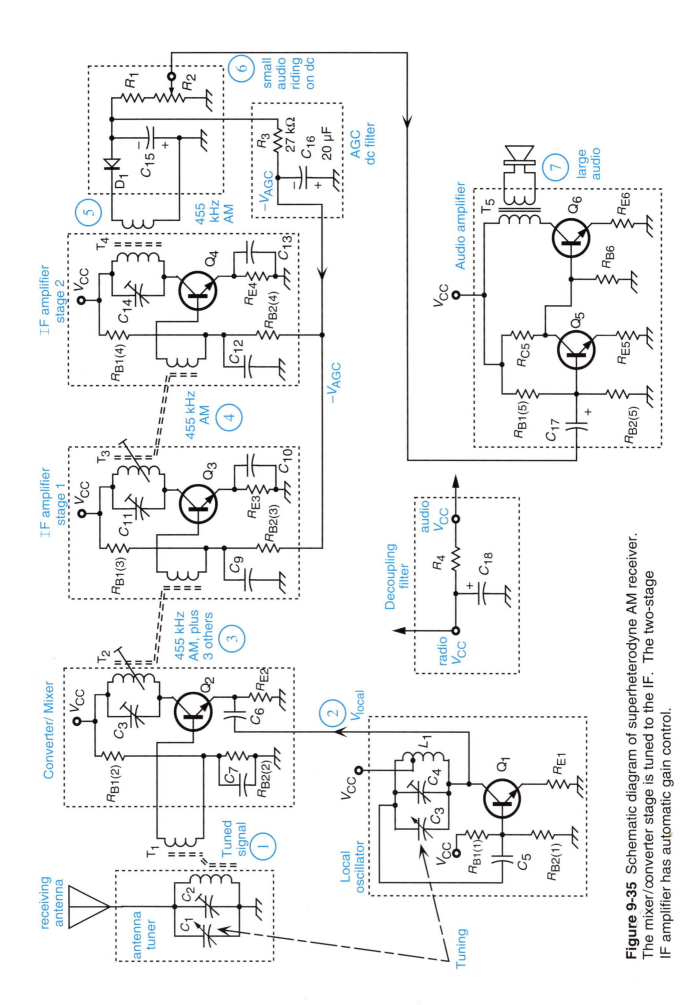

Figure 9-35 Schematic diagram of superheterodyne AM receiver. The mixer/converter stage is tuned to the IF. The two-stage IF amplifier has automatic gain control.

represents the magnitude of the audio signal. The R_3–C_{16} time constant is rather long, so normal audio loudness variations in music or speech do not affect V_{AGC}. Only long-term changes in voltage level that last more than several seconds can affect V_{AGC}.

The negative V_{AGC} is applied to the bottom of bias resistor R_{B2} in Fig. 9-36. It thereby controls the dc bias currents in transistor Q_4. As we know, dc bias current I_E determines the internal junction resistance r_{Ej}, by the formula $r_{Ej} \approx 30$ mV/I_E. Since emitter resistor R_E is ac-bypassed by C_{13}, voltage gain A_v depends heavily on internal junction resistance r_{Ej}. This is how $-V_{AGC}$ controls A_v.

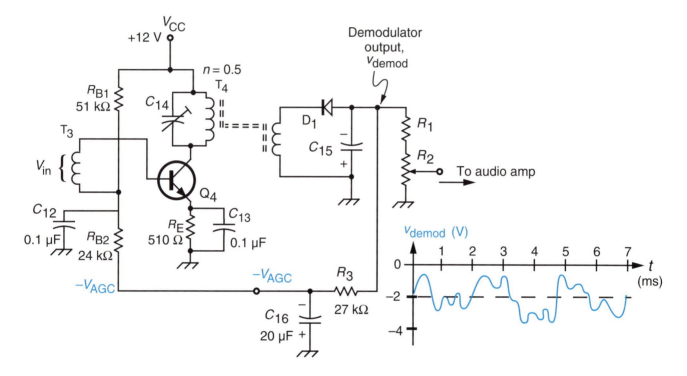

Figure 9-36 AGC – Automatic Gain Control. The output voltage is fed back in such a way that it actually changes the amplifier's voltage gain A_v.

For example, suppose the average audio signal from the demodulator is –2 V , as shown in Fig. 9-37(A). R_3 and C_{16} remove the audio, producing $V_{AGC} = -2.0$ V. This voltage is applied to R_{B2}, giving the following voltage division:

$$\frac{V_{RB1}}{V_{total}} = \frac{R_{B1}}{R_{B1} + R_{B2}} = \frac{51 \text{ k}\Omega}{51 \text{ k}\Omega + 24 \text{ k}\Omega}$$

$$
\begin{aligned}
V_{total} &= 12 \text{ V} - (-2 \text{ V}) \\
&= 14 \text{ V}.
\end{aligned}
$$

$$V_{RB1} = 14 \text{ V} \left(\frac{51 \text{ k}\Omega}{75 \text{ k}\Omega} \right) = 9.52 \text{ V}$$

Therefore, V_B is given by

$$V_B = V_{CC} - V_{RB1}$$
$$= 12 \text{ V} - 9.52 \text{ V} \approx 2.5 \text{ V}$$

V_E is found by subtracting the 0.7-V B-E drop from V_B, for

$$V_E = V_B - 0.7 \text{ V} = 2.5 \text{ V} - 0.7 \text{ V} = 1.8 \text{ V}$$

Applying Ohm's law to R_E gives

$$I_E = \frac{V_E}{R_E} = \frac{1.8 \text{ V}}{510 \text{ } \Omega} = 3.5 \text{ mA}$$

So r_{Ej} can be estimated as

$$r_{Ej} \approx \frac{30 \text{ mV}}{3.5 \text{ mA}} = 8.5 \text{ } \Omega$$

Let us assume that the resistance appearing in Q_4's collector circuit is 1700 Ω, taking into account the T_4 turns ratio of $n = 0.5$. There is no net reactance present, since the LC parallel circuit is resonant at 455 kHz. Therefore the Q_4 stage's voltage gain is given by

$$A_v = \frac{r_C}{r_{Ej}} (n) = \frac{1700 \text{ } \Omega}{8.5 \text{ } \Omega} (0.5) = 100$$

R_E does not appear in the A_v formula because it is bypassed by C_{13}.

Now suppose that the antenna signal in Fig. 9-35 gets much stronger, perhaps due to a change in atmospheric conditions. Imagine that the carrier voltage delivered to the mixer increases by a factor of 4. This would cause the T_2 secondary voltage, the input voltage to the IF amplifiers, to increase by a factor of 4 also. Such a great increase in V_{in} will cause the demodulator output voltage to increase by a more reasonable factor. This is because the demodulator increases the magnitude of AGC voltage, which causes a drastic reduction in the A_v of both amplifier stages. The situation is shown in Fig. 9-37(B), for stage-2.

[20] ✔

Let us begin by assuming that the demodulator voltage increases by a factor of only 1.5, as shown on the right of Fig. 9-37(B). This will boost the AGC magnitude from –2 V to –3 V. Applying voltage-division to R_{B1} and R_{B2}, we get

$$V_{RB1} = 15 \text{ V} \left(\frac{51 \text{ k}\Omega}{51 \text{ k}\Omega + 24 \text{ k}\Omega} \right) = 10.2 \text{ V}$$

Base voltage V_B is given by

$$V_B = V_{CC} - V_{RB1} = 12 \text{ V} - 10.2 \text{ V} = 1.8 \text{ V}$$

so V_E is

$$V_E = V_B - 0.7 \text{ V} = 1.8 \text{ V} - 0.7 \text{ V} = 1.1 \text{ V}$$

Applying Ohm's law to 510-Ω resistor R_E gives

$$I_E = \frac{V_E}{R_E} = \frac{1.1 \text{ V}}{510 \text{ } \Omega} = 2.16 \text{ mA}$$

so r_{Ej} is

$$r_{Ej} \approx \frac{30 \text{ mV}}{I_E} = \frac{30 \text{ mV}}{2.16 \text{ mA}} = 13.9 \text{ } \Omega$$

Therefore, for the same 1700 Ω value of r_C, the new voltage gain of this stage is

$$A_v = \frac{r_C}{r_{Ej}} (n) = \frac{1700 \text{ } \Omega}{13.9 \text{ } \Omega} (0.5) = 61.2$$

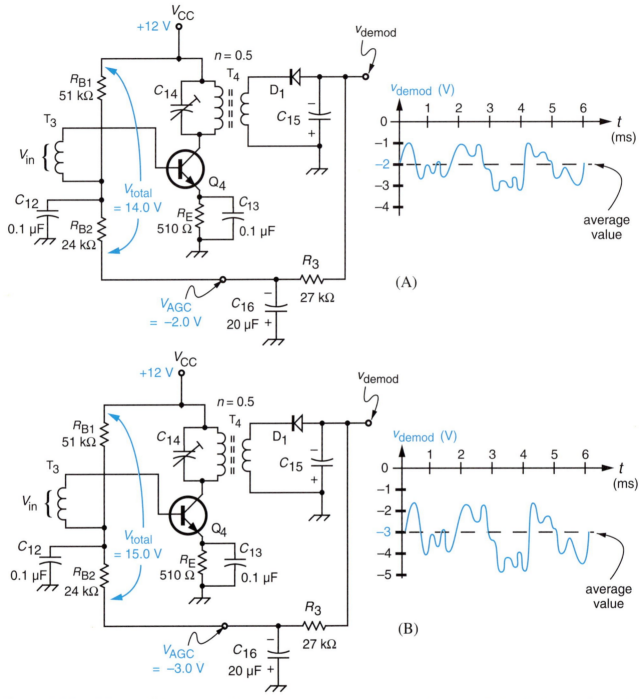

Figure 9-37 AGC conditions. (A) Weak antenna signal produces $V_{AGC} = -2$ V, which sets $A_v = 100$ for IF amplifier stage 2. (B) A much stronger antenna signal produces $V_{AGC} = -3$ V, which lowers A_v to only about 61 for IF amplifier stage 2.

If we assume that the same change in A_v occurs in stage-1, then the combined IF stages have a new value of A_v given by

$$A_{v \, (\text{total IF})} = (61.2) \times (61.2) \approx 3750$$

This is in contrast to the previous IF voltage gain of

$$A_{v \, (\text{total IF})} = (100) \times (100) = 10\,000$$

So here is what has happened:

1 V_{in} increased by a factor of 4.

2   Amplifier A_v decreased to a new value equal to 0.375 times its previous value.

The new value of A_v is 3750. The old value was 10 000. The ratio of the new to the old is $3750 \div 10\,000 = 0.375$.

Therefore V_{out} increases by a factor of

$$(4)\,(0.375)\ =\ 1.5$$

This agrees with our beginning assumption that V_{demod} increased its average magnitude from 2 V to 3 V, since 2 V× (1.5) = 3 V.

SELF-CHECK FOR SECTION 9-7

42. (T-F) One of the advantages of AGC is that the sound volume stays relatively constant as the receiver is tuned between stronger and weaker stations.
43. (T-F) The purpose of AGC is to stabilize the IF amplifiers' voltage gain, preventing $A_{v\,(\text{total IF})}$ from changing.
44. (T-F) AGC responds within a few milliseconds.
45. In Fig. 9-37(B), suppose the average demodulator voltage were –2.5 V rather than –3 V. Find:

a) V_{AGC} d) V_E g) A_v (stage-2, for the same n and r_C)
b) V_{RB1} e) I_E h) $A_{v\,(\text{total IF})}$
c) V_B f) r_{Ej}

9-8 FREQUENCY MODULATION

No matter how much we improve our receivers, there is one problem with AM radio communication that we cannot avoid. AM is sensitive to electrical noise interference. Electrical noise is an unwanted voltage burst that appears on top of the desired signal. Stray capacitance can couple a noise burst onto the wire carrying the signal, or noise can be magnetically coupled, by Faraday's law. In radio communication, noise is sometimes radiated through the air and picked up by the antenna, and sometimes is coupled directly into the receiver circuitry. No matter how the electrical noise appears, the outcome is always the same: It produces unwanted sound from the speaker. This is unavoidable, because it is voltage *amplitude* that contains the radio message.

Although electrical noise can distort voltage amplitude, it usually cannot affect frequency. To take advantage of this, we have devised a method of coding a radio message by frequency variation, rather than amplitude variations. This approach is called **frequency modulation**, or FM.

It is easy to see the difference between frequency modulation and amplitude modulation by comparing their time-domain waveform graphs. Figure 9-38 shows this comparison.

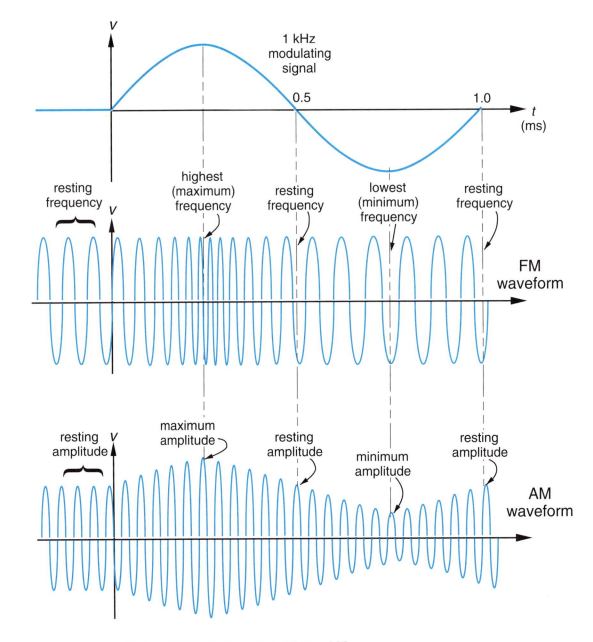

Figure 9-38 Comparing FM to AM.
FM has a constant amplitude, but varying frequency.
AM has a constant frequency, but varying amplitude.

The resting frequency is also called the center frequency, or the carrier frequency.

At the left of Fig. 9-38, the modulating signal is zero. In the FM waveform, this causes the frequency to hold at a steady value, called the **resting frequency**. When the modulating signal becomes positive, it causes the frequency to increase, as indicated by the closer-together oscillations on the FM waveform. As the instantaneous modulating voltage becomes more positive, the frequency becomes higher. Compare this to the AM wave, where the more positive modulating voltage makes the amplitude larger.

At the time instant when the modulating voltage returns to zero, the FM wave is back at its resting frequency. This occurs at t = 0.5 ms in Fig. 9-38.

When the modulating signal becomes negative, it causes the frequency to decrease, as indicated by the farther-apart oscillations on the FM waveform. As the modulating voltage goes more negative, the frequency becomes lower. Compare this to the AM wave, where the more negative modulation voltage makes the amplitude smaller.

The FM waveform in Fig. 9-38 shows 18 carrier oscillations occurring during the 1-ms period of the audio signal. In reality, there will be a much larger number of carrier oscillations. Therefore the frequency variations are much smoother, less abrupt, than shown here. For our example carrier frequency of 3 MHz, there would be 3000 total carrier cycles during the 1-ms time duration of Fig. 9-38.

There would be more than 1500 oscillations during the audio positive half cycle, and less than 1500 oscillations during the negative half cycle. The increase during the positive half cycle is balanced by an equal decrease during the negative half cycle.

Amount of Modulation – Frequency Deviation

The magnitude of the modulating signal determines how much the frequency changes from the resting frequency. The maximum frequency change is called **frequency deviation**. Figure 9-39 shows that a larger audio modulating signal produces greater frequency deviation.

Figure 9-39(B) shows a greater amount of frequency modulation than Fig. 9-39(A). The highest frequency in Fig. 9-39(B) is higher than the highest frequency in Fig. 9-39(A). And the lowest frequency in part (B) is lower than the lowest frequency in part (A).

Producing an FM Waveform

An FM wave can be produced by a variable-frequency oscillator or VFO. One approach is pictured in Fig. 9-40(A).

In the Hartley oscillator of Fig. 9-40(A), the tank circuit contains a **varicap diode**, D_1. This is a diode that is specially built to have variable capacitance that depends on the reverse bias voltage that is applied to it. Larger amounts of reverse bias voltage cause the varicap's depletion region to become wider, which lowers its capacitance, C_1. Smaller reverse-bias voltage narrows the depletion region, raising the diode's capacitance. A typical capacitance-versus-voltage relation is graphed in Fig. 9-40(B).

Varicap diodes are also called *varactor* diodes, or *voltage-variable capacitors*.

The oscillation frequency is set by the total capacitance of the oscillator's tank circuit, $C_T = C_1 \cdot C_2 / (C_1 + C_2)$. Since C_2 is more than 10 times as large as C_1, total capacitance C_T is determined mostly by C_1. As the audio modulating signal varies the reverse bias on D_1, its capacitance C_1 varies the oscillation frequency. D_1's reverse bias voltage is controlled by the audio transformer secondary voltage in combination with the dc voltage across resistor R_2 in the three-resistor voltage divider.

The oscillator's output voltage is amplified and applied to a transmitting antenna, to broadcast an FM radio wave. This arrangement is shown in Fig. 9-41.

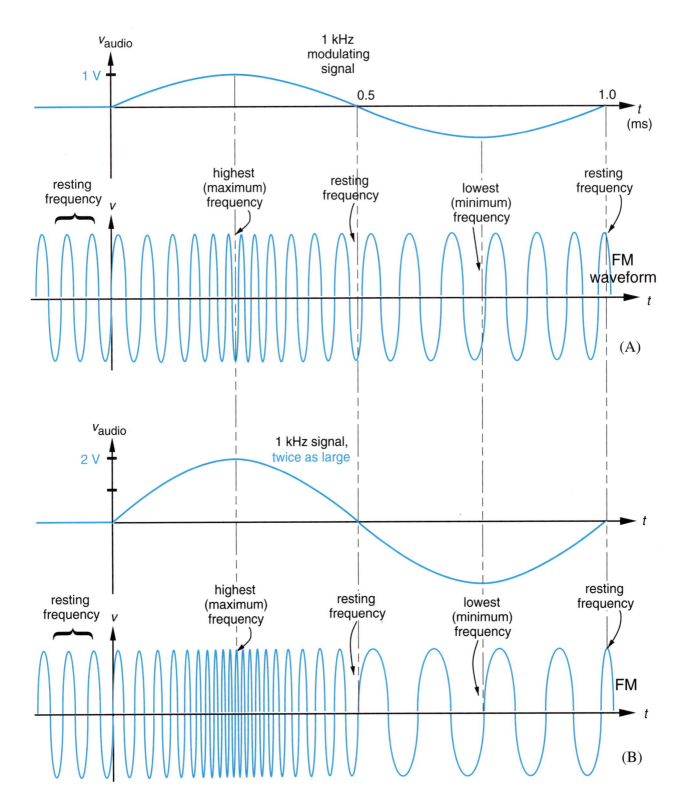

Figure 9-39 Different amounts of frequency modulation.
(A) A small modulation signal produces a certain amount of frequency deviation.
(B) A larger modulating signal produces greater frequency deviation.

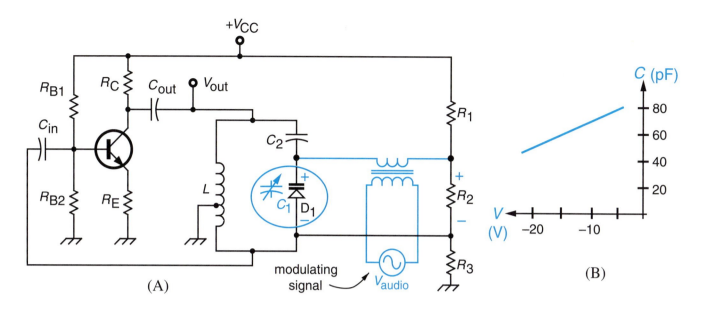

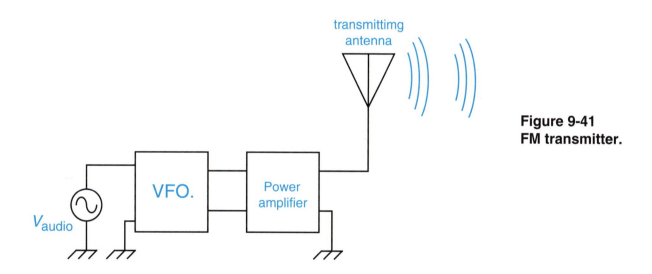

Figure 9-40 Hartley variable-frequency oscillator using a varicap diode. As the audio transformer voltage goes positive on the left, the reverse bias voltage across D_1 increases. This reduces capacitance C_1, which raises the oscillation frequency. As the audio transformer voltage goes negative on the left, D_1's reverse bias is reduced. C_1 increases and f decreases.
(B) *C*-versus-*V* curve for a varicap diode.

22 ✔

Figure 9-41
FM transmitter.

Frequency Spectrum of an FM Waveform

The frequency spectrum of an FM wave is quite different from that of an AM wave.

When a carrier wave is frequency-modulated by an audio signal of a single frequency, say 1 kHz, the resulting spectrum is *not* limited to only three radio frequencies.

For example, if a 3-MHz carrier (the resting frequency of the oscillator) is frequency-modulated by a 1-kHz sine wave, we do not get a simple spectrum containing just the three radio frequencies 2.999 MHz, 3 MHz, and 3.001 MHz. Instead, the FM spectrum is described as follows:

Get This

Remember the definition of harmonic. For a repeating waveform, the harmonics are a collection of sine waves with frequencies that are given by 1 times the basic frequency, 2 times the basic frequency, 3 times, 4 times, and so on.

The upper sideband contains all the frequencies gotten by adding the audio modulating frequency and all of its harmonics to the carrier frequency. The lower sideband contains all the frequencies gotten by subtracting the audio modulating frequency and all of its harmonics from the carrier frequency.

In other words, if we symbolize the carrier (resting) frequency as f_c and the audio modulating frequency as f_{mod}, the USB contains $f_c + f_{mod}$, $f_c + 2f_{mod}$, $f_c + 3f_{mod}$, and so on. The LSB contains $f_c - f_{mod}$, $f_c - 2f_{mod}$, $f_c - 3f_{mod}$, and so on.

This idea is illustrated in Fig. 9-42, for a 3-MHz carrier modulated by a 1-kHz audio signal. This figure stops at the sixth harmonic. In reality, the frequency components go on forever — there are an infinite number or them in the USB. However, their relative magnitudes keep getting smaller so that they eventually become insignificant.

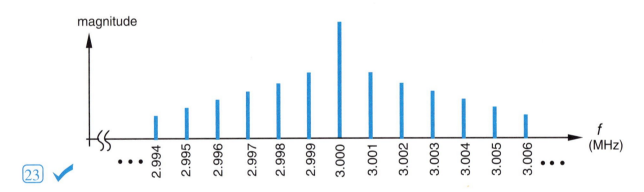

[23] ✓

Figure 9-42 FM spectrograph. For an audio modulating frequency $f_{mod} = 1$ kHz (0.001 MHz), the audio harmonics are 2 kHz, 3 kHz, 4 kHz, and so on. The sum of every harmonic appears in the USB; the difference of every harmonic appears in the LSB. The higher harmonics have weaker (lower-magnitude) components.

Facing Page

Figure 9-43   Showing how the magnitude of the modulating signal affects the frequency spectrum in FM. (A) With a small modulating voltage, the 1st and 2nd harmonics of the modulation frequency are significant. The 3rd, 4th, and higher harmonics are insignificant.
(B) For a larger modulating voltage, the 3rd harmonic grows so it becomes significant. The 4th harmonic is still insignificant.
(C) Raising the modulation voltage even higher (louder sound) causes the 4th and 5th harmonics to become significant.

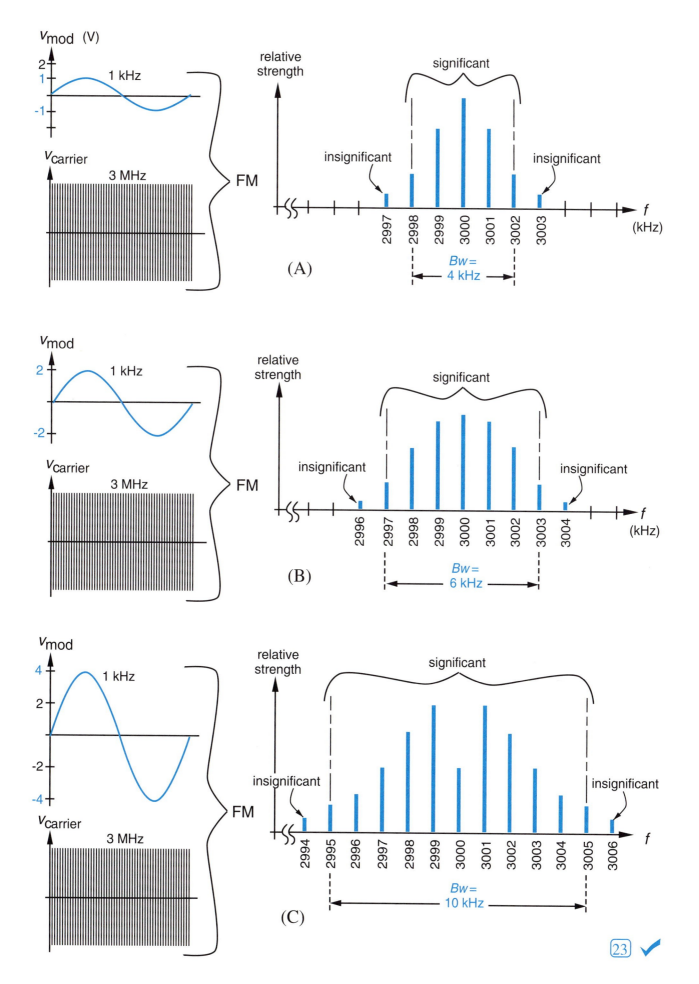

An important issue in FM radio communication is this: We must know which harmonic frequency components are large enough in magnitude so that they are significant. The transmitter and receiver filter circuits must have bandwidths wide enough to pass all these significant components. The insignificantly small frequency components do not have to be passed by the filter circuits. We don't mind if they are blocked, because they contribute very little to the fidelity of the original message.

As a general rule:

If the magnitude of the modulating signal is larger, a greater number of harmonics become significant.

This idea is demonstrated in Fig. 9-43, parts (A), (B), and (C), for three different magnitudes of modulation signal.

As the number of significant harmonics increases, the bandwidth of the frequency spectrum gets wider. In Fig. 9-43(A), Bw is 4 kHz. Bandwidth increases to 6 kHz in part (B), and to 10 kHz in part (C). The FM radio receiver's filter circuits must have a bandwidth at least as wide as the frequency spectrum, for high-fidelity reproduction of the message.

The foregoing discussion dealt with the effect of the audio modulation magnitude on the FM spectrum and bandwidth. But of course, audio magnitude is not the only thing that matters. The audio modulating *frequency* also has a direct effect on an FM frequency spectrum, just like it did for an AM frequency spectrum.

As things turn out, a higher audio modulation frequency has fewer significant harmonics than a lower audio modulation frequency, all other things being equal. This idea is illustrated in Fig. 9-44, which compares audio modulating frequencies of 1.5 kHz, 1 kHz and 500 Hz, all with the same 2-V magnitude. In Fig. 9-44(A), f_{mod} = 1.5 kHz. The 1st and 2nd harmonics are significant, but the 3rd is insignificant.

Figure 9-44(B) repeats the same information that was presented in Fig. 9-43(B). The 1st, 2nd, and 3rd harmonics are significant.

The modulating frequency is lowered to 500 Hz (0.5 kHz) in Fig. 9-44(C). With f_{mod} reduced by a factor of 2, the number of significant harmonics doubles, from 3 to 6.

Commercial FM Radio Bandwidths: When the modulating signal's magnitude and frequency effects are both taken into account, here is the final arrangement for commercial FM radio broadcasting:

Get This

The maximum allowed audio modulating frequency is 15 kHz. At this maximum f_{mod}, the magnitude is kept low enough to limit the number of significant harmonics to 5 in each sideband. Therefore the frequency spectrum extends 75 kHz (5 x 15 kHz) above the carrier in the USB, and 75 kHz below the carrier in the LSB. The bandwidth is 150 kHz. In the time domain, this is equivalent to saying that the maximum frequency deviation from the resting frequency is 75 kHz.

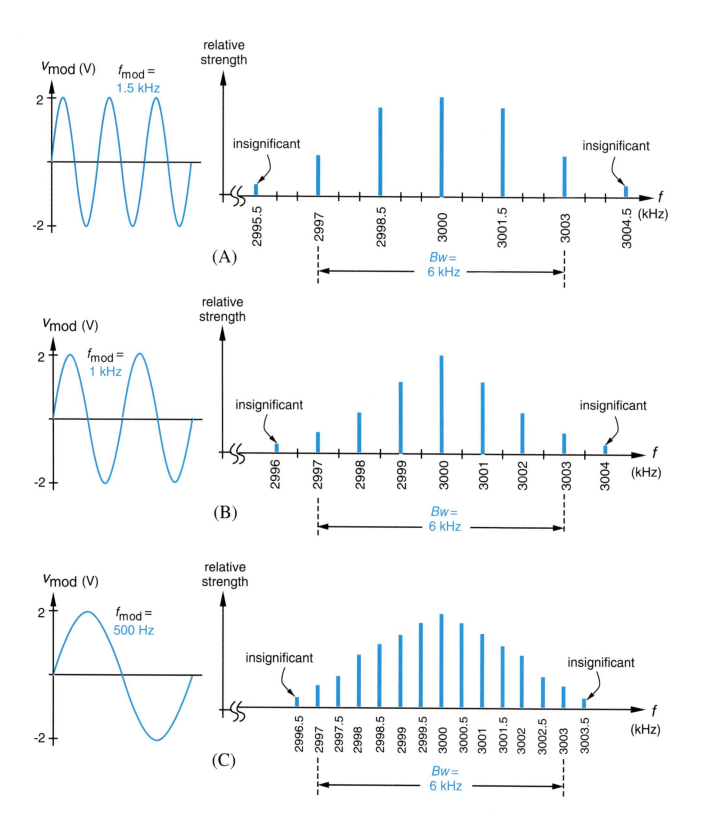

Figure 9-44 Showing how the frequency of the modulating signal affects the frequency spectrum in FM. (A) With a high modulating frequency, few harmonics are significant. (B) With a medium f_{mod}, more harmonics are significant. (C) With a low f_{mod}, even more harmonics become significant. The bandwidth stays constant because when more harmonics are significant, they are spaced closer together.

At lower audio modulating frequencies, the modulating magnitude is allowed to be greater, so there are more than 5 significant harmonics in each sideband. But as described in the explanations for Figs. 9-43 and 9-44, this does not change the spectrum bandwidth. It remains 150 kHz.

In North America, the commercial FM radio broadcast carrier frequencies start at 88.1 MHz and go in steps of 200 kHz (0.2 MHz) to 107.9 MHz. The 200-kHz-wide steps allow room for the 150-kHz signal bandwidth, plus two 25-kHz guard bands, one on each side of the bandwidth. This arrangement is shown in Fig. 9-45.

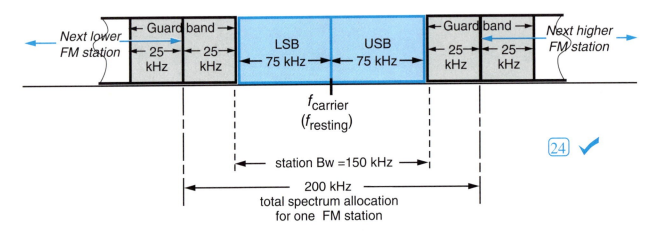

Figure 9-45 Frequency allocation for commercial FM radio. Since every station has a 25-kHz guard band at both ends of its spectrum, there is actually a 50-kHz unused band separating adjacent stations, to prevent their interfering with each other.

Because a commercial FM transmitter waveform has a frequency bandwidth of 150 kHz, a real FM radio receiver's filter circuits must also have $Bw = 150$ kHz.

SELF-CHECK FOR SECTION 9-8 IS ON PAGE 287

9-9 FM RECEIVERS

Conceptually, an FM heterodyne radio receiver is almost the same as an AM heterodyne receiver. The block diagram of Fig. 9-46 shows this overall similarity. That diagram also points out that there are two conceptual differences in an FM receiver — the amplitude limiting circuit and the different FM demodulator.

Facing Page
Figure 9-46 In a commercial FM radio receiver, the tuned carrier frequency will be between 88.1 and 107.9 MHz. The local oscillator is set to run at $f_{osc} = f_{carrier} + 10.7$ MHz. This gives a difference IF frequency of 10.7 MHz, so the IF stages are permanently tuned to 10.7 MHz. Their bandwidth is 150 kHz, as called for in Fig. 9-45. The limiter clips the IF FM output at a certain maximum voltage, thereby eliminating any amplitude noise that may have gotten in. The varying-frequency wave is applied to the FM demodulator, which converts the frequency variations into audio voltage variations.

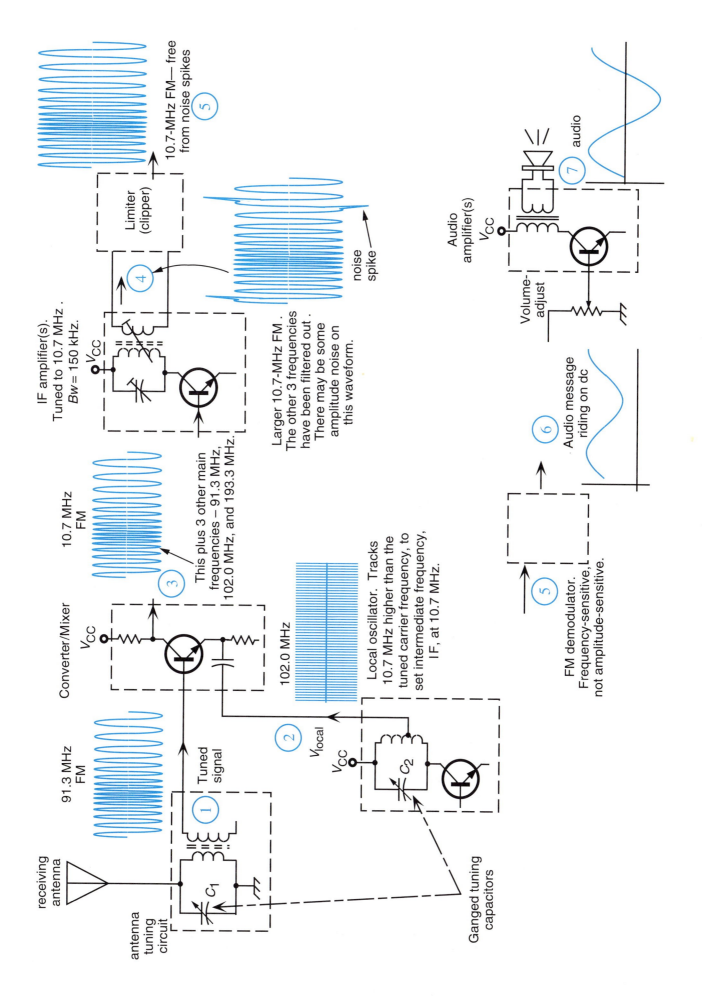

receiving antenna

antenna tuning circuit

91.3 MHz FM

① Tuned signal

② V_{local}

102.0 MHz

Local oscillator. Tracks 10.7 MHz higher than the tuned carrier frequency, to set intermediate frequency, IF, at 10.7 MHz.

C_1

C_2

V_{CC}

Ganged tuning capacitors

Converter/Mixer

V_{CC}

③ This plus 3 other main frequencies – 91.3 MHz, 102.0 MHz, and 193.3 MHz.

10.7 MHz FM

IF amplifier(s). Tuned to 10.7 MHz. $Bw = 150$ kHz.

V_{CC}

④

noise spike

Larger 10.7-MHz FM. The other 3 frequencies have been filtered out. There may be some amplitude noise on this waveform.

Limiter (clipper)

⑤ 10.7-MHz FM— free from noise spikes

FM demodulator. Frequency-sensitive, not amplitude-sensitive.

⑤

⑥ Audio message riding on dc

Volume-adjust

Audio amplifier(s)

V_{CC}

⑦ audio

In an FM receiver, the output of the IF amplifier is passed through a voltage limiting stage, or **limiter**. The voltage-limited output is then demodulated by a circuit which is sensitive to frequency deviations, not amplitude variations.

There are several different designs for FM demodulating circuits. The design that is easiest to understand is the **double-tuned discriminator** circuit. Its schematic diagram is shown in Fig. 9-47(A).

26 ✓

In Fig. 9-47(A), L_1 and L_2 have identical values because each inductance is exactly half of the secondary winding of the coupling transformer. But C_2 is sized slightly larger than C_1. This causes the resonant frequency of L_2C_2 to be slightly lower than f_r of L_1C_1. As pictured in Fig. 9-47(B), C_2 is chosen to produce an L_2C_2 resonant frequency of 10 625 kHz, which is 75 kHz lower than the 10.7 MHz intermediate frequency. Capacitance C_1 is chosen to produce an L_1C_1 resonant frequency of 10 775 kHz, which is 75 kHz higher than 10.7 MHz.

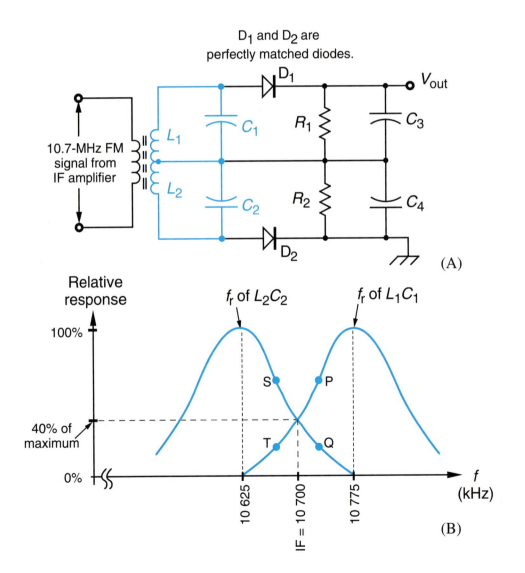

Figure 9-47 **Double-tuned discriminator. (A) Schematic diagram. (B) Frequency-response curves of the *LC* circuits.**

When the IF signal is at its resting frequency, 10.7 MHz, both *LC* circuits will have the same response. Figure 9-47(B) shows each circuit producing about 40% of maximum response. The voltage developed across L_1C_1 is rectified by diode D_1, and applied to the R_1–C_3 combination, **+** on top, **−** on bottom. This voltage is symbolized V_1 in Fig. 9-48(A), with an assumed value of 3 V.

At the same time, the equal voltage developed across L_2C_2 is rectified by diode D_2 and applied to the R_2–C_4 combination. In Fig. 9-48(A), this produces $V_2 = 3$ V across R_2–C_4, **+** on bottom, **−** on top. With equal magnitudes and opposing polarities, V_1 and V_2 combine to produce $V_{out} = 0$ V, at the resting frequency.

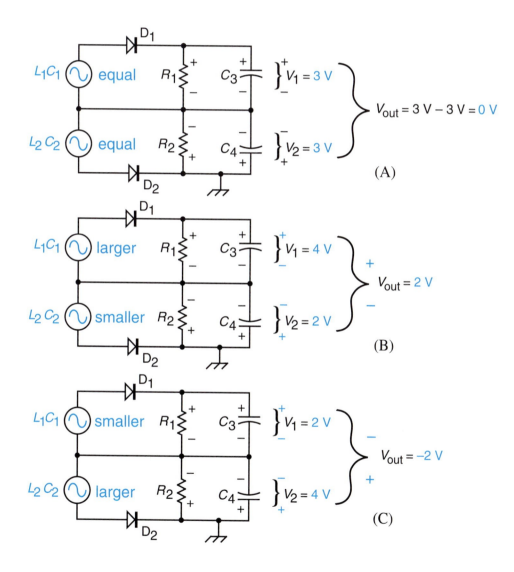

Figure 9-48 Demodulating the FM signal from the IF amplifiers.
(A) At a moment when *f* = resting frequency, $V_{out} = 0$.
(B) At a moment when *f* deviates above resting frequency, say to 10 720 kHz, V_{out} goes positive.
(C) When *f* deviates below resting frequency, say to 10 680 kHz, V_{out} goes negative.

When the IF signal deviates above 10.7 MHz, L_1C_1 develops a greater voltage than L_2C_2. This is indicated by points P and Q in Fig. 9-47(B). Therefore L_1C_1 delivers a V_1 value greater than 3 V in Fig. 9-48(B). L_2C_2 delivers a V_2 value less than 3 V. The output is produced by the series combination of V_1 and V_2, giving $V_{out} = +2$ V.

When the IF signal deviates below 10.7 MHz, the opposite happens. As shown by points S and T on Fig. 9-47(B), L_2C_2 now dominates L_1C_1. In Fig. 9-48(C), V_2 becomes larger than 3 V and V_1 becomes smaller than 3 V, producing a negative value of V_{out}.

The greater the frequency deviation, the greater the imbalance between V_1 and V_2 becomes. Thus, greater frequency deviation produces larger magnitude of V_{out}. In this way, the frequency-modulated signal is demodulated and the original FM message is converted to an audio signal.

The double-tuned discriminator is not often used in a modern FM receiver, because other demodulating circuit designs can outperform it. We will not discuss these other designs because they are more difficult to understand.

Today's FM receivers, like AM receivers, have ICs that contain most of the transistor, resistor and diode components of their circuits. The discrete filtering components, capacitors, inductors and transformers, are connected to the proper IC terminals.

Automatic Frequency Control: It is difficult to get perfect tracking between the local oscillator frequency and the tuned frequency through the entire 20-MHz-wide FM broadcast band. To solve this tracking problem, and to eliminate temperature-related frequency drift of the local oscillator, **automatic frequency control** (AFC) can be used. It works as follows.

If the local oscillator frequency is perfectly correct, the IF frequency will also be perfect – 10.7 MHz. In that case, the two tuned circuits in Fig. 9-47 will be treated equally, on average. The discriminator/demodulator output will be pure ac, with zero dc component.

On the other hand, if the local oscillator is slightly higher than it ought to be, the IF signal will be slightly higher than 10.7 MHz. In Fig. 9-47(B), this will give L_1C_1 an advantage over L_2C_2. The demodulator output will then contain a **+** dc component. If the reverse is true, local oscillator too low, V_{out} from the demodulator will ride on a **−** dc component. An AFC circuit has a low-pass filter that extracts the dc component from the demodulator output waveform. This dc voltage is fed back to the local oscillator in such a way that it tends to correct the frequency error. A varicap diode can be used as the voltage-sensitive frequency-controlling element.

SELF-CHECK FOR SECTION 9-9

52. (T-F) To solve the tracking problem and to maintain consistent amplifier bandwidth, a modern FM receiver uses the same heterodyning idea as an AM receiver.

53. In an FM receiver, the IF output has its oscillations clipped to a constant amplitude by the _____ circuit.
54. In a receiver for commercial FM broadcasts, what is the relationship between the local oscillator frequency and the tuned carrier frequency?
55. (T-F) For an FM demodulator, the output voltage is proportional to the frequency deviation.

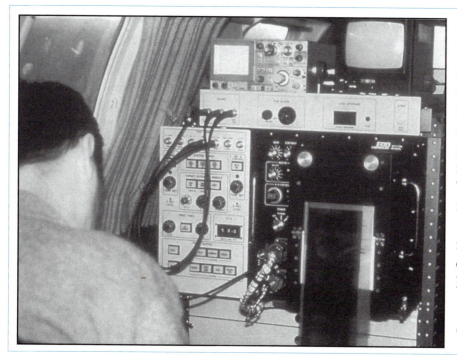

This equipment is carried aboard U.S. Forest Service aircraft for detecting just-started forest fires that might not be detectable on the ground. It senses infrared radiation from a small fire. Then it identifies the fire's map coordinates to within 150 meters, by radio communication with the earth-orbiting Navstar Global Positioning Satellite.

Courtesy of U.S. Forest Service

SELF-CHECK FOR SECTION 9-8 (FROM PAGE 282)

46. In frequency-modulation, the waveform's _____ is constant and its _____ varies.
47. At an instant when $v_{mod} = 0$, the frequency is equal to the _____ frequency.
48. (T-F) When a carrier is frequency-modulated by a sine-wave audio signal with frequency $= f_{mod}$, the resulting spectrum contains three radio frequencies: $f_{carrier}$, $f_{carrier} + f_{mod}$, and $f_{carrier} - f_{mod}$.
49. In commercial FM radio broadcasting, the signal bandwidth is _____ kHz.
50. When a carrier is frequency-modulated by a sine-wave audio signal with frequency $= f_{mod}$, the resulting spectrum contains many radio frequencies. Make a partial list of these frequencies.
51. In commercial FM radio broadcasting, the maximum frequency deviation from the resting frequency is _____ kHz in the time domain.

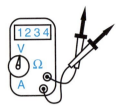

9-10 RECEIVER TROUBLESHOOTING

To troubleshoot a radio receiver, think of the receiver in block diagram fashion, as shown in Figs. 9-33 and 9-46. Your goal is to locate the block, or stage, that is breaking the signal path. To do this, follow the same procedure that was described in Sec. 6-5 for troubleshooting a multistage amplifier. Move your oscilloscope test probe from output toward input, always looking for a bad output but a good input. That identifies the particular block that is breaking the signal path. The trouble is most likely to be in that block.

EXAMPLE 9-9

Suppose the AM receiver of Figs. 9-33 and 9-35 produces no sound. The speaker is known to be good because you have tested it as shown in Fig. 9-49, or by the simpler ohmmeter "touching" test shown in Fig. 5-25. With the speaker connected, the scope reveals 0 V, no waveform, at the audio amp output, point ⑦ in Fig. 9-33. You now place your scope probe at the volume-control pot terminal, point ⑥ in Fig. 9-33.

a) Suppose you see a proper audio waveform, several volts in magnitude. What is your conclusion?

b) Suppose you see 0 V, no waveform. What is your conclusion?

SOLUTION

a) This is a case of the audio amplifier's output being bad, even though its input is good. The trouble is in the audio amp block.

b) If the input to the audio amp is bad, the trouble probably is not in that block. You must look earlier in the signal path for the trouble.

〔27〕 ✔

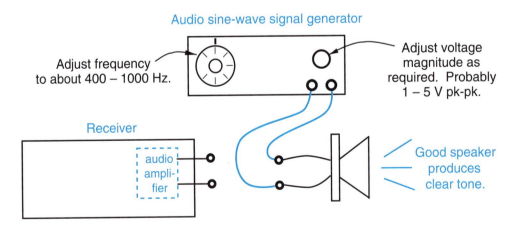

If one of the speaker connections is chassis ground, break the other, nonground, connection.

Figure 9-49 A speaker is tested by breaking open one or both of its connections to the audio amplifier and applying an audio sine-wave voltage.

EXAMPLE 9-10

Continuing the troubleshooting process from part (b) of the previous Example, you place your scope at the input of the demodulator, point ⑤.
a) If the scope reveals a proper waveform, namely a high-frequency signal enclosed in an audio-frequency "envelope", what is your conclusion?
b) If the scope reveals 0 V, no waveform, what is your conclusion?

SOLUTION

a) Here, the demodulator block has a good input but a bad output. The trouble is probably in the demodulator. Look for an open diode D_1. The trouble could possibly be with capacitor C_{15}, or the volume control pot, but these malfunctions would probably distort the output of the IF amplifier.
b) With 0 V at the input, the trouble is probably not in the demodulator. You must look earlier in the signal path.

However, as explained in Sec. 5-7, it is possible for a failure in a driven stage (the demodulator, here) to cause the output of the driving stage (the IF amplifier, here) to go bad.

Signal Injection: The troubleshooting process can be speeded up by injecting a correct signal at various locations in the signal path. The signal injection idea is illustrated in Fig. 9-50.

Figure 9-49 is a simple example of signal injection.

Figure 9-50 Signal injection at various points in a receiver's signal path.

The local oscillator can be tested with an oscilloscope or frequency counter. For an FM receiver, the oscillator frequency is in the 100 MHz range, so you must be sure that your instrument's frequency response goes that high. The test instrument's input impedance must be high enough to guarantee negligible loading effect on the oscillator circuitry.

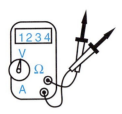

Whenever trouble is isolated to a particular block, check all the dc power supply and ground terminals within that block, as described in Sec. 5-7. Dc bias measurements of the transistors can reveal specific circuit faults, as outlined in Sec. 5-7.

Automatic gain control operation can be checked by measuring the dc voltage produced by the AGC circuit as the tuner is adjusted from no station to a strong station. Compare the actual measured variation in AGC voltage to the range specified in the receiver's service manual.

Typical homes contain many sources of radio frequency interference (RFI). Microwave ovens, commutator-type motors on power tools, lamp-dimmers, ceiling-fan speed controllers, and some engine ignition systems all produce electromagnetic radiation with a wide spectrum of frequency components. It is almost impossible to prevent antenna pickup of this radiation, other than by turning off the offending device.

FORMULAS

For AM:

$$\text{modulation percentage} = \frac{\text{maximum voltage change}}{\text{unmodulated voltage}} \times 100\%$$

$$\text{USB width} = f_{\text{carrier}} + f_{\text{audio(max)}}$$

$$\text{LSB width} = f_{\text{carrier}} - f_{\text{audio(max)}}$$

$$\text{RF amplifier bandwidth} = 2\,[f_{\text{audio(max)}}]$$

$$Bw = \frac{f_r}{Q_T} \qquad\qquad \text{Eq. (9-3)}$$

$$f_{\text{local}} = f_{\text{carrier}} + 455 \text{ kHz} \qquad\qquad \text{Eq. (9-4)}$$

For FM:

$$f_{\text{local}} = f_{\text{carrier}} + 10.7 \text{ MHz}$$

$$\text{RF amplifier bandwidth} = 150 \text{ kHz}$$

SUMMARY

● Radio waves are a combination of an electric field and a magnetic field. They are electromagnetic waves.

● A dipole antenna radiates radio waves in all directions, not just in a particular direction.

● The range of radio frequencies is about 20 kHz to 300 GHz.

● The simplest possible radio receiver contains four components: An antenna, a diode, a capacitor and a headset speaker.

● The diode and capacitor demodulate the incoming radio waves. Together, they detect the message.

● Sensitivity is the ability of a radio receiver to receive weak radio waves.

● Selectivity is the ability of a radio receiver to receive radio waves at one frequency and ignore radio waves at other frequencies.

● Sensitivity is improved by using electronic amplifiers to boost the received radio signal.

● Selectivity is improved by using band-pass filter circuits.

● Amplitude modulation is the continuous variation in amplitude of a high-frequency carrier signal.

● Modulation percentage refers to how drastic the variations in amplitude are. It affects the relative loudness of the sound.

● An AM waveform can be thought of as a combination of sine waves at different frequencies, namely the carrier frequency and the sideband frequencies. This view is called the frequency-domain view.

● In AM, the upper sideband frequency is equal to the modulating frequency added to the carrier frequency; the lower sideband frequency is the modulating frequency subtracted from the carrier frequency.

● Sidebands are usually not isolated frequencies, but bands of frequencies, representing all the individual component frequencies in the modulating signal.

● The bandwidth of an AM radio signal is 2 times the highest frequency component in the modulating signal. The bandwidth is centered on the oscillator carrier frequency.

● In a radio receiver, the bandwidth of the band-pass filters and TRF amplifiers should be at least as wide as the bandwidth of the AM signal that is being received.

● Flattened frequency-response is preferred to bell-shaped frequency-response in a receiver's filters and RF amplifiers.

● Music and speech waveforms can be thought of as a mixture of many sine waves of different frequencies, ranging from about 20 Hz to about 10 kHz. (There are different opinions on this high-end frequency.)

● The carrier frequencies of commercial AM radio stations start at 540 kHz. They go in 10-kHz increments to 1700 kHz.

● Each AM station has a frequency bandwidth of 10 kHz, so adjacent-frequency stations just touch each other's sidebands. This gives them the potential to interfere with one another.

- An original sound wave is converted into an electrical voltage wave by a microphone. The voltage wave can be used as the modulating signal in a radio transmitter. Or it can be recorded for later playback.

- The heterodyning process involves mixing a tuned carrier frequency with a local oscillator frequency. This produces an output hat has 4 radio frequencies in its spectrum. They are: $f_{carrier}$, f_{local}, the sum of $f_{local} + f_{carrier}$, and the difference between f_{local} and $f_{carrier}$.

- The difference between f_{local} and $f_{carrier}$ is called the intermediate frequency, IF. A superhet receiver's amplifiers are permanently tuned to the IF.

- The IF arrangement eliminates the TRF problems of mistracking and nonconstant bandwidth.

- Automatic gain control (AGC) is a circuit design that automatically decreases the IF amplifier gain when a strong radio signal is being received, and increases the IF gain for a weak signal.

- Frequency modulation (FM) causes the frequency of the carrier wave to deviate slightly above and below the resting frequency. The amplitude stays constant. This is the time-domain view.

- A larger audio modulating signal produces a greater deviation in frequency (time domain).

- The frequency spectrum of an FM signal contains all the harmonics of the modulating frequency, both added to the resting frequency and subtracted from the resting frequency. This is the frequency-domain view.

- In FM, most of the higher-harmonic components are so small that they are insignificant (frequency domain).

- A larger modulation signal voltage causes more harmonics to become significant.

- For good sound fidelity, the carrier must be allowed to deviate by 75 kHz from the resting frequency (time domain). This produces a 150-kHz bandwidth (frequency domain).

- Commercial FM radio broadcasting is done at carrier frequencies ranging from 88.1 MHz to 107.9 MHz, in 0.2-MHz steps.

- An FM superhet radio receiver has an intermediate frequency of 10.7 MHz. It is conceptually the same as an AM superhet receiver, except for the demodulation process.

- An FM demodulator produces an instantaneous output voltage that is proportional to the frequency deviation.

- During troubleshooting, you should take the block diagram view of a radio receiver. By testing a block's output signal and input signal, you isolate the block that is interrupting the signal path.

QUESTIONS AND PROBLEMS

1. Radio messages are carried on _____ waves. These waves consist of a varying _____ field in one dimension and a varying _____ field in the other, perpendicular, dimension.

2. The device that radiates the waves that carry radio messages is called a(n) _____ .

3. Electromagnetic waves travel at what approximate speed? About how much time does it take an electromagnetic wave to travel across the U.S.? Assume a coast-to-coast distance of 4000 km (≈ 2440 miles).

4. A simple-as-possible radio system contains the following parts: 1) Oscillator; 2) Modulator; 3) Transmitting antenna; 4) Receiving antenna; 5) Demodulator; 6) Speaker. Draw a diagram showing how these parts are interconnected, and explain the function of each part.

5. In a radio receiver, we include filter circuit(s) that respond to a particular frequency band in order to give the receiver _____ .

6. In a radio receiver, we include amplifiers that boost the power of the antenna signal in order to give the receiver _____ .

7. Draw the block diagram of a tuned radio frequency (TRF) receiver, assuming 2 stages of RF amplification and 2 stages of audio amplification. Show the tuning devices.

8. Referring to your block diagram from Problem 7, which of the following would improve the receiver's selectivity?
   a) Adding an additional tuned RF amplifier stage.
   b) Adding an additional audio amplifier stage.
   c) Changing to a higher-efficiency speaker.

9. Repeat Question 8, but substitute the word sensitivity for selectivity.

10. Draw the time-domain waveform of an amplitude-modulated carrier. Show about 50% modulation percentage.

11. Repeat Problem 10 for a greater magnitude of modulating voltage.

12. Repeat Problem 10 for a higher frequency modulating signal.

13. A 4-MHz carrier is amplitude-modulated by a 2-kHz sine-wave audio signal. List the frequencies that are present in the waveform's spectrum.

14. A 4-MHz carrier is amplitude-modulated by a signal that is a combination of a small-magnitude 2-kHz sine wave and a larger-magnitude 3-kHz sine wave. Draw the resulting frequency spectrum graph (spectrograph). Show relative strengths.

15. A 4-MHz carrier is amplitude-modulated by a complex audio waveform whose highest frequency component is 5 kHz. The upper sideband is from _____ MHz to _____ MHz. The lower sideband is from _____ MHz to _____ MHz.

16. In Problem 15, the total bandwidth is _____ Hz.

17. For the AM signal of Problems 15 and 16, the receiver must contain filter circuits with Bw = _____ Hz, in order to reproduce the original audio waveform.

18. For a TRF receiver that meets the condition specified in Question 17, its amplifiers' low cutoff frequency will be _____ Hz and the high cutoff frequency will be _____ Hz.

19. Generally speaking, which frequency-response shape is preferred for a filter circuit in a receiver, bell-shaped or square-shaped? Explain why this is so.

20. Draw the structure of a crystal microphone. Describe its principle of operation.

21. Repeat Question 20 for a dynamic microphone.

22. Which of the following are advantages of the superheterodyne receiver over the TRF receiver?
    a) Better sensitivity.
    b) Consistent receiver bandwidth as different stations are tuned in.
    c) Elimination of the image-frequency problem
    d) Better frequency tracking as different stations are selected because fewer circuits are being tuned.

23. A superhet AM receiver is tuned to receive a station at 1450 kHz. What is the frequency of the local oscillator?

24. For the situation in Problem 23, list the four frequencies that appear at the output of the mixer.

25. Of the four frequencies in Problem 24, which one passes through the IF amplifier stages? Why?

26. What AM superhet circuit tends to keep the volume reasonably constant as reception conditions change?

27. In a TRF receiver, the demodulator's time constant is designed for optimum performance in the middle of the AM band, about 1100 kHz.
    a) At the lower AM frequencies, near 540 kHz, the demodulator's RF ripple will _____ . (Increase or decrease). Explain why.
    b) At the higher AM frequencies, over 1500 kHz, the demodulator will do a poorer job of following the _____ - frequency components of the audio modulation signal (Answer high or low). Explain.

28. Does a superhet AM receiver have the problems described in Question 27? Explain.

29. (T-F) AGC is just like standard negative feedback, in that it stabilizes the amplifier's voltage gain against change.

30. In Fig. 9-37(B), suppose the average demodulator voltage were –4 V rather than –3 V Find:
    a) V_{AGC} d) V_E g) A_v (stage-2, for the same n and r_C)
    b) V_{RB1} e) I_E h) A_v (total IF)
    c) V_B f) r_{Ej}

31. Draw the time-domain waveform of a frequency-modulated carrier.

32. Repeat Problem 31 for a greater magnitude of modulating voltage.

33. Repeat Problem 31 for a lower frequency of modulating voltage.

34. (T-F) Frequency modulation by a steady 1-kHz audio sine wave produces exactly 2 sideband frequencies, one of them 1 kHz higher that the carrier and the other 1 kHz lower than the carrier, just like AM.

35. A 4-MHz carrier is frequency-modulated by a 1-kHz sine wave. Draw the resulting frequency spectrum.

36. Repeat Problem 36, assuming a larger magnitude for the 1-kHz sine-wave modulating signal.
37. In commercial FM broadcasting, the bandwidth of one station is _____ kHz.
38. In an FM superhet receiver, the filter circuits have Bw = _____ kHz.
39. In an FM superhet receiver, the IF amplifiers have a tuned frequency of _____ MHz.
40. For the IF amplifiers of Question 39, the low cutoff frequency is _____ MHz and the high cutoff frequency is _____ MHz.
41. Explain how the limiter circuit in an FM receiver makes the receiver free from amplitude noise interference.
42. An FM demodulator produces an instantaneous voltage in response to _____ from the resting frequency.
43. If the IF signal holds steady at exactly 10.7 MHz for several seconds, a dc meter connected to the demodulator output in Fig. 9-47(A) will measure _____ V.
44. Which of the following test signals would be useful for troubleshooting an AM superhet receiver?
    a) 500-Hz sine wave.
    b) Amplitude-modulated 10.7 MHz.
    c) Frequency-modulated 455 kHz.
    d) Amplitude-modulated 455 kHz.
    e) Amplitude-modulated 1.2 MHz.
45. Which of the following test signals would be useful for troubleshooting an FM superhet receiver?
    a) 500-Hz sine wave.
    b) Frequency-modulated 10.7 MHz.
    c) Frequency-modulated 455 kHz.
    d) Amplitude-modulated 10.7 MHz.
    e) Frequency-modulated 98 MHz.

This machine is the *Ambler*, a six-legged 4-meter-tall mobile robot that can step over large crevices or large boulders. It may someday be used for remote-controlled exploration of the moon and planets.

Courtesy of Carnegie-Mellon University and NASA

CHAPTER 10

HIGH-POWER AMPLIFIERS

An outdoor aircraft engine test site. This model CF6-80C turbofan engine is manufactured by GE Aircraft Engines for commercial aircraft.

At takeoff, 800 kilograms per second of air pass through the engine, producing a maximum thrust of 282 kilonewtons (63 500 pounds).

With its sophisticated electronic subsystems, its overall fuel efficiency is about 40% greater than that of a comparable-power turbofan jet engine of 25 years ago.

Courtesy of GE Aircraft Engines

OUTLINE

Amplifier Efficiency
Class-A Operation Compared to Class B
Combining two Class-B Transistors — The Push-Pull Circuit
Practical Push-Pull Circuit — Class-AB Operation
Complementary Symmetry Amplifiers
   IC Power Amplifiers
Class-C Operation
Decibels
   Applying Decibels to Multistage Systems
Decibels Applied to Voltage Gain

NEW TERMS TO WATCH FOR

amplifier efficiency
class A
class B
crossover distortion
class AB
complementary symmetry
class C
frequency multiplying

decibel
common logarithm
attenuator
half-power condition
down 3 dB
audio decibel
characteristic impedance

All the transistor amplifiers in Chapters 5 through 9 were designed to handle small signals. They were low-power amplifiers. Typically, we saw output voltages of a few volts across load resistances of a few hundred ohms. This combination produced output powers in the milliwatt range. With power levels so small, we were not much concerned about the amplifier's efficiency, or its wasted power.

In this chapter, we will study transistor amplifiers that are designed for higher output powers. In general, these amplifiers can range from several watts to many kilowatts. For such amplifiers, the questions of efficiency and wasted power are of major concern to us.

After studying this chapter, you should be able to:

1. Find an amplifier's efficiency, given its signal input power, its load output power, and its dc supply power.
2. Explain the meaning of class-A operation of a transistor. Show where the Q-point is located along a class-A amplifier's ac load-line.
3. For resistance-coupled class-A amplifiers, state the theoretical maximum efficiency and the approximate range of real practical efficiencies.
4. Do the same for transformer-coupled class-A amplifiers.
5. Explain the meaning of class-B operation of a transistor. Show where the Q-point is located along a class-B amplifier's ac load-line.
6. Explain why class-B amplifiers have higher efficiencies than class-A amplifiers, by referring to the location of the Q-point.

7. Describe the operation of a class-B transformer-coupled push-pull amplifier. Explain why such an amplifier has a crossover distortion problem.
8. For class-B amplifiers, state the theoretical maximum efficiency and the approximate range of real efficiencies.
9. Explain the meaning of class-AB operation of a transistor. Show where the Q-point is located along a class-AB amplifier's ac load-line.
10. Describe the operation of a class-AB transformer-coupled push-pull amplifier. Explain why such an amplifier solves the problem of crossover distortion.
11. Given a detailed schematic diagram of a transformer-coupled push-pull amplifier, analyze it to find its voltage gain, A_v.
12. Describe the operation of a class-AB complementary-symmetry push-pull amplifier.
13. Describe the operation of a quasi complementary-symmetry amplifier. Explain the advantages of quasi complementary symmetry over plain complementary symmetry.
14. Explain the meaning of class-C operation of a transistor.
15. Referring to the waveforms of collector voltage and collector current for a class-C amplifier with an *LC* tank, explain why its efficiency is so high.
16. State a typical practical efficiency for a class-C amplifier with an *LC* tank in the collector circuit.
17. Describe the process of frequency multiplication using a class-C amplifier.
18. Given power gain A_P as a numeric ratio, express that power gain in decibels, using a hand-held scientific calculator.
19. Given power gain in decibels, convert it to a numeric power ratio using a calculator.
20. Explain the meaning of positive decibels, zero decibels, and negative decibels.
21. Convert power attenuation from numeric ratio to negative decibels, and vice-versa, using a calculator.
22. Use decibel addition and subtraction to find the overall power gain of a multistage system.
23. Given numeric-ratio power data, plot a frequency-response curve in decibels.
24. Give the conditions that allow voltage gain to be expressed in decibels.

10-1 AMPLIFIER EFFICIENCY

Every amplifier boosts power. Figure 10-1 shows the overall situation. A small amount of average input power, P_{in}, is delivered to the amplifier's input by the signal source. The amplifier produces an output signal that duplicates the waveshape of the input signal, but has larger voltage and/or current. This output signal causes a larger average output power, P_{out}, to be delivered to the load, which converts it into a useful product. P_{out} is larger than P_{in} because the amplifier circuitry receives power from the dc power supply. This power is symbolized as $P_{\text{(dc supply)}}$ in Fig. 10-1.

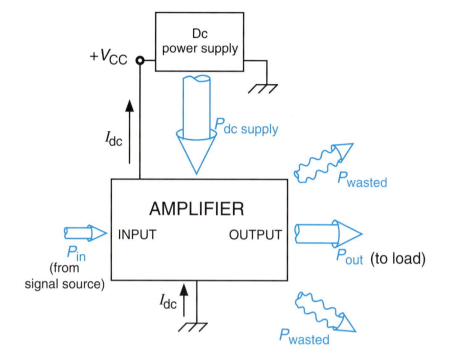

Figure 10-1
The various powers in an amplifier.

We wish that we could build amplifiers in which *all* of $P_{\text{(dc supply)}}$ went into P_{out}. But this is impossible. Instead, only a portion of $P_{\text{(dc supply)}}$ is converted to P_{out}; the rest is converted into heat and is dissipated into the surrounding air. This is symbolized as P_{wasted} in Fig. 10-1.

There are two reasons why we resent P_{wasted}.

1 It means that some of the energy that was available to be used from the dc supply never got properly used. Some of the earth's natural resources were consumed, and our money was spent to convert that energy and make it available. Both the natural resources and the money have been wasted because we did not obtain a useful product.

2 Before it can be dissipated into the air, the wasted energy appears as concentrated heat in the body of the transistor. This raises the transistor's temperature, which risks damaging it.

For these two reasons, we want to minimize P_{wasted}.

An **amplifier**'s **efficiency** is defined as the ratio of its output power to its total power consumption. Referring to Fig. 10-1,

$$\text{efficiency} \;=\; \eta \;=\; \frac{P_{\text{out}}}{P_{\text{(dc supply)}} + P_{\text{in}}} \qquad \text{Eq. (10-1)}$$

Get This

Efficiency is symbolized by the Greek letter η (eta), shown in Eq. (10-1). This equation gives efficiency as a decimal ratio. To express efficiency as a percentage, multiply by 100%.

EXAMPLE 10-1

A certain amplifier receives input power $P_{in} = 0.1$ W. It provides power gain $A_P = 300$. The amplifier's dc supply voltage is $V_{CC} = 25$ V and its average dc supply current is $I_{DC} = 7.0$ A.

a) Calculate $P_{(dc\ supply)}$, the power consumed by the amplifier from the dc power supply.

b) Find P_{out}, using $A_P = P_{out} \div P_{in}$.

c) Find the amplifier's efficiency, η.

d) Calculate P_{wasted} for this amplifier.

SOLUTION

a) In Fig. 10-1, $P_{(dc\ supply)}$ is given by the standard power equation, $P = VI$.

$$P_{(dc\ supply} = V_{CC} \cdot (I_{DC})$$

$$= 25\ V\ (7.0\ A) = \textbf{175 W}$$

b)

$$P_{out} = A_P \cdot (P_{in})$$

$$= 300\ (0.1\ W) = \textbf{30 W}$$

c) As stated in Eq. (10-1), the total power consumption is the sum of $P_{(dc\ supply)}$ and P_{in}.

$$P_{(dc\ supply)} + P_{in} = 175\ W + 0.1\ W = \textbf{175.1 W}$$

Applying Eq. (10-1) gives

$$\eta = \frac{P_{out}}{P_{(dc\ supply)} + P_{in}} = \frac{30\ W}{175.1\ W} = 0.171 \text{ or } \textbf{17.1\%}$$

This is a poor efficiency.

d) All power must be accounted for. So we can say

$$P_{(dc\ supply)} + P_{in} = P_{out} + P_{wasted} \qquad \textsf{Eq. (10-2)}$$

Rearranging gives

$$P_{wasted} = P_{(dc\ supply)} + P_{in} - P_{out}$$

$$= 175.1\ W - 30\ W = \textbf{145.1 W}$$

Poor efficiency implies that a great deal of power is wasted.

> Efficiencies of 10–20% are typical of real *RC*-coupled class-A amplifiers.

In this example, the input power is negligible compared to $P_{(dc\ supply)}$. This is usually the case. Therefore, we often ignore P_{in} in the efficiency equation, writing simply

$$\eta \approx \frac{P_{out}}{P_{(dc\ supply)}} \qquad \textsf{Eq. (10-3)}$$

Another efficiency expression is obtained by plugging Eq. (10-2) into Eq. (10-1), giving us

$$\eta = \frac{P_{out}}{P_{out} + P_{wasted}} \qquad \textsf{Eq. (10-4)}$$

10-2 CLASS-A OPERATION COMPARED TO CLASS-B

Operating a transistor with its Q-point in the center is called **class-A** operation. As Fig. 10-2(A) makes clear, class-A operation involves a substantial dc bias collector current I_C, and a substantial dc bias collector-emitter voltage V_{CE}. These two combine for a substantial dc bias power consumption, given by

The amplifier is called a **class-A amplifier**.

 ✔

$$P_{(dc\ supply)} = V_{CE} \cdot I_C$$

dc bias values

This power is consumed regardless of the size of the input signal, or regardless of whether an input signal is even present. This is the reason for the poor efficiency of class-A amplifiers, as pointed out in Example 10-1.

It can be shown that a class-A amplifier with resistor-capacitor output coupling has a maximum possible efficiency of only 25%. With transformer output coupling, the maximum possible efficiency is 50%.

Get This

These are theoretical maximums for the condition where the transistor is moving back and forth on its ac load-line from one extreme position to the other. Nonideal behavior of real transistors and transformers causes actual amplifier efficiencies to be even less than these theoretical maximums.

③ ✔

④ ✔

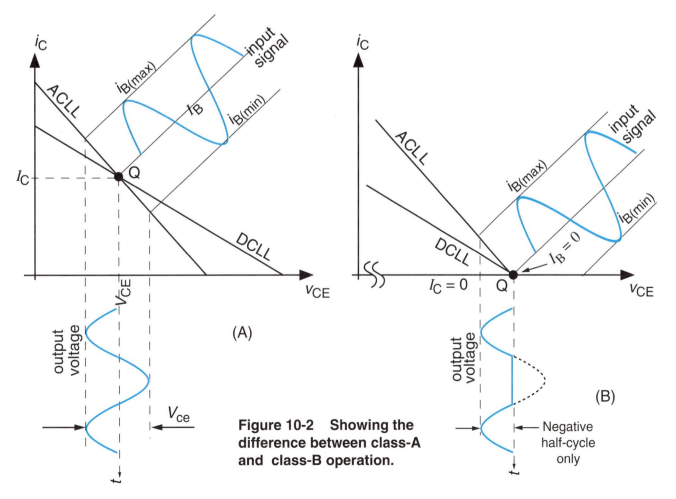

(A)

(B)

Figure 10-2   Showing the difference between class-A and class-B operation.

It is possible to design an amplifier so that the transistor's Q-point is at $I_C = 0$, just on the verge of conduction. This is called **class-B** operation. It is illustrated in the load-line diagram of Fig. 10-2(B). The advantage is that the dc bias power consumption drops to zero, since

$$P_{(\text{dc bias})} = I_C \cdot V_{CE} = (0 \text{ A}) (V_{CE}) = 0 \text{ W}$$

In other words, the amplifier consumes power only when an input signal is being amplified. The background dc power consumption has been eliminated. This dramatically improves the efficiency of the amplifier.

Of course, the great disadvantage of class-B operation is that only the positive half cycle can be amplified, not the negative half cycle. Figure 10-2(B) makes this clear.

Note that class-B versus class-A has nothing to do with the basic configuration of the amplifier circuit, whether it is common-emitter, common-base, or common-collector. Class of operation deals strictly with the location of the bias Q-point.

10-3 COMBINING TWO CLASS-B TRANSISTORS – THE PUSH-PULL CIRCUIT

We can bring in a second class-B-biased transistor to amplify the negative half cycle of the input signal. This gives us back the complete undistorted sine-wave output, like class-A, combined with the efficiency advantage of class-B. The circuit schematic is shown in Fig. 10-3.

Figure 10-3(A) shows the bias situation with no input signal present. It is easy to see that both base-emitter junctions have no forward bias, since the transformer secondary voltage is zero. Therefore both Q_1 and Q_2 are operating with their quiescent points at $I_C = 0$ mA, $V_{CE} = V_{CC}$. They are both class-B. The zero-signal background power consumption is 0 W, since $I_{C1} = I_{C2} = 0$.

When v_{IN} goes positive, the situation is shown in Fig. 10-3(B). The input coupling transformer produces voltage on its top half that is **+** on top, **−** at the center. This forward-biases the B-E junction of Q_1, causing it to conduct current. It moves up the ac load-line in Fig. 10-2(B). Collector current i_{C1} energizes the top half of the primary winding of T_2, the output coupling transformer. The T_2 secondary then induces the negative half cycle of the output signal.

Meanwhile, the bottom half of the T_1 secondary is **+** on the center tap, **−** on the bottom. This reverse-biases the B-E junction of Q_2, keeping it cut off. On the load-line diagram of Fig. 10-2(B), transistor Q_2 has hit its lower limit.

Figure 10-3(B) shows the situation when v_{IN} goes negative. Now Q_2 conducts and Q_1 is cut off. Current flows in the bottom half of the T_2 primary winding, in the reverse direction from Fig. 10-3(B). The T_2 secondary induces the positive half cycle of the output signal.

The events in Fig. 10-3 are shown synchronized with the circuit waveforms in Fig. 10-4.

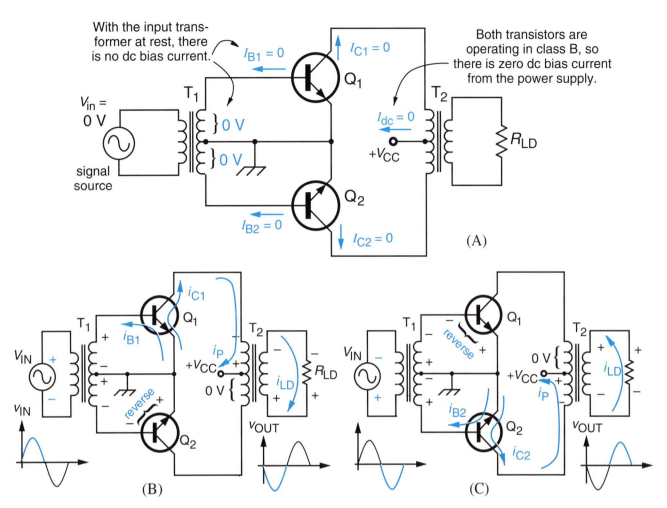

With the input transformer at rest, there is no dc bias current.

$I_{B1} = 0$

$I_{C1} = 0$

Q_1

Both transistors are operating in class B, so there is zero dc bias current from the power supply.

$V_{in} = 0\ V$

T_1

}0 V

}0 V

signal source

$I_{dc} = 0$

$+V_{CC}$

T_2

R_{LD}

Q_2

$I_{B2} = 0$

$I_{C2} = 0$

(A)

Figure 10-3
Push-pull amplifier. (A) Basic schematic arrangement. Both transistors have zero dc bias – class B.
(B) During the input's positive half-cycle, transistor Q_1 conducts and Q_2 remains cut off.
(C) During the input's negative half-cycle, Q_2 and Q_1 reverse their roles.

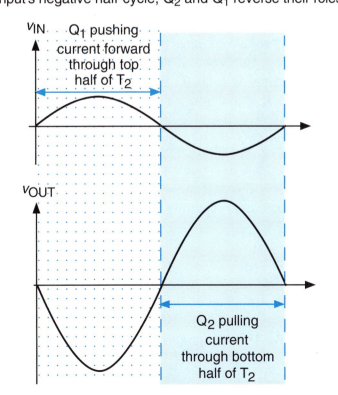

v_{IN} Q_1 pushing current forward through top half of T_2

v_{OUT}

Q_2 pulling current through bottom half of T_2

Figure 10-4
Handling the complete sine wave by alternating between two class-B transistors.

It can be shown that a class-B amplifier like this push-pull design has a theoretical maximum efficiency of 78.5%.

⑧ ✓

This theoretical maximum is for the transistor moving from one extreme to the other extreme along the ac load-line [from the cutoff Q-point to the saturation point in Fig. 10-2(B)]. As usual, nonideal component behavior causes actual efficiencies to be always less than the theoretical maximum.

SELF-CHECK FOR SECTIONS 10-1, 10-2 AND 10-3

1. A certain amplifier has $V_{CC} = 30$ V, $I_{DC\ (average)} = 8.5$ A. Its output power to the load is 50 W. The input power from the signal source is negligible.
   a) Calculate the amplifier's efficiency.
   b) What amount of waste power does the amplifier throw off?
2. In class-___ operation, the transistor is biased somewhere near the center of the load-line.
3. In class-___ operation, the transistor's Q-point is at cutoff, at the bottom of the load-line.
4. In class-B operation, the transistor responds to a(n) _____ in base drive, but cannot respond to a(n) _____ in base drive. (Increase or decrease)
5. The fundamental advantage of class-B operation over class-A operation is the higher _____ .
6. A _____ amplifier contains two transistors, both operating in class B, with transformer input coupling and transformer output coupling.

10-4 PRACTICAL PUSH-PULL CIRCUIT — CLASS-AB OPERATION

Figure 10-3 is good for understanding the basic functioning of a push-pull circuit, but it is not practical as it stands. The reason is that the transistors do not start conducting until the T_1 driving voltage reaches 0.6 V, enough to forward-bias the B-E junction. This causes the output voltage waveform to have a flat segment as it crosses through zero. Figure 10-5 illustrates this **crossover distortion** problem.

Inductive effects in the output coupling transformer make actual crossover distortion less abrupt than shown in Fig. 10-5. An oscilloscope photograph of a real crossover-distorted waveform is shown in Fig. 10-6.

The solution to the crossover distortion problem is to strike a compromise between class-A and class-B operation. We design a dc bias circuit that keeps the transistor just barely conducting. This places the Q-point just slightly above cutoff, as shown in Fig. 10-7. Now the operation is not quite class-B, but neither is it truly class-A. This compromise is called **class-AB** operation.

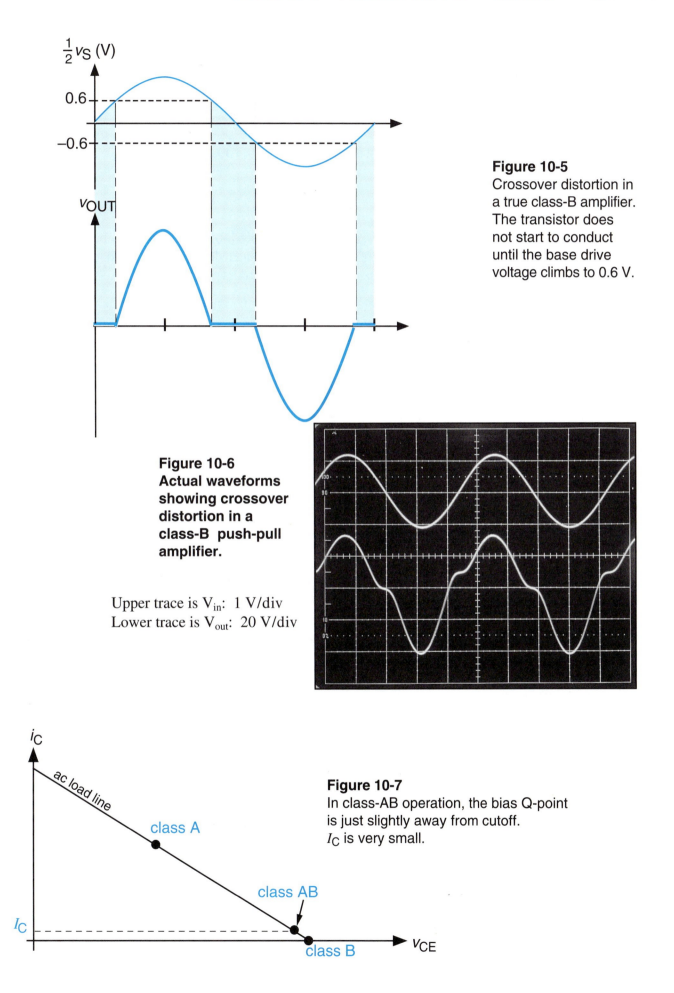

Figure 10-5
Crossover distortion in a true class-B amplifier. The transistor does not start to conduct until the base drive voltage climbs to 0.6 V.

Figure 10-6
Actual waveforms showing crossover distortion in a class-B push-pull amplifier.

Upper trace is V_{in}: 1 V/div
Lower trace is V_{out}: 20 V/div

Figure 10-7
In class-AB operation, the bias Q-point is just slightly away from cutoff.
I_C is very small.

The schematic diagram of a practical class-AB push-pull amplifier is shown in Fig. 10-8(A). In that circuit, R_1 and R_2 form a voltage divider driven by the 15-V V_{CC} supply. The voltage developed across R_2 applies a small forward bias to each B-E junction. Applying the voltage-division formula gives

$$\frac{V_{R2}}{15 \text{ V}} = \frac{R_2}{R_1 + R_2} = \frac{47 \text{ }\Omega}{1000 \text{ }\Omega + 47 \text{ }\Omega}$$

$$V_{R2} = 15 \text{ V} \left(\frac{47 \text{ }\Omega}{1047 \text{ }\Omega}\right) = 0.67 \text{ V}$$

This 0.67-V dc voltage is applied to each transistor's base-emitter path through its 1-Ω emitter resistor and one half of the T_1 secondary winding. The bias conditions are shown in Fig. 10-8(B).

In Fig. 10-8, 500-μF capacitor C_1 enables the ac base current to bypass R_2. Emitter resistors R_{E1} and R_{E2} raise the transistors' input resistances and help keep the voltage gains matched for both half cycles.

⌊10⌋ ✔

**Figure 10-8
Practical push-pull
circuit, with transistors
operating in class AB.**

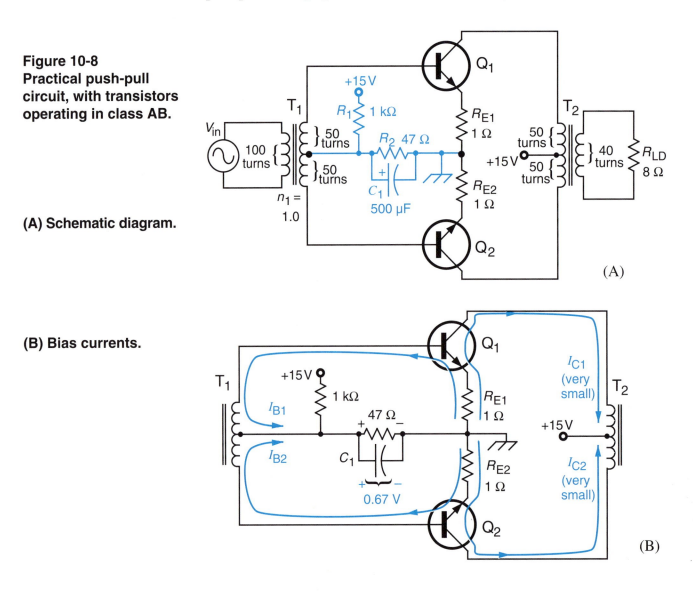

(A) Schematic diagram.

(B) Bias currents.

EXAMPLE 10-2

In Fig. 10-8(A), suppose the matched transistors have $\beta_1 = \beta_2 = 100$.
a) Draw the dc load-line of either transistor (say Q_1) on the collector characteristic curves. Consider transformer T_2 to be ideal.
b) Show the approximate Q-position on the dc load-line.
c) Find the equivalent ac resistance in the collector lead.
d) Find the slope of the ac load-line by applying Ohm's law to the collector-emitter ac path. Draw the proper slope on the collector characteristic curves.
e) Transfer the slope to the Q-point, thereby making the actual ac load-line.

SOLUTION

a) Figure 10-9 shows the family of characteristic curves for a $\beta = 100$ transistor (2 mA x 100 = 200 mA).

> With transformer-coupled output, there is ideally 0 Ω of dc resistance in the collector lead. Therefore, the slope of the dc load-line is very steep, nearly vertical (infinite).

Get This

A nearly vertical dc load-line is shown in Fig. 10-9, for $V_{CE} = V_{CC} = 15$ V.
b) The bias situation is shown in Fig. 10-8(B). Since 0.67 V barely exceeds the B-E turn-on voltage, we can't really make a reliable prediction of I_C. Let us just show it a bit larger than 0 mA, as indicated by the Q- point in Fig. 10-9.
c) Figure 10-8(A) shows T_2 having 40 turns in its secondary winding and 50 turns in each half of its primary winding. Since only half the primary winding is working at any moment, the effective turns ratio is

The dc load-line is not absolutely vertical due to the small amount of emitter resistance, here just 1 Ω.

$$n_{2\text{ (effective)}} = \frac{40 \text{ T}}{50 \text{ T}} = 0.8$$

Therefore the ac resistance seen in the primary circuit is given by

$$r_C = \frac{1}{n^2} R_{LD}$$

$$= \frac{1}{(0.8)^2} \, 8 \, \Omega = \mathbf{12.5 \, \Omega}$$

d) The total equivalent ac resistance is the sum of $r_C + R_E$. That is,
$$r_T = r_C + R_E = 12.5 \, \Omega + 1 \, \Omega = 13.5 \, \Omega$$
To find the slope of the ac load-line, we calculate the ac current that would cause a voltage of 15 V across the total ac resistance. This is the procedure that we learned in Sec. 7-7.

$$I_{max} = \frac{V_{CC}}{r_T} = \frac{15 \text{ V}}{13.5 \, \Omega} = 1.11 \text{ A or } 1110 \text{ mA}$$

Drawing a line between $V_{CE} = 15$ V and $I_C = 1.11$ A gives the proper slope, as shown in Fig. 10-9.
e) We slide the slope line slightly to the right so that it passes through the Q-point. This has been done in Fig. 10-9, to produce the ac load-line.

During the "push" half cycle when transistor Q_1 is conducting, its instantaneous collector current and collector-to-emitter voltage will move along this ac load-line. During the "pull" half cycle, transistor Q_2 will move along an identical load-line.

Figure 10-9
Showing the dc load-line, the approximate Q-point, and the ac load-line for one of the transistors in Fig. 10-8.

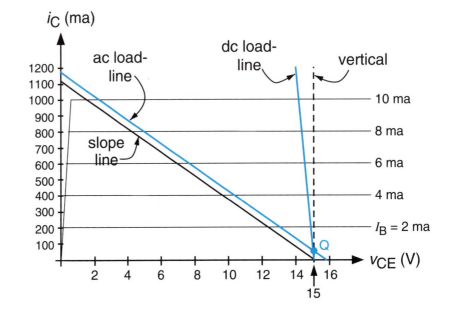

EXAMPLE 10-3

Now that we have the ac load-line, we can describe the specific operation of this push-pull class-AB amplifier.

Figure 10-8(A) shows transformer T_1 with 50 turns in each half of its secondary winding, and 100 turns in its primary winding. Since only half the secondary is working at any moment, the effective turns ratio is

$$n_{1 \text{ (effective)}} = \frac{50 \text{ turns}}{100 \text{ turns}} = 0.5$$

Let us assume that V_{in} has a peak value of 2 V.

a) Find the peak value of the sine-wave voltage that drives each transistor's base-emitter circuit.

b) Calculate the peak value of the emitter current, by applying Ohm's law to the base-emitter circuit. Neglect the effect of internal ac resistance r_{Ej} (since the peak value of I_E is very large), and assume that the transformer winding has zero resistance.

c) Locate the farthest point that each transistor moves up its ac load-line. Mark that point.

d) Using the farthest point on the load-line, use Ohm's law with r_C to find the peak voltage across the T_2 primary winding.

e) Find the peak value of voltage across the load.

f) Calculate the overall voltage gain $A_{\tilde{v}}$ of this amplifier.

SOLUTION

a) Since T_1 has an effective turns ratio of $n = 0.5$, we get

$$V_{S(pk)} = n\, V_{P(pk)} = (0.5)\,(2\text{ V}) = \mathbf{1\ V}$$

b) Since the transformer has $0\ \Omega$, $r_{Ej} = 0\ \Omega$ (we're neglecting it), and X_{C1} is assumed to be very low, the base-emitter circuit has only one current-limiting element, namely $R_E = 1\ \Omega$. Therefore we say

$$I_{e\,(pk)} = \frac{V_{S\,(pk)}}{R_E} = \frac{1\text{ V}}{1\ \Omega} = \mathbf{1\ A} \approx I_{c\,(pk)}$$

c) The ac load-line is reproduced in Fig. 10-10. $I_{c(pk)} = 1$ A, so the farthest point, P, is shown at 1 A.

d) Applying Ohm's law to r_C gives a T_2 primary voltage of

$$V_{P(pk)} = I_{c(pk)}\, r_C$$
$$= (1\text{ A})\,(12.5\ \Omega) = \mathbf{12.5\ V}$$

e) T_2 has an effective turns ratio of 0.8, as pointed out in Example 10-2. Therefore

$$V_{S(pk)} = (n_{2\,(effective)})\, V_{P(pk)}$$
$$= (0.8)\,(12.5\text{ V}) = \mathbf{10\ V} = V_{ld(pk)}$$

During the push half-cycle, Q_1 produces a sine-shaped load-voltage waveform of 10-V peak. During the pull half-cycle, Q_2 produces an opposite-polarity sine-shaped voltage waveform of 10-V peak. The overall load voltage waveform is a complete sine wave of 20 V peak-to-peak (7.07 V rms).

Figure 10-11 pictures the complete sine wave of primary current.

f) We know the peak values of the input voltage and the output (load) voltage. Voltage gain can be found from

$$A_v = \frac{V_{ld\,(pk)}}{V_{in\,(pk)}} = \frac{10\text{ V}}{2\text{ V}} = \mathbf{5.0}$$

⑪ ✔

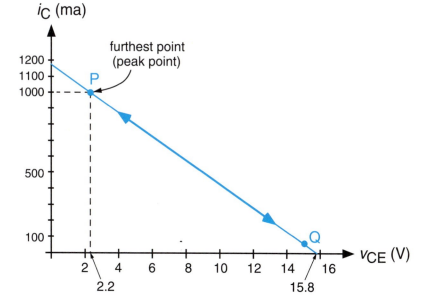

Figure 10-10
The ac load-line for either transistor in the push-pull amplifier. During a half cycle, the instantaneous transistor operating point moves from the Q-point to the peak point, then back to the Q-point.

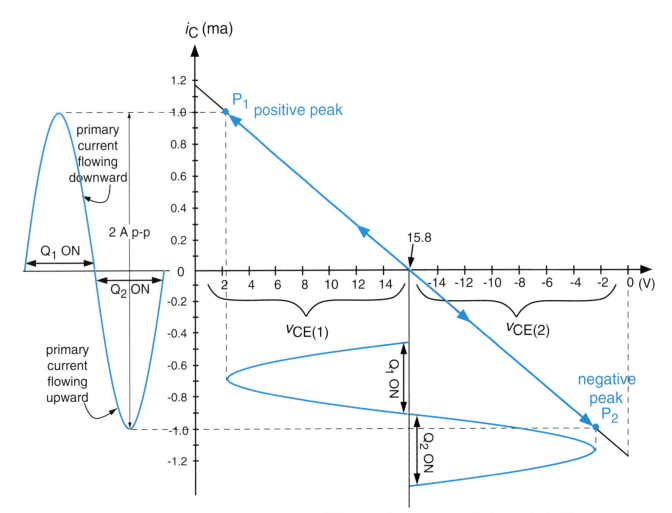

Figure 10-11 Showing how the push-pull amplifier produces a complete cycle in T_2.

If we continued analyzing this amplifier, we would find that it has a current gain of

$$A_i = 250$$

Finding this A_i value has been given as a problem at the end of the chapter.

The power gain is

$$A_P = A_v \cdot A_i = (5.0)(250) = 1250$$

We could make an estimate of the amplifier's efficiency as follows:

1 Calculate the average output (load) power as

$$P_{LD} = \frac{\left[V_{ld(rms)}\right]^2}{R_{LD}} = \frac{\left[(0.707)V_{ld(pk)}\right]^2}{R_{LD}} = \frac{[7.07 \text{ V}]^2}{8 \ \Omega}$$

$$= 6.25 \text{ W}$$

2 Estimate the average power wasted in the 1-Ω emitter resistors as

$$P_{wasted(RE)} = [I_{e(rms)}]^2 R_E$$
$$= [(0.7071)(1 \text{ A})]^2 (1 \ \Omega) = 0.5 \text{ W}$$

3 Estimate the average power wasted in the transistor during the run up and then back down the load-line in Fig. 10-10. By advanced math, this can be calculated as 3.3 W. It could be approximated by just multiplying the average

value of collector current i_C times the average value of transistor voltage v_{CE}. The average value of i_C is about 0.5 A. The average value of v_{CE} is about 6.8 V. This gives an average wasted power of

$$(0.5 \text{ A}) (6.8 \text{ V}) = 3.4 \text{ W} \text{ (approximate)}$$

Average v_{CE} is about $1/2 (15.8 \text{ V} - 2.2 \text{ V}) = 6.8 \text{ V}$. Look closely at Fig. 10-11 to see the 2.2 V.

$\boxed{4}$ Add together the wasted powers from steps $\boxed{2}$ and $\boxed{3}$ to get an estimate of total P_{wasted}.

$$P_{\text{wasted}} = 0.5 \text{ W} + 3.4 \text{ W} = 3.9 \text{ W}$$

$\boxed{5}$ Apply Eq. (10-4).

$$\eta = \frac{P_{\text{out}}}{P_{\text{out}} + P_{\text{wasted}}}$$

$$= \frac{6.25 \text{ W}}{6.25 \text{ W} + 3.9 \text{ W}} = \frac{6.25 \text{ W}}{10.15 \text{ W}} = 62\%$$

This is a big improvement over a transformer-coupled class-A amplifier, which would give an efficiency calculation of about 40%, assuming ideal transformers.

SELF-CHECK FOR SECTION 10-4

7. In Fig. 10-8(A), imagine that the T_2 secondary has 60 turns.
   a) Find the equivalent ac resistance seen in the collector lead.
   b) For the same $I_{c(pk)}$ value of 1 A, calculate the primary voltage.
   c) Find the peak value of load voltage.
8. In Example 10-3, we ignored r_{Ej}. In this push-pull amplifier, even if we tried to take r_{Ej} into account, we couldn't do a proper job of it. Explain why this is so.
9. An ideal class-B amplifier has a maximum theoretical efficiency of $\eta = 78.5\%$. What can you say about the maximum theoretical efficiency of a class-AB amplifier like Fig. 10-8?
10. In Example 10-3, the input signal was $V_{\text{in(pk)}} = 2$ V and our estimated amplifier efficiency was 62%. If the input signal increased to $V_{\text{in(pk)}} = 2.3$ V, the amplifier efficiency would _____ . (Increase or decrease). Explain why.

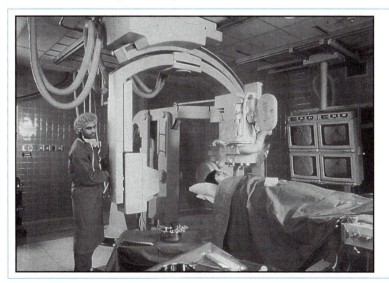

A cardiac video imaging system similar to the one shown on page 224. This particular system provides simultaneous viewing of the heart in two different planes.

Courtesy of Philips Medical Systems North America Co.

10-5 COMPLEMENTARY SYMMETRY AMPLIFIERS

Another high-power amplifier that operates in class-AB uses **complementary symmetry** design. The phrase complementary symmetry means that there are two transistors with identical betas (symmetry) but one is *npn* and the other is *pnp* (complementary). A schematic diagram is shown in Fig. 10-12.

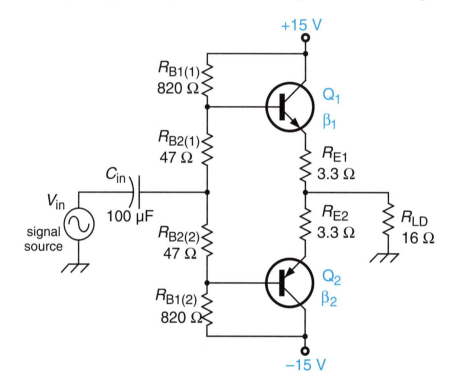

**Figure 10-12
Complementary
symmetry amplifier.
β_1 and β_2 must be
nearly identical.**

For ease in speaking, we often abbreviate the words complementary symmetry to comp-sym.

The comp-sym amplifier has no transformers. It uses capacitive coupling at the input and direct coupling at the output. Therefore it can operate at lower frequencies than a standard push-pull amplifier that uses audio transformers.

Because it does away with the transformers, it also is lighter in weight and less expensive. However, the complementary symmetry amplifier does require a dual-polarity power supply, if the output voltage is to be centered on ground (0-V dc level). Here is how the amplifier works.

Bias Conditions: Look at the 4-resistor series voltage-divider on the left side of the schematic. The top 2 resistors drop 15 V and the bottom 2 resistors also drop 15 V. Therefore the voltage at the junction of the 47-Ω resistors is 0 V relative to ground, as Fig. 10-13 shows.

The voltage across the 47-Ω $R_{B2(1)}$ resistor can be found by voltage division.

$$\frac{V_{RB2(1)}}{15\ V} = \frac{R_{B2(1)}}{R_{B2(1)} + R_{B1(1)}} = \frac{47\ \Omega}{47\ \Omega + 820\ \Omega}$$

$$V_{RB2(1)} = 15\ V \left(\frac{47\ \Omega}{867\ \Omega}\right) = 0.8\ V$$

This 0.8 V appears across the B-E junction of Q_1 in series with the 3.3-Ω R_{E1} resistor. Figure 10-13 shows where this voltage appears.

The very same bias conditions occur on the bottom half of the divider circuit. The 0.8 V developed across $R_{B2(2)}$ is used to drive the series combination of the Q_2 B-E junction and R_{E2}. As Fig. 10-13 points out, this works because the dc voltage at the junction of 3.3-Ω resistors R_{E1} and R_{E2} is also 0 V, the same as at the junction of $R_{B2(1)}$ and $R_{B2(2)}$.

For each transistor, a bit more than 0.6 V appears across the B-E junction. This leaves a bit less than 0.2 V across each R_E resistor. Let us assume 0.15 V. Then applying Ohm's law to R_E gives

The I_E calculation is the same for both transistors because they are selected to have identical B-E junction characteristics (V_{BE} is the same for both).

$$I_E \approx \frac{0.15 \text{ V}}{3.3 \ \Omega} = 45 \text{ mA}$$

This comparatively small current establishes the Q-point of each transistor. Thus, both transistors are operating in class-AB, like the push-pull amplifier in Sec. 10-4.

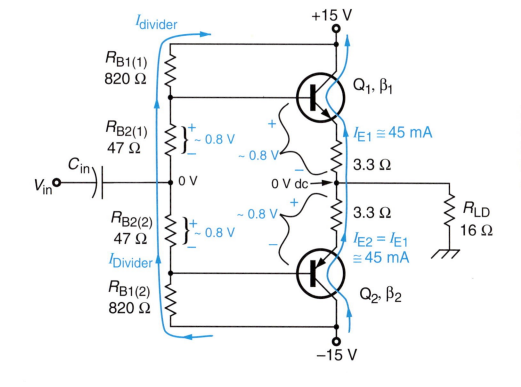

Figure 10-13
Dc bias conditions. Both Q_1 and Q_2 are biased barely on – class AB.

Ac Operation: When the input signal goes positive, the ac situation is shown in Fig. 10-14(A). The **+** v_{IN} polarity drives the Q_1 base in the forward direction, through 47-Ω resistor $R_{B2(1)}$. Therefore Q_1 turns on harder, moving up the load-line from its Q-point. The transistor has a common-collector configuration. It is driving a total emitter resistance of 19.3 Ω. It provides no voltage gain, but plenty of current gain. The ac current passes through R_{LD} from bottom to top and then through Q_1 to the +15-V supply. This forms the positive half-cycle of V_{ld}.

$R_{E1} + R_{LD} = 3.3 + 16$
$= 19.3 \ \Omega$

Meanwhile, the +v_{IN} polarity attempts to drive the Q_2 base through 47-Ω resistor $R_{B2(2)}$. But this transistor is *pnp*, so its B-E junction becomes reverse-biased, not forward biased. Q_2 moves slightly down its load-line into cutoff condition. This explains why R_{E2} does not appear in the emitter circuit of the turned-on transistor, Q_1.

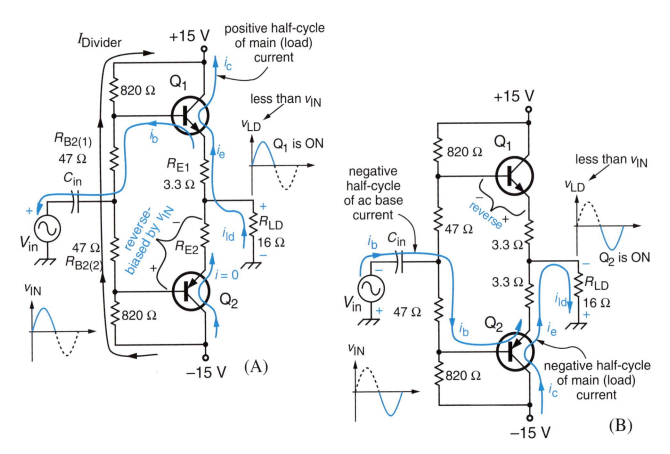

Figure 10-14 Ac operation of complementary symmetry amplifier. (A) Positive half-cycle. (B) Negative half-cycle. A comp-sym amplifier is common-collector, so $A_v < 1$.

When the input signal goes into its negative half cycle, the ac situation is shown in Fig. 10-14(B). The roles of the transistors reverse. *pnp* transistor Q_2 becomes forward-biased and it turns on, while *npn* transistor Q_1 becomes reverse-biased and moves into cutoff. Now the main current path is from the –15-V supply, through Q_2, and through R_{LD} from top to bottom. This forms the negative half-cycle of V_{ld}.

In Figs. 10-12, 13, and 14, the 0.8-V forward bias voltage is established by a resistive voltage divider. In a practical comp-sym amplifier, each 47-Ω resistor would be replaced with a diode and a lower-value resistor, as shown in Fig. 10-15. This provides better temperature-stability of the transistors' Q-points. Any temperature-related change in the transistors' V_{BE} values would be compensated for by an equal change in the V_F values of diodes D_1 and D_2.

To make sure that the diodes temperature–track the transistors, they are usually mounted on the same heat sink with the transistors.

Figure 10-15 illustrates another complementary symmetry feature. It is not really necessary to have dual-polarity dc supplies. However, operating with a single-polarity 30-V supply causes the center points to be at +15 V, not 0 V dc. If this dc voltage must be blocked from the load, then the circuit must have a very large output coupling capacitor C_{out}.

X_{Cout} must be negligible in comparison to the low resistance values 3.3 Ω and 16 Ω. Therefore C_{out} must be quite large.

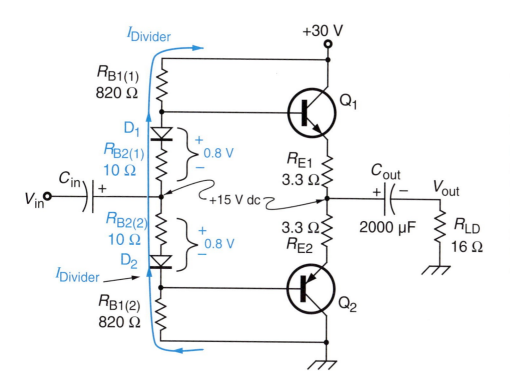

Figure 10-15
Comp-sym amp with temperature-stabilized bias and single-polarity supply. The peak ac input current is less than the dc current $I_{divider}$. Therefore D_1 and D_2 never become reverse-biased. They remain as conductive paths to ac.

A variation on the plain complementary symmetry design is **quasi** complementary symmetry. In this design, two Darlington pairs are used in the top and bottom halves of the circuit, as shown in Fig. 10-16. The bottom Darlington pair has a *pnp* transistor, Q_3, driving an *npn* transistor, Q_4.

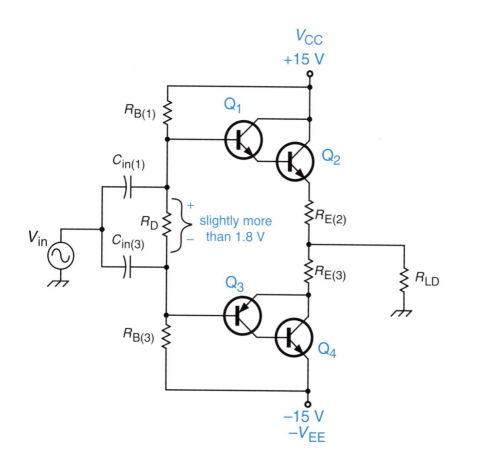

Figure 10-16
Quasi complementary-symmetry amplifier. Q_1 must be symmetrical with Q_3 and Q_2 must match Q_4.

Quasi complementary symmetry has two advantages over straight comp-sym.

[1] It is easier to match the betas and B-E junction characteristics of *pnp* and *npn* transistors if they are low-power models. The quasi circuit of Fig. 10-16 allows Q_1 and Q_3 to be low-power, rather than high-power models.

[2] Q_1 and Q_3 have very low dc bias currents, since the Darlington pairs have such high betas. Therefore the voltage divider current can also afford to be very low, improving the amplifier's overall efficiency.

Figure 10-16 shows another feature sometimes seen on comp-sym amplifiers – dual input capacitors. They couple the input signal straight to the base terminals of the transistors, avoiding all the components in the voltage-divider circuit. Such two-capacitor coupling could have been used on any of the amplifiers in Figs. 10-12 through 10-15.

Another wrinkle: In Figs. 10-12 through 10-15, it might have been possible to eliminate C_{in} altogether, since the right side of C_{in} connects to a point that is *already* at 0 V dc.

IC Power Amplifiers

Bipolar transistor integrated circuit power amplifiers are available with power output ratings in the 5 to 20 W range. Such IC amplifiers have a final output stage that uses quasi complementary symmetry, or a variation on that idea. A photo of an 8-W, 2-channel IC audio amplifier is shown in Fig. 10-17(A). A simplified schematic diagram of one channel is shown in Fig. 10-17(B). A complete stereo amplifier using a type No. LM2896P IC is shown in Fig. 10-17(C).

(A)

Figure 10-17 (A) Type No. LM2896P audio power amplifier that uses a variation of the quasi complementary symmetry idea. The large metal tab of the IC is in thermal contact with the collectors of transistors Q_2 and Q_4. It will be attached to a heat sink that is mounted on the printed-circuit board, with protruding heat-radiating fins.

SELF-CHECK FOR SECTION 10-5

11. A complementary symmetry amplifier contains one _____ transistor and one _____ transistor.
12. In a comp-sym amplifier, the transistors operate in class-_____ .
13. In a comp-sym amplifier, the transistors are in common-_____. configuration.
14. For a comp-sym amplifier, the voltage gain A_v is _____ .
15. Describe the basic advantage of a complementary symmetry amplifier over the amplifier of Fig. 10-8.
16. Draw the schematic symbol for an integrated circuit amplifier.

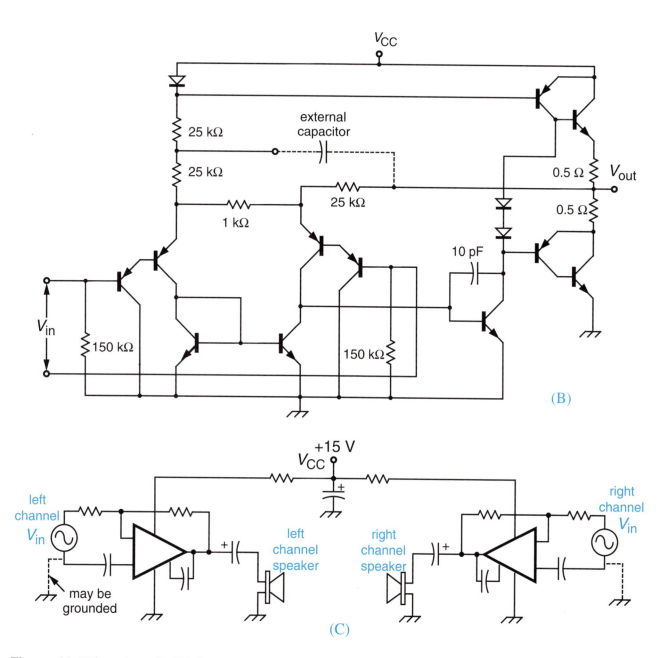

Figure 10-17(continued) (B) Simplified schematic diagram of one channel (one half) of the IC. (C) Complete stereo power amplifier. Each IC power amp is symbolized by a large triangle.

10-6 CLASS-C OPERATION

In class-C operation, the transistor is biased beyond cutoff. A simple example circuit is shown in Fig. 10-18(A).

The dc bias conditions are given in Fig. 10-18(B). With V_B actually reverse-biasing the B-E junction, the transistor cannot even get out of cutoff unless it has help from the ac input. This is what we mean when we say that class-C operation is biased beyond cutoff.

The Q-point is shown in Fig. 10-18(C). It appears the same as class-B operation. However, the transistor does not begin to move up the ac load-line at the instant that its ac input signal becomes 0.6-V positive, like class-B. Instead, it remains stuck at the Q-point until the ac input signal gets close to its positive peak.

Get This

A class-C-operated transistor conducts for only a short time duration, near the positive peak of the input signal. The transistor circuit itself does not attempt to duplicate the input waveshape. Instead, it produces only short pulses at the output.

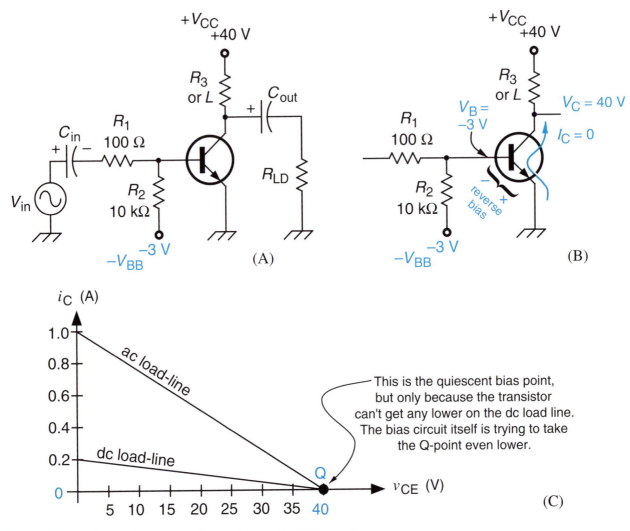

Figure 10-18 Class-C-operated transistor. (A) Schematic.
(B) Dc bias conditions. The B-E junction is reverse-biased, not zero-biased.
(C) Load-lines and Q-point.

A class-C transistor circuit is an amplifier only in the sense that it produces greater output power than input power. It is not an amplifier like all the others we have studied, that duplicate the input waveform.

In Fig. 10-18(C), the transistor moves up and then back down the ac load-line as the input signal passes through a narrow range near the positive peak (assuming a sine-wave input). The waveform graphs are shown in Fig. 10-19.

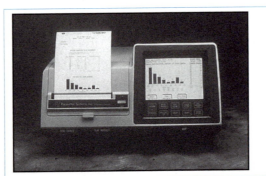

This instrument is used to reprogram an implantable heart pacemaker by radio, if the patient's condition varies. It avoids the need for surgery to get access to the pacemaker.

Courtesy of Siemens Pacesetter, Inc.

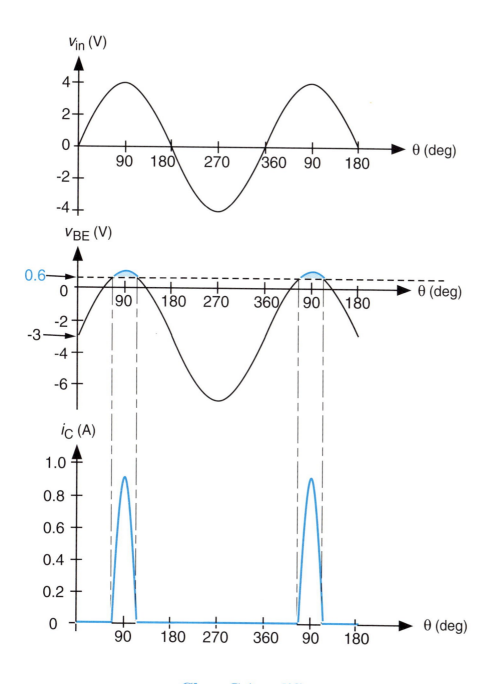

Figure 10-19
Waveforms for the class-C "amplifier" of Fig. 10-18. The narrow current pulses centered on the 90° points contain much greater power than the input signal.

Class-C Amplifier
with *LC* Tank Circuit in the Collector

Real class-C circuits almost never have a collector resistor like Fig. 10-18. Instead, they contain an *LC* circuit that is tuned to the frequency of the input signal. This is shown schematically in Fig. 10-20. The amplifier functions as follows:

The *LC* tank circuit oscillates with a peak voltage of 40 V, as shown in the waveform of Fig. 10-21(B). The oscillations synchronize themselves with the pulses of collector current shown in Fig. 10-21(D). The i_C pulses, of course, replenish the energy that was lost on the previous cycle of oscillation. This is the same idea that was put forward in Sec. 8-1, regarding oscillator circuits. Here, though, the power level is much greater. This is because when a load is transformer-coupled to the *LC* tank, as shown in Fig. 10-22, a great deal of

Class-C amplifiers share an idea with oscillators — the idea of replacing lost energy in order to keep the oscillations going.

v_{LC} is near its positive peak value, + on top and − on bottom, at the time of the narrow i_C pulse.

energy is delivered to the load. This energy is lost to the LC tank, requiring a large injection of makeup energy from the transistor amplifier.

Figure 10-20
Class-C amplifier withan LC tank circuit. The LC tank is tuned (its resonant frequency is adjusted) to match input frequency f_{in}.

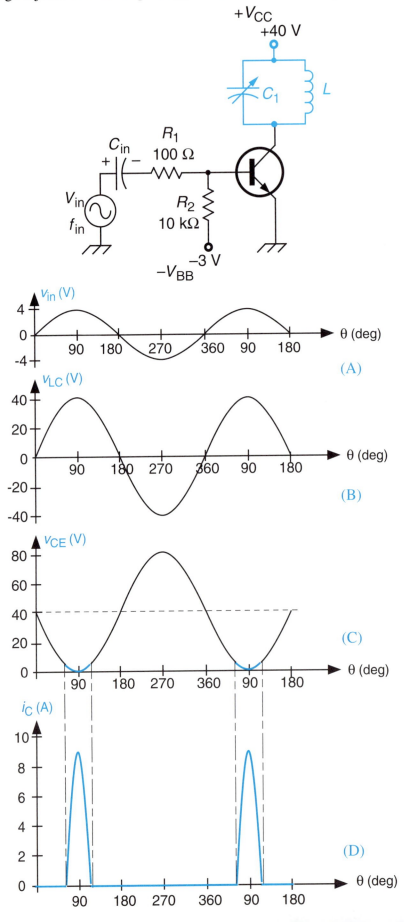

Figure 10-21
Waveforms for class-C amplifier of Fig. 10-20.

Radio transmitters almost always use a class-C circuit as the final amplifier stage to drive the transmitting antenna. For FM transmission, you can think of the C_1–L_P tank circuit as resonating at the radio carrier frequency in Fig. 10-22. Think of V_{in} as a frequency-modulated sine-wave signal, deviating around the carrier frequency. The load is the transmitting antenna.

Class-C Efficiency: A transformer-coupled class-C amplifier is even more efficient than a class-B amplifier. This can be understood by carefully studying the waveforms in Fig. 10-21. The waveform of collector-emitter voltage, v_{CE}, oscillates between +80 V ($2 \times V_{CC}$) and 0 V, as shown in Fig. 10-21(C). v_{CE} dips down to 0 V when the v_{CE} waveform in part (B) is at its positive peak. By Kirchhoff's voltage law,

$$v_{CE} = V_{CC} - v_{LC}$$
$$= 40\text{ V} - 40\text{ V} = 0\text{ V}$$

Modern class-C amps for radio transmission are switched abruptly from cutoff to saturation, shortly after the start of the positive half cycle of v_{IN}. Their collector current and voltage waveforms are nearly square, not sine-shaped as shown in Figs. 10-19 and 20. The sine-wave shape of the input is then restored by filtering out the square-wave's harmonics before application to the antenna.

Therefore, when collector current pulses occur, the transistor's v_{CE} voltage value is near zero. The transistor's wasted power is given by the product of voltage and current, or

$$p_{wasted} = (i_C) \cdot v_{CE} = i_C \cdot (\approx 0\text{ V}) \approx 0\text{ W}$$

Get This

The near-zero power waste in the transistor makes a transformer-coupled class-C amplifier nearly 100% efficient, ideally.

⑮ ✔

EXAMPLE 10-4

In Fig. 10-22, assume that the transformer has a turns ratio of 2.0 and $R_{LD} = 50\ \Omega$.
a) Referring to the waveforms of Fig. 10-21, find the output power to the load.
b) Assume that this class-C amplifier is in conduction for 80 degrees in each cycle. That is, the i_C pulse begins at $50°$ and ends at $130°$. This is an $80°$ difference, centered on the $90°$ point. Calculate the value of v_{CE} at the beginning and ending instants of conduction. Refer to Fig. 10-21(B) and (C).
c) Find an appropriate "average" value of v_{CE} during the conduction angle. Then multiply the average value of v_{CE} by the "average" value of i_C (say 8 A) to find the power wasted in the transistor during conduction.
d) Find the average power wasted in the transistor over the full cycle, by proportioning the conduction angle.
e) Use the results from parts (a) and (d) to estimate the amplifier's efficiency.

SOLUTION

a) The transformer's primary voltage is 40 V peak, the value of V_{LC}. This is an rms value of

$$V_{P(rms)} = (0.707)(40\text{ V}) = 28.3\text{ V}$$

By the transformer voltage law,

$$V_S = V_P n = 28.3 \text{ V} (2.0) = 56.6 \text{ V} = V_{\text{ld(rms)}}$$

The load power is

$$P_{LD} = \frac{V_{\text{ld}}^2}{R_{LD}} = \frac{(56.6 \text{ V})^2}{50 \, \Omega} = \textbf{64.0 W}$$

b) At 50°, v_{LC} is given by

$$v_{LC}\big|_{50°} = (40 \text{ V}) (\sin 50°) = 40 \text{ V} (0.766)$$
$$= 30.6 \text{ V}$$

Collector-emitter voltage v_{CE} is given by

$$v_{CE} = V_{CC} - v_{LC} = 40 \text{ V} - v_{LC}$$

At 50°, this gives

$$v_{CE}\big|_{50°} = 40 \text{ V} - v_{LC}\big|_{50°} = 40 \text{ V} - 30.6 \text{ V} = \textbf{9.4 V}$$

The very same value occurs at 130°.

c) Since v_{CE} dips from 9.4 V to 0 V, then back to 9.4 V during the conduction angle, we can approximate its average value as simply half of 9.4 V, or

$$v_{CE \text{ (avg)}} \approx \textbf{4.7 V}.$$

The approximate transistor power waste is then

$$P_{\text{wasted}} \approx [v_{CE(\text{avg})}] \cdot [i_{C(\text{avg})}]$$
$$= (4.7 \text{ V}) \, 8 \text{ A} = \textbf{38 W}$$

d) The transistor conducts for only 80° out of 360°. Therefore, the average wasted power over the complete cycle is given by proportion as

$$P_{\text{wasted}} = \left(\frac{80°}{360°}\right) 38 \text{ W} = (0.222) \, 38 \text{ W} = \textbf{8.4 W}$$

e) From Eq. (10-4),

$$\eta = \frac{P_{\text{out}}}{P_{\text{wasted}} + P_{\text{out}}} = \frac{64.0 \text{ W}}{8.4 \text{ W} + 64.0 \text{ W}} = \frac{64.0 \text{ W}}{72.4 \text{ W}} = \textbf{88\%}$$

This efficiency is better than any class-B amplifier can achieve.

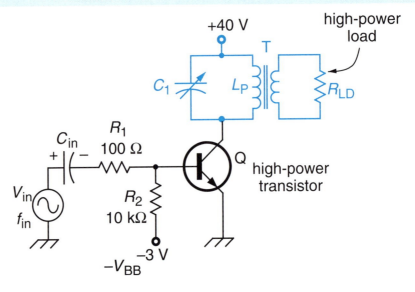

**Figure 10-22
Class-C amplifier
with transformer-
coupled output.**

Frequency Multiplying: It is not absolutely necessary that a class-C amplifier's *LC* tank circuit be tuned to the input frequency f_{in}. It can be tuned to resonate at 2 times or 3 times f_{in}. For example, if f_{in} is 1 MHz, and the resonant frequency of the tank is designed for 3 MHz, then the waveform situation is shown in Fig. 10-23.

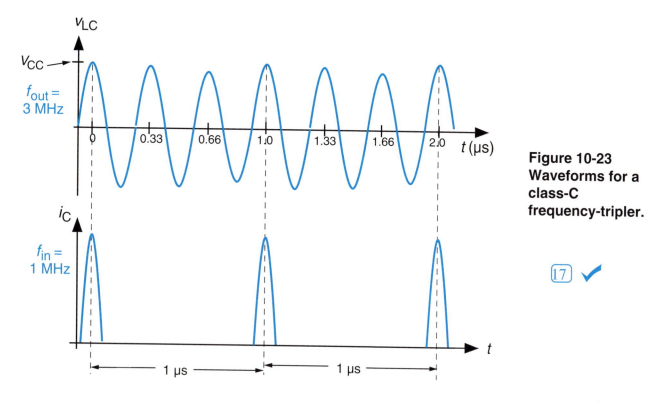

**Figure 10-23
Waveforms for a class-C frequency-tripler.**

17 ✔

Now, rather than the collector current pulse replenishing the lost energy on every cycle of the *LC* tank, it replenishes the energy on every 3rd cycle. Of course, the amount of energy injected per i_C pulse has to be 3 times larger than if the pulses came once every oscillation cycle. Therefore the magnitude of the i_C pulse must be 3 times as great.

This **frequency-multiplying** method cannot generally be used for larger multiplication factors, like 4, 5 or 10. Too much energy would be lost during the coasting cycles.

Since the *LC* tank oscillations must "coast" for 2 cycles, they have some loss in magnitude, as Fig. 10-23 shows.

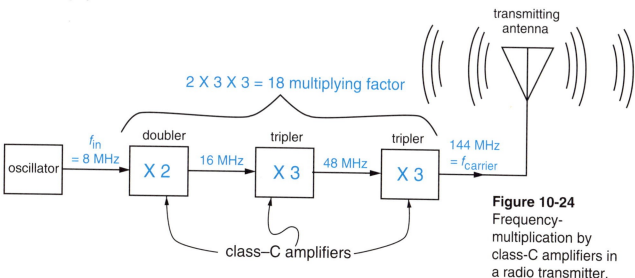

**Figure 10-24
Frequency-multiplication by class-C amplifiers in a radio transmitter.**

By cascading several stages of class-C frequency-doublers and triplers, we can obtain large multiples of the original input frequency. Figure 10-24 illustrates the cascade idea. Radio transmitters commonly use this method to develop the final carrier frequency.

SELF-CHECK FOR SECTION 10-6 IS ON PAGE 327

MAGLEV TRANSPORTATION

Courtesy of Railway Technical Research Institute of Japan.

Magnetic levitation, MAGLEV, will become the premier technology for medium- to long-distance overland transportation at some point in the next century, as the earth's petroleum reserves become exhausted. The photo shows a MAGLEV railway vehicle, which has a cruising speed of 420 km/hr (about 260 mi/hr).

This vehicle contains extremely strong dc electromagnets that require no electric driving source whatsoever. The electromagnets, once energized, are able to achieve their extremely strong magnetic flux densities even in the absence of a power source because their winding coils are made of zero-resistance superconducting material. With $R = 0 \ \Omega$, Ohm's law calls for zero voltage to produce large current through the winding coil, which does not overheat because its I^2R power dissipation is also zero. The difficulty is that the electromagnet coils must be held at a very cold temperature, below −269 °Celsius (−452 °F), in order to maintain their superconducting ability. This is accomplished by housing them in specially insulated enclosures, constantly supplied with high-pressure helium. Compressed liquid helium is carried in the vehicle and must be continually refrigerated. The refrigeration compressor accounts for the only on-board energy consumption.

Magnets made of newly discovered superconductive materials are now under development. They will operate in the −120 °Celsius (−185 °F) temperature range. These new magnets will be even cheaper and easier to refrigerate on-board, using liquid nitrogen rather than helium.

The vehicle's dc supermagnets route flux both vertically and horizontally, through the vehicle's floor and through its sides. The horizontal sideways flux provides the forward propulsion, as explained in Figs. 1 and 2. The vertical floor flux provides the levitation.

Propulsion Figures 1 and 2 are views from above. They show the railway's sidewall electromagnets that have to do with forward propulsion. These two figures ignore the railway's on-the-ground magnets, that have to do with levitation. There are thousands and thousands of railway magnets, spaced close together, as you can tell by a close look at the photograph.

Each sidewall magnet has a pair of electric leads that are controlled by the railway's electronic power circuitry. The circuitry switches each magnet's current off and on, and controls the current's

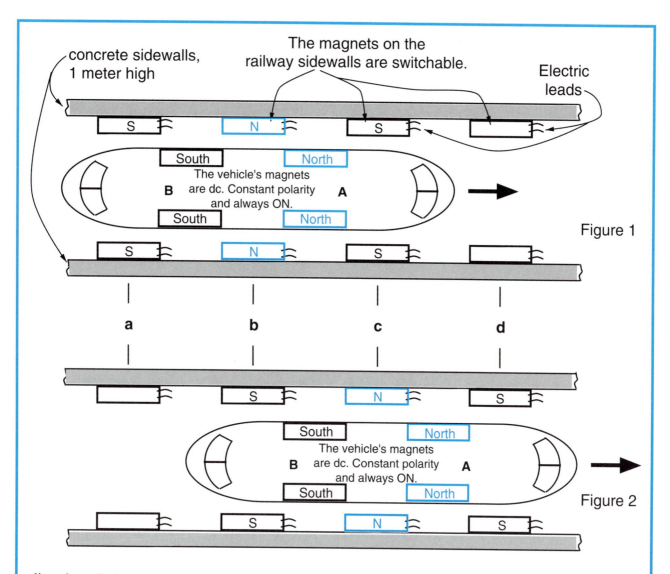

Figure 1

Figure 2

direction. It therefore is able to switch an individual magnet's polarity, changing it between north and south. Each magnet is electrically in parallel with the one directly across from it on the other sidewall, so these two are always alike in polarity (if they are turned ON).

Figure 1 shows the situation at time instant t_1. Figure 2 shows the situation a short time later at instant t_2. We have identified four sidewall magnet positions, labeled **a**, **b**, **c** and **d**. At time instant t_1, the magnets at position **d** are turned OFF by the power-control circuitry. The magnets at position **c** are turned ON, polarized south. The magnets at position **b** are turned ON, polarized north. And those at position **a** are ON, south. Inside the vehicle, the dc supermagnet positions are labeled **A** and **B**. At this instant, the north supermagnets at **A** are attracted to the south sidewall magnets at **c**. This attractive force tends to pull the vehicle forward, to the right. Also, the **A** supermagnets are repelled by the like-polarity (north) sidewall magnets at **b**. This repulsion tends to push the vehicle forward.

Meanwhile, the south supermagnets at **B** are attracted to the unlike-polarity (north) sidewall magnets at **b**, producing more forward propulsion. And the south supermagnets at **B** are repelled away from the like-polarity south sidewall magnets at position **a**.

As the vehicle moves forward, its **A** supermagnets pass by sidewall position **c**. At that moment, the electronic power control circuitry reverses the current direction through sidewall magnets **b** and **c**. Position **c** becomes north and position **b** becomes south. Figure 2 shows this reversal. At the same moment, the power control circuitry switches ON the sidewall magnets at **d**, with a south polarity. And it switches OFF the sidewall magnets at **a**, since the vehicle is now past that position.

All these conditions are indicated in Fig. 2, which shows the position of the vehicle at time t_2, a short time after the switching moment. By checking the magnetic polarities in Fig. 2, you can satisfy yourself that all forces are still propelling the vehicle forward to the right.

For simplicity of explanation, these figures show the vehicle with just two magnetic poles (one N and one S) per side. Actually, the 22-meter-long vehicle has 6 magnetic poles per side (3 N and 3 S).

Levitation When the MAGLEV vehicle starts from a standstill, it has its wheels lowered onto the railbed. Once it reaches a speed of about 100 km/hr, it begins to levitate. Then it retracts its wheels, like an airplane. Figure 3 shows how levitation occurs.

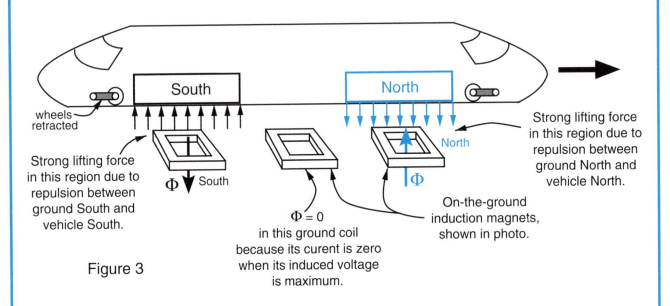

Figure 3

Some of the flux from the supermagnets passes through the floor of the vehicle, as mentioned before. As this vertical flux passes at high speed over any one of the on-the-ground magnets, the rapid rate of change of flux induces voltage in the ground magnet's coil, in accordance with Faraday's law and Lenz's law. The induced voltage is maximum at a moment when the leading or trailing edge of a supermagnet is just above it. The voltage is minimum (zero) at a moment when a supermagnet is centered directly above the ground coil. The inductive reactance of the coil causes the coil's current to be phase-shifted by 1/4-cycle. Therefore the ground-coil's current is maximum at a moment when it is centered beneath a supermagnet, as shown in Figure 3. This maximum current produces the magnetic flux that opposes (points in the opposite direction to) the supermagnet flux. Therefore the ground coil repels the supermagnet, providing the lifting force on the vehicle.

Figure 3 shows only a single ground coil beneath each on-board supermagnet. Actually, the ground coils are physically much smaller than the vehicle's supermagnets, as you can see in the photograph. Thus, there are really several ground coils interacting with a single supermagnet.

As the vehicle rises, the distance between the two flux surfaces increases. This weakens the magnetic repulsion-lifting force. When the vehicle reaches a height where the lifting force is exactly equal to its weight, it will maintain that levitation. This vehicle is designed to rise 10 cm above the railbed.

The MAGLEV vehicle has guide-arms, visible in the photo, that ride in channels in the concrete side walls. Rubber wheels on the guide-arms rub against the sidewalls if the vehicle strays too far off-center. This rarely happens because there is a natural sideways restoring force exerted on the vehicle by the sidewall magnets. We will not try to explain the origin of this restore-to-center force.

If any system should fail, the vehicle simply settles down with its guide-arms dragging on the bottom surfaces of the channels. Its safety advantage is obvious.

SELF-CHECK FOR SECTION 10-6

17. In Fig. 10-18(A), suppose $V_{BB} = -5$ V. The dc voltage across C_{in} will be _____ V. Therefore V_{in} must swing to about + _____ V in order to drive the transistor out of its cutoff state.

18. (T-F) A class-C transistor amplifier spends 360° in the conducting state, the same as a class-A amplifier.

19. (T-F) A class-C-operated transistor with an RC collector load circuit does a good job of reproducing the waveshape of the input signal.

20. (T-F) A class-C amplifier has higher efficiency than either class-B or class-A.

21. (T-F) Most practical class-C amplifiers have a resistor in the collector lead.

22. In Fig. 10-24, if the input frequency were 1.5 MHz, the output frequency to the antenna would be _____ Hz.

23. The reason for the great efficiency of a class-C amplifier with an LC collector load is that during the conduction angle when i_C is very large, the collector-to-emitter voltage v_{CE} is _____ .

10-7 DECIBELS

Until now, we have expressed amplifier gain as a plain numeric value, a ratio. For example, when we have made a statement like $A_v = 7.5$ for an amplifier, we meant that V_{out} is larger than V_{in} by a factor of 7.5, or equivalently, that the ratio of V_{out} to V_{in} is 7.5:1. The same for power gain. The statement $A_P = 120$ means that P_{out} is larger than P_{in} by a factor of 120. The ratio of P_{out} to P_{in} is 120:1.

Sometimes, there are advantages to stating voltage gain and power gain in a different way, by using the **decibel** system. We will see later what these advantages are.

Let us look first at using decibels to express power gain, A_P. Then, in Sec. 10-8, we will use decibels with voltage gain, A_v.

An amplifier's power gain in decibels is given by the equation

$$\text{dB power gain} = 10 \ \log \left(\frac{P_{out}}{P_{in}} \right)$$
$$= 10 \ \log (A_P) \qquad \text{Eq. (10-5)}$$

Get This

where log (A_P) means "the base-10 logarithm of A_P".

The base-10 logarithm is called the **common logarithm**.

Understanding What Logarithms Mean

You may be familiar with base-10 logarithms from your math courses. If not, here is a simple explanation of the meaning of a base-10 logarithm:

A numbers's logarithm is the power (exponent) that you must raise 10 to, in order to get that number. Thus, the logarithm of 100 is 2, because you must raise 10 to the power 2 in order to get 100. That is,

$$10^2 = 100$$

so $$\log 100 = 2.$$

It is not hard to understand logs for the numbers 10, 100, 1000, 10 000, and so on. We have

$$\log 10 = 1 \quad \text{since } 10^1 = 10$$
$$\log 100 = 2 \quad \text{since } 10^2 = 100$$
$$\log 1000 = 3 \quad \text{since } 10^3 = 1000$$
$$\log 10\,000 = 4 \quad \text{since } 10^4 = 10\,000$$

However, it is harder to understand logs for numbers that are *not* integer powers of 10. For example, the logarithm of 31.6 is approximately 1.5. Let us try to understand what this means.

If log 31.6 = 1.5, then raising 10 to the power 1.5 will give us 31.6. We can write

$$10^{1.5} = 31.6$$

Since 1.5 is equivalent to the fraction $\frac{3}{2}$, we can write

$$10^{3/2} = 31.6$$

The expression $10^{3/2}$ tells us to do two things.

①Raise 10 to the power 3. This gives $10^3 = 1000$.

②Find the 2 root (square root) of the result from step ①. This gives

$$\sqrt[2]{1000} \approx 31.6$$

Any logarithm that is a mixed number (a whole number combined with a fractional part) can be understood by the above process. Of course, the root that we must find in step ② will not always be the 2nd root (square root). It can be any root.

Dealing with logarithms used to be difficult and time-consuming. But modern hand-held scientific calculators have made them quite easy. If you inspect your calculator, you will find a key labeled $\boxed{\log}$ or $\boxed{\log x}$. The 2nd function of that key is labeled $\boxed{10^x}$ or $\boxed{\log^{-1}}$. These two functions enable us to handle logarithm equations with ease.

EXAMPLE 10-5

We demonstrated the hard way that log 31.6 ≈ 1.5.

a) Use the log function of your calculator to get the logarithm of 31.6.

b) Use the 10^x function of your calculator to reverse the process, showing that $10^{1.5} \approx 31.6$

SOLUTION

a) Enter the number 31.6 into the display, then press the ⃞log key. The keystroke sequence is

⃞3 ⃞1 ⃞. ⃞6 ⃞log

which gives a displayed result of **1.499**, or similar.

b) Enter the logarithm 1.5, then use the ⃞2nd key to get the ⃞10^x function. The keystroke sequence is

⃞1 ⃞. ⃞5 ⃞2nd ⃞10^x

which gives a display of **31.62**, or similar.

EXAMPLE 10-6

Use your calculator to find the logarithms of the following numbers:

a) 75   d) 825
b) 100   e) 1000
c) 138   f) 2600

SOLUTION

In each case, the keystroke sequence is the same. Just enter the number, then press the ⃞log key.

a) log 75 = **1.875**. The logarithm is between 1 and 2 because the number 75 is between 10 and 100.

b) log 100 = **2.0**. We knew this already.

c) log 138 = **2.14**. The log is between 2 and 3 because the number 138 is between 100 and 1000.

d) log 825 = **2.92**. The log is still between 2 and 3 because 825 is between 100 and 1000, but the log is close to 3 because 825 is close to 1000.

e) log 1000 = **3.0**. We knew this already.

f) log 2600 = **3.41**. The log is between 3 and 4 because the number 2600 is between 1000 and 10 000.

Using the Formula for dB Power Gain

If we know an amplifier's power gain A_P as a numeric ratio, then we can apply Eq. (10-5) with our calculators to express the power gain in decibels (dB). For example, the transformer-coupled push-pull amplifier of Fig. 10-8 had a power gain of $A_P = 1250$ (following Example 10-3). Figure 10-25(A) symbolizes this power gain. To express the same information in decibels, we write Eq. (10-5) as

$$\text{dB power gain} = 10 \log (A_P) = 10 \log (1250)$$

The keystroke sequence is

First, enter the numeric value of A_P into the calculator. Then get the logarithm. Last, multiply by 10. You're moving from right to left through the equation.

which gives a displayed result of 30.97 or similar. Rounding the result to 31.0, we could write

dB power gain = 31.0.

Figure 10-25(B) symbolizes this information.

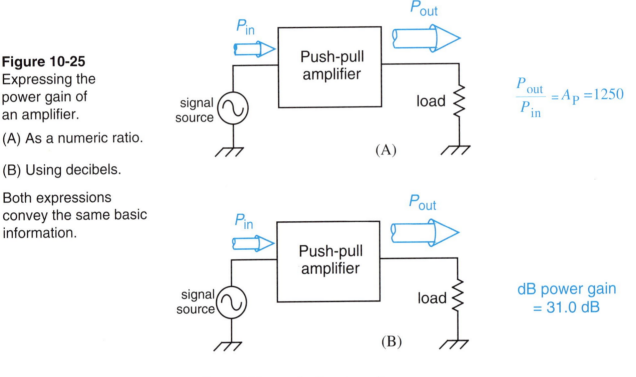

Figure 10-25
Expressing the power gain of an amplifier.

(A) As a numeric ratio.

(B) Using decibels.

Both expressions convey the same basic information.

It would be equivalent to write

power gain = 31.0 dB

or A_P (dB) = 31.0 dB

or dB power gain = 31.0 dB

or G_{dB} = 31.0 dB

There are several ways of writing the value of the dB power gain. These ways all mean the same thing.

EXAMPLE 10-7

For each amplifier power gain A_P expressed as a numeric ratio of P_{out}/P_{in}, express the power gain in decibels.

a) A_P = 325   c) A_P = 5800   e) A_P = 1

b) A_P = 54   d) A_P = 8.8

SOLUTION

In each case, the keystroke sequence is the same. Just enter the numeric ratio, press the log key, then multiply by 10.
a) dB power gain = 10 log (325) = **25.1 dB**. A_P is 325 which is between 100 and 1000. This causes the dB power gain to be between 20 dB and 30 dB. Think carefully about Eq. (10-5) and try to understand why this happens.
b) dB power gain = 10 log (54) = **17.3 dB**. When A_P is between 10 and 100, the dB power gain is between 10 dB and 20 dB. You can understand this by thinking carefully about Eq. (10-5).

c) dB power gain = 10 log (5800) = **37.6 dB**. When A_P is between 1000 and 10 000, the dB power gain is between 30 dB and 40 dB.
d) dB power gain = 10 log (8.8) = **9.44 dB**. When A_P is between 1 and 10, the dB power gain is between 0 dB and 10 dB.
e) dB power gain = 10 log (1.0) = **0 dB**. This points up an important fact that must be understood.

0 dB means that there is no *change* in power — the output power is the same as the input power. The circuit is not actually amplifying, or gaining; but it is not losing, either.

Get This

The 0-dB condition means that the circuit is neither gaining nor losing.

Table 10-1 summarizes the relationship between power gain expressed as a numeric ratio and power gain expressed in decibels.

POWER GAIN

Numeric Ratio	Decibels
1	0 dB
10	10 dB
100	20 dB
1000	30 dB
10 000	40 dB

Table 10-1
The relation between the two ways of expressing power gain.

Transposing the Formula for dB Power Gain

So far, we have used Eq. (10-5) to find decibels if we already know the numeric power ratio. That same formula can be rearranged and used the other way around. Then we can find the numeric ratio if we already know the decibel figure.

The mathematical rearrangement of Eq. (10-5) proceeds as follows:

$$\text{decibels} = 10 \log (A_P)$$
$$10 \log (A_P) = \text{decibels}$$
$$\log (A_P) = \frac{\text{decibels}}{10}$$
$$A_P = \log^{-1}\left(\frac{\text{decibels}}{10}\right) \qquad \text{Eq. (10-6)}$$

which just means

$$A_P = 10^{(\text{decibels} \div 10)} \qquad \text{Eq. (10-7)}$$

The $\log^{-1}$ function is also called the *antilog*.

Equations (10-6) and (10-7) mean exactly the same thing. They are just different ways of writing it.

EXAMPLE 10-8

The amplifier of Fig. 10-26 has a dB power gain of 24 dB. Its input power is $P_{in} = 20$ mW.

a) Use Eq. (10-7) to find its numeric power ratio, A_P.

b) Calculate its output power, P_{out}.

SOLUTION

a) With decibels = 24, Eq. (10-7) gives

$$A_P = 10^{(24 \div 10)} = 10^{2.4}$$

The calculator keystroke sequence is

which gives a displayed result of about 251. Therefore, $A_P = \mathbf{251}$.

b) $P_{out} = A_P \cdot P_{in} = 251 (20 \times 10^{-3} \text{ W}) = \mathbf{5.02 \text{ W}}$

⑲ ✔

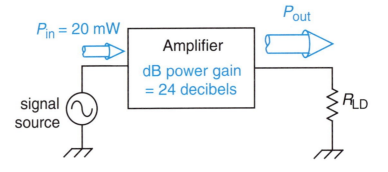

Figure 10-26
If we know
an amplifier's
dB power gain,
we can find its
numeric power
ratio, A_P.

Negative Decibels

⑳ ✔

Positive decibels mean that A_P is greater than 1 — output power is greater than input power. Zero decibels mean that A_P is equal to 1 — output power is equal to input power. Continuing this line of thought, we have:

> Negative decibels mean that A_P is less than 1 — output power is less than input power. Therefore negative decibels mean that the circuit is not an amplifier at all. Instead, it is a power reducer.

Get This

A power-reducer circuit is called an **attenuator**. It is the opposite of an amplifier. Rather than speaking about its power *gain*, we speak about its power *attenuation*. We will symbolize power attenuation as

$$\alpha_P$$

EXAMPLE 10-9

The attenuator in Fig. 10-27 has an attenuation ratio of
$$\alpha_P = 0.5$$
Express the attenuation in decibels.

SOLUTION

For an attenuator, we can rewrite Eq. (10-5) as

$$\boxed{\text{dB power attenuation} = 10 \log \left(\frac{P_{out}}{P_{in}} \right)} \qquad \text{Eq. (10-8)}$$

For Fig. 10-27, this gives
$$\text{dB power attenuation} = 10 \log (0.5)$$
By the usual keystroke sequence, the displayed result is –3.01.
Therefore,
$$\text{dB power attenuation} = \mathbf{-3.01 \ dB}$$

21 ✔

It is worthwhile memorizing that –3 dB is a numeric ratio of 0.5. We call the -3 dB condition the **half-power condition**.

This condition is often verbalized as "down 3 dB" or "**3 dB-down**".

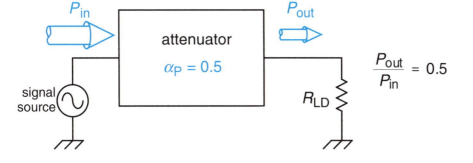

$$\frac{P_{out}}{P_{in}} = 0.5$$

Figure 10-27
An attenuator is the opposite of an amplifier.
This attenuator has a numeric ratio of $\alpha_P = 0.5$. Its dB attenuation is –3 dB.

Applying Decibels to MultiStage Amplifiers and Other Electronic Systems

One of the advantages of the decibel measurement system is its ability to handle multiple stages. For example, consider the 3-stage amplifier in Fig. 10-28. In part (A) of that figure, the individual stages have their A_P values expressed as numeric ratios. To find the overall total power gain, $A_{P(T)}$, we must multiply the three A_P values. Doing that gives

$$A_{P(T)} = A_{P(1)} \times A_{P(2)} \times A_{P(3)} = 4.5 \times 125 \times 60 = 33 \ 750$$

When individual stages have their power gains expressed in decibels, we can find the total power gain by simply *adding* the dB values, rather than multiplying. As a formula,

Get This

$$\boxed{\text{dB power gain}_{(total)} = \text{dB power gain}_{(1)} + \text{dB power gain}_{(2)} + \text{dB power gain}_{(3)}} \qquad \text{Eq. (10-9)}$$

For example, in part (B) of Fig. 10-28, the same amplifier stages have their power gains expressed in decibels. To find the total dB power gain, we use Eq. (10-9).

Adding decibels is equivalent to multiplying numeric ratios.

$$\text{dB power gain}_{(total)} = 6.5 \text{ dB} + 21.0 \text{ dB} + 17.8 \text{ dB} = 45.3 \text{ dB}$$

To check that 45.3 dB is equivalent to a ratio of 33 750, apply Eq. (10-4).

$$\text{dB power gain} = 10 \log(A_P) = 10 \log(33\ 750)$$
$$= 10 \times (4.528) = 45.3 \text{ dB} \quad \text{(checks)}$$

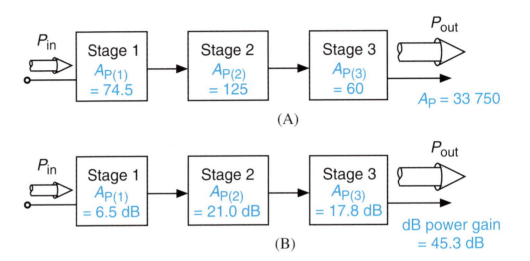

Figure 10-28
Adding decibels is the same as multiplying numeric ratios.

Thus, for multiple stages, using decibels makes the problem into an addition problem. This is an advantage, because addition is easier than multiplication.

The decibel advantage becomes even clearer when we have amplifier stages (gainers) cascaded with attenuator stages (reducers). For example, consider the 5-stage system shown in Fig. 10-29. The gains and attenuations are expressed as numeric ratios in part (A). They are expressed in decibels in part (B). The dB values are **+** for gainers and **−** for reducers.

Transporting an electronic signal by wire or cable always involves some power loss. Therefore, in a system with lengthy interconnecting wires, the wires act as attenuators.

In part (A) we must multiply the numeric ratios, giving

$$A_{P(T)} = 12 \times 0.73 \times 170 \times 0.2 \times 25 = 7446$$

In part (B) we add and subtract decibels, giving

$$\text{dB power gain} = +10.8 \text{ dB} - 1.4 \text{ dB} + 22.3 \text{ dB} - 7.0 \text{ dB} + 14.0 \text{ dB} = 38.7 \text{ dB}$$

We can check that these results express the same basic information. With $A_{P(T)} = 7446$,

checks

$$\text{dB power gain} = 10 \cdot \log(A_{P(T)}) = 10 \cdot (3.87) = 38.7 \text{ dB}$$

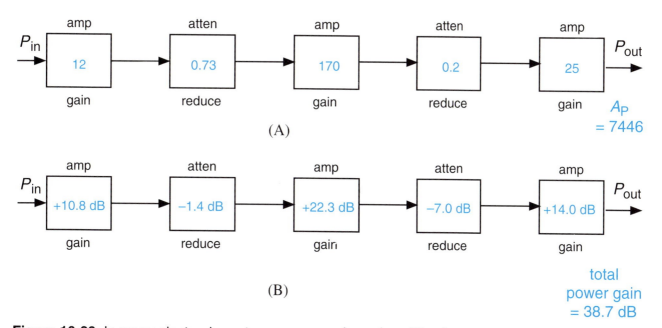

Figure 10-29 In many electronic systems, power-gainers (amplifiers) are cascaded with power-losers (attenuators). (A) We can multiply the A_P and α_P ratios to get total power gain. (B) We can add and subtract the dB values to get the same information.

Frequency-Response Curves Using Decibels

We are familiar with frequency-response curves for frequency-selective amplifiers and filter circuits. Our frequency-response curves have always been drawn with the vertical axis representing voltage — either percentage of maximum voltage, or absolute voltage gain, A_V. It is also possible to have the vertical axis represent power, and this is commonly done. Figure 10-30(A) shows an example frequency-response curve of A_P versus frequency, for an amplifier with a midband power gain of $A_P = 5000$ at a midband frequency of 7 kHz. Notice that the cutoff frequencies are at the 50% points on the power curve, namely $A_P = 2500$.

The 50% power points correspond to the 70.71% voltage points since $P = V^2 \div R_{LD}$. Squaring 70.71% (decimal 0.7071) gives 50% (decimal 0.50).

The power gain on the vertical axis of Fig. 10-30(A) could be expressed in decibel units. This has been done in Fig. 10-30(B). Notice how the low numeric values of A_P, that were unreadable in Fig. 10-30(A), now become easily readable on the decibel graph of Fig. 10-30(B). This is the second advantage of the decibel measurement system.

The decibel measurement system allows a very great range of variation in numeric ratio to be conveniently expressed by a reasonable range of dB values.

Get This

The same idea that is illustrated for an amplifier in Fig. 10-30 would hold for an attenuator. The only difference is that the dB values would be negative.

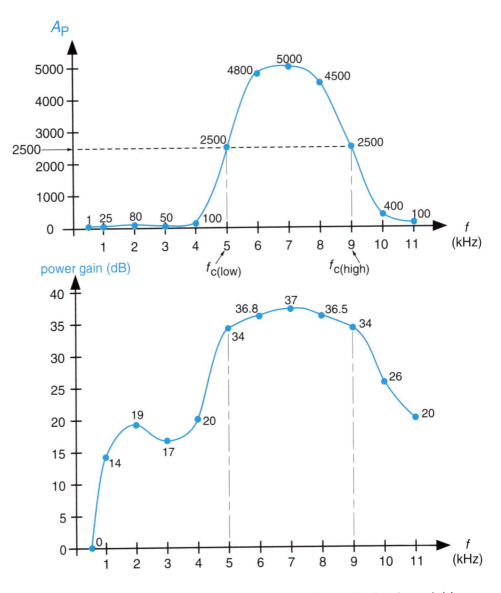

Figure 10-30 Frequency-response curves with power as the vertical-axis variable.
(A) Power gain measured as a numeric ratio.
(B) Power gain measured in decibels. The A_P values that could not be read in part (A) are easily readable in part (B).

EXAMPLE 10-10

A certain amplifier has the following performance over the frequency range from 1 kHz to 13 kHz.

f (kHz)	1	2	3	4	5	6	7	8	9	10	11	12	13
A or α ratio	0.3	2	16	100	1000	2000	800	63	4	1	0.5	0.1	0.01

Plot the frequency-response curve of this amplifier with power gain expressed in decibels.

SOLUTION

Each of the numeric ratios must be converted to decibels. The numeric ratios greater than 1 convert to **+** dB values. The numeric ratios less than 1 convert to **−** dB values.

Using the decibel conversion formulas, Eqs. (10-5) and (10-8), we get the results in Table 10-2.

f (kHz)	1	2	3	4	5	6	7	8	9	10	11	12	13
ratio	0.3	2	16	100	1000	2000	800	63	4	1	0.5	0.1	0.01
decibels	−5	+3	12	20	30	33	29	18	+6	0	−3	−10	−20

Table 10-2

Since our dB values span the range from -20 to +33, we scale the vertical axis as shown in Fig. 10-31. Plotting the dB values gives the frequency-response curve.

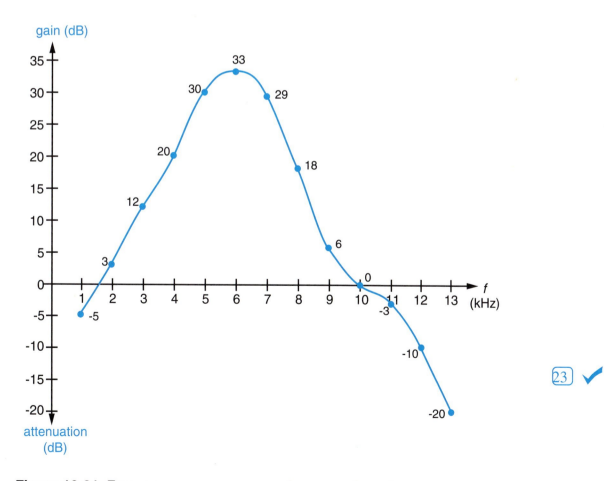

Figure 10-31 Frequency-response curve of power gain and power attenuation, measured in decibels.

The decibel system of measurement is used in other areas of technology, besides electronic circuit power ratios. In these applications, people agree on a reference value of the measured variable. This reference value is considered to have a numeric ratio of 1.0, which is expressed as 0 dB.

For example, you may be familiar with loudness of sound expressed in decibels. People who work in the field of sound have agreed that a sound power of 1×10^{-12} watts per square meter of surface area (1 pW/m^2) is the reference value for sound intensity. This is the value of sound power density that is barely detectable by normal human hearing.

Louder sounds are then expressed in **audio decibels**, symbolized dBA, by the equation

$$\text{sound loudness in dBA} = 10 \log \left(\frac{\text{sound power}}{1 \times 10^{-12}} \right) \qquad \text{Eq. (10-10)}$$

Using this measurement system, we can describe the human hearing experience by the list in Table 10-3.

Table 10-3

Description of Sound	Audio Decibels
Barely audible	0 dBA
Whisper	30 dBA
Normal conversation	60 dBA
City street at rush hour	80 dBA
Rock band, at stage level	100 dBA
Nearby thunder clapp (threshold of feeling)	120 dBA
Threshold of pain	130 dBA
Jet airplane at takeoff	140 dBA
More intense sound does not seem to be any louder because our ear's hearing mechanism saturates	150 dBA

SELF-CHECK FOR SECTION 10-7

24. Convert to decibels:
    a) $A_P = 70$ d) $A_P = 1500$
    b) $A_P = 180$ e) $A_P = 5$
    c) $A_P = 600$ f) $A_P = 1.0$
25. When power gain A_P is a numeric ratio between 10 and 100, decibel power gain is between _____ dB and _____ dB.
26. When power gain A_P is a numeric ratio between 100 and 1000, decibel power gain is between _____ dB and _____ dB.
27. Convert the following dB power gains into numeric ratios:
    a) 3 dB   d) 1 dB
    b) 16 dB   e) 30 dB
    c) 20 dB   f) 40 dB
28. Negative decibels mean that output power is _____ than input power. The numeric ratio is _____ than 1.0.
29. Convert the following attenuation ratios to decibels:
    a) 0.67   d) 0.05
    b) 0.5   e) 0.01
    c) 0.1   f) 0.001
30. When amplifiers are cascaded, we find the overall power gain as a numeric ratio by multiplying the A_P values. We find the overall decibel power gain by _____ the dB values.
31. When amplifiers and attenuators are cascaded, how do we find the overall decibel power gain?
32. A 5-stage system of amplifiers and attenuators has blocks with the following dB values: +10 dB, –4 dB, +20 dB, –8 dB, +3 dB.
    a) Find the overall decibel power gain.
    b) Express A_P as a numeric ratio.
33. When sound intensity (loudness) is measured in dBA, the quietest audible sound is referred to as _____ dBA.

10-8 DECIBELS APPLIED TO VOLTAGE GAIN AND LOSS

Some electronic systems have a **characteristic impedance**, which we can think of as a characteristic resistance. Simplified, this means that every amplifier and every attenuator has input resistance R_{IN} that is the same ohmic value as the load resistance R_{LD}. For example, Fig. 10-32 shows a radio signal-distribution system with a characteristic resistance of 75 Ω. The three radio receivers on the far right are the "loads" on the distribution system. When every block in a system presents a resistance of 75 ohms, we say that we are "working in a 75-Ω environment".

A radio or television signal-processing system usually has a characteristic resistance of either 50 Ω, 75 Ω, or 300 Ω. High-performance laboratory instrumentation systems usually have a characteristic resistance of 50 Ω. Studio audio systems and telephone systems usually have a characteristic resistance of 600 Ω.

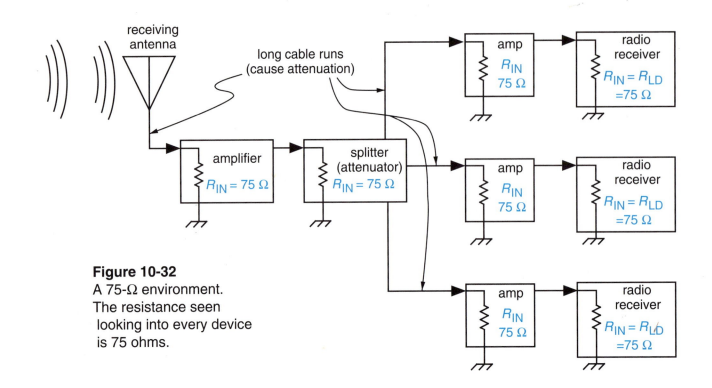

Figure 10-32
A 75-Ω environment.
The resistance seen
looking into every device
is 75 ohms.

Systems with a characteristic resistance, symbolized R_c, have an important feature:

Get This

Input voltage V_{in} appears across the same value of resistance that output voltage V_{out} appears across. This is because $R_{in} = R_{ld} = R_c$.

Therefore, for every amplifier or attenuator in the system, we can write

$$P_{in} = \frac{(V_{in})^2}{R_c}$$

and

$$P_{out} = \frac{(V_{out})^2}{R_c}$$

Dividing the output equation by the input equation gives

$$\frac{P_{out}}{P_{in}} = \frac{\dfrac{(V_{out})^2}{R_c}}{\dfrac{(V_{in})^2}{R_c}}$$

R_c cancels, since it is guaranteed to be equal for input and output.

$$\frac{P_{out}}{P_{in}} = \left(\frac{V_{out}}{V_{in}}\right)^2 \qquad \text{Eq. (10-11)}$$

In words, Eq. (10-11) tells us that the power gain is the square of the voltage gain. Remember, this is true only if the system guarantees equal-value input resistance and load resistance.

When Eq. (10-11) is true, we can express the voltage gain in decibels. Here is how:

$$\text{dB voltage gain} = 10 \log \left(\frac{V_{out}}{V_{in}}\right)^2$$

$$\boxed{\text{dB voltage gain} = 20 \log \left(\frac{V_{out}}{V_{in}}\right)} \qquad \text{Eq. (10-12)}$$

Equation (10-12) is valid only when R_{in} equals R_{ld}.

The number 20 appears because the logarithm of a number squared is 2 times the logarithm of the number.

EXAMPLE 10-11

Suppose you are working in a system with a 600-Ω characteristic resistance. While checking an IC power amplifier, you measure V_{in} to be 160 mV and V_{out} to be 7.2 V.
   a) Find the voltage gain A_v as a numeric ratio.
   b) Express the voltage gain in decibels.

SOLUTION

a)
$$A_v = \frac{V_{out}}{V_{in}} = \frac{7.2 \text{ V}}{160 \times 10^{-3} \text{ V}} = \mathbf{45}$$

b) Because the system guarantees equal-value R_{in} and R_{ld}, Eq. (10-12) can be applied.

$$\text{dB voltage gain} = 20 \log (45)$$
$$= 20 \,(1.65) = \mathbf{33.1 \ dB}$$

   It often happens that people try to use Eq. (10-12) when R_{in} and R_{ld} are not equal. This is incorrect.

It is wrong to use Eq. (10-12) when input resistance R_{in} and load resistance R_{ld} are unequal. People often make this mistake.

   The decibel voltage equation [Eq. (10-12)] is more popular than the decibel power equation [Eq. (10-5)] because voltages are easy to measure, while power is harder to measure. But if you want to express gain in decibels and voltage is the only variable you can measure, you have no choice. You must calculate both input and output power from $P = V^2/R$, then apply Eq. (10-5).

   If you don't know R_{in} and R_{ld}, then you are out of luck. You can't express gain in decibels, even though you know V_{in} and V_{out}.

SELF-CHECK FOR SECTION 10-8

34. In order to express voltage gain in decibels, the amplifier's input resistance must be equal to the _____ .
35. A certain amplifier is operating in a 600-Ω environment. It has V_{in} = 100 mV and V_{out} = 14 V.
   a) Is it possible to express its voltage gain in decibels? Explain why.
   b) What is its decibel voltage gain?

36. A certain amplifier has $R_{in} = 10$ kΩ and $R_{LD} = 24$ Ω. Measuring its input and output voltages gives $V_{in} = 200$ mV and $V_{out} = 9$ V.
    a) Is it possible to express its voltage gain in decibels? Explain why.
    b) With the given information, is it possible to express power gain in decibels?
    c) What is the value of decibel power gain?

FORMULAS

$$\eta = \frac{P_{out}}{P_{(dc\ supply)} + P_{in}}$$ Eq. (10-1)

$$P_{(dc\ supply)} + P_{in} = P_{out} + P_{wasted}$$ Eq. (10-2)

$$\eta \approx \frac{P_{out}}{P_{(dc\ supply)}} \quad (\text{ignoring } P_{in})$$ Eq. (10-3)

$$\eta = \frac{P_{out}}{P_{out} + P_{wasted}}$$ Eq. (10-4)

$$\text{dB power gain} = 10 \log\left(\frac{P_{out}}{P_{in}}\right) = 10 \log (A_P)$$ Eq. (10-5)

$$A_P = 10^{(decibels/10)}$$ Eq. (10-7)

$$\text{dB power attenuation} = 10 \log\left(\frac{P_{out}}{P_{in}}\right) \quad (\text{for } P_{out} < P_{in})$$ Eq. (10-8)

$$\text{dB}_{(total)} = \text{dB}_{(1)} \pm \text{dB}_{(2)} \pm \text{dB}_{(3)} \pm ..$$ Eq. (10-9)

$$\text{dB voltage gain or attenuation} = 20 \log\left(\frac{V_{out}}{V_{in}}\right)$$ Eq. (10-12)

SUMMARY

● An amplifier's efficiency describes its ability to convert power from the dc supply into output power, while wasting as little as possible.
● In class-A operation, the transistor's Q-point is near the middle of the characteristic curves. This allows the amplifier to faithfully reproduce the entire input waveform, but it causes efficiency to be low.
● In class-B operation, the transistor's Q-point is exactly at cutoff, with $I_C = 0$. It can amplify one half cycle of the input signal, but not the other half cycle. However, class-B efficiency is better than for class-A.
● To amplify the complete input signal, we can arrange for one class-B transistor to handle one half cycle and a second class-B transistor to handle the other half cycle. Amplifiers that use this approach are called push-pull amplifiers.

● Class-B push-pull amplifiers tend to distort the crossover points, where one half-cycle ends and the other begins.

● To eliminate class-B crossover distortion, the transistor can be biased in a barely conducting state, with I_C equal to some small value. This is called class-AB operation.

● The complementary symmetry design is a popular amplifier design that uses the push-pull idea but contains no transformer.

● The complementary symmetry design combines an *npn* transistor with a *pnp* transistor. The two transistors must be closely matched, (equal β and V_{BE} values).

● Quasi complementary symmetry uses two Darlington pairs. One Darlington pair contains 2 *npn* transistors, and the other contains one *pnp* and one *npn* transistor.

● Integrated circuit BJT high-power amplifiers use a variation of the quasi complementary symmetry design.

● In class-C operation, the transistor is biased beyond cutoff. The input signal must go far into its positive half cycle to make the transistor begin conducting.

● A class-C amplifier with a resistive collector circuit is unlike other amplifiers in that it does not duplicate the shape of the input waveform.

● Most practical class-C amplifiers have an *LC* collector circuit. Then class-C is the most efficient method of operation.

● Frequency-doublers or -triplers can be built using a class-C amplifier.

● An amplifier's power gain can be expressed in decibels, which are based on logarithms.

● There are two advantages to expressing power gain in decibels:
    1) When amplifier stages are cascaded, their dB power gains are simply added. There is no need to multiply A_P values.
    2) A very wide range of power gains can be graphed conveniently.

● Decibels are used to measure many physical variables that vary over great ranges. The most common example is the audio decibel system, that is used for measuring sound intensity in dBA units.

● An amplifier's voltage gain can be expressed in decibels, if its input resistance has the same value as its load resistance.

QUESTIONS AND PROBLEMS

1. A certain amplifier has $P_{in} = 1$ mW and $P_{out} = 1.5$ W. The average power taken from the dc supply is 8 W. What is the efficiency of the amplifier?

2. What is the most likely operating class for the amplifier in problem 1?

3. Draw an ac load-line with end points at $i_C = 100$ mA and $v_{CE} = 15$ V.
    a) Show the approximate position of the Q-point for class-A operation.
    b) Show the approximate position of the Q-point for class-B operation.
    c) Show the approximate position of the Q-point for class-AB operation.

4. Referring to the diagram that you made for problem 3, describe the location of the Q-point for class-C operation.

5. A certain amplifier has $P_{in} = 10$ mW and $P_{out} = 20$ W. The average power taken from the dc supply is 35 W.
   a) What is the efficiency of the amplifier?
   b) What is the likely operating class of the amplifier?
6. For the class-AB push-pull amplifier of Fig. 10-8, we calculated A_v as 5.0 in Example 10-3. Assume that both betas = 100. Now calculate the current gain A_i as follows:
   a) Find $I_{b(pk)}$, knowing that $I_{e(pk)} = 1$ A and $\beta = 100$.
   b) Apply the transformer current law to find the peak primary current in T_1. This is the amplifier's input current.
   c) Apply Ohm's law to the 8-Ω load to find the amplifier's peak output current.
   d) Take the ratio of $I_{out} \div I_{in}$ to get A_i. Compare to the value that was stated following Example 10-3.
7. Using your results from Problem 6 and the results from Example 10-3, convert the peak values of all currents and voltages to rms values.
   a) Find P_{out}.
   b) Find P_{in}.
   c) Calculate A_P. Compare to the value following Example 10-3.

   For Problems 8 through 17, redraw the push-pull amplifier of Fig. 10-8 with the following specifications: T_1 primary – 200 turns; T_2 secondary – 75 turns per half (150 turns total); $R_{E1} = R_{E2} = 0.56\,\Omega$; T_2 primary – 120 turns per half (240 turns total); T_2 secondary – 90 turns; $R_{LD} = 4\,\Omega$. The values of V_{CC}, R_1, R_2, C_1, and β remain unchanged. V_{in} remains 2 V peak.

8. Draw the dc load-line and show the approximate Q-point location .
9. a) Find the equivalent ac resistance in the collector leads, r_C.
   b) Find the total equivalent ac resistance in the main flow path of each transistor.
10. Find the slope of the ac load-line by applying Ohm's law to the total ac resistance from problem 9, part (b). Draw this slope line on the i_C-versus-v_{CE} characteristic graph. Scale the vertical i_C axis appropriately.
11. Transfer the slope to the Q-point to produce the ac load-line.
12. a) Find the peak secondary voltage from transformer T_1. Use the transformer voltage law.
    b) Neglecting r_{Ej} and X_{C1}, calculate the peak ac emitter current $I_{e(pk)}$. Use Ohm's law.
    c) Mark the peak point on the ac load-line.
13. a) Apply Ohm's law to the ac collector resistance r_C to get the peak primary voltage on T_2.
    b) Use the transformer voltage law to get the peak load voltage.
14. Calculate the amplifier's voltage gain A_v.
15. Repeat the steps outlined in Problem 6 to get the amplifier's current gain, A_i.
16. Calculate the amplifier's power gain, A_P.
17. Repeat the 5-step process in Sec. 10-4 for estimating the efficiency of the amplifier.

For Problems 18 through 27, redraw the complementary symmetry amplifier of Fig. 10-12 to have the following specifications: $R_{B1(1)} = R_{B1(2)} = 680\ \Omega$; $R_{B2(1)} = R_{B2(2)} = 36\ \Omega$; $R_{E1} = R_{E2} = 3.6\ \Omega$; $R_{LD} = 20\ \Omega$; $\beta_1 = \beta_2 = 140$. The values of V_{CC} and V_{EE} remain as they were, at ± 15 V. Let $V_{in(pk)} = 8$ V.

18. Assume that the V_{BE} value for each transistor is 0.65 V. Calculate the collector bias current I_C (the idling current).

19. Based on the Q-point calculated in problem 18, calculate what the value of r_{Ej} would be if the transistor were operating in class-A with a small signal.

20. (T-F) The r_{Ej} value calculated in Problem 19 is not meaningful in this class-AB amplifier because when a transistor moves up the ac load-line, its current far exceeds the Q-point idling current, on average. Therefore, we are justified in ignoring r_{Ej}.

21. With $V_{in(pk)} = 8$ V, calculate the peak base voltage $V_{b(pk)}$. Use voltage division between R_{B2} and R_{B1}. (We are assuming that $r_{b\,in}$ is large enough that its parallel effect can be ignored.)

22. Working with the V_b value from Problem 21, and ignoring r_{Ej}, calculate the peak output voltage $V_{out\,(pk)}$, appearing across R_{LD}. Remember that the configuration is common-collector.

23. Using the result from problem 22, find the amplifier's A_v.

24. Calculate the amplifier's ac input resistance R_{in}. Remember that there are always three parallel paths to ac ground: 1) $R_{B2(1)}$ and $R_{B1(1)}$; 2) $R_{B2(2)}$ and $R_{B1(2)}$; 3) Through the B-E junction of the conducting transistor.

25. Using the result from Problem 24, calculate the amplifier's peak ac input current by Ohm's law.

26. Using the result from Problem 22, apply Ohm's law to find the peak output current. Then combine that with the result from problem 25 to find the amplifier's current gain A_i.

27. Calculate the amplifier's power gain A_P.

28. All other things being equal, the dual-input capacitor arrangement shown in Fig. 10-16 causes the voltage gain to _____ .

29. (T-F) It is reasonableto expect the actual efficiency of a complementary symmetry amplifier to be in the 50-60% range.

30. In a class-C amplifier with an LC tank, the reason the efficiency is so high is that when the collector current surge occurs, the transistor voltage v_{CE} is very _____ .

31. Other things being equal, a smaller conduction angle tends to _____ the efficiency of a class-C amplifier.

32. A certain class-C amplifier contains a tank circuit with $f_r = 3.6$ MHz. In order to act as a frequency tripler, it must be driven by an input frequency of _____ MHz.

33. (T-F) Radio transmitting systems often use class-C frequency multipliers to generate the final carrier frequency.

34. For each of the following A_P numeric ratios, express power gain in decibels.

    a) 100   d) 2200

    b) 480   e) 7

    c) 63   f) 10 000

35. Convert each of the following dB power gains into an A_P numeric ratio.

    a) 20 dB   d) 30 dB

    b) 25 dB   e) 3 dB

    c) 29 dB   f) 0 dB

36. For each of the following attenuation (α_P) numeric ratios, express attenuation in negative decibels

    a) 0.6   d) 0.05

    b) 0.5   e) 0.01

    c) 0.1   f) 0.005

37. Convert each of the following dB attenuations into an α_P numeric ratio.

    a) –3 dB   c) –10 dB

    b) –6 dB   d) –20 dB

38. A certain system has 6 stages. The dB values of each stage are as follows: +12 dB, –6 dB, +23 dB, –9 dB, +4 dB, –2 dB.

a) What is the overall dB power gain of the system?

b) What is the overall system power gain, expressed as a numeric ratio?

39. A certain system has 5 stages. It is desired to make the overall power gain equal to 63 (numeric ratio). The first 4 stages have the following dB values: +7 dB, –2 dB, +10 dB, –5 dB. What dB value must the final stage have?

40. A certain amplifier has the following frequency-response data.

f (kHz)	30	35	40	45	50	55	60	65	70	75
A_p or α_P	0.4	1	80	500	1000	700	500	25	0.7	0.08

    Plot a frequency-response curve in decibels.

41. Amplifier voltage gain can be expressed in decibels if the amplifier's resistance (impedance) is equal to the _____ resistance (impedance).

42. A certain amplifier operates in a 50-Ω environment. It has $V_{in} = 0.5$ V and $V_{out} = 28$ V. Express its voltage gain in decibels

43. Suppose the amplifier in Problem 42 has switchable voltage gain, in 5-dB steps.

a) If A_v (dB) is switched to 25 dB, find V_{out}.

b) Repeat for A_v (dB) being switched to 15 dB.

CHAPTER 11

JUNCTION FIELD-EFFECT TRANSISTORS

This electronic instrument is a portable fluorometer. It is used by oceanographers to measure the concentration and activity of plankton, the microscopic plants that are at the bottom of the food chain in the sea. Like land-growing green plants, plankton perform the chemical reaction of photosynthesis, absorbing carbon dioxide and releasing oxygen. The reaction also produces fluorescence, the emission of electro-magnetic radiation. By measuring the intensity of radiation, the fluorometer gives an indirect indication of plankton productivity. Page 4 of the color section shows this instrument in use in the water.

Courtesy of Biospherical Instruments, Inc.

OUTLINE

JFET Structure and Operation
Common-Source JFET Amplifiers
FET Instability Problems
   Stabilizing the Bias Point
Ac Operation
Common-Drain FET Amplifier
   Common-Gate Amplifier
Troubleshooting JFET Amplifiers

NEW TERMS TO WATCH FOR

field-effect transistor
channel
source
drain
gate
shorted-source drain current

channel width
pinch-off voltage
transconductance
siemens

The bipolar junction transistors that we have studied in Chapters 4 through 10 are current-operated devices. The transistor's base input current controls the electrical action in the output circuit. In this chapter and in Chapter 12, we will study field-effect transistors. They are voltage-operated devices, not current-operated.

After completing this chapter, you should be able to:

1. Draw the schematic symbol, label the terminals, and show the proper dc supply polarity for an *n*-channel JFET. Do the same for a *p*-channel JFET.

2. Sketch the characteristic I_D-versus-V_{GS} curve for a JFET.

3. Define the shorted-source drain current, I_{DSS}, with reference to the FET's characteristic curves.

4. Define the gate-source cutoff voltage, $V_{GS(off)}$, also called pinch-off voltage, with reference to the FET's characteristic curves.

5. Explain how V_{GS} controls main-terminal current I_D, in terms of depletion region and channel width.

6. Sketch ideal and realistic families of I_D-versus-V_{GS} characteristic curves for a JFET. Describe the fundamental differences between FET and BJT curve families.

7. Given a complete schematic diagram of a common-source FET amplifier, and a specific I_D-versus-V_{GS} characteristic curve of the FET, use load-line analysis to describe the ac performance.

8. Explain why FET amplifiers tend to distort the input signal waveform, in class-A operation.

9. State the main advantage of an FET amplifier over a BJT amplifier, and explain why the FET has that advantage.

10. Describe the batch instability of JFETs, with reference to minimum, maximum and typical I_D-versus-V_{GS} characteristics.

11. Given the minimum and maximum values of I_{DSS} and $V_{GS(off)}$ for an FET type, and given a detailed amplifier schematic diagram, draw the line of possible dc bias Q-points.

12. Define an FET's transconductance g_m.

13. Given a specific FET characteristic curve, and given its Q-point and input signal voltage, find its transconductance g_m.

14. Knowing the value of g_m, and given a detailed amplifier schematic, find the amplifier's voltage gain A_v.

15. Interpret an I_D-versus-V_{GS} characteristic curve to explain why an FET's g_m is so variable, unlike a BJT's β.

16. Sketch the schematic diagram of a common-drain (source-follower) JFET amplifier.

17. Describe how a common-drain amplifier differs from a common-source amplifier in terms of bias stability, value of voltage gain, stability of voltage gain, dc supply voltage requirements, and input-output phase relation.

18. Troubleshoot a JFET amplifier stage.

11-1 JFET STRUCTURE AND OPERATION

The **junction field-effect transistor** (JFET or just FET) has a simpler structure than a BJT. Figure 11-1(A) shows that an **n-channel** JFET is simply a complete path of *n*-doped silicon surrounded by a collar of *p*-doped silicon bulging into the path. There are terminals at opposite ends of the *n*-doped channel. They are called the **source** (S) and **drain** (D), as shown in Fig. 11-1(A). The terminal on the *p*-doped collar is called the **gate** (G). Figure 11-1(B) shows the schematic symbol for an *n*-channel JFET.

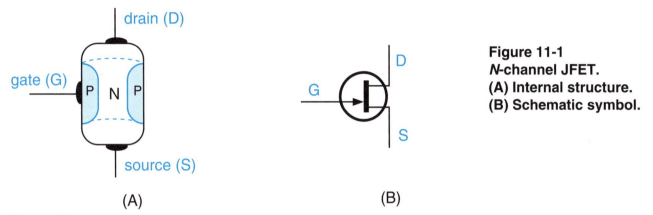

Figure 11-1
***N*-channel JFET.**
(A) Internal structure.
(B) Schematic symbol.

(A)   (B)

Operation

An *n*-channel JFET is connected in a circuit as shown in Fig. 11-2. The positive source voltage $+V_{DD}$ causes current I_D to flow from ground, through the FET's channel through resistor R_D, to the + source terminal. Because the channel is *n*-doped, all the charge carriers are free electrons. None of the current is due to hole flow.

① ✔

The gate-to-channel *p-n* junction must always be reverse-biased.
That is, the gate, G, must be negative relative to the source, S.

Get This

The negative polarity of gate-to-source voltage V_{GS} is shown in Fig. 11-2.

V_{GS} is allowed to be 0 V, because that maintains a depletion layer around the junction, keeping it reverse-biased. But V_{GS} must not become positive.

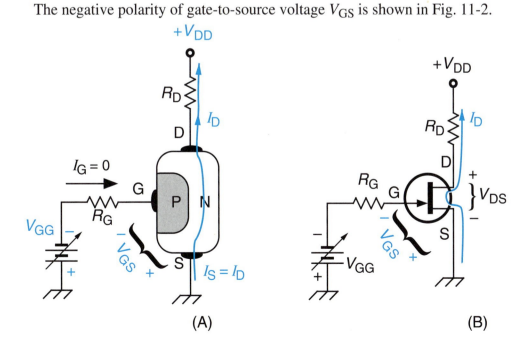

(A)   (B)

Figure 11-2 Basic circuit connection of an *n*-channel JFET. (A) Simplified structural diagram; only one side of the p-type collar is shown. (B) Schematic. In function, the drain D is similar to the collector C of a BJT. Source S is similar to emitter E, and gate G is similar to base B.

For an ideal FET, drain current I_D is not affected by V_{DS}, the main-terminal voltage from drain to source.

Drain current I_D is controlled by gate-to-source voltage V_{GS}. A typical I_D-versus-V_{GS} transfer curve is shown in Fig. 11-3. Ideally, the drain current is not affected by the main-terminal voltage V_{DS} in Fig. 11-2(B).

Figure 11-3
Typical transfer curve of I_D versus V_{GS} for a JFET.

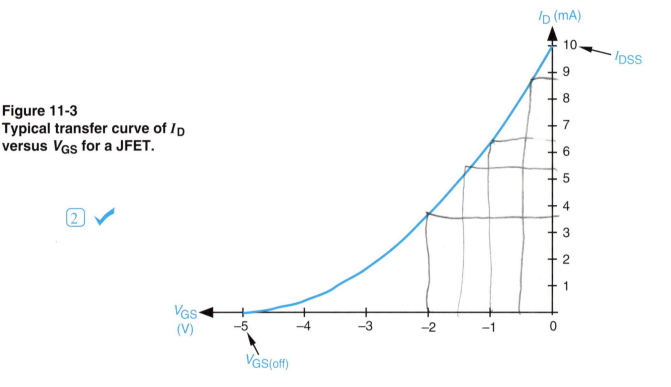

To understand the mechanism by which V_{GS} controls I_D, refer to the structural views in Figs. 11-4(A) through (D).

When V_{GS} is 0 V, it exerts no repulsion force on electrons trying to pass through the narrow part of the channel. Therefore electrons pass freely, and drain current I_D is as large as possible. This maximum value of I_D is called **shorted-source drain current**, symbolized I_{DSS}.

I_{DSS} = 10 mA, taken from Fig. 11-3.

When V_{GS} has a small magnitude, say –0.5 V, it causes opposition to electrons trying to pass through the narrow part of the channel. In solid-state terms, a small FET depletion region forms on both sides of the junction. On the channel side, the absence of free electrons means that the region is unable to pass current. Figure 11-4(B) shows this effect. We visualize the **channel width** as being reduced slightly, as indicated in Fig. 11-4(B). Therefore I_D is reduced to a value less than the maximum value I_{DSS}.

I_D = 8.1 mA, taken from Fig. 11-3.

When V_{GS} becomes larger in magnitude (more negative), it extends the depletion region farther into the channel. The channel width is reduced further, as shown in Fig. 11-4(C). Therefore I_D becomes smaller.

For V_{GS} = –2 V, I_D = 3.6 mA taken from Fig. 11-3.

There will be some value of V_{GS} that completely closes off the channel, causing I_D to become zero. This is pictured in Fig. 11-4(D). The value of V_{GS} that causes $I_D = 0$ is called the **pinch-off voltage**, symbolized $V_{GS(off)}$. In Fig. 11-4(D), $V_{GS(off)}$ = –5 V. This value is taken from Fig. 11-3.

Pinch-off voltage is sometimes symbolized V_P.

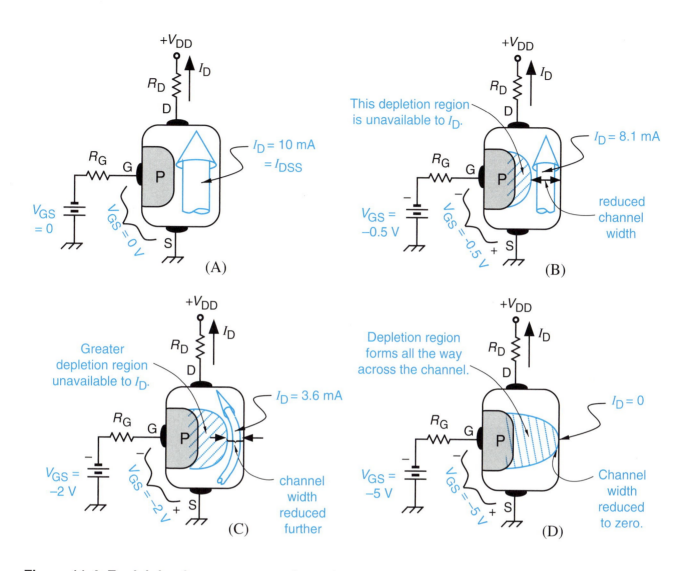

Figure 11-4 Explaining how more negative values of V_{GS} cause reduced values of I_D.

Interpreting the Characteristic Curves: The transfer curve of Fig. 11-3 shows a nonlinear relationship between I_D and V_{GS}. At the right of the curve, a slight change in control voltage V_{GS} produces a large change in I_D. At the left, even a large change in V_{GS} produces relatively little change in I_D. This non-proportional relation produces non-constant spacing in the ideal family of characteristic curves of I_D versus V_{DS}, shown in Fig. 11-5(A).

A realistic family of I_D-versus-V_{DS} curves is shown in Fig. 11-5(B). These curves indicate that I_D hardly increases at all for increasing V_{DS}, similar to a BJT. However, JFET saturation begins at larger voltage levels than BJT saturation. For instance, at $V_{GS} = -1$ V, saturation begins when V_{DS} is less than about 4 V, in Fig. 11-5(B). For $V_{GS} = 0$ V, saturation starts when V_{DS} decreases to 5 V. These voltages are much greater than the typical BJT saturation voltage of about 0.1-0.2 V. Because of this, saturation tends to be more of a problem for FET amplifiers than for BJT amplifiers.

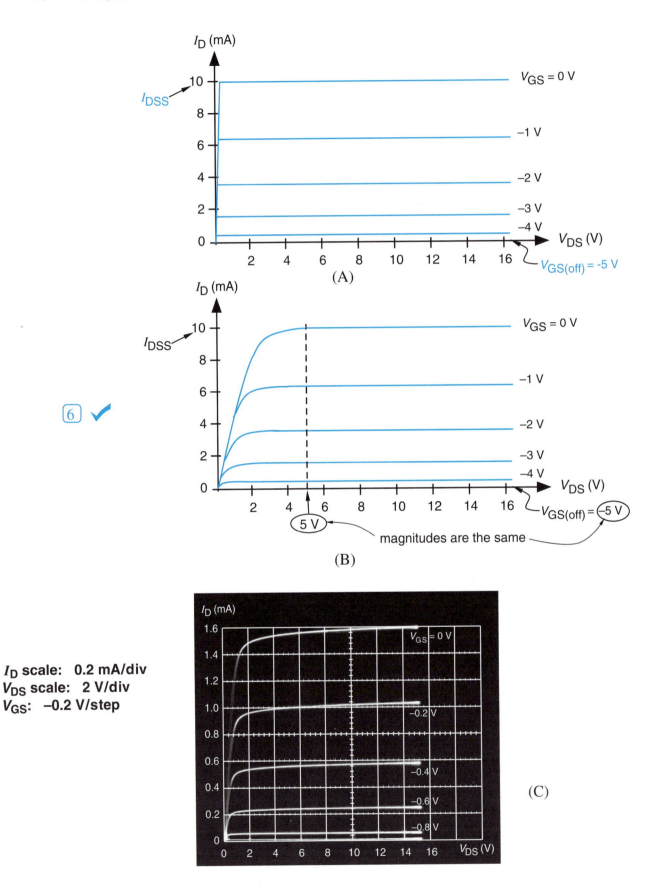

I_D **scale: 0.2 mA/div**
V_{DS} **scale: 2 V/div**
V_{GS}: **−0.2 V/step**

Figure 11-5 JFET I_D-versus-V_{DS} family of characteristic curves.
(A) Ideal curves. I_D is completely independent of main terminal voltage V_{DS}.
(B) Realistic curves. I_D is almost independent of V_{DS}, if V_{DS} is above the saturation value.
(C) Actual characteristics of a JFET with I_{DSS} = 1.6 mA and $V_{GS(off)}$ = −1 V.

Figure 11-5(B) shows that the saturation voltage for $V_{GS} = 0$ has the same magnitude as $V_{GS(off)}$. In this case the magnitude is 5 V. The magnitudes being the same is not a coincidence. It will always happen this way, for any FET. If a different FET were measured to have $V_{GS(off)} = -3$ V, then its saturation voltage level would be +3 V when $V_{GS} = 0$ V.

Figure 11-5(C) shows a photograph of an actual family of JFET characteristic curves.

P-Channel JFETs: The doping polarity can be reversed for an FET, like a BJT. The internal structure of a *p*-channel JFET is shown in Fig. 11-6(A). The V_{DD} power supply polarity must be reversed to $-V_{DD}$. Also, V_{GS} must be **+** on the gate in order to keep the *p-n* junction reverse-biased. The schematic symbol of a *p*-channel JFET has the gate arrow pointing outward, as shown in Fig. 11-6(B).

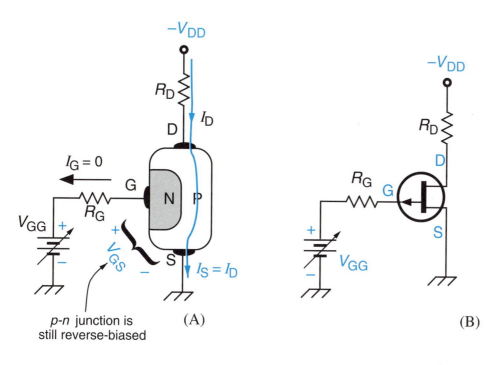

Figure 11-6
P-channel JFET.

(A) Internal structure simply has reversed doping. Main-terminal power supply polarity is like a *pnp* BJT.

(B) Schematic symbol.

SELF-CHECK FOR SECTION 11-1

1. For a field-effect transistor, the _____ is the controlling variable.
2. Draw the schematic diagram of an *n*-channel JFET connected to its two power supplies. Show power-supply polarities.
3. In the diagram from Problem 2, show the following:
   a) The polarity of gate-to-source voltage, V_{GS}.
   b) The direction of drain current I_D.
   c) The polarity of drain-to-source voltage, V_{DS}.
4. Repeat Problems 2 and 3 for a *p*-channel JFET.
5. For the FET in Fig. 11-3, if V_{GS} changes from -1.0 V to -0.5 V, I_D changes from _____ mA to_____ mA, for a net change of _____ mA.
6. In Fig. 11-3, if V_{GS} changes from -2.0 V to -1.5 V, I_D changes from _____ mA to_____ mA, for a net change of _____ mA.
7. (T-F) Change in I_D is always proportional to change in V_{GS}.

8. For an *n*-channel JFET, as V_{GS} gets larger in magnitude the channel width _____ . (increases or decreases)
9. (T-F) JFETs are like BJTs in this respect: The main current is not affected by the main terminal voltage, in the center part of the characteristic curves.
10. (T-F) JFETs are like BJTs in this respect: The milliampere-range main current remains unaffected by the main-terminal voltage until the main-terminal voltage decreases to a few tenths of a volt.

11-2 COMMON-SOURCE FET AMPLIFIER

An FET can serve as a common-source ac amplifier simply by biasing the drain and gate and connecting the input signal and load through coupling capacitors. Figure 11-7 shows the circuit schematic.

In Fig. 11-7, the dc bias gate current is zero, so there is zero dc voltage drop across R_G, even though its resistance is very large. The amplifier can be understood by the load-line analysis of the following example.

**Figure 11-7
Basic common-source FET amplifier with simple gate bias.**

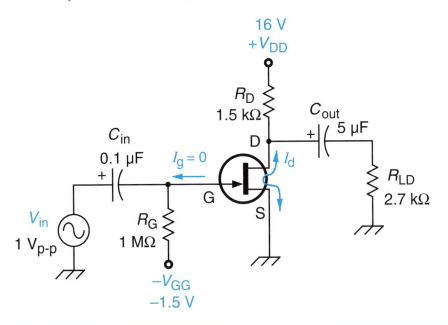

EXAMPLE 11-1

The FET in the amplifier of Fig. 11-7 has the characteristic curves of Fig. 11-3 and Fig. 11-5(B).
a) Draw the dc load-line on the I_D-versus-V_{DS} curves of Fig. 11-5(B).
b) Referring to the I_D-versus-V_{GS} curve of Fig. 11-3, find the I_D value of the Q-point. Locate the Q-point on the dc load-line.
c) Using the equivalent ac drain resistance in the drain line

$$r_D = R_D \| R_{LD},$$

draw the slope of the ac load-line.
d) Transfer the slope to the Q-point, to form the ac load-line.
e) Referring to the given value of V_{in}, locate the two extreme points (peak points) on the ac load-line. These are the points that the transistor swings between.

f) Find $V_{out(p-p)}$ by projecting down from the ac load-line to the v_{DS} axis.

g) Recalculate $V_{out(p-p)}$ by applying Ohm's law with $I_{d(p-p)}$ and r_D.
Compare to the result from part (f).

h) Calculate A_v for the amplifier.

i) Draw waveforms of V_{in} and v_{DS}.

SOLUTION

a) The i_D-versus-v_{DS} characteristic curves have been redrawn in Fig. 11-8(A). The dc load-line end-points are at $v_{DS} = 16$ V on the v axis and $i_D = 16$ V $\div 1.5$ kΩ = 10.7 mA on the i axis. The dc load-line is between these points.

b) From Fig. 11-3, $V_{GS} = -1.5$ V gives $I_D = 4.9$ mA. This characteristic curve has been added to the family in Fig. 11-8(A). The Q-point is located where this line intersects the dc load-line on Fig. 11-8(A).

c) The $+V_{DD}$ terminal is an ac ground, so R_D and R_{LD} are in parallel when viewed by ac current i_D emerging from the drain. Therefore

$$r_D = R_D \| R_{LD} = 1.5 \text{ k}\Omega \| 2.7 \text{ k}\Omega = 964 \ \Omega$$

If we try to identify the slope from Ohm's law with $v_{DS} = 16$ V, we get

$$i_D = \frac{v_{DS}}{r_D} = \frac{16 \text{ V}}{964 \ \Omega} = 16.6 \text{ mA}$$

which is way off the scale. So we prefer to choose a smaller test value of voltage v_{DS} for finding the slope. Choosing $v_{DS} = 6$ V gives

$$i_D = \frac{v_{DS}}{r_D} = \frac{6 \text{ V}}{964 \ \Omega} = 6.2 \text{ mA}$$

Therefore the slope line goes between 6 V on the v_{DS} axis and 6.2 mA on the i_D axis.

d) Transferring the slope to the Q-point gives the ac load-line shown in Fig. 11-8(A).

e) $V_{in(p-p)} = 1$ V, so V_{in} swings 0.5 V positive and 0.5 V negative. Coupled through capacitor C_{in}, this causes v_{GS} to oscillate by 0.5 V on either side of -1.5 V, namely between -2.0 V and -1.0 V. These v_{GS} values intercept the ac load-line at points P_1 and P_2 on Fig. 11-8(A).

f) The ac load-line has been redrawn in Fig. 11-8(B), with the important points marked. Projecting down from P_1 gives $v_{DS} = 7.2$ V. P_2 gives $v_{DS} = 9.9$ V.

So the ac voltage swing across the transistor is

$$9.9 \text{ V} - 7.2 \text{ V} = 2.7 \text{ V}$$

The ac voltage swing across the transistor is the same as the ac voltage swing across r_D (because the source lead is connected directly to ground). The ac voltage swing across r_D is the ac output voltage, $V_{out(p-p)}$. Therefore,

$$V_{out(p-p)} = \textbf{2.7 V}$$

g) From the vertical axis in Fig. 11-8(B), we have

$$I_{d(p-p)} = 6.4 - 3.6 = 2.8 \text{ mA}$$

Applying Ohm's law to r_D gives
$$V_{out(p-p)} = [I_{d(p-p)}] \, r_D$$
$$= (2.8 \text{ mA})(964 \, \Omega) = 2.7 \text{ V}$$

This agrees with the result from part (f), as it should.

h)
$$A_v = \frac{V_{out}}{V_{in}} = \frac{2.7 \text{ V}}{1 \text{ V}} = \mathbf{2.7}$$

i) Refer to Fig. 11-7. When V_{in} swings positive, v_{GS} moves closer to zero. This moves i_D higher on the ac load-line. Thus, the positive half-cycle of V_{in} moves the transistor to P_1 in Fig. 11-8(B). This is the negative half cycle of v_{DS} and of V_{ld}. We see that a common-source FET amplifier causes signal inversion, just like a common-emitter BJT amplifier.

During the positive half-cycle of V_{in}, v_{DS} swings from 8.6 V (the Q-point value) to 7.2 V (the P_1 value). During the negative half-cycle of V_{in}, v_{DS} swings from 8.6 V to 9.9 V (the P_2 value). These slightly unequal swings are clearly indicated on the v_{DS} waveform of Fig. 11-9.

Figure 11-8
Analyzing the FET amplifier by load-line techniques.

(A) Locating the dc load-line, the Q-point, the ac load-line, and the peak points on the ac load-line.

(B) Careful inspection of the operating points, P_1, P_2, and Q.

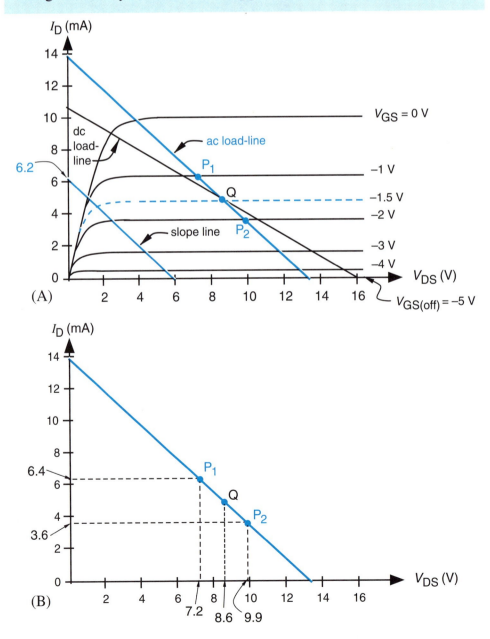

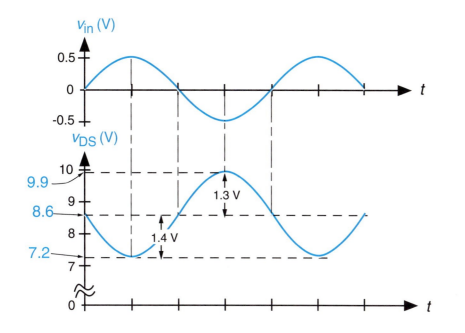

**Figure 11-9
Waveforms for the
FET amplifier.**

The v_{DS} waveform in Fig. 11-9 is a distorted sine wave. After the average value of v_{DS} is removed by coupling capacitor C_{out}, the resulting V_{ld} waveform will also show distortion. The distortion is unavoidable, because of the FET's nonlinear i_D-versus-v_{GS} curve (Fig. 11-3). However, it can be minimized by restricting the signal magnitudes to low levels. There are some applications where the JFET's distortion effect is not a serious disadvantage. In fact, in some RF applications, it can even be an advantage.

⑧ ✔

Input Resistance:

The major advantage of an FET amplifier is its very high input resistance. The transistor itself has infinite R_{in}, ideally. Thus, the only resistance seen by the ac signal source is the gate bias resistance.

Get This

For the Fig. 11-7 FET amplifier,
$$R_{in} = R_G = 1 \text{ M}\Omega$$
This R_{in} value is much higher than for any BJT amplifier that we studied in Chapters 4 through 10.

⑨ ✔

The International Telecommunications Satellite Organization (INTELSAT) is a consortium of satellite users from many nations. It operates and maintains these antennas / ground stations in Indonesia.

Courtesy of INTELSAT

Remember that open-circuit voltage (V_{oc}) is also called no-load voltage (V_{nl}).

EXAMPLE 11-2

Suppose that the Fig. 11-7 FET amplifier is driven by an input source with open-circuit voltage of $V_{oc} = 200$ mV, and internal resistance of $R_{int} = 100$ kΩ. This arrangement is shown in Fig. 11-10(A). We want to compare this FET circuit to a BJT common-emitter amplifier with $R_{in} = 5$ kΩ, which is a typical value. The BJT arrangement is shown in Fig. 11-10(B).

a) By voltage division, find the V_{in} value for the FET in Fig. 11-10(A).

b) Repeat for the BJT amplifier in Fig. 11-10(B).

c) Comment on the loading effect of an FET amplifier, compared to a BJT amplifier.

SOLUTION

a) The ideal portion of the signal voltage source sees R_{int} in series with the amplifier's R_{in}. Therefore,

$$\frac{V_{in}}{V_{oc}} = \frac{R_{in}}{R_{int} + R_{in}}$$

$$V_{in} = 200 \text{ mV} \left(\frac{1 \text{ MΩ}}{150 \text{ kΩ} + 1 \text{ MΩ}} \right) = \textbf{174 mV}$$

b) With $R_{in} = 5$ kΩ, we get

$$V_{in} = 200 \text{ mV} \left(\frac{5 \text{ kΩ}}{150 \text{ kΩ} + 5 \text{ kΩ}} \right) = \textbf{6.5 mV}$$

c) The loading problem on the high-resistance source is much less severe with the high-R_{in} FET amplifier.

Figure 11-10
Comparing performance of an FET amplifier and a BJT amplifier, driven by a high-resistance (high-impedance) source.

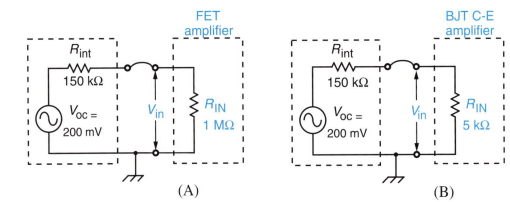

(A) (B)

11-3 FET INSTABILITY PROBLEMS

In Example 11-1, we were able to analyze the FET amplifier because we said that we knew the transistor's exact transfer curve. Our analysis relied entirely on the information contained in the I_D-versus-V_{GS} curve of Fig. 11-3, which is essentially the same information contained in the characteristic curve family of Fig. 11-5(B).

Real JFETs have great inconsistency in their characteristic curves. Individual FETs of a given type can differ from one another by a factor of 10 or more. In general, their batch instability is worse than that of BJTs.

EXAMPLE 11-3

A popular *n*-channel JFET is type No. 2N5457. This type has $V_{GS(off)}$ values that can range from –0.5 V to –6 V. It has I_{DSS} values that can range from 1 mA to 5 mA. These minimum and maximum values are summarized in Table 11-1.

	Minimum Value	Maximum Value
$V_{GS(off)}$	– 0.5 V	– 6 V
I_{DSS}	1 mA	5 mA

Table 11-1

Assuming that the minimum values go together and the maximum values go together, plot the minimum and maximum characteristic curves of I_D versus V_{GS} for the FET. Approximate the actual curvatures.

SOLUTION

The minimum curve descends from $I_D = 1$ mA to $V_{GS} = -0.5$ V, as shown in Fig. 11-11. The maximum curve descends from $I_D = 5$ mA to $V_{GS} = -6$ V.

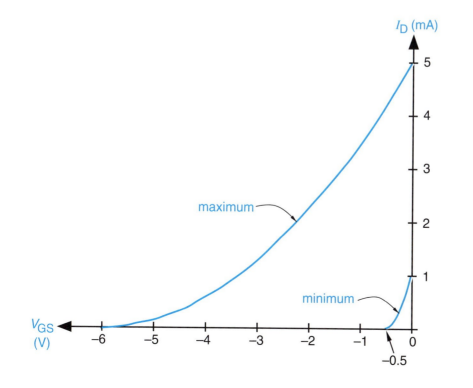

Figure 11-11 Showing that FETs have tremendous batch instability.

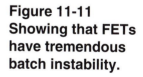

By inspecting Fig. 11-11, it is clear that a minimum unit requires a dc bias V_{GS} value between 0 and –0.5 V. But a maximum unit could not tolerate such a low V_{GS} bias value, because the drain current would be too large. Therefore, the simple gate bias method shown in Fig. 11-7 is not practical for mass-produced amplifiers.

Stabilizing the Bias Current

Figure 11-12
Voltage-divider bias-stabilizing design for FET amplifier. Large values are used for R_{G1} and R_{G2}, in order to make the amplifier's R_{in} very high.

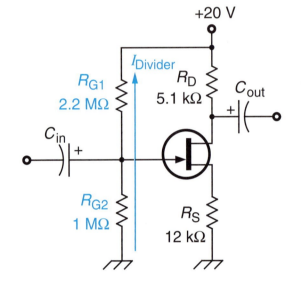

The most effective way to stabilize the bias current I_D is by using a voltage divider in the gate circuit and a large resistance in the source lead. The common-source amplifier in Fig. 11-12 shows typical component values. The working of this biasing circuit can be understood as follows.

$\boxed{1}$ Resistors R_{G1} and R_{G2} set the dc voltage at the gate, V_G, according to the voltage divider formula

$$V_G = V_{DD}\left(\frac{R_{G2}}{R_{G1} + R_{G2}}\right) \qquad \text{Eq. (11-1)}$$

R_{G1} and R_{G2} can afford to be very large because the FET gate has no loading effect at all on $I_{divider}$. Keeping them large raises the amplifier's R_{in}.

In this example, we have

$$V_G = 20\text{ V}\left(\frac{1\text{ M}\Omega}{2.2\text{ M}\Omega + 1\text{ M}\Omega}\right) = 6.3\text{ V}$$

which is specified in Fig. 11-13.

$\boxed{2}$ As drain current flows through the large source resistance R_S, it develops a voltage across R_S. This voltage, which is V_S, is larger than V_G. Voltage V_S must be larger than V_G in order to make V_{GS} negative. By Ohm's law,

I_D is equal to I_S.

$$V_S = I_D R_S$$

V_S is shown in Fig. 11-13.

3 The dc bias value of V_{GS} is equal to V_S subtracted from V_G. As a formula,

$$V_{GS} = V_G - V_S \qquad \text{Eq. (11-2)}$$

If the particular FET is closer to the minimum end of the range, I_D naturally tends to be small. This is suggested by the light line in Fig. 11-13(A). For instance, Fig. 11-13(A) shows $I_D = 0.6$ mA, which produces

$$V_S = (0.6 \text{ mA}) (12 \text{ k}\Omega) = 7.2 \text{ V}$$

Equation (11-2) gives

$$V_{GS} = 6.3 \text{ V} - 7.2 \text{ V} = -0.9 \text{ V}$$

which is a properly small value for a minimum-tending FET.

On the other hand, if the particular FET is closer to the maximum end of the range, I_D tends to be larger. This is suggested by the heavier line in Fig. 11-13(B). A larger I_D causes V_S to be larger. For instance, Fig. 11-13(B) shows $I_D = 0.8$ mA, which produces

$$V_S = (0.8 \text{ mA}) (12 \text{ k}\Omega) = 9.6 \text{ V}$$

Equation (11-2) gives

$$V_{GS} = 6.3 \text{ V} - 9.6 \text{ V} = -3.3 \text{ V}$$

which is a properly larger value for a maximum-tending FET.

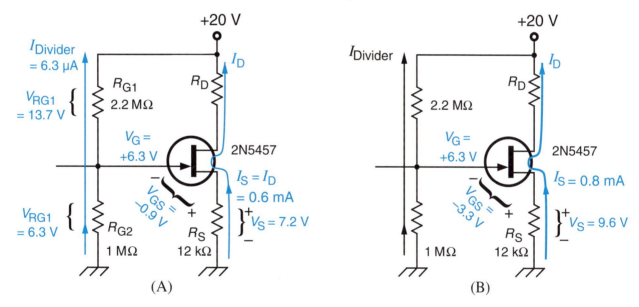

Figure 11-13 Explaining why the voltage-divider/R_S bias method stabilizes I_D reasonably well. (A) Minimum-tending FET.   (B) Maximum-tending FET.

The final outcome is that I_D is reasonably well stabilized. Its bias value might differ from the typical value by no more than 20%, perhaps, as we go from one transistor to another.

Notice that a large portion of the V_{DD} supply voltage must be dropped across R_S in order to get reasonable bias current stability. This takes away from the amount of voltage available to be split between R_D and V_{DS}, the transistor's main terminal bias voltages. In this regard, FETs are at a disadvantage compared to BJTs, since we can usually stabilize a BJT by spending only 2 or 3 volts across the emitter resistor R_E.

FET amplifiers are harder to stabilize than BJT amplifiers.

**Figure 11-14
Showing how to find the range of I_D bias values.**

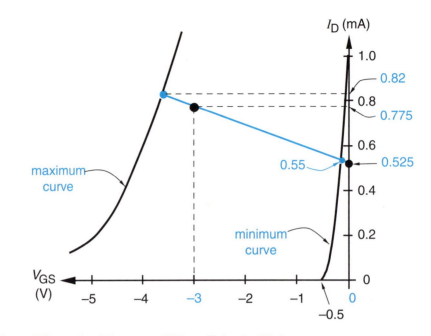

Graphing the Range of Possible I_D Values

Given a detailed schematic diagram of an FET amplifier, we can show the range where the I_D Q-point must be. Let us demonstrate this for the 2N5457 amplifier of Figures 11-12 and 11-13. Refer to the maximum and minimum characteristic curves that have been magnified in Fig. 11-14 for the region of $I_D < I_{DSS(min)}$, namely $I_D < 1$ mA.

We want to draw the line that the Q-point must lie on, between the minimum and maximum characteristic curves. Remember that we know the bias voltage V_G to be 6.3 V, from our previous analysis of the amplifier schematic.

$\boxed{1}$ To get started, we imagine $V_{GS} = 0$ V (this cannot really happen). Based on $V_{GS} = 0$ V, we can find V_S and I_D in the amplifier schematic by

$$V_S = V_G + V_{GS} = 6.3 \text{ V} + 0 \text{ V} = 6.3 \text{ V}$$

Applying Ohm's law to R_S gives

$$I_D = I_S = \frac{V_S}{R_S} = \frac{6.3 \text{ V}}{12 \text{ k}\Omega} = 0.525 \text{ mA}$$

We plot the point 0 V, 0.525 mA on the vertical axis of Fig. 11-14.

$\boxed{2}$ Now we choose a reasonable value of V_{GS}, some value near the maximum curve. Let us choose –3 V. We repeat our schematic diagram calculation from step 1.

$$V_S = V_G + V_{GS} = 6.3 \text{ V} + 3 \text{ V} = 9.3 \text{ V}$$

$$I_D = \frac{V_S}{R_S} = \frac{9.3 \text{ V}}{12 \text{ k}\Omega} = 0.775 \text{ mA}$$

We plot the point $V_{GS} = -3$ V, $I_D = 0.775$ mA on Fig. 11-14, as shown.

$\boxed{3}$ Draw a straight line between the plotted points, and extend it out to the maximum curve. The amplifier Q-point must lie somewhere on the line between the curves.

$\boxed{4}$ Project over to the I_D axis to find the range of I_D bias values. In this case we see that I_D must be between about 0.55 mA and 0.82 mA.

SELF-CHECK FOR SECTIONS 11-2 AND 11-3

11. In a common-source FET amplifier, the input voltage is applied between the _____ terminal and the _____ terminal. The output is taken between the _____ terminal and the _____ terminal.

12. In the common-source amplifier of Fig. 11-7, make the following changes: $V_{DD} = +22$ V; $R_D = 1.8$ kΩ; $R_{LD} = 3$ kΩ. Repeat Example 11-1, parts (a) through (h).

13. (T-F) Unless the ac signals are very small, a common-source FET amplifier distorts the input signal noticeably.

14. What is the main advantage of an FET amplifier over a BJT amplifier?

15. (T-F) JFETs tend to have closer manufacturing tolerances and closer (more stable) electrical specifications than BJTs.

16. The type 2N3819 n-channel JFET has these minimum and maximum ratings: $I_{DSS(min)} = 2$ mA; $I_{DSS(max)} = 20$ mA; $V_{GS(off)\,[min]} = -1$ V; $V_{GS(off)\,[max]} = -8$ V. Sketch the minimum and maximum characteristic curves of I_D versus V_{GS}.

17. Suppose the FET from Problem 16 is biased as shown in Fig. 11-13, with $R_{G1} = 2.2$ MΩ, $R_{G2} = 2.7$ MΩ, $R_S = 6.2$ kΩ and $V_{DD} = 20$ V. Graph the range of possible I_D values, using the method outlined on page 362.

11-4 AC OPERATION

Starting with the bias-stabilized JFET circuit of Fig. 11-12, we can build a complete common-source amplifier as shown in Fig. 11-15(A). The source-bypass capacitor C_S makes the S terminal an ac ground. Therefore the entire ac input voltage V_{in} appears between the G and S terminals, as V_{gs}.

Figure 11-15 Complete bias-stabilized common-source FET amplifier. (A) Schematic diagram. (B) Typical bias conditions and waveform appearance.

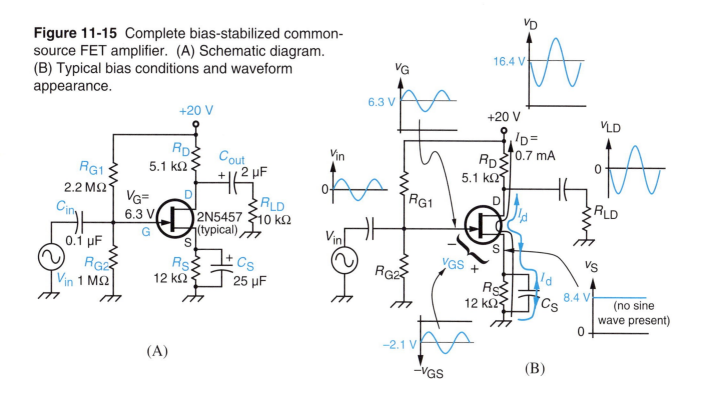

If we assume that the individual FET has typical specifications for the type No. 2N5457, its characteristic curve will be somewhere between the maximum and minimum extremes.

A typical I_D-versus-V_{GS} curve might be the one shown in Fig. 11-16(A). Then its dc bias current I_D will be somewhere between 0.55 mA and 0.82 mA. Here, we have $I_D = 0.7$ mA. Applying Ohm's law and Kirchhoff's voltage law to the drain-source circuit in Fig. 11-15(A), we get

$$V_S = I_D R_S$$
$$= (0.7 \text{ mA})(12 \text{ k}\Omega) = 8.4 \text{ V}$$

Figure 11-16
Performing ac analysis of the FET amplifier in Fig. 11-15.

(A) The I_D-versus-V_{GS} curve can be used to get transconductance g_m, as shown in Ex. 11-4 on page 366. Then g_m can be used to find A_v and V_{out} as shown in Example 11-5, page 367.

(B) Load-line analysis gives the same results. It is shown starting on page 367.

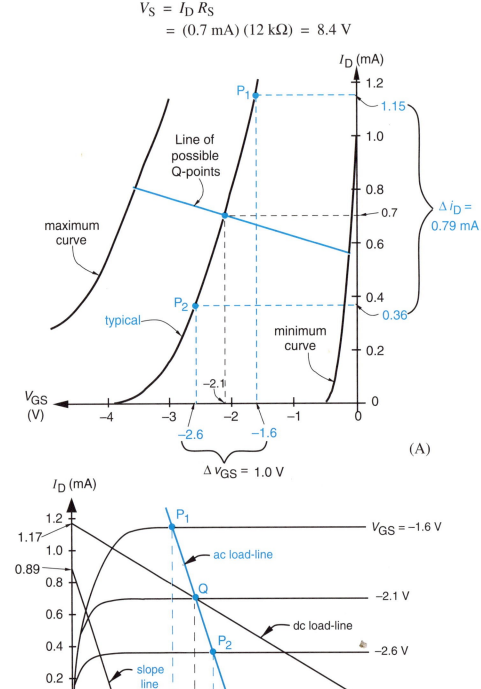

$$V_{RD} = I_D R_D = (0.7 \text{ mA}) (5.1 \text{ k}\Omega) = 3.6 \text{ V}$$

$$V_D = V_{DD} - V_{RD} = 20 \text{ V} - 3.6 \text{ V} = 16.4 \text{ V}$$

$$V_{DS} = V_{DD} - V_{RD} - V_S$$
$$= 20 \text{ V} - 3.6 \text{ V} - 8.4 \text{ V} = 8.0 \text{ V}$$

Then the ac waveforms riding on their dc bias levels will be as shown in Fig. 11-15(B), on page 363.

Input Resistance: Since the gate junction is reverse-biased, it acts like an open circuit to the ac signal. No ac current flows in the gate lead. Looking into the amplifier, the ac input source sees R_{G2} connected directly to ground and R_{G1} connected to ac ground at V_{DD}.. These resistors appear in parallel to the ac source, so the ac input resistance is given by

$$R_{in} = R_{G1} \| R_{G2} \qquad \text{Eq. (11-3)}$$

For the amplifier of Fig. 11-15, Eq. (11-3) gives

$$R_{in} = 2.2 \text{ M}\Omega \| 1 \text{ M}\Omega = 688 \text{ k}\Omega,$$

which is quite a high value.

Transconductance of an FET

Figure 11-16(A) shows the typical i_D-versus-v_{GS} curve, magnified in the region of $i_D < 1$ mA. Check for yourself that the dc bias Q-point is at $V_{GS} = -2.1$ V, $I_D = 0.7$ mA.

If we assume a peak-to-peak input voltage of 1 V, then v_{IN} swings positive to +0.5 V, and negative to –0.5 V in Fig. 11-15(B). This causes v_{GS} to oscillate by 0.5 V on either side of –2.1 V, namely between –1.6 V and –2.6 V. Therefore the transistor drain current moves up and down its transfer curve between points P_1 and P_2 in Fig. 11-16(A). Projecting over from point P_1 to the i_D axis gives $i_D = 1.15$ mA. P_2 has an i_D value of 0.36 mA, as Fig. 11-16(A) makes clear. The peak-to-peak change in current is given by

> The v_{GS} waveform's positive peak (P_1) is at -2.1 V + 0.5 V = –1.6 V. Its negative peak (P_2) is at –2.1 V – 0.5 V = –2.6 V.

$$\Delta i_D = 1.15 \text{ mA} - 0.36 \text{ mA} = 0.79 \text{ mA}$$

When an FET swings up and down its i_D-versus-v_{GS} transfer curve, the ratio of the current change to the voltage change is called the FET's **transconductance**, symbolized g_m. As a formula,

Get This

$$g_m = \frac{\Delta i_D}{\Delta v_{GS}} \qquad \text{Eq. (11-4)}$$

or equivalently,

$$g_m = \frac{I_{d \, (p\text{-}p)}}{V_{gs \, (p\text{-}p)}}$$

Transconductance is measured in the basic unit of **siemens**, symbolized S. Most JFETs have g_m values that are small fractions of a siemens, so we express g_m in millisiemens (mS) or microsiemens (µS).

> Transconductance is sometimes called mutual conductance (*g* represents conductance, the reciprocal [opposite] of resistance. *m* stands for mutual). It may be symbolized y_{fs} on some manufacturers' data sheets.

Another word for a siemens is **mho**, symbolized as an upside-down Ω. This usage is being eliminated.

13 ✔

Ac drain-line resistance for an FET is found in the same way as collector-line ac resistance for a BJT.

EXAMPLE 11-4

For the typical 2N5457 FET of Fig. 11-16(A), find the transconductance for the operating conditions shown in that figure.

SOLUTION

Applying Eq. (11-4) gives

$$g_m = \frac{\Delta i_D}{\Delta v_{GS}}$$

$$= \frac{1.15\ \text{mA} - 0.36\ \text{mA}}{-1.6\ \text{V} - (-2.6\ \text{V})} = \frac{0.79\ \text{mA}}{1.0\ \text{V}} = \textbf{0.79 mS} \ \text{or} \ \textbf{790 μS}$$

Knowing g_m enables us to predict the amplifier's voltage gain, A_v. Here is how:

1 As ac drain current I_d emerges from the transistor's D terminal, it sees an equivalent ac resistance given by

$$r_D = R_D \| R_{LD} \qquad \text{Eq. (11-5)}$$

This is because the top of R_D is at ac ground, and R_{LD} is grounded. It's the same situation as in the collector of a BJT. In Fig. 11-15,

$$r_D = R_D \| R_{LD} = 5.1\ \text{k}\Omega \| 10\ \text{k}\Omega = 3.38\ \text{k}\Omega$$

2 Also like a BJT, ac output voltage V_{out} is given by Ohm's law as

$$V_{out} = I_d\, r_D \qquad \text{Eq. (11-6)}$$

3 Equation (11-4) can b rearranged as

$$I_d = (V_{gs})\, g_m$$

Substituting this expression into Eq. (11-6) gives

$$V_{out} = (V_{gs})\, g_m\, (r_D)$$

which is the same as

$$\frac{V_{out}}{V_{gs}} = g_m\, r_D \qquad \text{Eq. (11-7)}$$

4 In the Fig. 11-15 amplifier, bypass capacitor C_S makes the S terminal an ac ground, so all of the ac input voltage V_{in} appears between the gate and source terminals, as V_{gs}. That is, $V_{in} = V_{gs}$. Therefore Eq. (11-7) can be written

$$\frac{V_{out}}{V_{in}} = A_v = g_m\, r_D \qquad \text{Eq. (11-8)}$$

EXAMPLE 11-5

For the common-source amplifier of Fig. 11-15 on page 363, containing the typical FET of Fig. 11-16 on page 364,
a) Calculate the voltage gain, A_v.
b) Calculate $V_{out(p-p)}$ for $V_{in(p-p)} = 1$ V, as shown in Fig. 11-16(A).
c) State the phase relation between V_{out} and V_{in}.

SOLUTION

a) From Eq. (11-8)

$A_v = g_m\, r_D$

$= (0.79 \text{ mS}) (3.38 \text{ k}\Omega) = (0.79 \times 10^{-3} \text{ S}) (3.38 \times 10^3 \ \Omega) = \textbf{2.67}$ $\boxed{14}$ ✔

b) $V_{out(p-p)} = A_v\, [V_{in(p-p)}]$

$= 2.67\, (1 \text{ V}) = \textbf{2.67 V}$

c) V_{out} is inverted with respect to V_{in}, like a common-emitter BJT amp.

Ac Load-Line Analysis

These amplifier results can be checked by load-line analysis. The typical i_D-versus-v_{DS} family of FET characteristic curves is drawn in Fig. 11-16(B), for v_{GS} values of –2.6 V, –2.1 V, and –1.6 V. This family contains the same basic information that is contained in the typical curve in Fig. 11-16(A).

The dc load-line's extreme points are gotten from the KVL equation

$V_{DD} = I_D R_D + V_{DS} + I_D R_S.$

$20 \text{ V} = I_D (R_D + R_S) + V_{DS}$

$20 \text{ V} = I_D (5.1 \text{ k}\Omega + 12 \text{ k}\Omega) + V_{DS}$

$20 \text{ V} = I_D (17.1 \text{ k}\Omega) + V_{DS}$

When $V_{DS} = 0$ V,

$$I_D = \frac{20 \text{ V}}{17.1 \text{ k}\Omega} = 1.17 \text{ mA}$$

This is marked on the I_D axis.

When $I_D = 0$, $V_{DS} = 20$ V. This is marked on the V_{DS} axis, and the dc load-line is formed by joining the two extreme points. The Q-point is located where the dc load-line intersects the curve for dc bias value $V_{GS} = -2.1$ V.

For the ac circuit, r_D is 3.38 kΩ as calculated earlier. The ac resistance in the source is zero, because 12-kΩ R_S is ac-bypassed by 25-μF C_S. Therefore, assuming a V_{ds} of 3 V, a reasonable value, the ac Ohm's law equation is

$$I_d = \frac{V_{ds}}{3.38 \text{ k}\Omega} = \frac{3 \text{ V}}{3.38 \text{ k}\Omega} = 0.89 \text{ mA}$$

This point is plotted on the i_D axis. Joining it to $V_{ds} = 3$ V forms the slope line in Fig. 11-16(B).

Then the slope line has been transferred to the Q-point to produce the ac load-line. This line intersects the curves for $V_{GS} = -1.6$ V and $V_{GS} = -2.6$ V at points P_1 and P_2, as shown. A projection down from P_1 hits the V_{DS} axis at 6.5 V. Projecting down from P_2, we hit at 9.2 V. The difference between these values is the peak-to-peak V_{ds} voltage, or

$$V_{ds} = 2.7 \text{ V (p-p)}$$

To graph-paper accuracy, this result agrees with the voltage calculated in Example 11-5.

Ac Instability (Transconductance Instability)

It is important to recognize this fact:

Transconductance g_m for a particular FET is extremely non-constant.

This is because the steepness of the I_D-versus-V_{GS} curve is extremely variable. The typical FET of Fig. 11-16(A) had $g_m = 790$ μS for the Q-point and input signal conditions in Example 11-4. But it is clear from Fig. 11-16(A) that g_m could be very different from 790 μS, at different Q-points.

15 ✔

The gain figure (g_m) for an individual FET is not nearly as constant as the gain figure (β) for an individual BJT.

The great variation in g_m for a particular FET is quite different from the BJT situation. A particular BJT has fairly constant β from one bias point to another.

Of course, transconductance also varies greatly from one individual FET to another. For example, Fig. 11-16(A) shows the line of possible Q-points for the amplifier of Fig. 11-15. It is clear by inspection that a minimum-tending FET has a steeper slope than a maximum-tending FET, at the Q-point's location on its curve.

Because g_m is so variable, Eq. (11-8) is not really very reliable for predicting voltage gain A_v. Unless we do a great amount of initial investigation to find out the specific value of g_m, we cannot confidently calculate A_v for a bypassed common-source FET amplifier.

Partially stabilizing A_v with unbypassed source resistance: By leaving part of the source resistance unbypassed, we can improve the *gm* instability problem, to a limited extent. Figure 11-17 shows this approach.

The ac input V_{in} will split between the gate-source junction and R_{S1}. This causes a reduction in V_{gs}, the voltage that actually makes the transistor work. Therefore A_v is lowered.

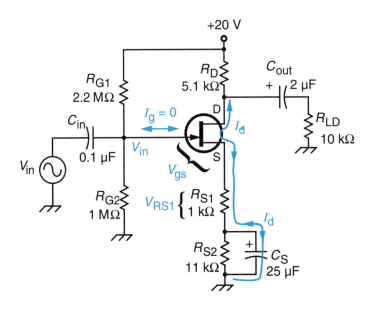

Figure 11-17 With unbypassed resistor R_{S1} in the source lead, A_v is partially stabilized.

You will recognize this as the same method that we used to stabilize A_v for a common-emitter BJT amplifier. It can be shown that the amplifier of Fig. 11-17 has a voltage gain given by

$$A_v = \frac{r_D}{\dfrac{1}{g_m} + R_{S1}} \qquad \text{Eq. (11-9)}$$

A careful analysis of Fig. 11-17 would show that A_v has been stabilized between about 1.1 and 1.8, for a type 2N5457 FET. Without R_{S1} stabilization, A_v would have varied between about 1.7 and 4.1. Thus, partial stability has been accompanied by a sacrifice in the amount of gain. ($A_v < 2.67$, the bypassed value from Example 11-5).

FET common-source amplifiers have naturally low A_v values. Source resistance stabilization makes their A_v even lower. Table 11-2 compares the results of gain-stabilizing common-source FET and common-emitter BJT amplifiers.

The quantity $1 / g_m$ is sometimes called the FET's internal *dynamic source resistance*, $r_{S(int)}$. It can be thought of as similar to emitter-junction resistance r_{Ej} of a BJT. Then Eq. (11-9) for a JFET reminds us of Eq. (5-14) for a BJT. See page 135.

As always, when we stabilize gain, we sacrifice gain.

	Common-emitter BJT amplifier	Common-source FET amplifier
Natural A_V (unstabilized)	High: $A_V > 100$ typically	Low; $A_V < 10$ typically
Reduced A_V (stabilized)	Moderate; $A_V = 10 - 50$ typically	Very low: $A_V < 3$ typically
Degree of stabilization possible	Very stable; typical variation $\pm 10\%$	Only limited stability; typical variation $\pm 50\%$

Table 11-2 Ac-stabilization of common-source FET amplifiers is not as satisfactory as it is for BJT amplifiers.

An FET amplifier's current gain A_i tends to be greater than that of a comparable BJT amplifier. This is due to the FET's very high R_{in}, which causes it to take very little current from the input signal source.

Restricting the Range of FET Characteristics

When mass-producing FET common-source amplifiers, we could take the trouble to test every individual FET. By rejecting those units that are too close to the minimum curve, it would be possible to guarantee a higher minimum value for I_{DSS}. For example, we could test every 2N5457 FET in a batch, and reject all those with $I_{DSS} < 2$ mA.

Guaranteeing a higher value of I_{DSS} would then allow us to redesign the bias circuitry for higher I_D values. For example, the reduced R_S values in Fig. 11-18(A) will raise the line of possible Q-points to the 1.5–1.9 mA range, as shown in Fig. 11-18(B).

Notice that the new Q-point line intersects the I_D-versus-V_{GS} curves at locations with steeper slopes . This provides greater g_m values, thereby increasing A_v.

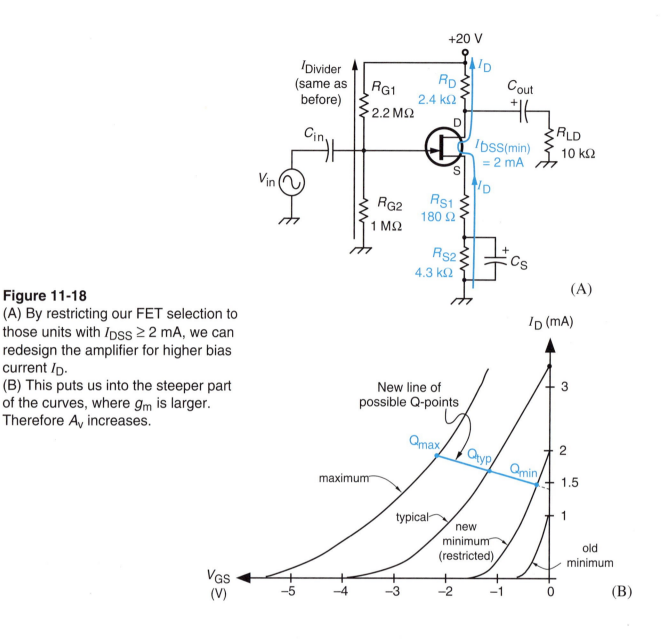

Figure 11-18
(A) By restricting our FET selection to those units with $I_{DSS} \geq 2$ mA, we can redesign the amplifier for higher bias current I_D.
(B) This puts us into the steeper part of the curves, where g_m is larger. Therefore A_v increases.

SELF-CHECK FOR SECTION 11-4

18. A typical 2N5457 FET characteristic curve is drawn in Fig. 11-16(A). The typical value of $V_{GS(off)}$ is _____ V.
19. For the amplifier of Fig. 11-15, suppose we investigate that particular FET, and find that it has g_m = 800 μS. (This g_m value is valid only for these specific conditions of bias and input magnitude.) Find the amplifier's A_v.
20. In Problem 19, suppose that the FET were replaced with a unit that is closer to the minimum curve [see Fig 11-16(A)]. Would g_m become greater or smaller?
21. In Question 20, A_v would become _____ .

    For Problems 22 and 23, redraw or photocopy and enlarge the typical i_D-versus-v_{GS} curve of Fig. 11-16(A).

22. Suppose the dc bias Q-point is at $V_{GS} = -2.2$ V, $I_D = 0.62$ mA. The peak-to-peak magnitude of V_{in} is 0.4 V.
    a) Locate peak point P_1. Project over to get the i_D value (approximate).
    b) Locate peak point P_2. Project over to get the i_D value.
    c) Use Eq. (11-4) to find g_m under these conditions.

23. Repeat Problem 22 for a Q-point at $V_{GS} = -2.8$ V, $I_D = 0.26$ mA. The peak-to-peak magnitude of V_{in} is 0.4 V.
    a) Locate peak point P_1. Project over to get the i_D value.
    b) Locate peak point P_2. Project over to get the i_D value.
    c) Use Eq. (11-4) to find g_m under these conditions.
24. Compare your results from Problems 22 and 23. What have you demonstrated about the constancy of g_m under different bias conditions?
25. If gate resistors R_{G1} and R_{G2} in Fig. 11-15 were changed to produce the bias conditions of Problem 22, the amplifier would have $A_v =$ _____ .
26. If gate resistors R_{G1} and R_{G2} in Fig. 11-15 were changed to produce the bias conditions of Problem 23, the amplifier would have $A_v =$ _____ .
27. In the stabilized common-source amplifier of Fig. 11-17, suppose gate resistors R_{G1} and R_{G2} are changed to produce the dc bias conditions of Problem 22. Then we would have $g_m =$ _____ μS, and $A_v =$ _____ . [Use Eq. (11-9).]
28. Repeat Problem 27, but for a change to the bias conditions of Problem 23. Then $g_m =$ _____ μS, and $A_v =$ _____ .

11-5 COMMON-DRAIN FET AMPLIFIER

A common-drain FET amplifier is shown in Fig. 11-19. Its circuit configuration is just like that of a common-collector BJT amplifier:

■ There is no resistance in the drain lead — the D terminal is connected directly to the V_{DD} power supply.

■ The ac output signal is developed across source resistance R_S.

■ The load is capacitively coupled between source S and ground.

Common-drain is also called **source-follower**. The phrasing is the same as for BJTs.

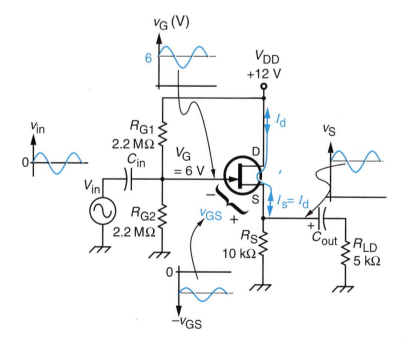

Figure 11-19
In a common-drain FET amplifier, the drain D is at ac ground. The ac input is applied between gate G and ground, and the output is taken between source S and ground. Therefore the ac-grounded drain is common to both input and output.

The common-drain amplifier offers improved performance over the common-source amplifier in two ways:

■ Its bias point is more stable

■ It can operate with lower values of dc supply voltage V_{DD}.

A common-drain amplifier is like a common-collector BJT amplifier in that $A_v < 1$.

In a common-drain amp, the resistor R_D is not present.

Source-line ac resistance is gotten in the usual way.

Again (as on page 369), the FET formula reminds us of the comparable BJT formula if $1/g_m$ is replaced with internal dynamic source resistance $r_{S(int)}$. Compare to Eq. (6-4) on page 164, for a common-collector amp.

Of course, the common-drain amplifier can't provide any voltage gain ($A_v < 1$). But the common-source amplifier has a rather low voltage gain anyway, so we're not giving up much. With proper design, it is possible to keep a common-drain's A_v value quite stable. This is not possible in the common-source mode.

The reason for the success of the common-drain amplifier is that there is no dc voltage wasted across R_D.

Because there is no voltage wasted, V_{DD} can be made smaller without running the risk of saturating the FET. By contrast, in the common-source amplifier there is considerable voltage spent on R_D. That makes less voltage available for V_{DS}, and it is more likely that v_{DS} will fall too low.

Also, now that a larger portion of the V_{DD} supply voltage can afford to be dropped across R_S, we can design the bias voltage V_G to be comparatively larger. This helps swamp out V_{GS} batch variations, holding I_D more stable.

Voltage-gain Formula: The ac-equivalent resistance in the source lead of Fig. 11-19 is

$$r_S = R_S \| R_{LD} \qquad\qquad \text{Eq. (11-10)}$$

A portion of V_{in} appears across the G-S junction, to operate the transistor. The rest of V_{in}, usually most of it, appears at the source S. This V_s voltage is V_{out}.

It can be shown that common-drain voltage gain is given by the formula

$$\frac{V_{out}}{V_{in}} = A_v = \frac{r_S}{\dfrac{1}{g_m} + r_S} \qquad\qquad \text{Eq. (11-11)}$$

EXAMPLE 11-8

For a typical 2N5457 FET in the common-drain amplifier of Fig. 11-19, the Q-point would be $V_{GS} = -1.95$ V, $I_D = 0.82$ mA. At this point, the FET's transconductance is $g_m = 860$ μS.

a) Find A_v.
b) For $V_{in} = 0.5$ V peak-to-peak, find $V_{out\,(p\text{-}p)}$. State their phase relation.
c) Find R_{in}.

SOLUTION

a) From Eq. (11-10),

$$r_S = 10\text{ k}\Omega \| 5\text{ k}\Omega = 3.33\text{ k}\Omega$$

Equation (11-11) gives

$$A_v = \frac{r_S}{\dfrac{1}{g_m} + r_S} = \frac{3330\ \Omega}{\dfrac{1}{860 \times 10^{-6}\text{ S}} + 3330\ \Omega}$$

$$= \frac{3330\ \Omega}{1160\ \Omega + 3330\ \Omega} = \mathbf{0.74}$$

b) $V_{out} = A_v \cdot V_{in}$

$= (0.74)(0.5\ V) =$ **0.37 V (p-p), in phase with V_{in}**

⟨17⟩ ✓

c) As always, the JFET itself has virtually infinite input resistance. The amplifier's R_{in} is

$$R_{in} = R_{G1}\,||\,R_{G2}\,||\,\infty = 2.2\ M\Omega\,||\,2.2\ M\Omega = \textbf{1.1 M}\boldsymbol{\Omega}$$

The common-drain amp, like the common-collector, is often used to match a high-resistance source to a low-resistance load.

Common-Gate Amplifier: A common-gate FET amplifier has the same circuit configuration as a common-base BJT amp. However, it has limited application because it gives away the chief advantage of an FET amplifier, namely high R_{in}.

Dual-Polarity Power Supplies

Any JFET amplifier configuration can be operated with dual-polarity power supplies. This is shown in Fig. 11-20(A) for a common-source amplifier, and in Fig. 11-20(B) for a common-drain amplifier. For the common-drain circuit, V_G is usually about half of the total power-supply voltage (look at Fig. 11-19). With equal-magnitude V_{DD} and V_{SS}, this can be accomplished with a single gate resistor R_G. It may not be necessary to set up a two-resistor voltage-divider bias circuit.

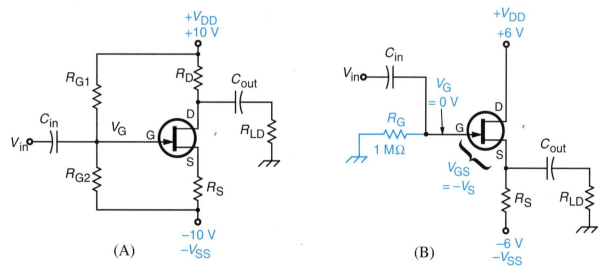

(A)   (B)

Figure 11-20 Dual-polarity power supply circuits. (A) Common-source. (B) Common-drain. There may or may not be a two-resistor voltage divider in the common-drain circuit.

SELF-CHECK FOR SECTION 11-5

29. In a common-drain amplifier, the output signal is _____ phase with the input signal.

30. A common-drain amplifier does provide _____ gain but does not provide _____ gain.

31. One of the drawbacks of common-source amplifiers is that they usually require rather high power supply voltage V_{DD}. Why don't common-drain amplifiers have that same problem?

32. In the common-drain amplifier of Fig. 11-19, suppose R_{LD} is changed to 3 kΩ.
    a) Will the bias conditions change?   c) Find A_v.
    b) Will transconductance g_m change?   d) Find R_{in}.

33. In the common-drain amp of Fig. 11-20(B), suppose the FET is a typical 2N5457, whose transfer curve is shown in Fig. 11-16(A) [and 11-18(B)]. Suppose R_S = 7.5 kΩ and R_{LD} = 1 kΩ. Let us find the Q-point.
    a) For an assumed V_{GS} = 0 V, find V_S, V_{RS}, and I_S. Use Ohm's law. Locate the V_{GS}, I_D point on the graph.   (Remember that I_D= I_S)
    b) Repeat part **a** for V_{GS} = –1 V.
    c) Connect the points from parts (a) and (b) and then extend the straight line to get the line of possible Q-points. Locate Q for the typical FET.

34. For the bias conditions in Problem 33, find the transconductance g_m, assuming $V_{gs\ (p\text{-}p)}$ = 0.4 V.

35. Referring to Problems 33 and 34, calculate the circuit's A_v from Eq. (11-11).

BIOFEEDBACK FOR VISION

Our eyes can focus on both far objects and near objects because the eye's ciliary muscle is able to adjust the curvature of the eye's lens. Specifically, when viewing a far object, the ciliary muscle relaxes, allowing the eye lens to spring into its normal shape, with a gentle radius of curvature (about 10 mm). This is shown in part A of Fig. 1 (for simplicity, only the front of the eye lens is shown).

When viewing a close object, the ciliary muscle contracts, forcing the eye lens away from its normal shape into a sharper radius of curvature (about 5 mm, at extreme condition). This is pictured in part B of Fig. 1.

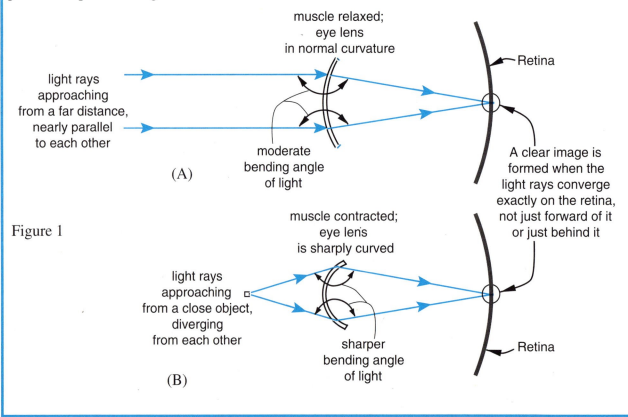

Figure 1

About 60% of all Americans have some vision defect that prevents the eye from performing this job properly. For many people, the problem is simply inability to flex the ciliary muscle strongly enough, or inability to relax the muscle. Recent research indicates that these problems sometimes can be solved by biofeedback training in ciliary muscle control.

The biofeedback arrangement is shown in the top photo. In a darkened room, the patient's head is held at a constant distance from the focusing tunnel. An infrared source inside the tunnel simulates light from a far object or from a close object, coming from the left or coming from the right. This is under the control of the optometrist manipulating the joystick. Some of the infrared radiation is reflected back from the eye, and reenters the tunnel through the aperture hole. By electronically detecting this reflected infrared radiation, the degree of lens curvature can be measured.

If the eye's lens is under-curved, indicating that the ciliary muscle is not contracted tightly enough, the electronic system produces a low-pitched tone in the patient's headset. The worse the lens' under-curvature, the lower the tone. On the other hand, for over-contraction and over-curvature, the system produces high-pitched audio tones, again proportional to the degree of misadjustment.

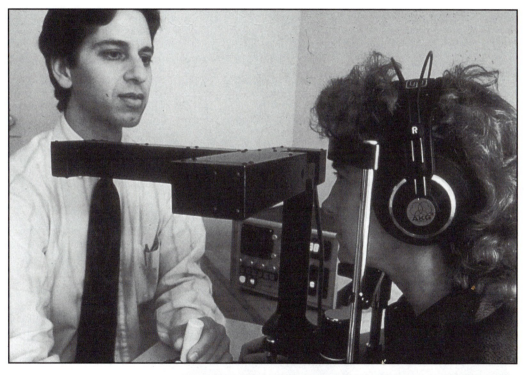

In repeated training sessions with this audible biofeedback, some patients learn to voluntarily control the contraction of their ciliary muscles—normally an involuntary function. If the training is successful, clear vision is restored without corrective lenses.

The vision system is shown by itself in the lower photo. Also see page 402.

Courtesy of Biofeed Trac, Inc.

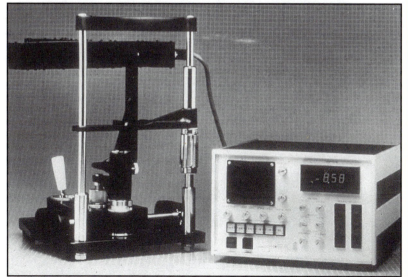

11-6 TROUBLESHOOTING JFET AMPLIFIERS

The standard troubleshooting method is used to test a JFET amplifier stage. This method has been described before, in Sections. 5-7 and 9-10. Look for bad output but good input. If a JFET stage is malfunctioning, there are two probable causes.

$\boxed{1}$ The gate-to-channel junction may be shorted.

$\boxed{2}$ The gate junction may be open-circuited.

Shorted junction: The shorted-junction problem is pictured in Fig. 11-21(a). With the *p-n* junction no longer acting as an open-circuit, a dc current flow path is established as follows: Up through R_{G2}, into the gate terminal, through the short circuit, then up through the channel and out the drain. This causes several measured conditions to occur that would never occur in a properly functioning circuit:

■ $I_G \neq 0$, and $I_{RG2} > I_{RG1}$. Since current is difficult to measure, this is best detected by measuring the dc voltages across R_{G1} and R_{G2} with a DMM. The voltage V_{RG2} will be much larger than the voltage-division formula predicts in Fig. 11-21(C), which is a properly functioning circuit shown for comparison purposes. Of course, V_{RG1} will be smaller than expected.

■ $V_{GS} \approx 0$ V. A short circuit between the gate and channel regions will cause V_{GS} to be close to 0 V, as Fig. 11-21(B) suggests. Therefore I_D is close to I_{DSS}, or else I_D will be the maximum limit value permitted by Ohm's law in the main-terminal path. That is,

$$I_{D\,(\text{limit})} = \frac{V_{DD} - V_{sat}}{R_D + R_S}$$

This larger-than-normal main-terminal bias current is best detected by measuring V_{RD} and then dividing by R_D.

The larger-than-normal main current (I_D and I_S) causes V_S to be much too large also. With the gate essentialy shorted to the main flow path, V_G is much larger than normal too, as indicated in Fig. 11-21(B).

In summary, the voltage symptoms of a shorted-gate FET are:

$\boxed{1}$ V_G is much too large.

$\boxed{2}$ V_{RG2} and V_{RG1} not in correct proportion. V_{RG2} is much too large.

$\boxed{3}$ V_{RD} and V_{RS} are both far too large, and they are not quite in correct proportion.

Open junction: The open-junction problem is pictured in Fig. 11-22(A). With a gap between the gate *p*-region and the channel *n*-region, no depletion layer forms, no matter how hard the junction is reverse-biased. The channel retains its entire width. Therefore I_D is equal to the maximum possible value, either I_{DSS} or $I_{D(\text{limit})}$.

The measured dc conditions are indicated in Fig. 11-22(B). In summary, the voltage symptoms of an open-junction FET are:

$\boxed{1}$ V_G is correct.

$\boxed{2}$ V_{GS} is not 0 V. In fact, V_{GS} tends to be greater than normal.

$\boxed{3}$ V_{RD} and V_{RS} are too large, although they are in correct proportion.

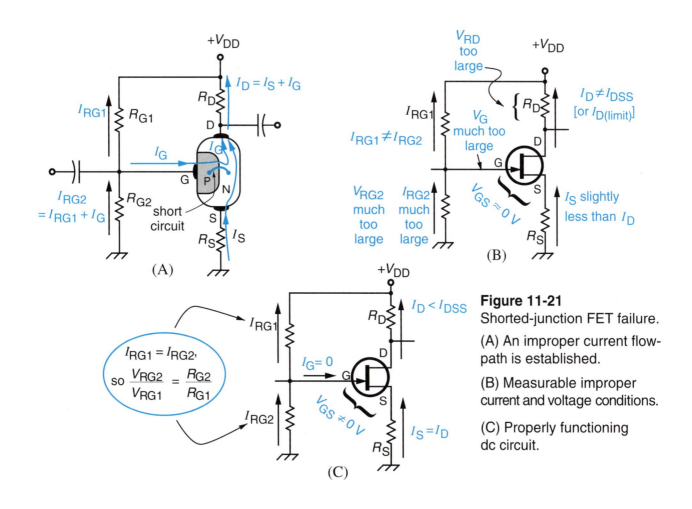

Figure 11-21
Shorted-junction FET failure.

(A) An improper current flow-path is established.

(B) Measurable improper current and voltage conditions.

(C) Properly functioning dc circuit.

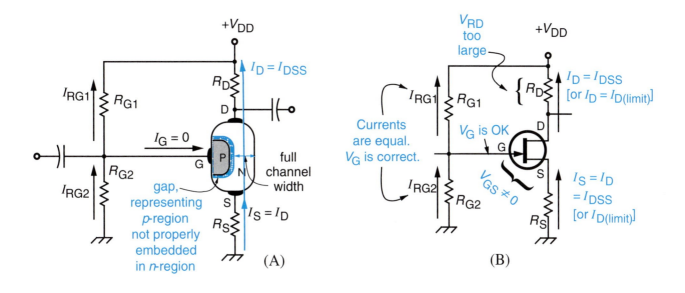

Figure 11-22 Open-junction FET failure. (A) No depletion layer forms, so the channel width and I_D are maximum. (B) Measurable improper voltage and current conditions.

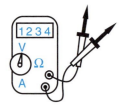

Component failures: Of course, a JFET amplifier stage can malfunction for other reasons besides the transistor itself. Any of the resistors can fail in the open condition, and any capacitor can fail either open or shorted.

Shorted capacitors can be detected by the fact that there is zero dc voltage across their terminals. Open coupling capacitors can be detected by the fact that the ac voltage across their terminals is not zero.

In a common-source amplifier, if any of the four resistors fails open, the symptoms and consequences are as follows: Use Fig. 11-21(C) as a reference diagram to understand these outcomes.

Open R_D: With no current in the drain lead, $V_{RD} = 0$ V. Transistor voltage V_{DS} becomes nearly zero, so $V_D \approx V_S \approx 0$ V. A current path is established through R_S, the p-n junction, and R_{G1}. This forward-biases the junction, so $V_{GS} \approx 0.7$ V. However, the junction may be destroyed.

Open R_S: No current can flow in either the source or drain leads. $V_{RD} = 0$ V so $V_D = V_{DD}$. Gate voltage V_G remains correct, but V_{GS} takes on a large negative magnitude ($V_{GS} = V_G - V_{DD}$). No harm is done to the transistor.

Open R_{G1}: $I_{divider}$ becomes zero, so $V_{RG2} = 0$ V and $V_G = 0$ V. V_{GS} becomes greater than normal, so I_D and I_S decrease to less than normal. V_S falls closer to 0 V and V_D rises closer to V_{DD}. The amplifier may still function if V_{in} is small.

Open R_{G2}: A current flow path is established through R_S, the p-n junction, and R_{G1}. The junction is forward-biased, so V_{GS} becomes slightly positive. However, the junction may be destroyed. If it survives, $I_D = I_{DSS}$ [or $I_{D(limit)}$].

Drain-Source mix-up: Here is a word of advice. It is possible for a JFET amplifier to function halfheartedly even if the drain and source terminals are reversed in the circuit. After all, the D and S terminals are not essentially different — they are just at opposite ends of the channel. It can happen that the person who installed the FET made the mistake of turning its body by 1/2 turn. This would reverse the D and S positions, which are often the two outside terminals of a TO-92-type package. If an amplifier is functioning poorly at high frequencies, check this possibility.

This same advice holds for BJT amplifiers, although reversing C and E usually degrades the amplifier's performance very badly.

SELF-CHECK FOR SECTION 11-6

36. In Fig. 11-23, it has been verified that the output was bad even though the ac input signal was OK. After disconnecting the input signal, the dc voltages are measured as shown. What is the trouble here?
37. In Fig. 11-24, the output was bad but the input was good. After disconnecting the input, the voltages are measured as shown. What appears to be the trouble here?

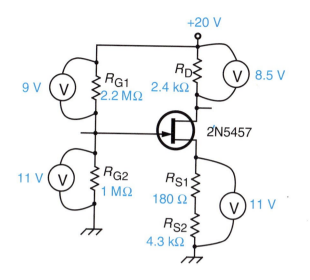

Figure 11-23
For Problem 36.

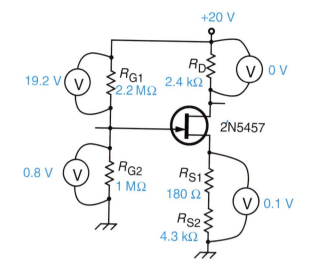

Figure 11-24
For Problem 37.

FORMULAS

$$V_{\mathrm{G}} = V_{\mathrm{DD}}\left(\frac{R_{\mathrm{G2}}}{R_{\mathrm{G1}} + R_{\mathrm{G2}}}\right) \qquad \text{Eq. (11-1)}$$

$$V_{\mathrm{GS}} = V_{\mathrm{G}} - V_{\mathrm{S}} \qquad \text{Eq. (11-2)}$$

$$R_{\mathrm{in}} = R_{\mathrm{G1}} \,\|\, R_{\mathrm{G2}} \quad \substack{\text{(common-source or} \\ \text{common-drain)}} \qquad \text{Eq. (11-3)}$$

$$g_{\mathrm{m}} = \frac{\Delta i_{\mathrm{D}}}{\Delta v_{\mathrm{GS}}} \qquad \text{Eq. (11-4)}$$

$$g_{\mathrm{m}} = \frac{I_{\mathrm{d\,(p\text{-}p)}}}{V_{\mathrm{gs\,(p\text{-}p)}}}$$

$$r_{\mathrm{D}} = R_{\mathrm{D}} \,\|\, R_{\mathrm{LD}} \quad \text{(common-source)} \qquad \text{Eq. (11-5)}$$

$$A_{\mathrm{v}} = g_{\mathrm{m}} r_{\mathrm{D}} \quad \substack{\text{(common-source,} \\ R_{\mathrm{S}}\ \text{bypassed})} \qquad \text{Eq. (11-8)}$$

$$A_{\mathrm{v}} = \frac{r_{\mathrm{D}}}{\dfrac{1}{g_{\mathrm{m}}} + R_{\mathrm{S1}}} \quad \substack{\text{(common-source,} \\ R_{\mathrm{S1}}\ \text{unbypassed})} \qquad \text{Eq. (11-9)}$$

$$r_{\mathrm{S}} = R_{\mathrm{S}} \,\|\, R_{\mathrm{LD}} \qquad \text{(common-drain)}$$

$$A_{\mathrm{v}} = \frac{r_{\mathrm{S}}}{\dfrac{1}{g_{\mathrm{m}}} + r_{\mathrm{S}}} \qquad \text{(common-drain)} \qquad \text{Eq. (11-11)}$$

SUMMARY

● A JFET has a complete-path channel of one doping polarity and a gate region of the opposite doping polarity. Therefore it has only one *p-n* junction. The two terminals at the ends of the channel are called source and drain.

● The gate-source junction is always reverse-biased, or zero-biased.

● The magnitude of gate-source voltage, V_{GS}, controls the channel current, I_D and I_S.

● Larger magnitude of V_{GS} causes smaller I_D. Smaller V_{GS} produces larger I_D. However, the relationship between these two is not proportional.

● The two important electrical specifications for a JFET are:

     a) I_{DSS}, the value of I_D when $V_{GS} = 0$ V.

     b $V_{GS(off)}$, the value of V_{GS} that causes $I_D = 0$.

● *P*-channel JFETs are biased oppositely from *n*-channel JFETs, in the same way that *pnp* BJTs are biased oppositely from *npn* BJTs.

● When used as large-scale ac amplifiers, JFETs tend to distort the input signal waveshape, since the I_D-versus-V_{GS} relationship is not proportional.

● The major advantage of an FET amplifier is its very high input resistance, r_{in}

● JFETs have very unstable batch characteristics. This makes it difficult to stabilize their bias point in an amplifier.

● Common-source FET amplifiers provide both voltage gain and current gain, like common-emitter BJT amplifiers. However, voltage gain A_v tends to be lower than for a BJT.

● Transconductance g_m is the gain figure for an FET, somewhat comparable to β for a BJT. g_m is the ratio of change in drain current i_D to change in gate-source voltage v_{GS}.

● Transconductance is extremely variable from one position to another on the FET's characteristic curves. This makes amplifier voltage gain A_v quite variable and difficult to predict.

● Common-drain (source-follower) FET amplifiers provide current gain but no voltage gain ($A_v < 1$), like common-collector BJT amplifiers.

● Common-drain amplifiers tend to be more bias-stable and A_v-stable ($A_v < 1$) than common-source amplifiers.

QUESTIONS AND PROBLEMS

1. Draw the schematic symbol for an *n*-channel JFET and name the terminals. Repeat for a *p*-channel JFET.

2. In an n-channel JFET, the gate voltage is _____ relative to the source (positive or negative).

3. As V_{GS} becomes larger in magnitude, the depletion layer around the *p-n* junction becomes _____. That causes the channel width to be-come _____. (Answer wider or narrower.)

4. (T-F) For a JFET, drain current I_D is proportional to the negative of gate-source voltage V_{GS}.

5. Define I_{DSS} for a JFET.

6. Define $V_{GS(off)}$ for a JFET.

7. (T-F) Ideally, I_D is completely independent of the main-terminal voltage V_{DS}.

8. (T-F) In reality, I_D is fairly independent of V_{DS} if the V_{DS} value is greater than the saturation value.

9. (T-F) For a JFET, the V_{DS} saturation value is a few tenths of a volt, like a BJT.

10. Make an approximate sketch of the I_D-versus-V_{GS} transfer curve for an *n*-channel FET with I_{DSS} = 6 mA and $V_{GS(off)}$ = –5 V.

11. Make an approximate sketch of the family of I_D-versus-V_{DS} characteristic curves for the FET of Problem 10. Show realistic saturation characteristics.

12. An FET has the characteristics shown in Figs. 11-3 and 11-5(B), with I_{DSS} = 10 mA and $V_{GS(off)}$ = –5 V. It is used in the single-resistor gate bias circuit of Fig. 11-7, with R_G = 1 MΩ, R_D = 1.8 kΩ, and R_{LD} = 1 kΩ. The dc supplies are V_{DD} = +24 V and V_{GG} = –1.0 V.
   a) Draw the dc load-line on the family of I_D-versus-V_{DS} curves. (You will have to photocopy Fig. 1-5(B) and then extend both axes.)
   b) Locate the Q-point on the dc load-line.
   c) Calculate the ac drain-line resistance r_D and draw the slope of the ac load- line.
   d) Transfer the slope to the Q-point to produce the ac load-line.

13. For the amplifier of Problem 12, suppose $V_{in(p\text{-}p)}$ = 2.0 V.
   a) Locate the two peak points, P_1 and P_2, on the ac load-line.
   b) Find $V_{ds(p\text{-}p)}$.
   c) Calculate the amplifier's voltage gain A_v.
   d) V_{out} is _____ phase with V_{in}.

14. For the amplifier of Problems 12 and 13, what is the value of input resistance R_{in} ?

15. (T-F) For the amplifier of Problems 12 and 13, the output waveform V_{out} would appear on an oscilloscope to be a perfectly shaped (symmetrical) sine wave.

16. A certain type of JFET has minimum characteristics of I_{DSS} = 2 mA, $V_{GS(off)}$ = –1.5 V, and maximum characteristics of I_{DSS} = 5 mA, $V_{GS(off)}$ = –6 V. These characteristic specifications are shown in Fig. 11-18(B). Photocopy and enlarge that figure. The transistor is used in a bypassed common-source amplifier like Fig. 11-15, with V_{DD} = 24 V, R_{G1} = 3.3 MΩ, R_{G2} = 2.4 MΩ, R_S = 5.6 kΩ, and R_D = 3.3 kΩ.
   a) Draw the line of possible dc bias points.
   b) Give the range of possible values of I_D.
   c) Give the range of possible values of V_D.
   d) Give the range of possible values of V_{DS}.

17. For the amplifier of Problem 16, suppose $V_{in(p\text{-}p)}$ = 0.2 V.
   a) Find the value of transconductance g_m for a minimum-characteristic FET.
   b) Find the value of transconductance g_m for a typical-characteristic FET.
   c) Find the value of transconductance g_m for a maximum-characteristic FET.

18. Referring to your results from Problem 17, comment on the stability of the gain figure g_m, for this FET type.

19. For the common-source amp of Problems 16 and 17.
   a) Calculate A_v for a minimum-characteristic FET.
   b) Calculate A_v for a typical-characteristic FET.
   c) Calculate A_v for a maximum-characteristic FET.

20. Referring to your results from problem 18, comment on the voltage gain-stability of this bypassed common-source amplifier.

21. For the common-source amplifier of Problem 16, replace 5.6-kΩ R_S with two source resistors: R_{S1} = 510 Ω and R_{S2} = 5.1 kΩ, with R_{S2} capacitor-bypassed. The total source resistance remains unchanged, so the dc bias conditions are the same as they were in Problems 16 and 17.
   a) Calculate the new value of A_v for a minimum-characteristic FET.
   b) Calculate the new value of A_v for a typical-characteristic FET.
   c) Calculate the new value of A_v for a maximum-characteristic FET.

22. Compare your results from Problem 21 and Problem 19. Comment on the voltage gain and the gain-stability of the two amplifier designs.

23. The same JFET type that was used in problems 16 through 22 is to be used in a common-drain amplifier. Referring to Fig. 11-19, suppose the component values are V_{DD} = 15 V, R_{G1} = 2.2 MΩ, R_{G2} = 2.7 MΩ, R_S = 5.1 kΩ, and R_{LD} = 7.5 kΩ.
   a) Draw the line of possible dc bias points.
   b) Give the range of possible values of I_D.
   c) Give the range of possible values of V_S.

24. For the amplifier of Problem 23, suppose $V_{in(p-p)}$ = 0.2 V.
   a) Find the value of transconductance g_m for a minimum-characteristic FET.
   b) Find the value of transconductance g_m for a typical-characteristic FET.
   c) Find the value of transconductance g_m for a maximum-characteristic FET.

25. For this common-drain amplifier,
   a) Calculate A_v for a minimum-characteristic FET.
   b) Calculate A_v for a typical-characteristic FET.
   c) Calculate A_v for a maximum-characteristic FET.

26. A common-drain amplifier is like a common-collector amplifier in that A_v is _____ 1.0.

27. The common-source JFET amplifier in Fig. 11-25 has bad output but good input. Disconnecting C_{in} gives the dc voltage measurements that are shown. What is the trouble?

28. The common-source JFET amplifier in Fig. 11-26 has bad output but good input. Disconnecting C_{in} gives the dc voltage measurements that are shown. What is the trouble?

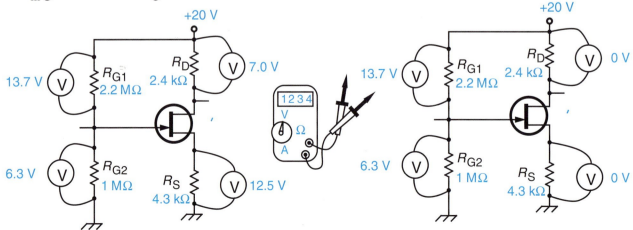

Figure 11-25 For Problem 27.   **Figure 11-26 For Problem 28.**

CHAPTER 12

MOSFETS

Courtesy of National Marine Mammal Laboratory

The tails of humpback whales have enough shape variation to enable photo-identification of individual whales. A computer data-base of about 800 humpback tail shapes has been compiled by the National Marine Mammal Laboratory, the College of the Atlantic, and other collaborating institutions. Using this data-base with a photo-matching computer program, the travels of individual whales can be monitored by marine biologists. This photo shows humpback whale Number 01315 (not Humphrey, the media-sensation humpback whale who mistakenly wandered up California's Sacramento river in 1985).

OUTLINE

Depletion-Type MOSFET Structure
    and Operation
MOSFET Amplifiers
Enhancement-Type MOSFETs
E-Type MOSFET Amplifiers
Popular Applications of MOSFETs
Proper Handling of MOSFETs

NEW TERMS TO WATCH FOR

depletion-type
enhancement-type
zero-biasing
normally-off
normally-on
gate-source threshold voltage

dual-gate MOSFET
input impedance
inverter
CMOS
diode-protected MOSFET

A MOSFET is a field-effect transistor containing a separate layer of insulating material that electrically insulates the gate from the channel semiconductor. Because the gate lead is insulated, the gate-to-source voltage V_{GS} is allowed to be either negative or positive. Removing the polarity restriction on V_{GS} gives MOSFETs some advantages over JFETs.

After studying this chapter, you should be able to:

1. Draw the schematic symbol for a depletion-type *n*-channel MOSFET and label the terminals. Do the same for a *p*-channel device.
2. Given a structural diagram of a D-type MOSFET, explain its operation in the depletion mode and its operation in the enhancement mode.
3. Draw the transconductance curve for a D-type MOSFET. Identify $V_{GS(off)}$ and I_{DSS}. Discuss the meaning of $g_{m(0)}$.
4. Given a schematic diagram of a common-source D-type MOSFET amplifier and the FET's transconductance curve, analyze the amplifier for A_v and R_{in}.
5. Draw the schematic symbol for an enhancement-type *n*-channel MOSFET and label the terminals. Do the same for a *p*-channel device.
6. Tell why E-type MOSFETs are called normally-off and D-type MOSFETs are called normally-on.
7. Given a structural diagram of an E-type MOSFET, explain how it operates in the enhancement mode.
8. Draw the transconductance curve for an E-type MOSFET. Identify $V_{GS(th)}$.
9. Given a schematic diagram and transistor transconductance curve, analyze an E-type MOSFET amplifier for A_v and R_{in}.
10. Draw the schematic symbol for a dual-gate MOSFET. State the reasons why MOSFETs are popular in radio-receiver RF and IF amplifiers.
11. Given the schematic diagram of a MOSFET digital inverter, explain its operation in detail.
12. Describe what is meant by complementary MOSFET circuitry, or CMOS.
13. Give the reasons why MOSFETs are popular in large-scale ICs.
14. List the precautions necessary for protecting standard MOSFETs during storage and handling.

12-1 DEPLETION-TYPE MOSFET STRUCTURE AND OPERATION

There are two different types of MOSFETs. They are called depletion-type (D-type) and enhancement-type (E-type). Let us concentrate on D-type MOSFETs first. We will tackle E-type MOSFETs in Section 12-3.

Structure

The structure of an *n*-channel D-MOSFET is shown in Fig. 12-1. It is like a JFET in that it has a complete current-flow path through an *n*-doped channel region. The terminal at the bottom end of the channel is called the source and the top terminal is called the drain, just like a JFET. When the dc supply is connected, electron current flows from bottom to top, from source to drain, just like a JFET. This action is shown in Fig. 12-1(B).

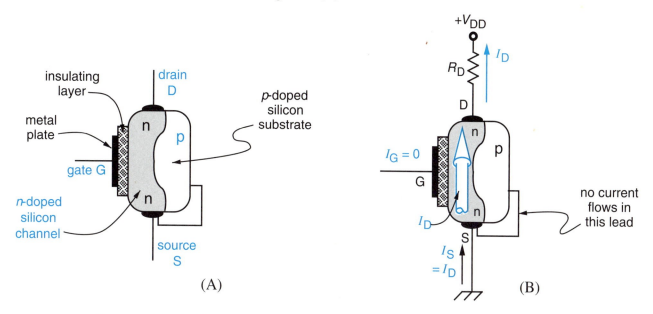

Figure 12-1 (A) Physical structure of *n*-channel depletion-type MOSFET. (B) Main terminal dc power supply polarity and current direction.

The important structural difference between a MOSFET and a JFET is that a MOSFET's gate is not a *p*-doped region of silicon. Instead, the gate is a metal plate, separated from the channel by an insulating layer of silicon oxide.

The insulating oxide layer is clearly labeled in Fig. 12-1(A).

Going from left to right in Fig. 12-1, we encounter <u>M</u>etal, then <u>O</u>xide, then <u>S</u>emiconductor. Thus, MOSFET stands for Metal Oxide Semiconductor Field Effect Transistor.

Often a MOSFET package has just three leads, namely source, gate, and drain. A fourth "lead" connects the substrate region to the source terminal internally. Figure 12-2(A) is the schematic symbol for a 3-lead *n*-channel D-type MOSFET. Notice that the arrow is not on the gate lead now. Instead, the arrow is on the substrate lead. Like a JFET symbol, the arrow points *in* toward the center for an *n*-channel device.

Sometimes the substrate lead is brought out of the package as a fourth lead. We externally connect the substrate to the source, in most cases. We may then draw the schematic symbol as shown in Fig. 12-2(B).

The schematic symbols suggest the electrical insulation of the gate, since the gate lead line turns upward without touching any other line.

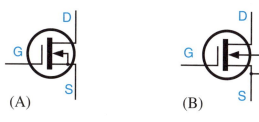

Figure 12-2 (A) Schematic symbol when the substrate is internally connected to the source. (B) Schematic symbol when the substrate lead is brought out of the package. This schematic distinction is not a universally recognized standard. We will use the (A) symbol always.

Operation

A depletion-type MOSFET operates much like a JFET when a zero or negative value of V_{GS} is applied.

As usual, the I_{DSS} value is quite variable from one individual MOSFET to another. The same is true for $V_{GS(off)}$ (next paragraph).

$V_{GS} = 0$ V: Figure 12-3(A) shows the situation when $V_{GS} = 0$ V. There is no depletion region in the n-doped channel, so the entire channel width is available for carrying current. The FET becomes a current source of value I_{DSS}.

The solid-state mechanism that forms the depletion region is different from the mechanism for a JFET, since now there is no p-n junction for charge carriers to diffuse across. In a MOSFET, the depletion region is formed by the electrostatic effect of the negative charge concentration on the metal gate electrode.

V_{GS} is negative: When a negative V_{GS} is applied, a depletion region forms in the n-channel near the gate, as shown in Fig. 12-3(B). Therefore the effective channel width is decreased, and I_D becomes less than I_{DSS}. As V_{GS} is made more negative, the depletion region becomes wider, the channel width becomes narrower, and I_D becomes smaller, the same as a JFET. If V_{GS} reaches the $V_{GS(off)}$ value, the channel width becomes completely pinched off and I_D decreases to zero.

V_{GS} is positive:

The important operating difference between MOSFETs and JFETs is that a MOSFET can operate with a positive V_{GS} value. This is possible because the gate is insulated from the channel.

Figure 12-3(C) shows what happens when V_{GS} becomes positive. A positive potential on the gate plate repels the holes in the nearby portion of the p-doped substrate material. It also pulls some bonded electrons out of the right side of the p-doped substrate, into this **enhancement region**, making them free electrons. Thus, the effective channel width is widened. With a wider usable channel width, drain current increases to a value larger than I_{DSS}.

When a MOSFET has its channel width widened to allow greater drain current, we say that it is operating in the enhancement mode.

You can see that a "depletion-type" MOSFET actually can operate in either the depletion mode *or* the enhancement mode. A better name for this type of MOSFET would be "depletion/enhancement-type" or "D/E-type".

Figure 12-4(A) shows the I_D-versus-V_{GS} transfer curve for a D-type MOSFET. The left half of the curve represents its depletion mode of operation and the right half represents its enhancement mode.

The resulting family of I_D-versus-V_{DS} characteristic curves is shown in Fig. 12-4(B).

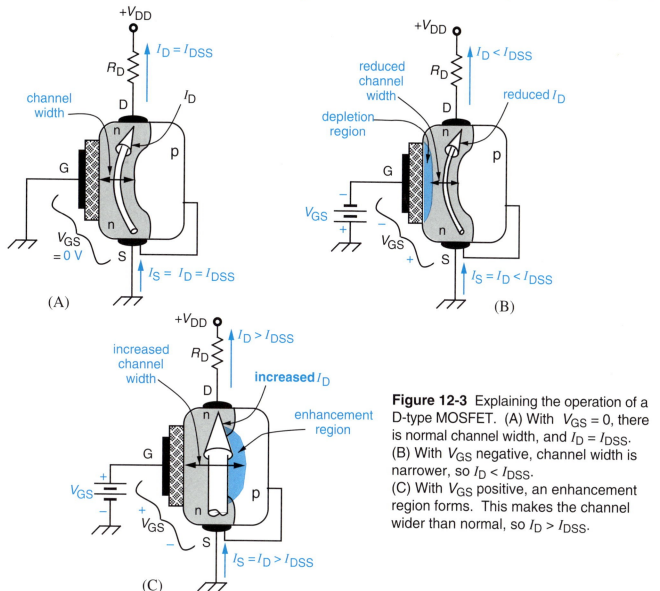

Figure 12-3 Explaining the operation of a D-type MOSFET. (A) With $V_{GS} = 0$, there is normal channel width, and $I_D = I_{DSS}$. (B) With V_{GS} negative, channel width is narrower, so $I_D < I_{DSS}$. (C) With V_{GS} positive, an enhancement region forms. This makes the channel wider than normal, so $I_D > I_{DSS}$.

Figure 12-4 Curves for a depletion-type (D-type) MOSFET. Remember that a "depletion-type" MOSFET is actually a depletion *or* enhancement device. (A) Drain current I_D versus V_{GS} transfer curve. This curve has the identical nonlinear shape as for a JFET. The only difference is that it keeps going into the positive part of the V_{GS} axis. (B) Family of characteristic curves of I_D versus V_{DS}. Saturation voltages are rather high (several volts), just like a JFET.

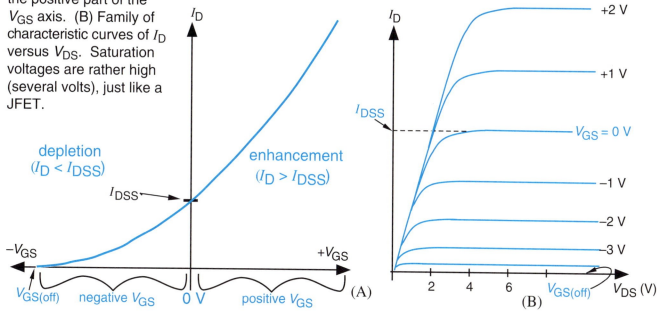

COAL-FIRED ELECTRIC UTILITY STACK CLEANUP
STEP 1: REMOVAL OF SOLID ASH PARTICLES

Coal is the most abundant fossil fuel on the earth. It has been estimated that the coal reserves of North America contain perhaps ten times as much energy as the oil reserves of the Persian Gulf region. The problems with coal are that it's expensive to dig, and it's dirtier than oil to burn.

Burning coal for electric power generation affects the environment in several ways:
1) It produces solid ash particulates that do not belong in the atmosphere or on fertile farmland.
2) It produces sulfur dioxide (SO_2) gas, which is the cause of acid rain. 3) It produces carbon dioxide (CO_2) gas, which contributes to the greenhouse effect.

While we're waiting for the fusion people to get their act together, the coal-burning industry is making big improvements in the first two areas—ash particulates and SO_2 removal. We'll describe the ash process now and talk about SO_2 in the next chapter.

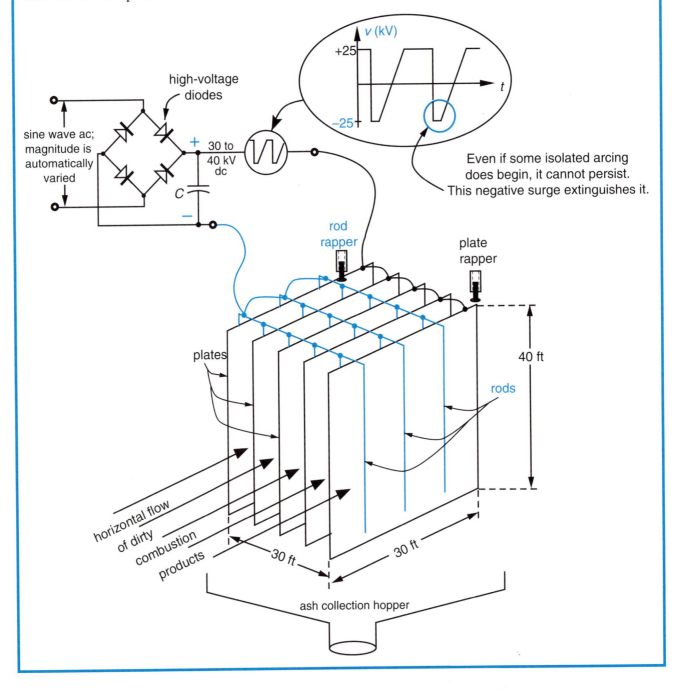

Solid ash particles are removed by passing the combustion products through an electrostatic precipitator. This is a group of large-area metal collecting plates with charged metal rods between them. The rods are sometimes called discharge electrodes.

All the collecting plates are electrically connected together and all the rods are electrically connected together, as shown in the figure.

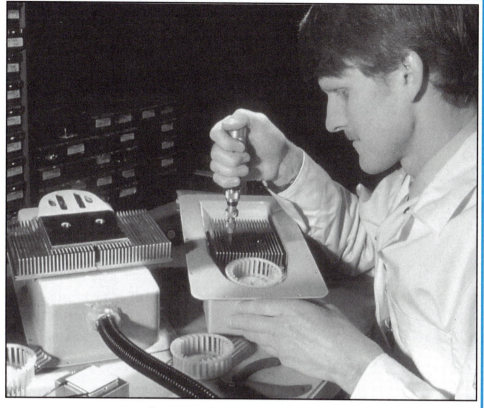

Courtesy of GE Environmental Systems

Very high voltage, about 40 kV dc with 50 kV p-p ac pulses superimposed, is applied between the plates and the rods, negative on the rods (in the coal-fired utility industry). As the ash particles pass over the rods, they themselves become negatively charged. They are then attracted to and captured by the positive collecting plates. Periodically, the plates must be mechanically rapped in order to knock the accumulated ash particles off their surface into the collection hopper below. This is accomplished by electromagnetic spring-loaded solenoids. Over time, some ash particles also cling to the rods, so they must be rapped occasionally too. The figure shows one plate rapper and one rod rapper, but really there are dozens of each.

A state-of-the-art "intelligent" precipitator can capture 99.98% of the ash in the exhaust flow stream. It can do such a good job because it is equipped with electronic controls that sense the buildup of dirt on the plates and automatically adjust the applied voltage. Both the baseline dc value and the magnitude and frequency of the ac pulses are automatically varied as the ash particles accumulate. Refer to the diagram on the facing page to see the dc and ac applied voltages. In general, the control circuit tries to adjust the voltages to the highest possible values, without allowing severe arcing between the rods and plates. Prolonged arcing would damage the metal surfaces.

The electronic controller also adjusts the rapping rate. For mild exhaust conditions the rapping period can be as long as 100 minutes, round-trip time. (The rappers are operated sequentially, not all at once.) For very dirty conditions, rapping round-trip time can be as quick as 1 minute.

The photo shows a technician servicing an electronic module used to control a precipitator with 55 000 square feet of collecting-plate area.

P-channel MOSFETs: There are both n-channel and p-channel MOSFETs, the same as JFETs. The schematic symbol of a p-channel MOSFET has the arrow pointing *out*, away from the center, as shown in Fig. 12-5. The main terminal bias is reversed, and so is the direction of I_D, as that diagram indicates. The gate bias is also reversed.

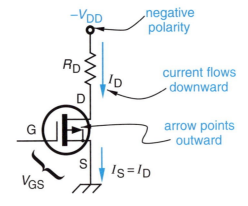

**Figure 12-5
Schematic symbol
and main terminal
voltage and current
of a *p*-channel MOSFET.**

SELF-CHECK FOR SECTION 12-1

1. Draw the schematic symbol and label the terminals for a D-type *n*-channel MOSFET. Use the internally connected-substrate symbol.
2. (T-F) A D-type *n*-channel MOSFET can operate either with negative V_{GS} or with positive V_{GS}.
3. Explain what the six letters M-O-S-F-E-T stand for.
4. When a D-type *n*-channel MOSFET is operating in the depletion mode, V_{GS} is _____ . When it is operating in the enhancement mode, V_{GS} is _____ . (answer positive or negative)
5. We call them depletion-type (D-type) MOSFETs, but a better name would be _____ .
6. (T-F) A D-type MOSFET is basically a voltage-controlled current-source, like a JFET.
7. A D-type MOSFET has a _____ relationship between input voltage and output current (answer linear or nonlinear). In this respect, it is like a _____ , and unlike a _____ (answer JFET or BJT).

12-2 MOSFET AMPLIFIERS

A MOSFET common-source amplifier is shown in Fig. 12-6. It is not always necessary to place a resistor R_S in the source lead. This is because it is not necessary to raise the S terminal to a positive potential, since we don't have to make the source more positive than the gate. The gate bias circuit can then be very simple — just a large-value gate resistor R_G connected directly to ground, as shown in Fig. 12-6. This arrangement is called **zero-biasing**.

However, to stabilize this bias point V_D and bias current I_D against batch variations and temperature variations, the standard voltage-divider/source-resistor bias arrangement shown in Fig. 12-6(B) is often used.

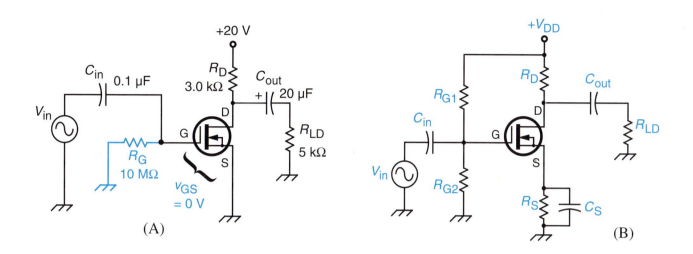

Figure 12-6 Common-source MOSFET amplifier schematics.
(A) Unstabilized zero-bias method. (B) Bias-stabilized design.

A low-power D-type MOSFET could have the following electrical characteristics:

	Minimum	Typical	Maximum
I_{DSS} (mA)	2	6	12
$V_{GS(off)}$ (V)	–2	–5	–8
$g_{m(0)}$ (mS)	1	2.4	4

MOSFETs are just as variable as JFETs in their electrical characteristics, I_{DSS}, $V_{GS(off)}$, and g_m. And their temperature instability tends to be even worse than JFETs'.

Table 12-1 Range of electrical specs for our example MOSFET.
$g_{m(0)}$ **is the transconductance at the V_{GS} = 0 point on the curve.**

Thus, a typical unit would have the transconductance (I_D-versus-V_{GS}) transfer curve shown in Fig. 12-7.

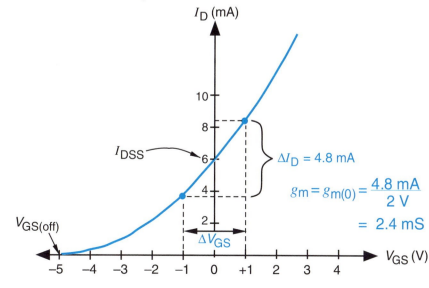

Figure 12-7
Typical
transconductance
curve for our
example MOSFET.

$$g_m = g_{m(0)} = \frac{4.8 \text{ mA}}{2 \text{ V}}$$

$$= 2.4 \text{ mS}$$

$\Delta I_D = 4.8$ mA

EXAMPLE 12-1

Suppose a MOSFET with the typical characteristics shown in Fig. 12-7 is used in the zero-biased amplifier of Fig. 12-6(A). Let $V_{in(p-p)} = 0.8$ V.
  a) Find the ac drain-line resistance r_D, and calculate the voltage gain A_v.
  b) Find $V_{out(p-p)}$. What is its phase relation to V_{in} ?
  c) Find the amplifier's input resistance R_{in}.

SOLUTION:

a) Ac drain-line resistance is given by the usual formula

$$r_D = R_D \| R_{LD} = \frac{R_D R_{LD}}{R_D + R_{LD}} \qquad \text{Eq. (12-1)}$$

$$r_D = \frac{3000 \ \Omega \ (\ 5000 \ \Omega)}{3000 \ \Omega + 5000 \ \Omega} = \mathbf{1880 \ \Omega}$$

With no resistance present in the source lead, voltage gain is given by the same formula that applied to a JFET amplifier.

This is the same as Eq. (11-8) for the JFET common-source amplifier.

$$A_v = g_m \, r_D \qquad \text{Eq. (12-2)}$$

As before, we must use the g_m value that applies at the bias point. Since this amplifier has $V_{GS} = 0$ V, $g_m = g_{m(0)} = 2.4$ mS, as shown in Fig. 12-7. Equation (12-2) gives

$$A_v = g_m \, r_D = 2400 \ \mu\text{S} \ (1875 \ \Omega) = \mathbf{4.5}$$

b) $V_{out(p-p)} = A_v \, V_{in(p-p)} = 4.5 \ (0.8 \ \text{V}) = \mathbf{3.6 \ V, \ inverted}$

The V_{out} waveform will have some visible amount of distortion due to the nonlinearity of the MOSFET's transconductance curve in Fig. 12-7.
  c) The input resistance of the MOS transistor itself is virtually infinite. In fact, R_{in} for a MOSFET is even higher than R_{in} for a JFET. Typical MOSFET input resistances are several hundred or several thousand gigohms. Thus, for the Fig. 12-6 amplifier as a whole,

④ ✔

$$R_{in} = R_G \| \infty = R_G = \mathbf{10 \ M\Omega}$$

All the analysis methods and formulas from Chapter 11 apply just as well to MOSFET amplifiers.

Common-drain and common-gate MOSFET amplifiers also have the same basic designs as the JFET versions.

SELF-CHECK FOR SECTION 12-2

8. (T-F) A D-type MOSFET amplifier can be biased very simply by just connecting a large resistor between gate and ground.
9. (T-F) MOSFETs are like JFETs in that their electrical characteristics vary greatly from one unit to another.

10. (T-F) MOSFET amplifiers have the same problems as JFET amplifiers, with regard to bias instability and A_v instability.

11. If a D-type MOSFET is biased farther to the right on the transconductance curve (Fig. 12-7), its transconductance _____ (increases or decreases).

12. In Question 11, the amplifier's voltage gain A_v would _____ .

12-3 ENHANCEMENT-TYPE MOSFETS

An enhancement-type (E-type) MOSFET can operate only in the enhancement mode. This means that V_{GS} must be positive (n-channel) in order for the transistor to conduct current.

Structure: The structure of an n-channel E-type MOSFET is shown in Fig. 12-8(A). It is different from a D-type MOSFET in this way:

An E-type MOSFET does not have a complete channel between the source and drain terminals. The substrate material touches the insulating oxide layer.

Get This

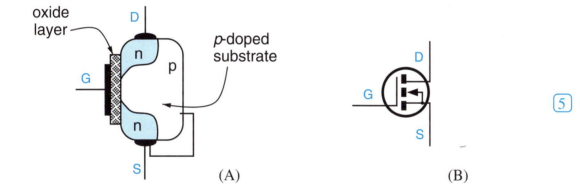

(A)   (B)

Figure 12-8 (A) Structure of an enhancement-type MOSFET. The n-channel is interrupted by the p-substrate. (B) Schematic symbol for n-channel E-type MOSFET. The arrow points *in* toward the center. The p-channel symbol would have the arrow pointing *out* away from the center.

Because the channel is not complete, no main-terminal current I_D can flow when the MOSFET has $V_{GS} = 0$ V. For this reason, E-type MOSFETs are sometimes called **normally-off** MOSFETs. D-type MOSFETs would then be called **normally-on**.

Normally-off compared to normally-on MOSFETs.

The schematic symbol for an n-type E-type MOSFET is shown in Fig. 12-8(B). The broken line is used to represent the incomplete channel between source and drain.

Operation: To get an E-type MOSFET to start conducting, we must force an enhancement region to form in the substrate. This enhancement region connects the two n-doped regions together, as shown in Fig. 12-9(A). The enhancement region is formed by the same method described in Sec. 12-1 for

D-type MOSFETs. Namely, a positive V_{GS} value attracts bonded electrons from the right side of the p-doped substrate region into the vicinity of the gate. This wipes out the holes and establishes an effective *n*-type connection from the S-region to the D-region. Then free electrons can pass from source to drain through that enhancement region.

As the positive V_{GS} is increased, the enhancement region becomes larger. This allows a greater amount of I_D to flow, as shown in Fig. 12-9(B).

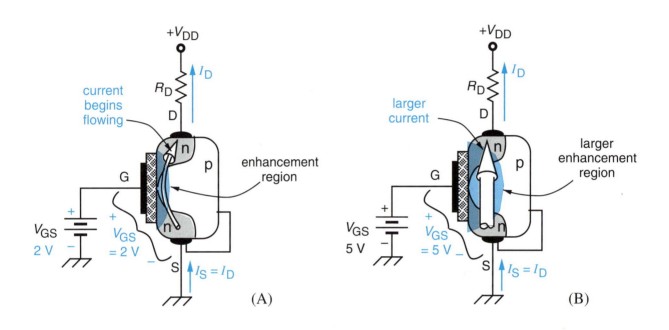

Figure 12-9 Explaining the operation of an E-type MOSFET. (A) V_{GS} is a positive value that is large enough to create an enhancement region, joining the Source-region to the Drain-region. A certain amount of current flows through the "channel". (B) With greater positive V_{GS}, the enhancement region becomes larger. The increased "channel" width allows current I_D to increase.

In order for current I_D to start flowing at all, the enhancement region must be large enough to touch the Source-region and the Drain-region. A very small enhancement region will not work. This means that V_{GS} must increase to a certain critical value, before the transistor begins to carry any drain current.

> **Get This**
>
> To start conduction in an E-type MOSFET, V_{GS} must increase to a critical value called the gate-source threshold voltage, symbolized $V_{GS(th)}$ by us.

$V_{GS(th)}$ is the amount of gate-source voltage needed to turn on an E-type MOSFET.

The I_D-versus-V_{GS} transfer curve for an E-type MOSFET is shown in Fig. 12-10(A). This curve clearly shows the effect of $V_{GS(th)}$. The resulting family of I_D-versus-V_{GS} curves is shown in Fig. 12-10(B).

E-type MOSFETs have the same variations in their electrical characteristics as other FETs. Thus, a given type will have a minimum, typical, and maximum $V_{GS(th)}$ rating.

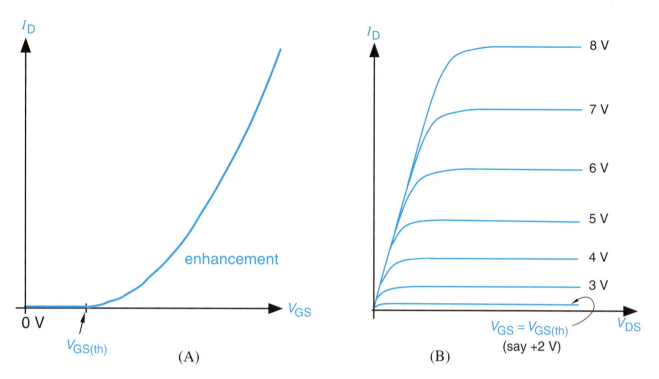

Figure 12-10 (A) Transconductance curve for an E-type MOSFET. The basic shape of the curve is the same as for the other field-effect transistors, JFETs and D-type MOSFETs. (B) Family of drain characteristic curves.

⑧ ✓

12-4 E-TYPE MOSFET AMPLIFIERS

The MOSFET market is dominated by E-type MOSFETs. Discrete (one transistor contained in one package) D-type MOSFETs are seldom used in new designs. So you may wonder why we discussed D-type MOSFETs in so much detail in Sections 12-1 and 12-2. We did so for two reasons: ① D-type MOSFETs help you to bridge the gap conceptually between depletion-mode JFETs and enhancement-mode (E-type) MOSFETs. ② D-type MOSFETs are used in certain integrated circuits (many interconnected transistors contained in one package).

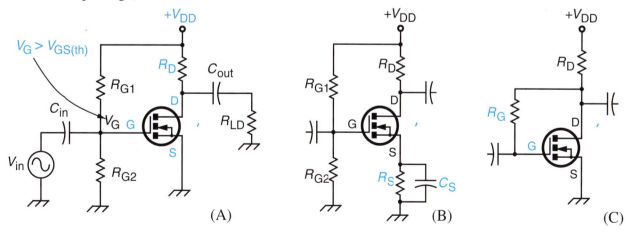

Figure 12-11 (A) Schematic of basic E-type common-source amplifier. (B) Stabilizing the bias point with R_S. (C) Stabilizing the bias point by tapping the D terminal. This forces V_{DS} to equal V_{GS}, since there is zero voltage drop across R_G with $I_G = 0$.

A common-source E-type amplifier is shown in Fig. 12-11(A). The voltage-divider bias circuit is used to set V_{GS} at a value greater than $V_{GS(th)}$. Bias-stabilized designs are shown in Figs. 12-11(B) and (C).

The usual FET analysis methods and formulas apply to E-type amplifiers.

EXAMPLE 12-2

A type number 2N4351 E-type MOSFET has a typical transconductance curve shown in Fig. 12-12. Redraw the unstabilized amplifier of Fig. 12-11(A), with the following component values: $V_{DD} = 17.5$ V, $R_{G1} = 10$ MΩ, $R_{G2} = 7.5$ MΩ, $R_D = 2.7$ kΩ, $R_{LD} = 5$ kΩ, and $V_{in(p-p)} = 2$ V.

a) Find the bias voltage $V_{GS} = V_G$, by voltage division.

b) Locate the Q-point for a typical unit.

c) Mark the negative peak point P_1 and the positive peak point P_2 on the transconductance curve.

d) Find the FET's transconductance g_m by $\Delta I_D / \Delta V_{GS}$.

e) Use Eqs. (12-1) and (12-2) to find r_D and A_v.

SOLUTION:

a)
$$V_{GS} = V_G = V_{DD}\left(\frac{R_{G2}}{R_{G1} + R_{G2}}\right)$$

$$= 17.5\text{ V}\left(\frac{7.5\text{ M}\Omega}{10\text{ M}\Omega + 7.5\text{ M}\Omega}\right) = \textbf{7.5 V}$$

b) Projecting up from the 7.5-V point on the voltage axis, the Q-point is at about 4.3 mA, as shown in Fig. 12-12.

c) With $V_{in(p-p)} = 2$ V, V_{GS} swings negative to 6.5 V and positive to 8.5 V. These two voltages identify P_1 and P_2 on the curve.

d) Projecting over to the I_D axis from P_1 gives $I_D = 3.0$ mA. Repeating for P_2 gives 5.9 mA. Transconductance g_m is given by

$$g_m = \frac{\Delta I_D}{\Delta V_{GS}} = \frac{5.9\text{ mA} - 3.0\text{ mA}}{8.5\text{ V} - 6.5\text{ V}}$$

$$= \frac{2.9\text{ mA}}{2\text{ V}} = \textbf{1.45 mS}$$

e) From Eq. (12-1),
$$r_D = 2.7\text{ k}\Omega \| 5\text{ k}\Omega = 1.75\text{ k}\Omega$$
Equation (12-2) gives
$$A_v = g_m r_D = (1.45 \times 10^{-3}\text{ S})(1.75 \times 10^3\ \Omega) = \textbf{2.54}\ \text{(inverting)}$$

Common-drain and common-gate amplifiers are built like their JFET and D-type counterparts. Of course, for class-A linear operation, the bias circuit must always be designed to cause V_{GS} to be greater than $V_{GS(th)}$.

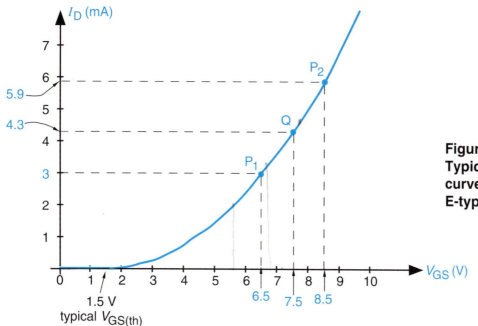

Figure 12-12
Typical transconductance curve for type No. 2N4351 E-type MOSFET.

Comparing JFET, D-Type, and E-Type MOSFET Amps

Table 12-2 summarizes the features of the three types of FETs. It should help you to keep your thoughts organized on these devices. Remember that only JFETs and E-type MOSFETs are common in discrete circuitry.

Name	JFET	D-type MOSFET	E-type MOSFET
Schematic symbol			
Transconduct-ance curve			
Operating mode	Depletion only	Depletion or enhancement	Enhancement only
Advantages	Very high input resistance	Extremely high input resistance	Extremely high input resistance
Disadvantages	Bias and gain affected by batch and temperature instability. Low A_v. Large-signal nonlinearity.	Same batch instability. Worse temperature instability. Low A_v. Large-signal nonlinearity.	Batch and temperature instability like D-type. Low A_v. Large-signal nonlinearity.

Table 12-2
Summarizing and comparing the three types of field-effect transistors.

SELF-CHECK FOR SECTIONS 12-3 AND 12-4

13. (T-F) An E-type MOSFET has basically the same internal construction as a D-type MOSFET.
14. An E-type MOSFET is sometimes described as normally off, in contrast to a D-type being normally on. Explain what this means.
15. Draw the schematic symbol for an *n*-channel E-type MOSFET. Repeat for a *p*-channel device.
16. In order to start an E-type MOSFET conducting current, V_{GS} must be greater than _____ .
17. Repeat Example 12-2 if R_{G2} is changed to 6.2 MΩ. Everything else remains the same.

12-5 POPULAR APPLICATIONS OF MOSFETS

The characteristics of MOSFETs make them very useful in certain electronics applications. We will describe two such applications — IF amplifiers and digital switching circuits.

IF Amplifier

The extremely high input resistance of a MOSFET makes it well-suited for the intermediate-frequency amplifiers of a radio receiver. Figure 12-13(A) shows an E-type MOSFET used for this purpose.

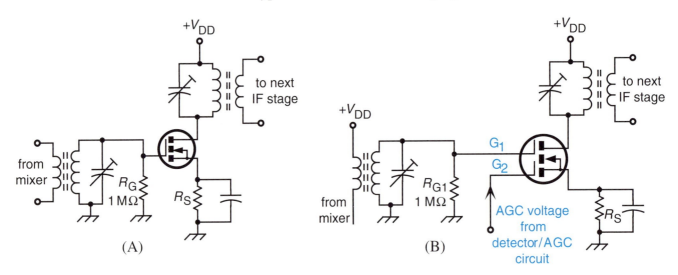

Figure 12-13 (A) MOSFET used in an IF amplifier of a receiver. (B) With a dual-gate MOSFET, G1 can handle the intermediate-frequency signal and G2 can be used for automatic gain control.

The amplifier's input resistance is 1 MΩ, which takes negligible resistive current from the mixer-circuit transformer. This helps keep the transformer's $I \times R$ winding loss as small as possible.

There is no reason why a MOSFET is limited to a single gate electrode. In Fig. 12-8, it would be possible to cut the metal plate in half, and have two gate leads, G_1 and G_2. When this is done we call the device a **dual-gate MOSFET**. Such a device is very useful in a radio receiver, because G_2 can be used to apply automatic gain control (AGC). This arrangement is shown in Fig. 12-13(B). The AGC voltage on G_2 combines with the zero-bias connection on G_1 to determine the bias Q-point. This varies the transistor's transconductance g_m, which varies the amplifier voltage gain.

In this automotive testing laboratory, the computer system gathers large amounts of data on engine, drive-train, and chassis performance. The test runs may be quite long – equivalent to 100 000 miles or more. Computer analysis of the assembled data helps engineers to identify the strengths and weaknesses of the vehicle's design. See page 10.

Courtesy of General Motors Corp.

Digital Switching Circuit

A *digital* circuit has only two voltage levels, usually +5 V and 0 V. These two voltages are referred to as HI and LO, or as logic 1 and logic 0.

The simplest digital circuit is called an **inverter**. When it is given a HI input, it produces a LO output. When it is given a LO input, it produces a HI output. This operation is summarized in Table 12-3.

HI and LO can also be spelled in the normal way, high and low.

Input	Output
HI (+5V)	LO (0V)
LO (0V)	HI (+5V)

**Table 12-3
The operation of a digital inverter.**

A digital inverter can be built with bipolar junction transistors, of course. But a digital inverter built with MOSFETs has some great advantages over a BJT inverter. Let us study the circuitry for a MOSFET inverter, to understand how it works. Then we can explain its advantages over BJT technology.

Figure 12-14
Digital inverter circuit using two E-type MOSFETs, one *p*-channel and one *n*-channel. There are no resistors, diodes, or capacitors needed.

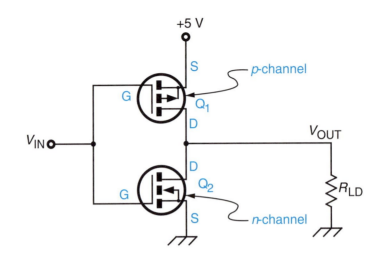

A MOSFET inverter is shown in Fig. 12-14. The MOSFETs are E-type — normally off. Note carefully that Q_1 is *p*-channel, and is "upside-down" (source S is on top, drain D is on bottom). Q_2 is *n*-channel, and is right-side-up. Here is how the circuit works:

V_{IN} HI: When V_{IN} is HI, +5 V, transistor Q_2 has positive V_{GS}, as shown in Fig. 12-15(A). It is *n*-channel, and it has $V_{GS} = +5$ V, so it tries to conduct. In fact, it tries to conduct so hard that it acts like a zero-ohm path, or a closed switch. The switch diagram in Fig. 12-15(B) shows SW_2 closed.

But transistor Q_1 has $V_{GS} = 0$ V, since both the gate G and the source S are at the same 5-V potential. This is shown clearly in Fig. 12-15(A). Because $V_{GS} = 0$ V, this E-type MOSFET remains off. It acts like an infinite-ohm path, or an open switch. The switch diagram in Fig. 12-15(B) shows SW_1 open.

With Q_2 turned on (like SW_2 closed) and Q_1 turned off (like SW_1 open), it is easy to see that the output terminal is connected to ground potential, 0 V. Thus, a HI input has produced a LO output.

Figure 12-15
Explaining a MOSFET inverter, with HI input.

(A) Conditions of the two transistors.

(B) Equivalent switch circuit. The output is LO.

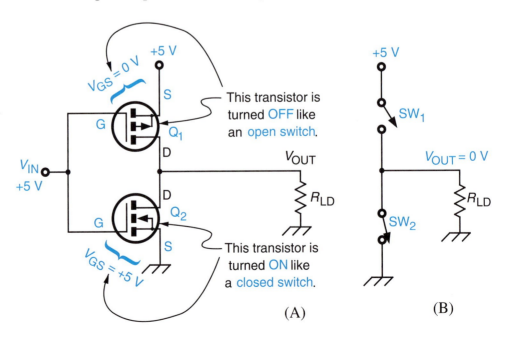

V_{IN} LO: When V_{IN} is LO, transistor Q_2 has both gate G and source S at 0 V. Therefore, $V_{GS} = 0$ V and the transistor remains off, as shown in Fig. 12-16(A). SW_2 is open in Fig. 12-16(B).

But transistor Q_1 has $V_{GS} = -5$ V, since the source is at +5 V and the gate is at 0 V. Negative V_{GS} turns on a p-channel MOSFET. Therefore Q_1 is turned on in Fig. 12-16(A), and SW_1 is closed in Fig. 12-16(B).

With Q_1 turned on and Q_2 turned off, the output terminal is connected to the +5-V supply. Thus, a LO input has produced a HI output.

In a p-channel device, all polarities are reversed. Therefore $V_{GS(th)}$ has gate G negative relative to source S.

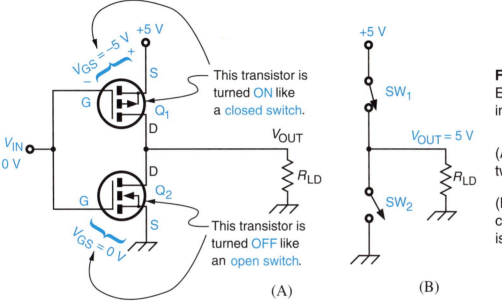

(A)   (B)

Figure 12-16
Explaining a MOSFET inverter, with LO input.

(A) Conditions of the two transistors.

(B) Equivalent switch circuit. The output is HI.

The arrangement shown in Figs 12-14 through 12-16 has a p-channel MOSFET on top and an n-channel MOSFET on bottom. This is like the complementary symmetry BJT arrangement of Sec. 10-5, where we had a pnp transistor on top and an npn on bottom. We use the letters COSMOS to refer to all circuits that use this <u>CO</u>mplementary <u>S</u>ymmetry <u>MOS</u>FET arrangement. Usually, COSMOS is shortened to CMOS, pronounced "see´–moss".

The CMOS digital inverter is extremely simple, containing just two transistors and no supporting components. A BJT digital inverter with similar input resistance and output current capabilities would require 4 transistors, 1 diode, and 4 resistors.

Advantage of CMOS over BJT circuitry.

> Therefore the CMOS design is cheaper to build, and it takes up much less space than a BJT design.

These advantages are very important when more complicated digital circuits are built as integrated circuits, ICs.

Also CMOS digital circuits have a power-consumption (efficiency) advantage over BJTs. This is because one or the other transistor is always off, so there is never a complete current flow path through the pair. And if R_{LD} is actually the input to another CMOS circuit, which is usually the case, then there is virtually zero current through that path as well. CMOS circuits operating **statically** (input changes are infrequent) take virtually zero current from the dc power supply. When they are operating **dynamically** (input changes occur at a rapid frequency), their short bursts of current during level transitions cause some non-negligible amount of average dc supply current to flow.

Not all digital circuits built with MOSFETs use the CMOS idea of mixing *p*-channel and *n*-channel transistors. The most complex digital ICs use only *n*-channel MOSFETs, because the manufacturing process is simpler than for CMOS. These <u>V</u>ery <u>L</u>arge <u>S</u>cale <u>I</u>ntegrated circuits (VLSI circuits) can contain over 10 000 *n*-channel E-type MOSFETs on a single silicon chip. You will cover these ideas thoroughly in a course devoted specifically to digital electronics.

13 ✔

The biofeedback vision-training system explained on pages 374-75 is being used by several professional sports franchises to improve their athletes' peripheral vision. Among them are the National Hockey League's Boston Bruins.

Courtesy of Biofeed Trac, Inc.

12-6 PROPER HANDLING OF MOSFETS

The oxide layer in a MOSFET is a nearly perfect insulator. However, it can be ruptured by moderately high voltages produced by static electricity. For this reason, there are certain precautions that must be taken to protect MOSFETs.

1 Store a MOSFET with its leads stuck into conductive foam, or with its leads all shorted together by a metal ring. MOSFETs are shipped to you in one or the other of these two conditions.

Styrofoam packing must never be used — it is insulating, not conductive.

2 Touch an earth-grounded object with your hand before you pick up a MOSFET. This will discharge any static charge that has accumulated on your body from walking on certain carpets and other surfaces.

3 For long-term handling of MOSFETs, use a wrist- or ankle-connected grounding strap made for that purpose. Such straps drain any accumulated charge off your body through a built-in 1-MΩ resistor.

4 Use only a high-quality earth-grounded soldering pencil when soldering MOSFET leads to a printed-circuit board.

5 Always turn the power off and wait for the filter capacitors to discharge before inserting or removing a MOSFET from its socket.

These precautions apply to both discrete MOSFETs and MOS ICs, such as digital CMOS.

14 ✔

Diode-Protected MOSFETs: Many MOSFETs have a series pair of diodes built in between the gate and source terminals. This is shown schematically in Fig. 12-17. The diodes are a special type, called zener diodes. They start conducting at a fairly low value of reverse voltage, usually 30-40 V for zeners built into MOSFETs. This arrangement protects the oxide insulating layer, because the zeners prevent the gate voltage from exceeding 30-40 V in either polarity. Diode-protected MOSFETs do not require the careful handling techniques listed above.

We sill study zener diodes in detail in Chapter 15.

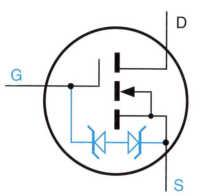

Figure 12-17
Diode-protected MOSFET.
Also called zener-protected
MOSFET, because the built-in
protective diodes are zener
diodes.

FORMULAS

$$r_D = R_D \| R_{LD} = \frac{R_D R_{LD}}{R_D + R_{LD}} \qquad \text{Eq. (12-1)}$$

$$A_v = g_m r_D \qquad \text{Eq. (12-2)}$$

$$g_m = \frac{\Delta i_D}{\Delta v_{GS}}$$

SUMMARY

- MOSFETs do not have a gate-source *p-n* junction like a JFET. Instead, they have a metal gate electrode that is perfectly insulated from the channel.
- Depletion-type (D-type) MOSFETs have a natural complete channel from source to drain. They are "normally-on" with $V_{GS} = 0$.
- A D-type MOSFET can have either positive or negative V_{GS}. As V_{GS} goes more negative, drain current I_D decreases. As V_{GS} goes more positive, I_D increases. (*n*-channel)
- MOSFET amplifiers are built in the same basic ways as JFET amplifiers. They are analyzed by the same methods and they have the same nonlinear behavior as JFET amplifiers.

● Enhancement-type (E-type) MOSFETs do not have a natural complete channel from S to D. They are "normally-off", with $V_{GS} = 0$.

● An E-type MOSFET must have positive V_{GS} larger than the threshold voltage $V_{GS(th)}$, in order to conduct. (*n*-channel)

● For an E-type MOSFET, as V_{GS} increases beyond $V_{GS(th)}$, I_D increases in the usual nonlinear way.

● For class-A operation, an E-type MOSFET amplifier must be biased with V_{GS} greater than $V_{GS(th)}$. (*n*-channel)

● E-type MOSFETs are much more common than D-type.

● The main advantage of a MOSFET amplifier is its very high input resistance R_{in}. It tends to be even higher than for a JFET amplifier.

● Dual-gate MOSFETs are especially useful in radio receivers for AGC and frequency-mixing purposes.

● MOSFETs have important advantages over BJTs in integrated circuits. They require fewer manufacturing steps and they take up less chip area.

● LSI and VLSI ICs use E-type MOSFETs rather than JFETs or BJTs.

● Standard MOSFETs require special storage and handling precautions to avoid electro-static damage (ESD) to the oxide insulating layer.

● Diode-protected MOSFETs are not so susceptible to oxide-layer damage from static electricity. They can be handled more normally.

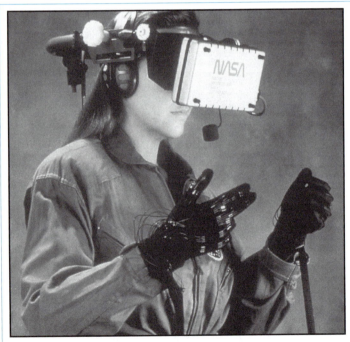

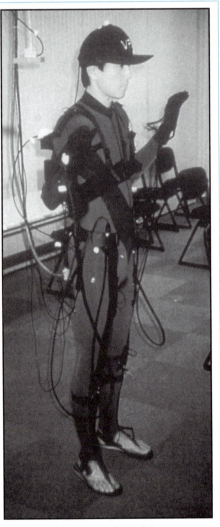

In the left photo, the technician is wearing a *virtual reality* headset. She sees a three-dimensional scene of a real environment that is either: 1) Presently being scanned by video cameras and transmitted into the headset; or 2) Has been scanned in the past, has been stored in computer memory, and is now being recreated in the headset. The imagery is so realistic that she feels as though she is actually present in the remote environment.

At right, with data-gloves and data-suit, the wearer's hand- and body-motions are duplicated in the artificial environment so realistically that it seems to him that he is actually moving within that remote environment (when he's wearing his headset).

The potential applications for virtual reality are very exciting.

Courtesy of NASA

QUESTIONS AND PROBLEMS

1. Draw the schematic symbol for an *n*-channel depletion-type MOSFET. Assume the internal substrate connection to the source. Show the proper voltage polarity between drain and source.

2. Repeat for *p*-channel.

3. (T-F) An *n*-channel D-type MOSFET is like an *n*-channel JFET in that the gate G must always be negative relative to the source S.

4. (T-F) A D-type MOSFET has a complete channel between source and drain.

5. (T-F) A D-type MOSFET can be described as normally-on.

6. Draw the I_D-versus-V_{GS} (transconductance) curve for a D-type MOSFET (*n*-channel). On the curve, label $V_{GS(off)}$ and I_{DSS}.

7. Referring to the transconductance curve in Problem 4, describe what is meant by $g_{m(0)}$.

8. For the MOSFET of Problem 6, g_m is _____ than $g_{m(0)}$ for $+ V_{GS}$ and g_m is _____ than $g_{m(0)}$ for $- V_{GS}$.

9. Draw the schematic symbol for an *n*-channel enhancement-type MOSFET. Assume the internal substrate connection to the source. Show the proper voltage polarity between drain and source.

10. Repeat Problem 9 for *p*-channel.

11. Describe in words the difference between the schematic symbols for a D-type and an E-type MOSFET. Explain why this symbol difference makes good sense.

12. (T-F) An E-type MOSFET has a complete channel between source and drain.

13. (T-F) An E-type MOSFET can be described as normally-off.

14. Draw the I_D-versus-V_{GS} (transconductance) curve for an E-type MOSFET (*n*-channel). On the curve, label $V_{GS(th)}$.

15. Draw the schematic diagram of a common-source amplifier using an *n*-channel E-type MOSFET with voltage-divider/source resistor-biasing. Show an ac-bypass capacitor around the source resistor.

16. Draw the schematic diagram of a common-source E-type amplifier with no bias-stabilizing R_S present.

17. For your schematic diagram of Problem 16, assume the following values: $V_{DD} = 20$ V, $R_{G1} = 1.5$ MΩ, $R_{G2} = 1.2$ MΩ, $R_D = 3.0$ kΩ, $R_{LD} = 2.5$ kΩ, and $V_{in(p-p)} = 2$ V. The FET is a typical type No. 2N4351, having the transconductance curve shown in Fig. 12-12.

a) Use voltage division to find $V_G = V_{GS}$.

b) Identify the Q-point on the transconductance curve.

c) Identify the negative peak point, P_1, and the positive peak point, P_2, on the curve.

d) Find the FET's transconductance g_m over this range, by $g_m = \Delta i_D / \Delta v_{GS}$.

e) Use Eqs. (12-1) and (12-2) to find r_D and A_v.

f) Calculate $V_{out(p-p)}$.

18. As an extra check on the results of Problem 17, do the following:
    a) Use the typical transconductance curve of Fig. 12-12 to create a detailed family of I_D-versus-V_{DS} characteristic curves on graph paper, like the family shown in Fig. 12-10(B). Show V_{GS} going from 2 V to 10 V in 1-V steps. Scale the V_{DS} axis to 20 V and I_D axis to 8 mA.
    b) Plot the dc load-line by joining the two extreme points $V_{DS} = 0$ V and $I_D = 0$ mA. Use Ohm's law and Kirchhoff's voltage law in the amplifier's schematic diagram.
    c) Locate the Q-point on the dc load-line.
    d) Using r_D from part (e) of Problem 16, plot the slope line. Do this by choosing a small value of V_{ds}, say 6 V, and calculating $\Delta i_D = I_d = 6\,V/r_D$. This is the same load-line analysis technique that was used in Sections 7-7 and 11-4.
    e) Transfer the slope line to the Q-point to produce the ac load-line.
    f) Identify the negative peak point, P_1, and the positive peak point, P_2, on the ac load-line. Remember that $V_{in(p-p)} = 2$ V.
    g) Project down to the v_{DS} axis from points P_1 and P_2 to find the change in v_{DS}. This is equal to V_{out}, since there is no resistance in the source lead, and thus no ac voltage drop there.
    h) Compare to the V_{out} result obtained in Problem 16. Does this prove that the ac load-line analysis method works just as well for MOSFETs as it does for JFETs and BJTs?

19. What is the main advantage of MOSFET amplifiers over BJT amplifiers ?

20. For the amplifier of Problem 17, find the value of R_{in}.

21. Besides their main advantage mentioned in Question 19, what other feature makes MOSFETs popular in radio and TV IF amplifiers?

22. In the CMOS inverter of Fig. 12-14:
    a) Q_1 turns on and Q_2 turns off when V_{IN} goes _____ . Explain this in detail.
    b) Q_2 turns on and Q_1 turns off when V_{IN} goes _____ . Explain this in detail.

23. What do the letter CMOS stand for?

24. (T-F) It is possible to build digital inverters and other more complex digital circuits using MOSFETs alone, with no supporting components.

25. (T-F) MOSFET circuits take up less space than comparable BJT circuits on an IC's silicon chip.

26. What is the main danger in storing and handling standard MOSFETs?

27. List some of the precautions that we take to avoid the danger of Question 26.

28. Describe what is meant by the term diode-protected MOSFET.

CHAPTER 13

OP AMPS

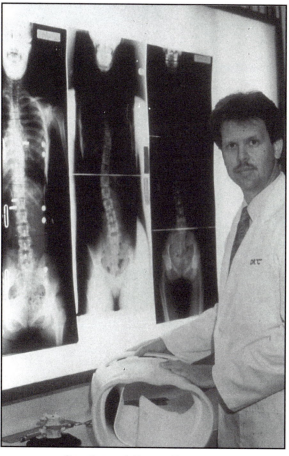

Courtesy of Copes Foundation

Scoliosis is a spinal disorder in which the spine curves sideways. This photo shows a recent innovation in scoliosis treatment. It is a body-brace containing an air-filled bladder that exerts corrective force on the spinal column. The bladder is the dark pouch on the left inside edge of the brace. An op-amp strain-gage circuit measures the pneumatic pressure in the bladder, which is periodically adjusted to match the patient's condition. Page 4 of the color section presents a picture of a ballerina whose dancing career was saved by treatment with the brace.

OUTLINE

Op Amp Schematic Symbol and Polarity Marks
Physical Appearance and Pin Configuration
Amplifier Specifications and Performance
Voltage-Comparers
The Op Amp Inverting Amplifier
Op Amp Noninverting Amplifier
The Offset Problem and Other Nonideal Effects
Troubleshooting Op Amp Circuits
Some Other Op Amp Applications

NEW TERMS TO WATCH FOR

operational amplifier

differential input

inverting input

noninverting input

voltage comparer

inverting amplifier

closed-loop voltage gain

open-loop voltage gain

virtual ground

noninverting amplifier

voltage follower

compensating resistor

null (offset) adjust

compensating capacitor

slew rate

summing amplifier

integrator

differentiator

passive filter

active filter

In our study of transistor amplifiers so far, we have dealt with *discrete* transistors, meaning individually packaged transistors. Figure 4-34 showed the appearance of several discrete transistors.

It is also possible to build transistor amplifiers as integrated circuits. In an integrated circuit, or IC, all the transistors, diodes, resistors and capacitors are created on a small "chip" of silicon. These microscopic components are interconnected by tiny layers of metal that are deposited directly onto the chip. In this way, the small silicon chip becomes a complete transistor amplifier. It is contained in a package that is no larger than the package that holds a single discrete transistor. Such a transistor amplifier that meets certain criteria is referred to as an **operational amplifier**, or **op amp**.

After studying this chapter, you should be able to:

1. Draw the schematic diagram of an op amp and show how it is connected to the power supply, signal source, and load.
2. Explain the meaning of differential input, as distinct from single-ended input.
3. Describe the input–output polarity relationship for an op amp. Given the voltage values at the inverting and noninverting input terminals, tell the polarity of the output voltage.
4. Interpret the pin diagram of an op-amp integrated-circuit package.
5. List the key electrical characteristics of an op amp, including typical values of voltage gain A_v and input resistance R_{in}.
6. Given the value of power-supply voltage, make an estimate of an op amp's maximum output voltage.
7. Explain the operation of an op-amp voltage comparer.
8. Given a complete schematic diagram of an op-amp inverting amplifier, find its voltage gain, its input resistance, current gain and power gain.
9. Use the virtual ground idea to explain how an inverting amplifier operates.
10. Explain the operation of a noninverting amplifier, using the idea of virtual zero differential input voltage.

11. Given the schematic diagram of an op amp noninverting amplifier, find its voltage gain; also describe its input resistance, current gain and power gain.
12. Describe the functioning of an op amp voltage follower.
13. Define what is meant by op-amp offset; state the two things that can be done to eliminate offset.
14. Explain the meaning of slew rate for an op amp.
15. Given complete information about an operational-amplifier circuit, find out whether the maximum operating frequency is set by the bandwidth limitation or by the slew rate.
16. Troubleshoot op-amp inverting and noninverting amplifiers by observation of the output waveform.
17. Explain the operation of an op-amp summing amplifier.
18. Given a complete schematic diagram of a summing circuit, and the values of the input voltages, calculate the output voltage.
19. Draw the basic schematic diagram of an op-amp integrator.
20. Given a complete schematic diagram of an integrator, and given the value of dc input voltage, describe the output waveform in detail.
21. Tell how a two-op-amp voltage-to-frequency converter works.
22. Draw the basic schematic diagram of an op-amp differentiator.
23. Describe how the output voltage of a differentiator depends on its input voltage.

13-1 OP AMP SCHEMATIC SYMBOL AND POLARITY MARKS

Figure 13-1(A) shows the schematic symbol for an op amp. It has 5 lines attached to a large triangle. The 5 lines correspond to 5 external pins on the IC package. We, the users, must make the connections to these 5 pins, or terminals.

To get an op amp working, here is what we must do.

● Connect a dual-polarity power supply to the $+V_{CC}$ and $-V_{EE}$ terminals. In Fig. 13-1(B), a ± 12-V dual-polarity power supply has been used. The $+12$ V power-supply terminal connects to the op amp's $+V_{CC}$ terminal. The -12-V power-supply terminal connects to the op amp's $-V_{EE}$ terminal. The center-ground terminal of the power supply does not connect directly to any one of the op amp's terminals.

An op amp's $+V_{CC}$ and $-V_{EE}$ terminals are often named simply $+V_S$ and $-V_S$.

An op amp does not have a ground terminal.

● Connect the load resistance between the op amp's output terminal and ground (that is, the center-ground terminal of the power supply). Figure 13-1(B) shows the load connected.

● Apply an input signal source between the op amp's two input terminals, called $-$IN and $+$IN, as shown in Fig. 13-1(B).

With these three things done, the op amp circuit becomes a working amplifier. It provides both voltage gain and current gain from source to load.

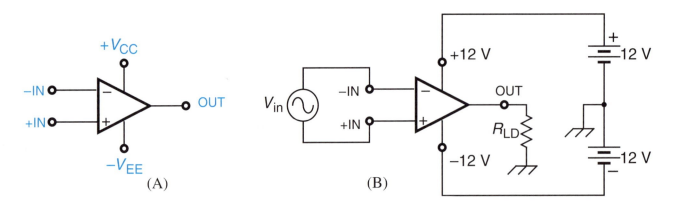

Figure 13-1 (A) Op amp schematic symbol. Usually, the input terminals are marked only with the − and + signs inside the triangle. The −IN and +IN symbols outside the triangle are usually not present in a real schematic diagram. (B) Completely wired op-amp circuit. ①✔

Standard amplifier input is called **single-ended** input.

The meaning of differential input.

②✔

Inverting and noninverting input terminals defined.

Differential Input: The input signal source does not have to be referenced directly to ground. In other words, it is not necessary for either the bottom input terminal (+IN) or the top input terminal (−IN) to be at ground potential. This is unlike the amplifiers that we have discussed up till now, in which the input voltage has always been measured between the single input terminal and ground. When an amplifier has two input terminals, neither of which has to be grounded, we say that it has **differential input**. All op amps have differential input.

Input Polarity: Figure 13-2 demonstrates the relationship between the op amp's input voltage polarity and its output voltage polarity. All parts of Fig. 13-2 show a true differential input. The ungrounded side of signal source v_1 connects to the op amp's **inverting input** (−IN). The ungrounded side of signal source v_2 connects to the op amp's **noninverting input** (+IN).

Because neither the +IN terminal nor the −IN terminal is connected directly to ground, the input voltage is differential. The op amp amplifies the *difference* between v_2 and v_1. We can symbolize this voltage difference as $v_{\text{IN(dif)}}$. As a formula,

$$v_{\text{IN(dif)}} = v_2 - v_1 \qquad \text{Eq. (13-1)}$$

$$v_{\text{OUT}} = (A_v)\, v_{\text{IN(dif)}} = A_v\,(v_2 - v_1) \qquad \text{Eq. (13-2)}$$

[A] In Fig. 13-2(A), $v_2 = +3$ V and $v_1 = +2$ V. Therefore, by Eq. (13-2),

$$v_{\text{IN(dif)}} = v_2 - v_1 = +3\text{ V} - (+2\text{ V}) = +1\text{ V}$$

A positive $v_{\text{IN(dif)}}$ means that the op amp's **+** terminal (+IN) is more positive than its **−** terminal (−IN), as Fig. 13-2(A) makes clear.

Get This

When $v_{\text{IN(dif)}}$ is positive, the op amp's output voltage, v_{OUT}, goes positive relative to ground.

Figure 13-2(A) shows v_{OUT} to be **+** on the top terminal of R_{LD}. That is, it has an actual polarity that is **+** on top, **−** on bottom.

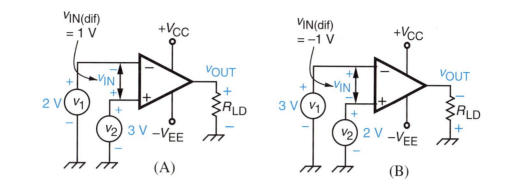

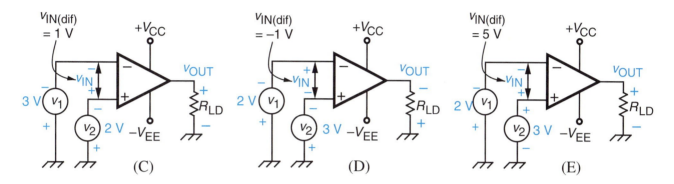

Figure 13-2 Showing the relationship between v_{OUT} polarity and $v_{IN(dif)}$ polarity.

As you study Fig. 13-2, be sure you understand that the input terminals' *schematic symbols* [the − sign for the inverting (−IN) terminal, and the + sign for the noninverting (+IN) terminal] do not necessarily indicate the *actual polarity* of the differential input voltage. In other words, it's quite possible for the actual polarity of the input to be more positive on the inverting (− schematic symbol) terminal, and more negative on the noninverting (+ schematic symbol) terminal. Pay special attention to parts B and D, to get this idea clear in your mind.

$\boxed{B}$ In Fig. 13-2(B), v_1 and v_2 are both positive, but now v_1 is more positive than v_2. Therefore Eq. (13-1) gives

$$v_{IN(dif)} = v_2 - v_1 = +2 \text{ V} - (+3 \text{ V}) = -1 \text{ V}$$

A negative value for $v_{IN(dif)}$ means that the op amp's − terminal (−IN) is more positive than its + terminal (+IN), as Fig. 13-2(B) makes clear.

> When $v_{IN(dif)}$ is negative, the op amp's output voltage, v_{OUT}, goes negative relative to ground.

Figure 13-2(B) shows v_{OUT} to be − on the top terminal of R_{LD}.

$\boxed{C}$ In Fig. 13-2(C), v_1 and v_2 are both negative relative to ground, with v_1 more negative than v_2. Therefore, from Eq. (13-1),

$$v_{IN(dif)} = v_2 - v_1 = (-2 \text{ V}) - (-3 \text{ V})$$
$$= -2 \text{ V} + 3 \text{ V} = +1 \text{ V}$$

This positive value of $v_{IN(dif)}$ produces + v_{OUT}, as Fig. 13-2(C) indicates.

$\boxed{D}$ In Fig. 13-2(D), v_2 is more negative than v_1. From Eq. (13-1),

$$v_{IN(dif)} = v_2 - v_1 = (-3 \text{ V}) - (-2 \text{ V})$$
$$= -3 \text{ V} + 2 \text{ V} = -1 \text{ V}$$

This negative value of $v_{IN(dif)}$ produces − v_{OUT}, as indicated in Fig. 13-2(D).

E There is no rule that says v_1 and v_2 must be the same polarity relative to ground. Figure 13-2(E) shows $v_1 = -2$ V and $v_2 = +3$ V. Applying Eq. (13-1) gives

$$v_{IN(dif)} = v_2 - v_1 = (+3 \text{ V}) - (-2 \text{ V})$$
$$= +3 \text{ V} + 2 \text{ V} = +5 \text{ V}$$

This positive $v_{IN(dif)}$ is shown to produce $+v_{OUT}$ in Fig. 13-2(E).

The input–output polarity relationship for an op amp is summarized in Table 13-1.

Table 13-1
When the differential input voltage polarity matches the schematic marks on the schematic symbol, the output voltage goes positive.

When the input voltage polarity does *not* match the schematic marks on the schematic symbol, v_{OUT} goes negative.

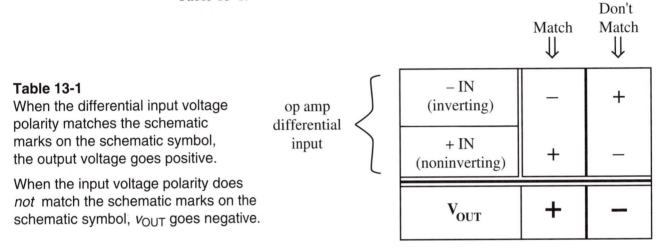

op amp differential input	Match ⇓	Don't Match ⇓
− IN (inverting)	−	+
+ IN (noninverting)	+	−
V_{OUT}	+	−

13-2 PHYSICAL APPEARANCE AND PIN CONFIGURATION

Op amps can be bought in any of the packages shown in Fig. 13-3. An 8-pin dual-in-line package (DIP) is shown in Fig. 13-3(A). The 8 pins can be soldered directly into a printed circuit board, or the DIP can be inserted into an 8-pin socket. The standard pin configuration for this package is shown in Fig. 13-3(B). The two **null-adjust** terminals are not shown in the basic schematic symbol that was presented earlier. We will discuss the purpose of those terminals later.

Sometimes a single op amp is contained in a 14-pin DIP, as indicated in Figs. 13-3(C) and (D). It is also common to have two op amps contained in one 14-pin DIP. In that case, the $+V_{CC}$ and $-V_{EE}$ power supply terminals are shared by both op amps. The round metal can package in Figs. 13-3(E) and (F) is less common than the DIP.

A popular general-purpose op amp is the type 741. It contains 17 BJTs, 4 diodes, 11 resistors, and 1 capacitor. All these components are integrated on a silicon chip that has dimensions about 0.2 cm by 0.2 cm (less than 1/8 inch square). Its schematic diagram is shown in Fig. 13-4.

The 741 contains all bipolar transistors, with no FETs. Other op amps contain a mixture of BJTs and FETs, sometimes JFETs and sometimes MOSFETs.

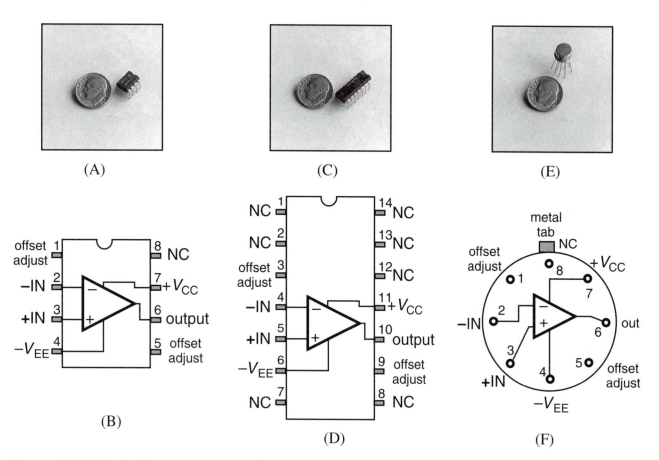

(A)   (C)   (E)

(B)   (D)   (F)

Figure 13-3 Physical appearances and standard pin configurations of op amps. The pin diagrams are top views. NC stands for *not connected*. ④ ✔

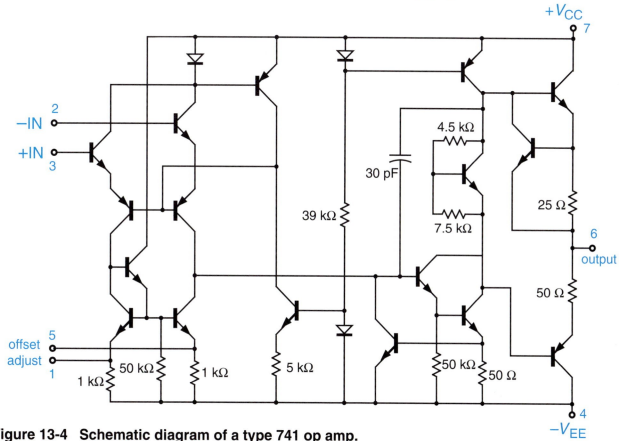

Figure 13-4   Schematic diagram of a type 741 op amp.

SELF-CHECK FOR SECTIONS 13-1 AND 13-2

1. Draw the schematic symbol for a basic op amp. Label all 5 terminals.
2. The input terminal marked **−** is called the _____ input.
3. The input terminal marked **+** is called the _____ input.
4. (T-F) An op amp can work successfully with neither one of its input terminals connected to ground.
5. The condition referred to in Question 4 is called _____ input.
6. A positive polarity of $v_{IN(dif)}$ is considered to be positive on the ____ input terminal and negative on the ____ input terminal. This causes v_{OUT} to be _____ relative to ground.
7. A negative polarity of $v_{IN(dif)}$ is considered to be positive on the ____ input terminal and negative on the ____ input terminal. This causes v_{OUT} to be _____ relative to ground.
8. (T-F) The center-ground terminal of the dual power supply must be connected to the ground pin of the op amp.
9. List and describe the three common kinds of IC packages that are used for packaging op amps.
10. When looking at an op amp IC package from the top, we count increasing pin numbers in the _____ direction. (clockwise or counterclockwise)
11. On an IC pin diagram, what do the letters NC stand for ?

13-3 AMPLIFIER SPECIFICATIONS AND PERFORMANCE

A modern op amp has the following general specifications:
● Very high voltage gain Typically, $A_v > 10\ 000$.
● Very high input resistance. Typically, $R_{in} > 1\ M\Omega$.
● Direct (dc) coupling throughout, all the way from input to output. There are no coupling capacitors in the signal path. Therefore an op amp can amplify dc signals as well as ac.
● Power supply voltage can be varied over a wide range. Typically from ±2 V to ±20 V.

Because its voltage gain is so extremely high, a plain op amp tends to go into saturation if the input voltage is greater than a fraction of a millivolt. For example, look at the plain amplifier of Fig. 13-5(A).

This circuit has −IN connected to ground, and +IN connected to the input source; the op amp's voltage gain is 50 000, which is typical. An op amp's maximum output voltage can be estimated by this rule:

*V*sat, the output saturation voltage of an op amp, can be estimated as 1 to 3 V less than the power-supply voltages ±*V*S, if the op amp is driving a moderate- or high-resistance load. For low-resistance loads (high output current), *V*sat is reduced.

In this case, we could estimate the output saturation voltage to be approximately ±10 V. With $A_v = 50\ 000$, a very small amount of input voltage will cause v_{OUT} to become 10 V. Using the voltage-gain formula, we get

$$v_{IN\,(sat)} = \frac{V_{sat}}{A_v}$$

$$= \frac{10\text{ V}}{50\,000} = 0.0002 \text{ V or } 0.2 \text{ mV}$$

In other words, a v_{IN} value of 0.2 mV will cause the output to saturate at 10 V. In Fig. 13-5(B), with a 0.4-mV peak sine-wave input, the v_{OUT} waveform drives back and forth between positive and negative saturation (+10 V and −10 V).

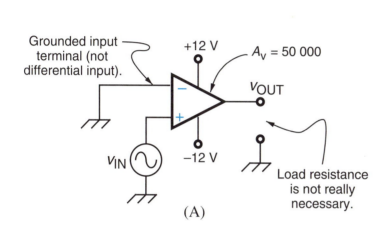

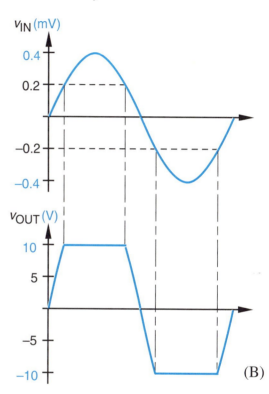

Figure 13-5 (A) Straightforward op-amp amplifier. Here, the inverting input is connected to ground. This is legal. Differential input means that neither input terminal *has* to be grounded. But if we want to, we can ground an input terminal.
(B) The huge voltage gain of an op amp usually causes the output to saturate, even for very small values of v_{IN}.

The amplifier's waveform distortion is obvious. In general, a plain op amp cannot be used for linear amplification of an input signal, unless the input signal is extremely small.

All the v_{OUT} values in Fig. 13-2 were saturated values (+V_{sat} or −V_{sat}).

Courtesy of Sky Vision, Inc.

This aerial video camera/recorder is used for low-altitude inspection of oil and gas pipelines, and electric power lines.

Mounted on the wing of a single-engine aircraft, the camera can be aimed and zoomed by a solo pilot, as he watches the image on a video monitor located where the right-side copilot seat would normally be. When the plane returns to base, the video film can be inspected frame-by-frame, if necessary, to study any problems along the line. Each frame has printed on it the airplane's exact latitude and longitude, accurate to within 15 meters. Such precise information is available anywhere on earth by exchanging radio signals with the Global Positioning System satellites (see page 287).

13-4 VOLTAGE-COMPARERS

Voltage-comparer with V_{IN} applied to the noninverting input.

The saturating tendency of an op amp makes it useful as a **voltage-comparer**, often called a comparator. The circuit of Fig. 13-6(A) compares V_{IN} to 0 V. If V_{IN} is even slightly greater than 0 V, (it is positive), then V_{OUT} = +10 V. If V_{IN} is even slightly less than 0 V (it is negative), then V_{OUT} = −10 V.

Figure 13-6(B) compares V_{IN} to +4.0 V. If V_{IN} is even slightly greater than +4.00 V (perhaps +4.01 V), then V_{OUT} = +10 V. If V_{IN} is even slightly less than +4.00 V (perhaps +3.99 V), then V_{OUT} = −10 V.

Any negative value of V_{IN} (perhaps −1.5 V) also is less than +4.0 V, so V_{OUT} = −10 V.

Voltage-comparer with V_{IN} applied to the inverting input.

We can reverse the positions of the comparison voltage and the input voltage V_{IN}. This is shown in Fig. 13-6(C). Now, if V_{IN} is slightly more positive than 4.00 V, then V_{OUT} goes to negative saturation, namely −10 V. If V_{IN} is any positive value less than +4.0 V, or any negative value, then V_{OUT} goes to positive saturation, namely +10 V.

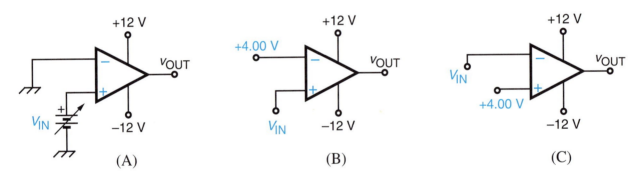

(A)   (B)   (C)

Figure 13-6 Voltage comparers. (A) Compare to 0 V. V_{IN} is applied to noninverting input so V_{OUT} goes in the same "direction" as V_{IN}. (B) Compare to +4.0 V. (C) Compare to +4.0 V, but V_{IN} is applied to inverting input. Therefore V_{OUT} goes in the opposite direction to V_{IN}.

EXAMPLE 13-1

Figure 13-7 shows a voltage comparer with adjustable comparison voltage. Either a positive or a negative value of comparison voltage can be selected. Suppose we adjust the pot to give a comparison voltage of −3.00 V.
a) Tell the range of values of V_{IN} that will cause V_{OUT} = +10 V.
b) Tell the range of values of V_{IN} that will cause V_{OUT} = −10 V.
Assume that the extreme values of V_{IN} are ±9 V.

SOLUTION:

a) V_{IN} is applied to the noninverting (+) input. Therefore, if V_{IN} is more positive than −3.0 V, then V_{OUT} will go to +10 V. This V_{IN} range can be described as **−2.99 V to +9 V**.

b) If V_{IN} is more negative than −3.00 V, then the actual differential input voltage polarity will be the opposite of the schematic marks, making V_{OUT} = −10 V. The V_{IN} range can be described as **−3.01 V to −9 V**.

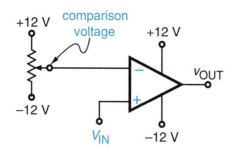

Figure 13-7
Comparer with adjustable comparison voltage.

There are many applications for voltage-comparers. One of the most common applications is in circuitry that converts analog signals to digital signals. Such circuits are called analog-to-digital converters, or ADCs. You will study the operation of ADCs in a course on digital electronics

SELF-CHECK FOR SECTIONS 13-3 AND 13-4

12. (T-F) Op amps have moderate voltage gains, with A_v typically between 10 and 50.
13. Op amps have high input resistances, with R_{in} typically greater than 1 MΩ.
14. The lowest frequency that an op amp can successfully amplify is _____ Hz.
15. (T-F) If an op amp is working all right in a circuit with $V_S = \pm12$ V, it will probably work all right if the power supply voltage is increased to $V_S = \pm15$ V.
16. (T-F) An op amp can work successfully with its inverting (−) input terminal connected to ground.
17. (T-F) An op amp can work successfully with its noninverting (+) input terminal connected to ground.
18. A plain op amp amplifier like the one in Fig. 13-5 contains an op amp with $A_v = 120\,000$. What value of v_{IN} will cause the output voltage to become saturated (±10 V).
19. (T-F) A plain op amp can give distortion-free amplification of a 1-V p-p sine wave. Explain your answer.
20. In Fig. 13-7, suppose the pot is adjusted to give +5.50 V as the comparison voltage. What range of values of V_{IN} will produce $V_{OUT} = +10$ V ? What range will produce $V_{OUT} = -10$ V ?

13-5 THE OP AMP INVERTING AMPLIFIER

By connecting external resistors to an op amp, we can get linear (distortion-free) amplification of an input signal. One resistor-connection method is shown in Fig. 13-8. This method produces an **op amp inverting amplifier**. It is called that because v_{OUT} will have the opposite polarity from v_{IN}. If v_{IN} is sine-wave ac, v_{OUT} will be an inverted sine wave — phase-shifted by 180 degrees. This inversion is shown in the waveform sketches in Fig. 13-8.

The defining features of an inverting amplifier are:
1. The **+** input is connected to ground.
2. Input signal voltage v_{IN} is applied to input resistor R_i, with the other side of R_i connected to the **−** input.
3. Feedback resistor R_f is connected between the output and the **−** input.

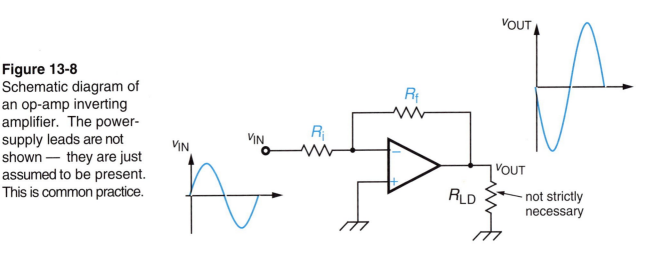

Figure 13-8
Schematic diagram of an op-amp inverting amplifier. The power-supply leads are not shown — they are just assumed to be present. This is common practice.

With resistors R_f and R_i in place, the overall amplifier circuit's voltage gain becomes much less than the voltage gain of the op amp itself. Therefore we need a way of distinguishing between the two voltage gains.

The overall voltage gain for an op amp inverting amplifier is called **closed-loop voltage gain**, symbolized $A_{v(CL)}$.
The very large voltage gain of the op amp itself is called **open-loop voltage gain**, symbolized $A_{v(OL)}$.

The term closed-loop comes from the fact that resistor R_f closes a feedback loop from the op amp's output back to its input.

An inverting amplifier's closed-loop voltage gain is given by

$$A_{v\,(CL)} = \frac{R_f}{R_i}$$
Eq. (13-3)

The inverting amplifier's overall (closed-loop) input resistance is simply

$$R_{in(CL)} = R_i$$
Eq. (13-4)

We will show later (on page 420) why Eqs. (13-3) and (13-4) are true.

EXAMPLE 13-2

For the circuit of Fig. 13-9(A),
a) Find the overall closed-loop voltage gain, $A_{v(CL)}$.
b) Calculate V_{out}.
c) Sketch waveform graphs of V_{in} and V_{out}.

SOLUTION

a) From Eq. (13-3),

$$A_{v(CL)} = \frac{R_f}{R_i} = \frac{39\ k\Omega}{3\ k\Omega} = \mathbf{13}$$

b)
$$V_{out} = A_{v(CL)}\ (V_{in})$$
$$= 13\ (1\ V\ \text{p-p}) = \mathbf{13\ V\ p\text{-}p}$$

c) V_{out} swings between –6.5 V and +6.5 V, out of phase with V_{in}. The waveforms are shown in Fig. 13-9(B).

⑧ ✓

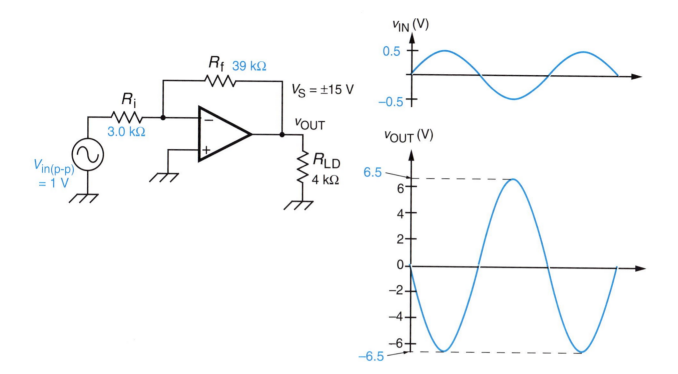

Figure 13-9   Op amp inverting amplifier.   (A) Schematic.   (B) Waveforms.

In this example, note carefully that input voltage V_{in} is the *overall* amplifier input voltage that the signal source applies to external resistor R_i. V_{in} is *not* the differential input voltage at the op amp itself. In fact, we will have an interesting thing to say about $V_{in(dif)}$ when we prove Eqs. (13-3) and (13-4).

V_{in} is different from $V_{in(dif)}$.

EXAMPLE 13-3

For the inverting amplifier of Example 13-2,
a) Find the amplifier's overall input resistance, $R_{in(CL)}$.
b) Calculate the peak-to-peak source input current, $I_{in(p-p)}$, by Ohm's law.
c) Calculate the peak-to-peak output load current, $I_{out(p-p)}$, by Ohm's law.
d) Divide I_{out} by I_{in} to find the amplifier's overall current gain, $A_{i(CL)}$.
e) Find the amplifier's overall power gain, $A_{P(CL)}$.

SOLUTION

a) Equation (13-4) applies to an inverting amplifier.

$$R_{in(cl)} = R_i = 3\ k\Omega$$

Note carefully that this is not the op amp's internal input resistance, $R_{in(OL)}$, which would be greater than 1 MΩ.

b)

$$I_{in(p-p)} = \frac{V_{in(p-p)}}{R_{in(cl)}} = \frac{1\ V}{3\ k\Omega} = \textbf{0.333 mA}$$

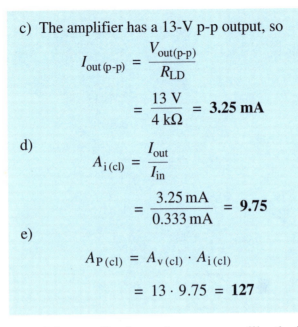

c) The amplifier has a 13-V p-p output, so

$$I_{out(p\text{-}p)} = \frac{V_{out(p\text{-}p)}}{R_{LD}}$$

$$= \frac{13\ V}{4\ k\Omega} = \mathbf{3.25\ mA}$$

d)

$$A_{i(cl)} = \frac{I_{out}}{I_{in}}$$

$$= \frac{3.25\ mA}{0.333\ mA} = \mathbf{9.75}$$

e)

$$A_{P(cl)} = A_{v(cl)} \cdot A_{i(cl)}$$

$$= 13 \cdot 9.75 = \mathbf{127}$$

⟨8⟩ ✔

Most op amp circuits contain a feedback resistor R_f.

Most applications of op amps are like the inverting amplifier in that they have feedback. That is, most applications are closed-loop, not open-loop. Because of this, we often omit the (CL) subscript, and adopt the following conventions:

When no subscript is written, it means closed-loop.

[1] When an op amp circuit specification has no subscript, we take it to mean a closed-loop specification.

[2] When we want to refer to an open-loop condition, we specifically write the subscript (OL).

When you mean open-loop, you must write the (OL) subscript.

Thus, Eq. (13-3) and (13-4) could be written simply as

$$A_v = \frac{R_f}{R_i}$$

Eq. (13-3)

$$R_{in} = R_i$$

Eq. (13-4)

The results from Example 13-3 would be written without subscripts as $R_{in} = 3\ k\Omega$, $A_i = 9.75$, and $A_P = 127$.

Back in Fig. 13-5, there was no feedback. The voltage gain of the plain open-loop op amp would be written as

$$A_{v(ol)} = 50\ 000.$$

Explaining Equations (13-3) and (13-4) — The Virtual Ground Idea:

Always remember than $A_{v(OL)}$ is huge for a modern op amp. Therefore we can understand the behavior of an op-amp amplifier by using the following fact:

Get This

For any *non-saturated* value of v_{OUT}, the differential input $V_{IN(dif)}$ must be extremely small, since $A_{v(OL)}$ is so huge. Therefore if one of the op amp's input terminals is connected to ground, then the other input terminal must be at a potential that is extremely close to ground. We say that the non-grounded input terminal is at **virtual ground**.

For example, the inverting amplifier of Fig. 13-9 would have dc output voltage $V_{OUT} = -6.5$ V for a dc input of $V_{IN} = +0.5$ V. This dc situation is pictured in Fig. 13-10(A). With $A_{v(OL)} = 100\,000$, the differential input voltage must be given by

$$V_{IN\,(dif)} = \frac{V_{OUT}}{A_{v(OL)}} = \frac{6.5\,V}{100\,000} = 0.065 \text{ millivolt}$$

or 0.000 065 V. This voltage is extremely close to zero, compared to the V_{IN} value of 0.5 V. Figure 13-10(B) stresses the idea that the op amp's –IN terminal is virtually at ground potential.

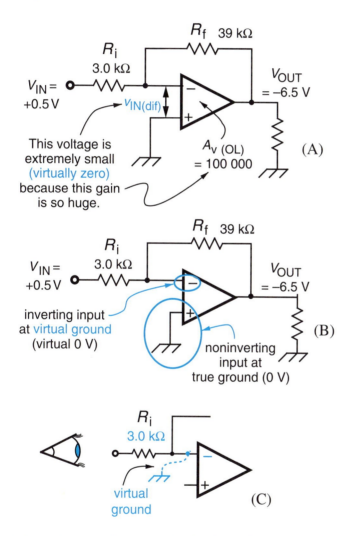

Figure 13-10
Explaining the virtual ground idea, and why $R_{IN} = R_i$.

(A) $v_{IN(dif)}$ is extremely small.

(B) The op amp's –IN terminal is at virtual ground.

(C) With R_i connected to virtual ground, the signal source sees only R_i.

When the signal source looks into the overall amplifier, it sees R_i connected to virtual ground, as pictured in Fig. 13-10(C). Therefore the source sees an equivalent overall input resistance (dc or ac) of

$$R_{IN} = R_{in} = R_i \qquad \text{Eq. (13-4)}$$

Now let us use the virtual ground idea to understand why A_v is given by Eq. (13-3). Look at Fig. 13-11(A). The –IN terminal is at virtual ground, and the op amp's internal (open-loop) input resistance is quite large. Therefore the current into the op amp is extremely small, virtually zero.

Assume that $R_{IN(OL)} = 2\,M\Omega$.
Then $I_{dif} = V_{IN(dif)} \div R_{IN(OL)}$
$= 0.065\,mV \div 2\,M\Omega =$
32 pA, or 0.000 032 μA.
This is extremely small.

Now look at Fig. 13-11(B). With $I_{dif} = 0$, we can say:

The current through R_f must be the same as the current through R_i. That is

$$I_{Rf} = I_{Ri} \qquad\qquad \text{Eq. (13-5)}$$

The left side of R_f is at virtual 0 V. Therefore the voltage across R_f is v_{OUT}, as Fig. 13-11(C) shows. I_{Rf} can be found by applying Ohm's law to R_f, giving

$$I_{Rf} = \frac{-V_{OUT} - 0 \text{ V}}{R_f}$$

$$I_{Rf} = \frac{-V_{OUT}}{R_f}$$

The right side of R_i is at virtual 0 V. Therefore the voltage across R_i is v_{IN}, as Fig. 13-11(C) shows. I_{Ri} can be found by applying Ohm's law to R_i, giving

$$I_{Ri} = \frac{0 - V_{IN}}{R_i}$$

$$I_{Ri} = \frac{-V_{IN}}{R_i} \qquad\qquad \text{Eq. (13-7)}$$

Figure 13-11

Using the virtual ground idea to understand why $A_v = R_f/R_i$.

(A) The differential input current is virtually 0.

(B) Because $I_{dif} \approx 0$, the two resistor currents must be equal to each other.

(C) With the −IN terminal at virtual 0 V, v_{OUT} appears across R_f and v_{IN} appears across R_i.

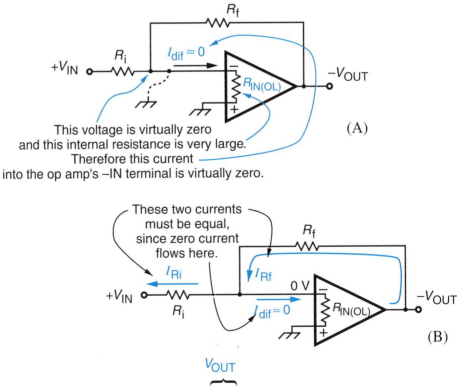

Plugging Eqs. (13-6) and (13-7) into Eq. (13-5), we get

$$\frac{-V_{OUT}}{R_f} = \frac{-V_{IN}}{R_i}$$

which can be rearranged to

$$\frac{V_{OUT}}{V_{IN}} = \frac{R_f}{R_i}$$

or $$\boxed{A_v = \frac{R_f}{R_i}}$$ Eq. (13-3)   ⑨ ✔

The foregoing was a rather formal explanation of the Equation $A_v = R_f/R_i$ for an inverting amplifier. Here is a more intuitive explanation, based on the simpler values shown in Fig. 13-12.

$\boxed{1}$ An input voltage $v_{IN} = +1$ V is applied to the amplifier. This immediately causes 1 mA to flow from right to left through 1-kΩ resistor R_i (since the right side of R_i is at 0 V).

$\boxed{2}$ The current flowing from right to left through R_f must also become 1 mA, since the two resistor currents must match. ($I_{dif} \approx 0$)

$\boxed{3}$ V_{OUT} must go to whatever negative value is necessary to cause 1 mA to flow through R_f. Because R_f is 10 times as large as R_i, V_{OUT} must become 10 times as large as V_{IN}. Here, V_{OUT} must become –10 V.

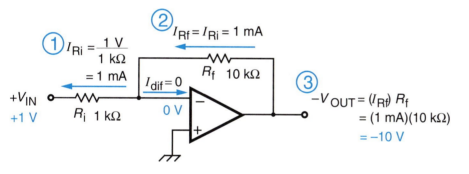

Figure 13-12
Explaining why v_{OUT} must be 10 times as large as V_{IN} if R_f is 10 times as large as R_i. This explains why, in general, $A_v = R_f / R_i$.

SELF-CHECK FOR SECTION 13-5

21. Draw the schematic diagram of an op amp inverting amplifier. Label the external resistors.
22. Explain what is meant by closed-loop voltage gain, as opposed to open-loop voltage gain.
23. For an op-amp inverting amplifier, write the formula for closed-loop voltage gain.
24. An inverting amplifier is called by that name because v_{OUT} is _____ with respect to v_{IN}.
25. An inverting amplifier has $R_i = 2.2$ kΩ, $R_f = 18$ kΩ, $R_{LD} = 5$ kΩ, and $V_S = \pm 15$ V.
    a) Find A_v.   c) If V_{in} is a 1-V peak sine wave, describe V_{out}.
    b) Find R_{in}.   d) If V_{in} is a 2-V peak sine wave, describe V_{out}.
26. Because the –IN op amp terminal is so close to ground potential in an inverting amplifier, it is called _____ ground.

COAL-FIRED ELECTRIC UTILITY STACK CLEANUP
STEP 2: REMOVAL OF SULFUR

Most coal contains a considerable amount of sulfur. When the coal is burned, the sulfur reacts with oxygen to form sulfur dioxide, a gas. If this gas is allowed to escape into the atmosphere, it reacts with water vapor to form sulfuric acid. Rain brings the acid back to the earth's surface with well-publicized harmful effects on the environment.

Removal of sulfur dioxide gas is accomplished in a water-spray tower, also called a scrubbing tower, illustrated in Fig. 1. The 150-ft-high, 50-ft-diameter tower holds thick liquid (called slurry) about 20 feet deep at its bottom. Powerful pumps lift the water-based slurry to a collection of spray nozzles near the top of the tower.

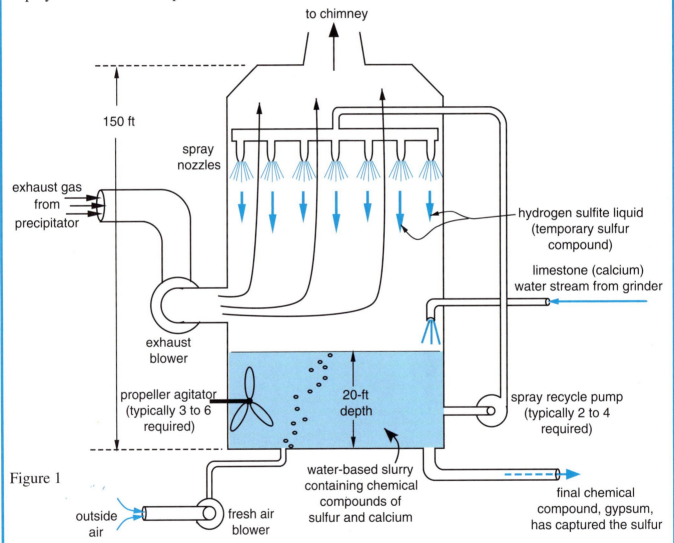

Figure 1

Having already passed through an intelligent precipitator, the SO_2-contaminated exhaust gas is blown into the tower below the nozzles. As the hot gas rises, it must pass through the fine-droplet water spray. As it does so, the initial capturing chemical reaction occurs: The sulfur dioxide, SO_2, reacts with the water, H_2O, to produce liquid hydrogen sulfite, H_2SO_3. Over 90% of the total sulfur in the exhaust is captured by this reaction with water. Less than 10% escapes up the chimney.

Hydrogen sulfite is not a stable chemical compound, so the sulfur cannot be permitted to remain in that form. It falls into the liquid slurry at the bottom, where it is mixed with pulverized limestone carried in on a water stream, as Fig. 1 shows. Of course, the limestone had to be mined or quarried and hauled to the site by freight train, just like the coal.

Limestone has a large concentration of calcium (Ca), in the form of calcium carbonate, $CaCO_3$. In the stirred-up slurry, the final capturing chemical reactions occur:

Courtesy of GE Environmental Systems

1) The limestone's $CaCO_3$ reacts with the H_2SO_3 to produce $CaSO_3$, calcium sulfite.

2) The $CaSO_3$ picks up one additional oxygen atom from the fresh air that is bubbling up through the slurry. This produces $CaSO_4$, calcium sulfate, also known as gypsum. It is a stable compound that forms solid crystals.

The solid gypsum is filtered out of the slurry (pipe leaving Fig. 1 at lower right). After the filtering operation, the liquid is returned to the slurry. The return pipe is not shown in Fig. 1.

Electronic sensors measure the chemical concentrations in the slurry. An electronic pump-control system automatically adjusts the rate of limestone delivery to maintain the concentrations in their proper range.

The photo above shows one of the spray recycle pumping stations. The 600-V, 500-HP ac drive-motor is on the right, closest to our view. The square unit in the center is a gear box for reducing the shaft speed from about 1750 rpm to about 300 rpm. The pump itself is the large-diameter unit closest to the tower.

13-6 OP AMP NONINVERTING AMPLIFIER

An op amp noninverting amplifier has external resistors R_i and R_f, like the inverting amplifier. But v_{IN} and ground are reversed, as shown in Fig. 13-13. Thus, the defining construction features of the noninverting amplifier are:

[1] Input signal voltage v_{IN} is applied to the + input of the op amp.

[2] One side of R_i is connected to ground, with the other side connected to the − input.

[3] Feedback resistor R_f is connected between the output and the − input, just like before.

The circuit is called noninverting because it produces an output voltage v_{OUT} that has the same polarity as v_{IN}. If v_{IN} is sine-wave ac, v_{OUT} will be sine-wave ac in phase with v_{IN}. Check the waveform sketches in Fig. 13-13.

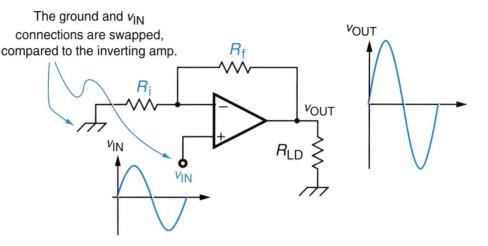

**Figure 13-13
Schematic diagram
of a noninverting
amplifier.**

Remember, an op-amp
circuit requires that both
dc power supplies be
connected, even though the
schematic diagram doesn't
show them.

For a noninverting amplifier, the closed-loop voltage gain and input resistance are given by:

$$A_v = \frac{R_f}{R_i} + 1$$

Eq. (13-9)

$$R_{in} \approx \infty \ \Omega$$

Eq. (13-10)

To understand why Eq. (13-9) is true, refer to the noninverting amplifier in Fig. 13-14(A), operating under dc conditions. Remember that the op amp itself has a huge open-loop gain. Therefore, if V_{OUT} is not saturated, the differential input voltage, $V_{IN(dif)}$, is extremely small. This means that:

$V_{IN(dif)}$ is virtually 0.

Get This

The voltage on the op amp's − input is virtually the same as the voltage on its + input, namely V_{IN}.

Figure 13-14(B) points out this fact.

Just as we did for the inverting amplifier, we can apply Ohm's law to the op amp's input terminals.

I_{dif} is virtually 0.

$$I_{dif} = \frac{V_{IN\,(dif)}}{R_{IN\,(OL)}} = \frac{\text{extremely small voltage}}{\text{several megohms}} \approx 0$$

Eq. (13-11)

Once again, we conclude that the differential input current, the current into the −IN terminal, is virtually 0. This is pointed out in Fig. 13-14(C).

By the same reasoning as before, if $I_{dif} = 0$, then R_i and R_f are virtually in series with each other, so

$$I_{Ri} = I_{Rf}$$

Eq. (13-12)

as pointed out in Fig. 13-14(C).

R_i has 0 V on its left and voltage v_{IN} on its right. Applying Ohm's law to R_i gives

$$I_{Ri} = \frac{V_{IN}}{R_i}$$

Eq. (13-13)

In this specific example,

$$I_{Ri} = \frac{+1\ V}{1\ k\Omega} = 1\ mA$$

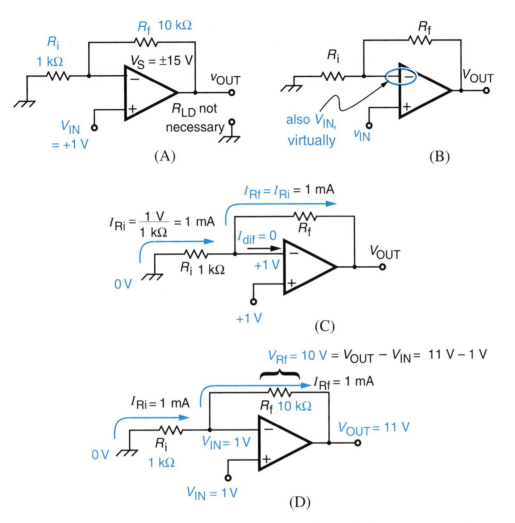

Figure 13-14 Understanding a noninverting amplifier. (A) Schematic. (B) The op amp's two input terminals have virtually the same voltage, V_{IN}. (C) The op amp's differential input current I_{dif} is extremely small, virtually zero. So the resistor currents must be equal. (D) With equal currents, R_f must have 10 times as much voltage as R_i because it has 10 times as much resistance as R_i.

Feedback resistor R_f has $+V_{IN}$ on its left and $+V_{OUT}$ on its right. Applying Ohm's law to R_f gives

$$I_{Rf} = \frac{V_{OUT} - V_{IN}}{R_f} \qquad \text{Eq. (13-14)}$$

Plugging Eq. (13-13) and (13-14) into Eq. (13-12) gives

$$\frac{V_{IN}}{R_i} = \frac{V_{OUT} - V_{IN}}{R_f} \qquad \text{Eq. (13-15)}$$

We can rearrange this equation by swapping the positions of V_{IN} and R_f, giving

$$\frac{R_f}{R_i} = \frac{V_{OUT} - V_{IN}}{V_{IN}}$$

Explaining why
$A_v = (R_f/R_i) + 1.$

$$\frac{R_f}{R_i} = \frac{V_{OUT}}{V_{IN}} - \frac{V_{IN}}{V_{IN}} = \frac{V_{OUT}}{V_{IN}} - 1$$

which we can rewrite as

⑩ ✔

$$A_v = \frac{V_{OUT}}{V_{IN}} = \frac{R_f}{R_i} + 1 \qquad \text{Eq. (13-9)}$$

In the specific example of Fig. 13-14,

$$A_v = \frac{R_f}{R_i} + 1 = \frac{10\ k\Omega}{1\ k\Omega} + 1 = 10 + 1 = 11$$

so $\qquad V_{OUT} = 11\ (V_{IN}) = 11(1\ V) = 11\ V$

Here is an intuitive summary of what happened in this noninverting amplifier.

1 An input voltage $V_{IN} = +1$ V was applied to the +IN terminal of the op amp. Virtually the same voltage, +1 V, had to appear on the –IN terminal, because their difference is virtually zero.

2 With 1 V across R_i, 1 mA immediately began flowing through R_i from left to right.

3 The current through R_f also became 1 mA from left to right, since the two resistor currents must match.

4 V_{OUT} had to go to whatever positive value was necessary to cause 1 mA to flow through R_f. Because R_i is 10 times as large as R_i, V_{OUT} had to go to the value that causes the *difference* between V_{OUT} and V_{IN} ($V_{OUT} - V_{IN}$) to be 10 times as large as V_{IN} alone. In this case, V_{OUT} had to become +11 V.

It is easy to see that the amplifier's closed-loop input resistance is virtually infinite. We already showed, in Eq. (13-11), that $I_{dif} \approx 0$. But this current, I_{dif}, is also the current in the +IN lead, besides being the current in the –IN lead [the two leads are in series with each other, connected together by the op amp's internal resistance, $R_{IN(OL)}$]. Therefore, from the input source's point of view,

Explaining why $R_{IN} = \infty\ \Omega$.

$$R_{IN(CL)} = R_{IN} = R_{in} = \frac{V_{IN}}{I_{dif}} = \frac{+1\ V}{\text{virtually}\ 0} = \text{virtually}\ \infty\ \Omega$$

EXAMPLE 13-4

For the noninverting amplifier of Fig. 13-15(A), driven by an ac source:
   a) Find the amplifier's closed-loop voltage gain A_v.
   b) Calculate V_{out}.
   c) Sketch the waveforms of V_{in} and V_{out}.
   d) Give some description of the amplifier's current gain, A_i.
   e) Give some description of the amplifier's power gain, A_P.

SOLUTION

a) From Eq. (13-9)

$$A_v = \frac{R_f}{R_i} + 1$$

$$= \frac{7.5\ k\Omega}{1.5\ k\Omega} + 1 = 5 + 1 = 6$$

b) $V_{\text{out(p-p)}} = A_{\text{v}} [V_{\text{in(p-p)}}]$

$$= 6 \,(3 \text{ V}) = \textbf{18 V}$$

c) V_{out} is in phase with V_{in}, as shown in Fig. 13-15(B).

d) Input current I_{in} is virtually zero, as Eq. 13-11 states. Therefore the amplifier's current gain can be described as

$$A_{\text{i}} = \frac{I_{\text{LD}}}{\text{virtually } 0} = \textbf{virtually } \infty$$

e) By the same reasoning, the input power to the amplifier from the signal source is virtually zero. So

$$A_{\text{P}} = \frac{P_{\text{out}}}{\text{virtually } 0} = \textbf{virtually } \infty$$

A realistic way of expressing this is:

⑪ ✔

A noninverting op-amp amplifier can deliver reasonable amounts of output power to the load, even though it requires virtually no input power from the source.

Get This

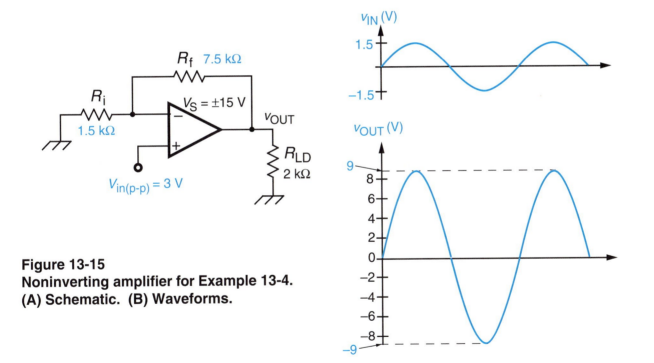

Figure 13-15
Noninverting amplifier for Example 13-4.
(A) Schematic. (B) Waveforms.

Voltage-Follower: A special case of the noninverting amplifier is shown in Fig. 13-16(A). This circuit, called a **voltage-follower**, is equivalent to having $R_{\text{f}} = 0 \text{ } \Omega$ and $R_{\text{i}} = \infty \text{ } \Omega$. These resistances are suggested in Fig. 13-16(B). Applying Eq. (13-9) gives

$$A_{\text{v}} = \frac{R_{\text{f}}}{R_{\text{i}}} + 1 = \frac{0 \,\Omega}{\infty \,\Omega} + 1 = 0 + 1 = 1$$

As with any noninverting op-amp amplifier,

$$R_{in} = \text{virtually } \infty \ \Omega$$

Thus, the voltage-follower provides no voltage gain, but lots of current gain. It is similar to a common-collector (emitter-follower) discrete transistor amplifier. But it outperforms a common-collector amplifier.

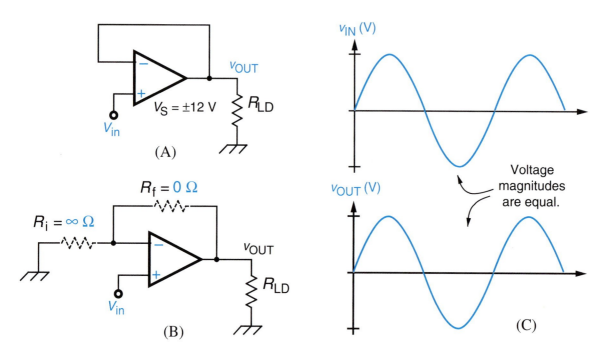

Figure 13-16 Voltage follower. (A) Schematic diagram. (B) Equivalent way of viewing the circuit. The direct connection of the output to –IN is like $R_f = 0 \ \Omega$. The absence of a resistor between –IN and ground is like $R_i = \infty \ \Omega$. (C) Waveforms showing V_{out} identical to V_{in}.

Explaining the operation of a voltage-follower.

Here is an intuitive explanation of the performance of a voltage-follower:

☐1☐ Input voltage v_{IN} is applied to the noninverting (+) input of the op amp. Because of the op amp's huge open-loop gain, the voltage existing on the inverting (–) input must be virtually the same as v_{IN}. (There is virtually no difference between the two voltages.)

☐2☐ v_{OUT} must become whatever voltage is necessary, so that the feedback path makes voltage value v_{IN} appear at the –IN terminal.

☐3☐ The feedback path is nothing but a direct (0-Ω) wire connection. So v_{OUT} must become equal to v_{IN}, to make v_{IN} appear at the –IN terminal.

EXAMPLE 13-5

A real ac signal source has an open-circuit (no-load) terminal voltage of 2.5 V rms, and an output resistance of 30 kΩ, as shown in Fig. 13-17. We wish to use this signal source to drive a load device that has $R_{LD} = 500 \ \Omega$.

a) If the source is connected directly to the load, calculate the load's voltage, current, and power.

b) If a voltage-follower is placed between the source and the load, calculate the load's voltage, current, and power.

SOLUTION

a) The direct connection is shown in Fig. 13-17(B). By voltage division,

$$\frac{V_{ld}}{2.5\ V} = \frac{500\ \Omega}{30\ 000\ \Omega + 500\ \Omega}$$

$$V_{ld} = \textbf{0.0410 V}$$

This very small load voltage occurs because of the mismatch between the load resistance and the source's output resistance.

Using Ohm's law and the $P = IV$ formula, the current and power values are found to be

$$I_{ld} = \textbf{82.0 } \mu\textbf{A}$$

$$P_{ld} = \textbf{3.36 } \mu\textbf{W}$$

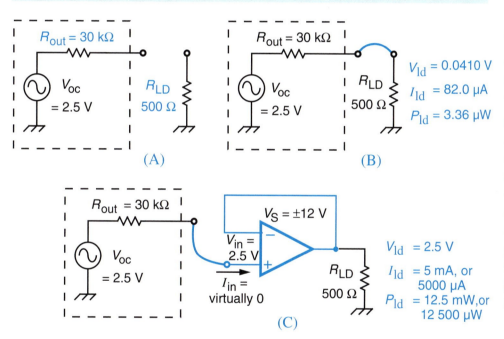

Figure 13-17

(A) Bad impedance mismatch between source and load.

(B) With nothing to *buffer* the load, the mismatch results in poor load power. Most of the source's voltage is lost across its internal resistance.

(C) With a voltage-follower acting as a buffer, load power is much better.

b) A voltage-follower has been installed in Fig. 13-17(C). Now the virtually infinite input resistance of the voltage-follower causes virtually zero current to be taken from the signal source. Therefore zero voltage is lost across R_{out}, and $V_{in} = 2.5$ V.

$$V_{ld} = A_v\ V_{in}$$

$$= (1)\ (2.5\ V) = \textbf{2.5 V}$$

The entire 2.5-V open-circuit value appears at the amplifier input.

The load's current and power can be calculated from Ohm's law and from $P = V \times I$ as

$$I_{ld} = \textbf{5 mA}$$

and $$P_{ld} = \textbf{12.5 mW}$$

The voltage-follower does a great job of matching a high-resistance (or impedance) source to a low-resistance (or impedance) load.

The voltage-follower improved the load power from 3.36 μW to 12 500 μW, a factor of almost 4000.

SELF-CHECK FOR SECTION 13-6

27. Draw the schematic diagram of an op amp noninverting amplifier.
28. In your schematic diagram, suppose $V_{CC} = \pm 12$ V, $R_i = 10$ kΩ,
    $R_f = 47$ kΩ, and $R_{LD} = 1.5$ kΩ.
    a) Find the amplifier's voltage gain A_v.
    b) Find the amplifier's dc input resistance, R_{IN}.
29. For the noninverting amplifier of Problem 28:
    a) If $V_{IN} = +1.2$ V dc, find V_{OUT}.
    b) If $V_{IN} = -1.5$ V, find V_{OUT}.
    c) If $V_{IN} = -2.5$ V, find V_{OUT}.
    d) If $V_{in} = 2.0$ V p-p, describe V_{out}.
30. (T-F) All op amp noninverting amplifiers are very effective at matching a high-impedance source to a low-impedance load.
31. The _____ - _____ is a special case of the noninverting amplifier, having $A_v = 1$.

13-7 THE OFFSET PROBLEM
AND OTHER NONIDEAL EFFECTS

Ideally, an op amp amplifier should produce $V_{OUT} = 0$ V when $V_{IN} = 0$ V. Figure 13-18 shows this ideal behavior for both the inverting and noninverting circuits.

> **Get This**
>
> In reality, simple inverting and noninverting amplifiers usually will *not* produce a 0-V output when a 0-V input is applied. This nonideal behavior is called the **offset** problem.

Figure 13-18
Ideal amplifiers would have zero offset. That is, when $V_{IN} = 0$, $V_{OUT} = 0$. (A) Ideal inverting amplifier. (B) Ideal noninverting amplifier.

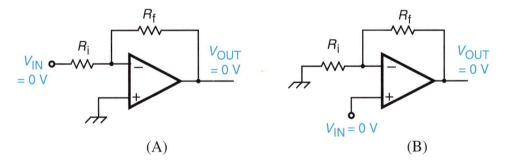

(A)   (B)

The precise reasons for this offset problem are studied in an advanced course. Right now, the only thing that concerns us is how to correct the problem.

To correct offset, we do the following:

1 Place a **compensating resistor** in the lead going to the op amp's +IN terminal. This resistor is usually symbolized R_B, as shown in Figs. 13-19(A) and (B).

2 Use the **offset adjust** terminals that were identified in the IC pin diagrams of Fig. 13-3. A potentiometer is connected to these two terminals, with the adjustable tap wired to either the $+V_{CC}$ or the $-V_{EE}$ dc supply voltage. Figure 13-20 shows this.

Compensating resistor R_B should have a value given by

$$R_B = R_i \,\|\, R_f = \frac{R_i R_f}{R_i + R_f} \qquad \text{Eq. (13-17)}$$

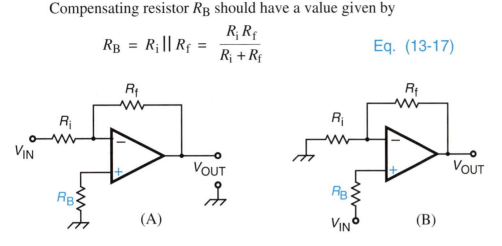

(A)

(B)

Figure 13-19
Placing compensating resistor R_B in the +IN lead to help eliminate offset.
(A) Inverting amp
(B) Noninverting amp.

The offset-adjust pot is usually $10\,\text{k}\Omega$ to $50\,\text{k}\Omega$. The op amp manufacturer recommends the proper pot value. To eliminate the offset, here is the procedure:

⟨1⟩ Power the op amp with $\pm V_S$, connect the load, and wait a minute or so for the op amp to come up to its normal operating temperature.

⟨2⟩ Make $V_{IN} = 0\,\text{V}$, by temporarily connecting the input terminal directly to ground.

⟨3⟩ With a sensitive voltmeter connected to the output, adjust the potentiometer until V_{OUT} becomes $0\,\text{V}$. It may not be possible to get exactly $0.0000\,\text{V}$, if your voltmeter has 0.1-millivolt resolution.

Once the offset has been adjusted to zero, it will hold steady in the short term if the supply voltage, load resistance, and temperature conditions remain unchanged. Long-term variations are always possible, of course.

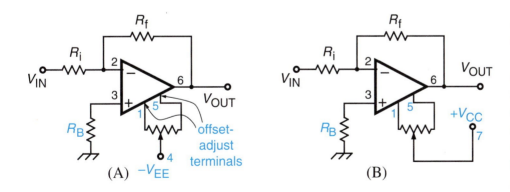

(A) $-V_{EE}$ offset-adjust terminals

(B)

Figure 13-20
Combining an offset-adjust pot with a compensating resistor.
(A) For some op amps, the pot wiper is tied to $-V_{EE}$.
(B) For other op amps, it is tied to $+V_{CC}$.
(Not all op amps have offset-adjust capability.)

Frequency Compensation

Op amp amplifiers are liable to break into high-frequency self-oscillation. This is more likely for low-A_v amplifiers (low values of R_f) than for high-A_v amplifiers. Most op amps have an internal **compensating capacitor** right in the integrated circuit to prevent such self-oscillation. But some op amps have no internal compensating capacitor. Or they may have some internal capacitance, but not enough to prevent oscillation under all operating conditions. For these op amps, the manufacturer provides a special compensation pin on the IC package. The user then connects an external *RC* series circuit between this compensation terminal and the output terminal, as shown in Fig. 13-21. The manufacturer gives detailed instructions on how to select the *R* and *C* values for the external frequency-compensation circuit.

Figure 13-21
In some op amp applications, an external *RC* frequency-compensation circuit is needed to prevent self-oscillation.

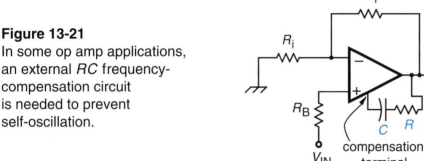

Slew Rate

Slew rate *SR* is how fast the op amp can change its output voltage.

Any op amp amplifier has a certain frequency bandwidth. The bandwidth spans from 0 Hz (dc) to the high-cutoff frequency value, f_{high}, that causes A_v to decrease to 0.707 of the normal A_v value given by Eq. (13-3) or Eq. (13-9).

An op amps's frequency bandwidth is valid only for small values of output voltage. This is because every op amp also has a certain maximum rate of change for its output voltage. This maximum rate is called the op amp's **slew rate**, which we will symbolize *SR*.

In formula form,

$$SR = \frac{\Delta v_{OUT}}{\Delta t} \Big|_{(maximum)}$$ Eq. (13-18)

Slew rate is measured in units of volts per microsecond (V/μs).

Get This If an op amp is producing large swings in v_{OUT}, it can happen that a sine-wave frequency less than f_{high} would require the voltage to change at a rate greater than the slew rate. The op amp is unable to satisfy this requirement, so it is unable to operate at that high a frequency. This is true even though the frequency is well within the small-signal bandwidth.

Look at Fig. 13-22 to understand this problem

The op amp's bandwidth (f_{high}) is 300 kHz.

For purposes of explanation, let us assume that we have an op amp with slew rate = 10 V/μs. Also suppose that the op amp's small-signal cutoff frequency (bandwidth) is 300 kHz, and that the input signal is at a frequency of 250 kHz.

The sine waves in Fig. 13-22 have *f* = 250 kHz. Check this for yourself by noting that 1/2 cycle takes 2 μs. Therefore, one full cycle has a period *T* = 4 μs, and

$$f = \frac{1}{T} = \frac{1}{4 \times 10^{-6} \text{ s}} = 250 \text{ kHz}$$

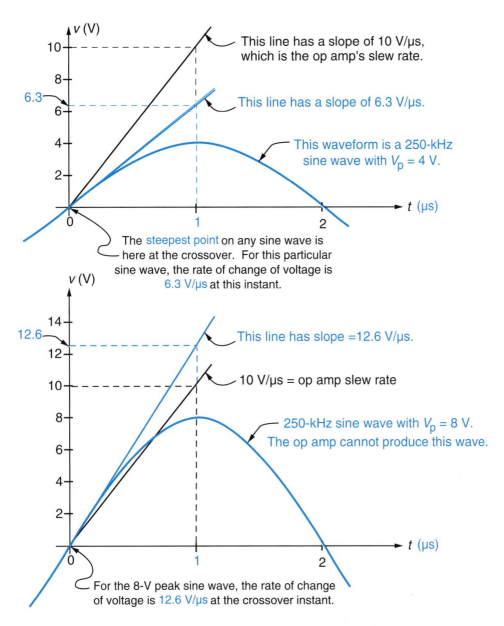

Figure 13-22

(A) Small-magnitude output voltage at a certain frequency can be handled by an op amp, because the waveform never exceeds the op amp's slew rate.

This line has a slope of 10 V/μs, which is the op amp's slew rate.

This line has a slope of 6.3 V/μs.

This waveform is a 250-kHz sine wave with V_p = 4 V.

The steepest point on any sine wave is here at the crossover. For this particular sine wave, the rate of change of voltage is 6.3 V/μs at this instant.

(B) But a larger output voltage at that same frequency cannot be handled by the op amp, because the waveform would exceed the slew rate at the crossover instant.

This line has slope =12.6 V/μs.

10 V/μs = op amp slew rate

250-kHz sine wave with V_p = 8 V. The op amp cannot produce this wave.

For the 8-V peak sine wave, the rate of change of voltage is 12.6 V/μs at the crossover instant.

Figure 13-22(A) shows the situation if the op amp's output voltage has a peak value of 4 V. It can be shown mathematically that the maximum rate of change of a sine-wave voltage is given by the formula

$$\frac{\Delta v_{OUT}}{\Delta t}\bigg|_{(maximum)} = \text{steepest slope (at crossover instant)}$$

$$= (2\pi) V_{pk} f \qquad \text{Eq. (13-19)}$$

For the sine wave in Fig. 13-22(A), Eq. (13-19) gives

$$\text{steepest slope (at crossover instant)} = 2\pi (4 \text{ V}) (250 \times 10^3 \text{ Hz})$$

$$= 6.3 \times 10^6 \text{ V/s or } 6.3 \text{ V/μs}$$

A straight line with slope = 6.3 V/μs has been drawn through the zero crossover point in Fig. 13-22(A). As the output voltage sine wave crosses through zero, its maximum (fastest) rate of change lies exactly on that straight line. The black line in Fig. 13-22(A) is the op amp's slew rate of 10 V/μs. Comparing these two straight lines makes it clear that the v_{OUT} waveform's fastest rate of change does not exceed the slew rate. Therefore the op amp can successfully produce this output waveform.

Doubling the value of V_{pk} from 4 V to 8 V causes the maximum rate to double from 6.3 V/μs to 12.6 V/μs.

The same comparison is made in Fig. 13-22(B), for a larger voltage magnitude. Now Eq. (13-3) gives

$$\text{maximum rate of change} \Big|_{\text{(at crossover instant)}} = 6.28 \, (8 \text{ V}) \, (250 \times 10^3 \text{ Hz})$$
$$= 12.6 \text{ V/μs}$$

This rate of change does exceed the op amp's 10 V/μs slew rate, as Fig. 13-22(B) clearly shows. Therefore the op amp cannot produce this output waveform, even though the 250-kHz operating frequency is within the 300-kHz small-signal bandwidth.

In summary:

The maximum operating frequency of an op amp amplifier is given by the *lower* of these two frequencies:

1 f_{high}. This is the small-signal cutoff frequency (bandwidth frequency). Its value is found from the manufacturer's data sheet. It depends on the amplifier's closed-loop voltage gain, A_v. And it also depends on the size of the external compensating capacitor, if there is one.

2 $f_{\text{(slew rate limit)}}$. This frequency is found by rearranging Eq. (13-19), to give

$$f_{\text{(slew rate limit)}} = \frac{SR}{2\pi \, V_{\text{out (pk)}}} \qquad \text{Eq. (13-20)}$$

EXAMPLE 13-6

The inverting amplifier of Fig. 13-23 uses a type 741 op amp. The amplifier has $A_v = 22$. No external compensating capacitor is used with a type 741. The manufacturer's data sheet indicates that the small-signal cutoff frequency is 45 kHz under these conditions. The 741 has a typical slew rate of 0.7 V/μs. (The 10 V/μs value in Fig. 13-22 was chosen for graphing convenience. General-purpose op amps like the 741 do not have slew rates that high.)
   a) If $V_{\text{in(pk)}} = 0.05$ V (50 mV), what is the amplifier's frequency limit?
   b) If $V_{\text{in(pk)}} = 0.25$ V, what is the amplifier's frequency limit?

SOLUTION

a) $V_{\text{out(pk)}} = A_v \, V_{\text{in(pk)}} = (22) \, (0.05 \text{ V}) = 1.1 \text{ V}$

$$f_{\text{(slew rate limit)}} = \frac{SR}{2\pi \, V_{\text{out (pk)}}}$$

$$= \frac{0.7 \text{ V}/(1 \times 10^{-6} \text{ s})}{6.28 \, (1.1 \text{ V})} = 101 \text{ kHz}$$

Since f_{high} is less than $f_{\text{(SR limit)}}$, the amplifier's frequency limit is f_{high}, **45 kHz.**

b) With the larger value of V_{in},

$$V_{out(pk)} = (22)(0.25 \text{ V}) = 5.5 \text{ V}$$

$$f_{(SR \text{ limit})} = \frac{0.7 \text{ V}/(1 \times 10^{-6} \text{ s})}{6.28(5.5 \text{ V})} = 20.3 \text{ kHz}$$

Since $f_{(SR \text{ limit})}$ is less than f_{high}, the amplifier's frequency limit is $f_{(SR \text{ limit})}$, **20.3 kHz**.

15 ✔

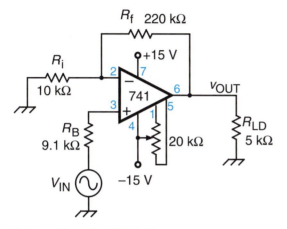

Figure 13-23

Complete op-amp inverting amplifier with offset correction. With $A_v = 22$, a type 741 op amp has $Bw = 45.5$ kHz, per the manufacturer's specification sheet.

SELF-CHECK FOR SECTION 13-7

✔

32. Explain what is meant by the offset problem for a real op amp amplifier.
33. Describe the two steps that can be taken to correct the offset problem.
34. (T-F) It is possible for some low-gain op amp amplifiers to break into high-frequency self-oscillation.
35. What step is sometimes taken to prevent the self-oscillation problem?
36. Define the term slew rate for an op amp.
37. (T-F) Slew rate limitation tends to be more of a problem for low-voltage signals than for higher-voltage signals.
38. In Fig. 13-23, suppose that the input voltage increases to $V_{in(pk)} = 0.5$ V. Everything else stays the same. Calculate the new slew rate limit frequency, $f_{(SR \text{ limit})}$.
39. In Fig. 13-23, suppose that we replace the type 741 op amp with a type 453 unit. This is a much faster op amp, with typical slew rate of 13 V/μs. Its high cutoff frequency is also higher. For $A_v = 22$, $f_{high} = 180$ kHz.
    a) If $V_{in(pk)} = 0.05$ V (50 mV), what is the amplifier's frequency limit?
    b) If $V_{in(pk)} = 0.25$ V, what is the amplifier's frequency limit?
    c) If $V_{in(pk)} = 0.55$ V, what is the amplifier's frequency limit?

A laser beam is a group of electromagnetic waves (light waves) that are all the same frequency and all in phase with one another. This is unlike natural light, in which the electromagnetic waves have differing frequencies,

all in random phase relationship to each other.

A laser being used for precise measurement of fluid flow-rate is shown on page 5 of the color section, with a second experimental laser apparatus.

Courtesy of Dantec Measurement Technology, Inc.

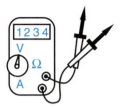

13-8 TROUBLESHOOTING OP-AMP CIRCUITS

In an op amp amplifier, there are only 5 external resistances. They are R_i, R_f, R_{LD}, R_B, and sometimes an offset-adjust pot. If something goes wrong with an op amp amplifier, it is because one of these 5 resistors is either open-circuited or shorted, or else the op amp itself has failed internally. Of course, this assumes that the $\pm V_S$ and ground power-supply connections are known to be good.

As stressed in Sec.5-7, when an amplifier has trouble, you should always check the power supply connections first. Figure 13-24 identifies the check-points.

**Figure 13-24
Check all power- supply
connections first.**

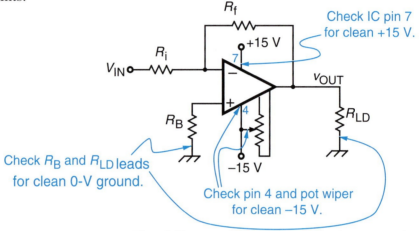

Feedback Resistor R_f: If R_f fails open in either an inverting or noninverting amplifier, the amplifier's voltage gain becomes the same as the op amp's open-loop gain $A_{v(OL)}$. Any reasonable amount of v_{IN} causes the amplifier to go to saturation. This is illustrated in Fig. 13-25, for an ac input signal to an inverting amplifier.

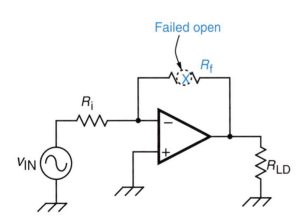

Figure 13-25 If R_f fails open, it's as if $R_f = \infty \ \Omega$. The gain formula gives $A_v = R_f/R_i = \infty \ \Omega/R_i = \infty$. Therefore v_{OUT} saturates. A noninverting amplifier would also saturate.

If R_f is shorted, the output is connected to the –IN terminal. For an inverting amplifier, shown in Fig. 13-26(A), the +IN terminal is referenced to ground, so the –IN terminal is at virtual ground. Therefore v_{OUT} is also frozen at 0 V. However, this circuit might break into high-frequency self-oscillation because of the strong feedback through the 0-Ω short. Then the v_{OUT} oscillations will be low-magnitude, a few millivolts.

For a noninverting amplifier, shown in Fig. 13-26(B), a shorted R_f turns the circuit into a voltage-follower, $A_v = 1$.

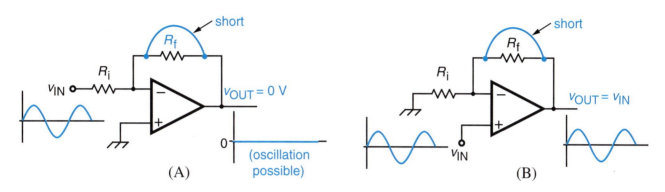

Figure 13-26 Effects of shorted R_f.   (A) Inverting amplifier: $A_v = R_f/R_i = 0\ \Omega/R_i = 0$. (B) Noninverting amplifier: $A_v = (R_f/R_i) + 1 = (0\ \Omega/R_i) + 1 = 1$.

Input Resistor R_i: If R_i fails open in an inverting amplifier, the input source is out of the picture. With R_i disconnected, R_f is placed in series with the op amp's internal resistance, $R_{IN(OL)}$, as Fig. 13-27(A) shows. This makes the output virtually connected to the –IN terminal because R_f is negligible compared to the several million-ohm $R_{IN(OL)}$. This is similar to the shorted-R_f situation shown in Fig. 13-26(A). Therefore the output is frozen at 0 V. However, this circuit is also liable to self-oscillation.

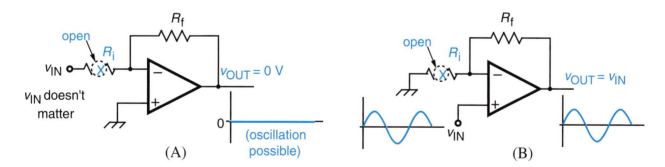

Figure 13-27 Effects of open R_i.   (A) Inverting amplifier: $A_v = R_f/R_i = R_f/\infty\ \Omega = 0$. (B) Noninverting amplifier: $A_v = (R_f/R_i) + 1 = (R_f/\infty\ \Omega) + 1 = 0 + 1 = 1$.

If R_i fails open in a noninverting amplifier, R_f is again placed in series with internal resistance $R_{IN(OL)}$. Since R_f is likely to be negligible compared to $R_{IN(OL)}$, the circuit becomes essentially a voltage-follower. We get $A_v = 1$, and v_{OUT} is a duplicate of v_{IN}, as shown in Fig. 13-27(B).

If R_i is short-circuited in either amplifier, the entire input voltage v_{IN} appears across the differential input terminals. Thus v_{IN} gets amplified by the op amp's $A_{v(OL)}$. This drives the output to saturation, like the waveforms shown in Fig. 13-25.

Bias Resistor R_B: If R_B fails open in a noninverting amplifier, input voltage v_{IN} is disconnected, as Fig. 13-28 shows. Any noise voltage that is capacitively coupled onto the +IN lead simply gets amplified by the factor $A_v = R_f /R_i + 1$.

Also, with R_B failed, the output offset problem can reappear. Therefore the output is an amplified noise waveform superimposed on a nonzero dc level. This is indicated in the v_{OUT} waveform sketch of Fig. 13-28.

Figure 13-28
Effects of open R_B.
Noninverting amplifier amplifies the noise and may offset from 0 V.

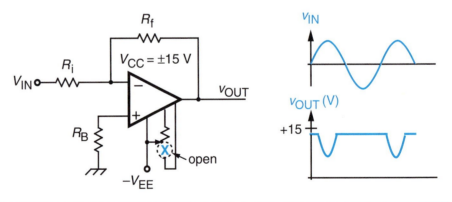

If R_B fails open in an inverting amplifier, a similar thing happens. The output amplified noise waveform may show a tendency to track the v_{IN} waveform.

Offset-Adjust Pot: If the offset-adjust pot fails in any way, the output offset problem reappears. It may be worse than if the pot had never been installed in the first place. Figure 13-29 shows a possible consequence.

Figure 13-29
Failed offset-adjust pot.

This man is using a portable gas chromatograph to check for the presence of hazardous materials. The chromatograph separates a gaseous mixture into its chemical components, then electronically identifies any dangerous components and measures their concentrations. It is useful for ensuring workplace safety and checking compliance with pollution laws.

Courtesy of Microsensor Technology, Inc.

Load Resistance R_{LD}: If R_{LD} opens, the amplifier doesn't even notice. R_{LD} is not necessary for the proper operation of an op amp, as we know.

If R_{LD} is shorted, the output terminal is shorted to ground, and $v_{OUT} = 0$ V.

Table 13-2 summarizes the amplifier symptoms resulting from the possible external resistor faults.

OUTPUT SYMPTOM	POSSIBLE CAUSES	
	Inverting Amplifier	Noninverting Amplifier
Saturated V_{out}	R_f is open R_i is shorted	R_f is open R_i is shorted
Zero V_{out} (with possible self-oscillation)	R_f is shorted R_i is open R_{LD} is shorted	Can't be caused by external resistor fault. Problem must be dc supply or internal IC failure.
$V_{out} = V_{in}$	Can't be caused by a single external resistor fault.	R_f is shorted R_i is open
V_{out} is noisy (either centered on 0 V or containing dc offset).	R_B is open	R_B is open
V_{out} is clean, but dc offset is present	R_B is shorted. Offset pot is open or shorted.	R_B is shorted. Offset pot is open or shorted.

Table 13-2
Trouble in op-amp amplifiers.

All 5 external resistors can be checked easily with an ohmmeter, without disconnecting them from the circuit. Just turn off the power, disconnect the signal source, and start measuring resistances.

IC Connections: If the op amp is soldered into a printed circuit board, it is a good idea to check the resistance between every IC pin and its copper track. As shown in Fig. 13-30, place one ohmmeter lead right on the IC pin and the other lead on the copper track, a little distance away from the solder joint. You expect every resistance measurement to be near 0 Ω, if all the solder joints are OK.

If the op amp is mounted in an IC socket, do the same kind of resistance test between every IC pin and its socket pin. This may reveal an IC pin that is not inserted into the socket, because it has been bent underneath the IC package. Another possibility is that the metal surfaces of the socket and/or the IC pin may have become oxidized, and are not making proper electrical contact. This problem can usually be fixed by spraying each socket hole with contact cleaner and reaming it several times with a newly stripped piece of AWG #24 solid wire.

If every external check is OK, then the problem must be internal in the op amp. A failed IC cannot be repaired — you replace it.

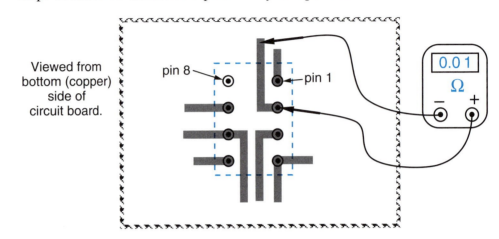

Figure 13-30 Checking for bad solder joints with an ohmmeter. Here the ohmmeter leads are placed onpin 2 of the IC (or socket) and the copper track that is supposed to be soldered to pin 2. If the solder joint were bad, we would measure a nonzero resistance value.

SELF-CHECK FOR SECTION 13-8

40. An inverting amplifier is designed to have $A_v = 5$. However, with $V_{in} = 1$ V p-p, the output is a square-looking waveform switching between positive and negative saturation. What is the likely cause?

41. An inverting amplifier is designed to have $A_v = 5$. However, with $V_{in(p-p)} = 1$ V, the output terminal has low-magnitude high-frequency oscillation, centered on 0 V. What are two possible causes for this trouble?

42. A noninverting amplifier is designed to have $A_v = 6$. However, with $V_{in(p-p)} = 1$ V, the output is also a 1-V p-p waveform, in-phase with V_{in}. Give two possible explanations for this.

43. A noninverting amplifier is designed to have $A_v = 6$. However, with $V_{in(p-p)} = 1$ V, the output has an unstable waveform with a fundamental frequency of 60 Hz. Give two possible explanations for this.

13-9 SOME OTHER OP-AMP APPLICATIONS

Summing Amplifier

One common op application is the **summing amplifier**. This circuit receives two or more input signals and gives an output this is proportional to the sum of the input voltages. Figure 13-31(A) shows a summing amplifier with three dc inputs, V_1, V_2, and V_3.

We can understand the operation of the summing amplifier by using the virtual ground idea again. With the –IN terminal at virtual ground (0 V), and all three input voltages positive, there are three currents leaving the –IN node flowing from right to left. They are

$$I_1 = \frac{V_1}{R_1}, \quad I_2 = \frac{V_2}{R_2}, \quad \text{and } I_3 = \frac{V_3}{R_3}$$

By Kirchhoff's current law, these three dc currents must be balanced by a single dc current entering the –IN node through R_f. This current is given by Ohms law applied to R_f,

$$I_f = \frac{-V_{OUT}}{R_f}$$

Therefore $I_f = I_1 + I_2 + I_3$

$$\frac{-V_{OUT}}{R_f} = \frac{V_1}{R_1} + \frac{V_2}{R_2} + \frac{V_3}{R_3} \qquad \text{Eq. (13-21)}$$

Rearranging Eq. (13-21), we get

$$-V_{OUT} = R_f \left(\frac{V_1}{R_1} + \frac{V_2}{R_2} + \frac{V_3}{R_3} \right) \qquad \text{Eq. (13-22)}$$

If R_1, R_2, and R_3 are all equal to each other, symbolized R_i, then Eq. (13-22) simplifies to

$$-V_{OUT} = \frac{R_f}{R_i} \left(V_1 + V_2 + V_3 \right) \qquad \text{Eq. (13-23)}$$

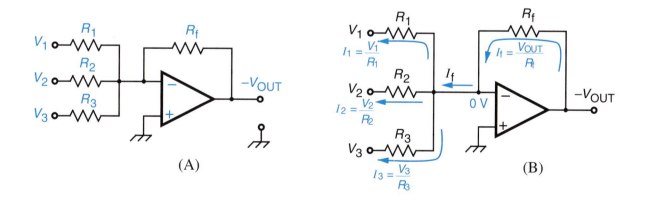

Figure 13-31 Op amp summing circuit. (A) Basic schematic. (B) Virtual ground and current balance at –IN.

The minus sign in Eq. (13-23) means that V_{OUT} has a negative polarity if the sum of $V_1 + V_2 + V_3$ is positive. On the other hand, if negative dc input values are present, the sum of $V_1 + V_2 + V_3$ could be negative. Then V_{OUT} would be positive.

EXAMPLE 13-7

For the summing amplifier of Fig. 13-32:
a) Suppose $V_1 = +2$ V, $V_2 = +1.4$ V, and $V_3 = +2.5$ V. Find V_{OUT}.
b) Suppose V_1 and V_2 remain the same, but V_3 changes to -1.2 V. Find V_{OUT}.
c) Suppose V_3 changes to -4.5 V. Find V_{OUT}.

SOLUTION

a) From Eq. (13-23),

$$-V_{OUT} = \frac{R_f}{R_i}\left(V_1 + V_2 + V_3\right)$$

$$-V_{OUT} = \frac{20\text{ k}\Omega}{10\text{ k}\Omega}\left(2\text{ V} + 1.4\text{ V} + 2.5\text{ V}\right)$$

$$V_{OUT} = -2\,(5.9\text{ V}) = \textbf{–11.8 V}$$

b)
$$-V_{OUT} = 2\,(2\text{ V} + 1.4\text{ V} - 1.2\text{ V})$$

$$V_{OUT} = -2\,(2.2\text{ V}) = \textbf{–4.4 V}$$

⟨18⟩ ✓

c)
$$-V_{OUT} = 2\,(2\text{ V} + 1.4\text{ V} - 4.5\text{V}) = -2\,(-1.1\text{ V}) = \textbf{+2.2 V}$$

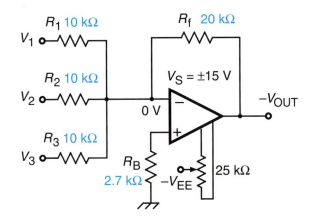

Figure 13-32
Complete summing amplifier with gain of 2.
~~Bias resistor R_B is calculated from $R_B = R_1 \| R_2 \| R_3 \| R_f$.~~

Integrator

An op-amp **integrator** is a circuit whose output voltage is dependent on two things:

1. The magnitude of the input voltage V_{IN}.
2. The amount of *time* that the input voltage has been present.

The basic integrator schematic is shown in Fig. 13-33(A). The main feature is that the feedback resistor is replaced by a feedback capacitor, C_f. Here is how the integrator works.

⟨19⟩ ✓

● When a constant-value dc input voltage V_{IN} is applied, it produces a current I_{Ri} through the input resistor, by the usual Ohm's law relation

A careful description of how an integrator works.

$$I_{Ri} = \frac{V_{IN} - 0\text{ V}}{R_i} \longleftarrow \text{virtual ground at } -\text{IN}$$

$$= \frac{V_{IN}}{R_i} \qquad\qquad \text{Eq. (13-24)}$$

● An equal-value dc current I_f must flow through the feedback capacitor C_f, because Kirchhoff's current law must balance at the –IN node. Therefore, in Fig. 13-33(A),

$$I_f = I_{Ri} \qquad \text{Eq. (13-25)}$$

● The constant dc current I_f causes charge Q_{Cf} to accumulate on the plates of capacitor C_f, as indicated in Fig. 13-33(B). At any instant in time, the overall charge that has accumulated is given by

$$Q_{Cf} = (I_f)\, t \qquad \text{Eq. (13-26)}$$

which can be rewritten as

$$I_f = \frac{Q_{Cf}}{t} \qquad \text{Eq. (13-27)}$$

where t in Eq. (13-27) stands for the amount of time that I_f has been flowing, which is the same as the amount of time that V_{IN} has been present at the input. Equations (13-27) and (13-26) are nothing more than the standard relationships among charge, current, and time, namely $I = Q/t$ or $Q = It$.

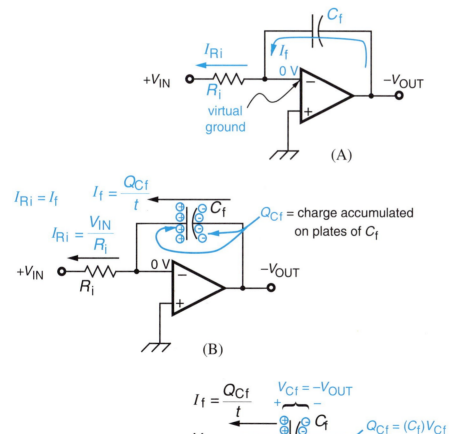

Figure 13-33 Explaining the operation of an op-amp integrator.

(A) Basic schematic, showing –IN at virtual ground. The feedback path has a capacitor, not a resistor.

(B) As usual, the feedback current I_f must be the same as the input current, I_{Ri}. But now I_f is given by a charge / time (Q/t) equation, instead of an Ohm's law equation.

(C) The capacitor voltage V_{Cf} is the same as the op amp's output voltage, $-V_{OUT}$. The thing that finally ties together V_{OUT} with V_{IN} is the charge / voltage (Q/V) equation for a capacitor.

● Accumulated charge Q_{Cf} is related to the capacitance and the capacitor voltage V_{Cf} by the standard capacitance formula

$$Q_{Cf} = (C_f)(V_{Cf})$$ Eq. (13-28)

But V_{Cf} is the same as V_{OUT}, because the left side of C_f is connected to the 0-V virtual ground, as Fig. 13-33(C) makes clear. Therefore, Eq. (13-28) can be written as

$$Q_{Cf} = (C_f)(-V_{OUT})$$ Eq. (13-29)

● By plugging the Ohm's law equation, Eq. (13-24) and the charge equation, Eq. (13-27), into the current-balance equation, Eq. (13-25), we get

$$I_{Ri} = I_f$$

$$\underset{\text{Eq. (13-24)}}{\frac{V_{IN}}{R_i}} = \underset{\text{Eq. (13-27)}}{\frac{Q_{Cf}}{t}}$$ Eq. (13-30)

Then plugging the capacitance equation, Eq. (13-29), into Eq. (13-30) gives us

$$\frac{V_{IN}}{R_i} = \frac{(C_f)(-V_{OUT})}{t} \quad \leftarrow \text{Eq. (13-29)}$$ Eq. (13-31)

Rearranging Eq. (13-31) gives the final integrator equation

$$\boxed{-v_{OUT} = \left(\frac{1}{R_i\,C_f}\right)(V_{IN})\,t}$$ Eq. (13-32)

in which $-v_{OUT}$ has been written with a lower-case v to signify that it is not steady dc, but changes continuously as time t passes.

In words, Eq. (13-32) tells us that

For an op-amp integrator, v_{OUT} is proportional to V_{IN} and is proportional to the time t that V_{IN} has been present. The proportionality factor is $\frac{1}{R_i\,C_f}$.

Summarizing the overall operation of an integrator.

EXAMPLE 13-8

The op-amp integrator of Fig. 13-34(A) receives a +1-V input at time $t = 0$.

a) Write the equation that predicts output voltage v_{OUT} in terms of input voltage, V_{IN}, and elapsed time, t.

b) Draw detailed waveform graphs of V_{IN} and v_{OUT}.

SOLUTION

a) Applying Eq. 13-32, we get

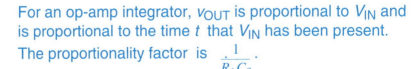

$$-v_{OUT} = \left(\frac{1}{R_i\,C_f}\right)(V_{IN})\,t$$

$$= \left(\frac{1}{(200 \times 10^3 \, \Omega)(10 \times 10^{-6} \, \text{F})} \right) (V_{\text{IN}}) \, t$$

$$= \left(\frac{1}{2 \times 10^0} \right) (V_{\text{IN}}) \, t$$

or $-v_{\text{OUT}} = 0.5 \, (V_{\text{IN}}) \, t$

b) With $v_{\text{IN}} = +1$ V, the result from part (a) becomes

$$-v_{\text{OUT}} = (0.5) \, (1) \, t = 0.5 \, t$$

In words, the output voltage goes negative 1/2 volt during every 1 second of time that passes. That is, v_{OUT} is a negative-going straight-line graph, called a **ramp**, changing at a rate of $\frac{1}{2}$-volt per second. This ramp result is drawn in Fig. 13-34(B). Of course, the ramp must stop when it hits the op amp's saturation voltage, assumed to be -12 V in this example.

With a steady dc input, an integrator produces an output ramp.

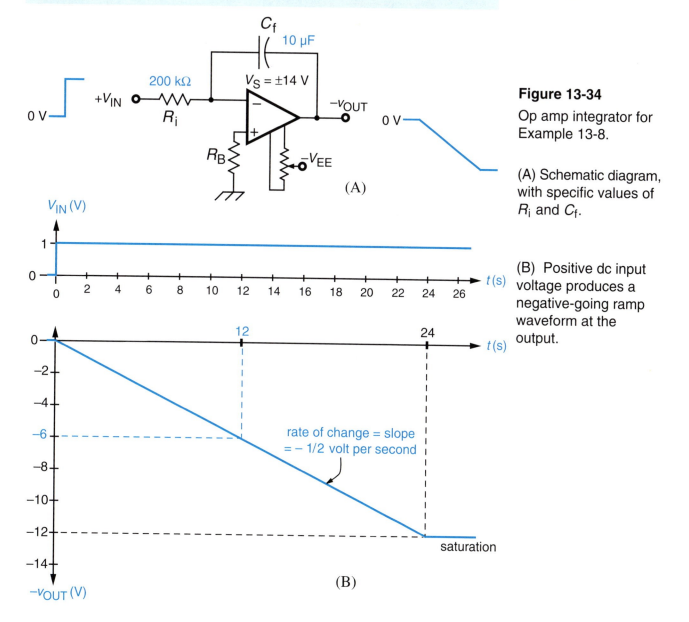

Figure 13-34

Op amp integrator for Example 13-8.

(A) Schematic diagram, with specific values of R_i and C_f.

(B) Positive dc input voltage produces a negative-going ramp waveform at the output.

Voltage-to-Frequency Converter

Op amp integrators have many applications. They are used in digital measuring instruments, signal generators, and industrial process-control circuitry. A digital measurement application example is shown in Fig. 13-35(A). This circuit combines two op amps, one used as an integrator and one used as a comparer. The purpose of the circuit is to produce a train of output pulses at a frequency that is proportional to the input voltage V_{IN}. That is,

$$f_{OUT} = k_P V_{IN} \qquad \text{Eq. (13-33)}$$

k_P is the constant number that is multiplied by input voltage (in volts) to give the output frequency (in hertz).

where k_P is the proportionality constant between input voltage and output frequency.

A separate digital counting circuit (not shown in Fig. 13-35) counts the number of pulses that occur in a fixed amount of time, say 1 second. That counted number of pulses then represents the magnitude of input voltage V_{IN}. In turn, V_{IN} could represent the value of some physical variable, like temperature. Thus, the overall system becomes a very accurate temperature-measuring system, or thermometer. This basic idea of converting an input voltage to an output frequency is useful in many areas.

Voltage-to-frequency conversion idea.

Here is how the circuit of Fig. 13-35 works. With a positive V_{IN}, the op amp$_1$ integrator will be ramping in the negative direction. For purposes of explanation, let us assume that the op amp$_2$ comparer is in negative saturation, with $V_{OUT(2)} = -10$ V. Then the R_5-R_6 voltage divider has -10 V applied to the top terminal of R_5. That voltage is halved by the voltage divider, producing -5 V as the comparer's reference voltage. That is, $V_{ref} = -5$ V. As long as the integrator output, $v_{OUT(1)}$, is not more negative than -5 V, the comparer remains in negative saturation. This is the state of affairs as we begin our explanation of the circuit.

As $v_{OUT(1)}$ ramps downward toward -5 V, the comparer keeps $v_{OUT(2)}$ at -10 V.

The -10-V saturation voltage is also applied to the top of the R_7-R_8 voltage divider in Fig. 13-35(A). With a 2-V difference across that divider, R_7 drops 1 V and R_8 drops 1 V. The $+1$-V voltage V_{R8} is applied through R_9 to the gate of Q_1. That value of voltage is below the threshold value $V_{(th)}$ of the E-type MOSFET. Therefore Q_1 remains cut off, which means that the 330-Ω R_{10} path is an open circuit. Therefore that R_{10} path has no effect on op amp$_1$ at this time.

$v_{OUT(2)}$ at -10 V results in Q_1 being cut off.

Eventually, the $v_{OUT(1)}$ negative ramp will reach -5 V. When it goes slightly below -5 V, the op amp$_2$ comparator sees a differential input voltage that is more negative on its $-IN$ terminal relative to its $+IN$ terminal. Therefore it switches to positive saturation, $+10$ V. This action is shown by the waveforms of Fig. 13-35(B).

When the downward ramp hits -5 V, the circuit switches.

Now the R_5-R_6 voltage divider produces a positive reference voltage, namely $V_{ref} = +5$ V. This sudden change in V_{ref} reinforces the positive saturation of op amp$_2$.

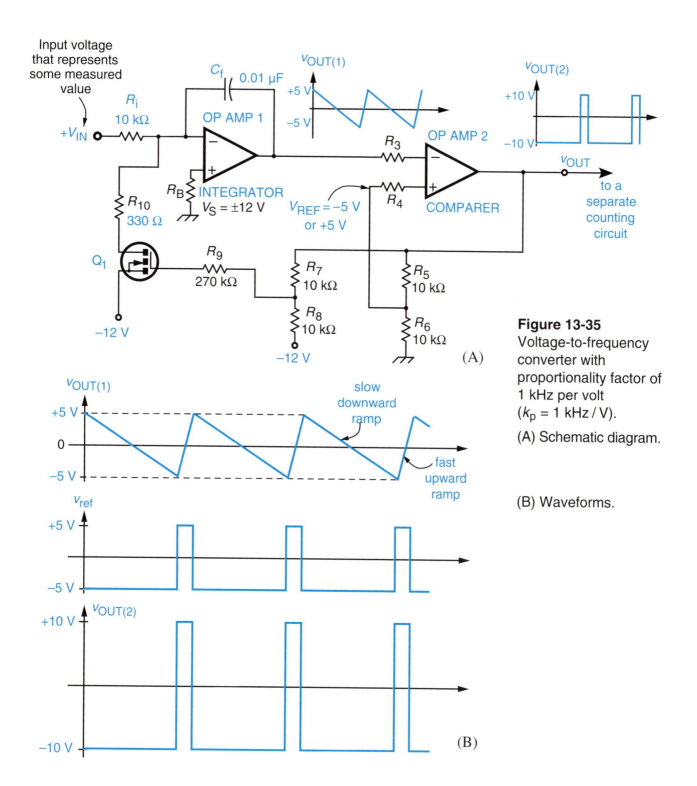

Figure 13-35
Voltage-to-frequency converter with proportionality factor of 1 kHz per volt ($k_p = 1$ kHz / V).

(A) Schematic diagram.

(B) Waveforms.

The +10-V $v_{OUT(2)}$ value is applied to the R_7-R_8 voltage divider. That places 22 V across the divider, so R_8 drops an amount given by

$$V_{R8} = \left(\frac{10\ k\Omega}{10\ k\Omega + 10\ k\Omega}\right) 22\ V = \left(\frac{1}{2}\right) 22\ V = 11\ V$$

This 11 volts is applied to the gate circuit of transistor Q_1. It much exceeds the threshold voltage, driving Q_1 hard into saturation. Taking into account the FET's V_{DS} saturation voltage of about 3 V, this leaves about –9 V connected to the bottom of 330-Ω resistor R_{10}.

Now, with $v_{OUT(2)} = +10$ V, Q_1 is driven into saturation.

Now look at the input situation to the integrator. It sees a relatively large negative voltage, –9 V, fed through a small resistance, 330 Ω; this is combined with a small positive voltage V_{IN} fed through a much larger resistance, 10 000 Ω. The negative input overwhelms the positive input. It is as if the integrator were seeing the new input circuit drawn in Fig. 13-36.

The Q_1 switching causes the integrator to reverse its ramping, to upward.

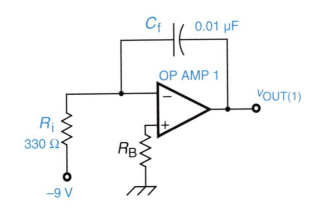

Figure 13-36
When the comparer output goes to +10 V, it turns ON transistor Q_1, which overwhelms the input of op amp 1. It is as if R_i = 330 Ω with V_{IN} = –9 V.

In the circuit of Fig. 13-36, the very small R_i value and the large-magnitude input voltage cause the integrator to produce a very rapid positive ramp. Thus, $v_{OUT(1)}$ begins ramping steeply upward in Fig. 13-35(B). In a very short time, $v_{OUT(1)}$ ramps up to +5 V, which is now the V_{ref} value for the comparer. When $v_{OUT(1)}$ rises slightly above +5 V, the comparer's differential input polarity reverses. Therefore it switches back to negative saturation, with $v_{OUT(2)} = -10$ V, as graphed in Fig. 13-35(B). Now V_{ref} returns to –5 V, and Q_1 returns to cutoff condition. The circuit has returned to where our discussion began, and the negative-going integrator ramp starts all over again. This overall circuit action repeats indefinitely, producing the output pulse train that is counted by the counting circuit.

The upward ramp is very fast.

When the $v_{OUT(2)}$ upward ramp hits +5 V, the circuit switches back.

We can find the voltage-to-frequency conversion proportionality constant by assuming $V_{IN} = +1$ V. Then, from Eq. (13-32), we have

$$-v_{OUT} = \left(\frac{1}{R_i\, C_f}\right) (V_{IN})\, t$$

$$= \left(\frac{1}{(10 \times 10^3\ \Omega)(0.01 \times 10^{-6}\ F)}\right) (1\ V)\, t$$

$$-v_{OUT(1)} = (10 \times 10^3)\, (1\ V)\, t$$

The integrator must go from +5 V to –5 V, which is a 10-V change. Therefore the elapsed time for the downward ramp is found by

$$10\ V = (10 \times 10^3)\, (1\ V)\, t$$

which we rearrange to

$$t = \frac{10\ V / (1\ V)}{10 \times 10^3} = 1 \times 10^{-3}\ s \quad \text{or} \quad 1\ ms$$

The upward ramp is so fast that its elapsed time is negligible. So an entire cycle takes 1 ms. Therefore the output frequency is

$$f = \frac{1}{T} = \frac{1}{1\text{ ms}} = 1\text{ kHz}$$

The upward ramp is actually much faster than Fig. 13-35(B) suggests.

We conclude that the frequency-converter's proportionality constant is 1 kilohertz per volt (1 kHz/V).

If V_{IN} were increased to 2 V, the downward integration rate would be twice as fast. By Eq. (13-32),

$$-v_{OUT} = \left(\frac{1}{R_i\,C_f}\right)2\text{ V})\,t = (10 \times 10^3)\,(2\text{ V})\,t$$

$$t = \frac{10\text{ V}/(2\text{ V})}{10 \times 10^3} = 0.5 \times 10^{-3}\text{ s}\quad\text{or}\quad 0.5\text{ ms}$$

$$f = \frac{1}{T} = \frac{1}{0.5\text{ ms}} = 2\text{ kHz}$$

Doubling V_{IN} would cause the downward ramp to occur in half the time. Therefore frequency would be doubled.

Differentiator

An op-amp **differentiator** uses a resistor and a capacitor, but their positions are reversed compared to an integrator. A basic differentiator circuit is shown in Fig. 13-37.

> A differentiator produces an output voltage v_{OUT} that is proportional to *how quickly* the input voltage is changing.

A thorough explanation of differentiator operation would involve the same basic ideas that were used to explain integrator operation. We will not go through the explanation for a differentiator.

The waveforms of Fig. 13-37(B) illustrate the differentiator idea.

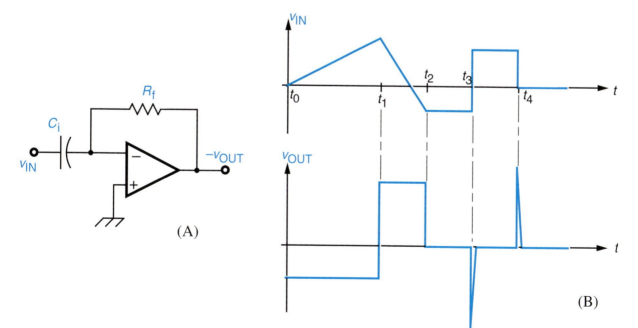

Figure 13-37 Op-amp differentiator. (A) Basic schematic. The feedback component is a resistor, but the input component is now a capacitor, C_i. (B) Input/output voltage waveforms for a differentiator. The magnitude of v_{OUT} depends on the rate of change of v_{IN}.

● During the time interval from t_0 to t_1, v_{IN} is changing at a moderate rate. Therefore v_{OUT} has a moderate value. v_{OUT} is negative, because of the op amp's inversion from –IN to output.

● During the interval from t_1 to t_2, v_{IN} is changing at a more rapid rate. Therefore v_{OUT} has a larger value. v_{OUT} is positive because the op amp inverts the negative-going input ramp.

● During the interval from t_2 to t_3, v_{IN} is not changing at all. It is steady. Therefore v_{OUT} is zero.

● At the t_3 instant, v_{IN} makes a very fast change in the positive direction. Therefore v_{OUT} goes to a large-magnitude negative value, probably its saturated value.

● From t_3 to t_4, v_{IN} does not change, so v_{OUT} again is zero.

● At the t_4 instant, v_{IN} makes a very rapid change in the negative direction. Therefore v_{OUT} is a short-lived positive pulse, probably $+V_{sat}$.

Single-supply Operation

It is not absolutely necessary to have a dual-polarity dc supply for powering an op amp. The single dc supply arrangement of Fig. 13-38(A) will also work successfully. The R_3-R_4 voltage-divider holds the +IN terminal at +6 V dc, but input capacitor C_{in} also charges to +6 V dc, so the dc component of $v_{IN(dif)}$ is still 0 V. The ac input V_{in} is amplified by $A_{v(CL)}$, producing an ac output waveform that is centered on +6 V dc. This is shown in the output waveform sketch of Fig. 13-38. However, with the load coupled through capacitor C_{out}, the final V_{ld} waveform is restored to 0-V ground reference.

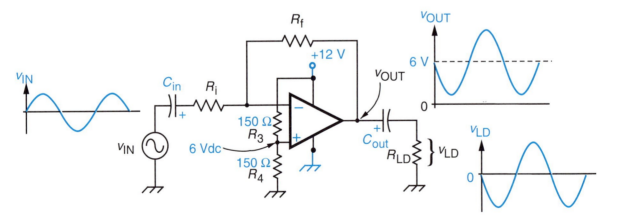

Figure 13-38 An op amp can be operated from a single dc supply, but coupling capacitors are then needed. This sacrifices the dc and low-frequency capability of the amplifier.

SELF-CHECK FOR SECTION 13-9

44. A summing amplifier like Fig. 13-32 has $R_1 = 10$ kΩ, $R_2 = 10$ kΩ, $R_3 = 10$ kΩ, and $R_f = 50$ kΩ. The input voltages are $V_1 = -1.5$ V, $V_2 = +2.6$ V, $V_3 = -0.7$ V. Find V_{OUT}.

45. In Problem 44, if V_3 changes to –3.1 V, find V_{OUT}.

46. If V_3 changes to –5.0 V, find V_{OUT}.

47. Draw the schematic of an inverting amplifier configuration with a resistor as the input component and a capacitor as the feedback component. This circuit is called a(n) _____ .
48. Reverse the positions of the resistor and capacitor in your schematic from Problem 47. Now the circuit is called a(n) _____ .
49. Of the two preceding circuits, which one produces a ramp output waveform if the input voltage is a steady dc value?

Op Amp Active Filters

You have studied the frequency-filtering idea in your basic electricity course. Recall that there are four basic filtering configurations: low-pass, high-pass, band-pass, and band-stop. We have seen filter circuits constructed with resistor-capacitor (*RC*) combinations, resistor-inductor (*RL*) combinations, and resistor-inductor-capacitor (*RLC*) combinations. All such filters are called **passive filters**, because resistors, inductors and capacitors are passive devices.

All four kinds of frequency discrimination can be achieved by op amps combined with external resistor/capacitor combinations. Such filters are called **active filters**, because the op amp, being an amplifier, is an active device. Often active filters can be designed to have better selectivity than passive filters. This means that their frequency-response curves are steeper, giving the filter a finer ability to discriminate between frequencies that are close to one another.

Active filters use op amps and *RC* networks. They don't contain inductors.

In an advanced analog electronics course, you will study the construction details and response of various active filters.

FORMULAS

For an inverting amplifier:

$$A_{v\,(CL)} = A_v = \frac{R_f}{R_i}$$ Eq. (13-3)

$$R_{in\,(CL)} = R_{in} = R_i$$ Eq. (13-4)

For a noninverting amplifier:

$$A_v = \frac{R_f}{R_i} + 1$$ Eq. (13-9)

$$R_{in} \approx \infty \ \Omega$$ Eq. (13-10)

For a voltage-follower:

$$A_v = 1$$

$$\text{slew rate } (SR) \;=\; \frac{\Delta v}{\Delta t}\bigg|_{\text{(maximum allowable)}} \qquad \text{Eq. (13-18)}$$

$$f_{\text{(slew rate limit)}} \;=\; \frac{SR}{2\pi\, V_{\text{out(pk)}}} \qquad \text{Eq. (13-20)}$$

For a summing amplifier:

$$-V_{\text{OUT}} \;=\; \frac{R_{\text{f}}}{R_{\text{i}}}\Big(V_1 + V_2 + V_3 + \ldots\Big) \qquad \text{Eq. (13-23)}$$

For an integrator (with dc input):

$$-v_{\text{OUT}} \;=\; \Big(\frac{1}{R_{\text{i}}\,C_{\text{f}}}\Big)(V_{\text{IN}})\,t \qquad \text{Eq. (13-32)}$$

SUMMARY

- An op amp is a complete amplifier circuit integrated on a small chip of silicon. All you have to do is connect the dc power supplies, the input source, and the load.
- The inputs to an op amp are called inverting (marked −) and noninverting (marked +). In the schematic symbol, we always show the − input on top and the + input on bottom.
- An op amp has differential inputs. This means that neither input needs to be connected to circuit ground. The op amp amplifies the difference between the two voltages appearing at +IN and −IN.
- When the actual differential input voltage polarity matches the schematic marks, then v_{OUT} goes positive with respect to ground. When the actual input polarity is the opposite of the schematic marks, v_{OUT} goes negative relative to ground.
- For a plain op-amp amplifier, the op amp's voltage gain is so great ($A_v > 10\,000$) that even a millivolt input signal can drive the output to saturation. For this reason, plain (open-loop) op-amp amplifiers are never used for linear amplification of a signal waveform.
- An op amp's ability to suddenly switch from one saturated polarity to the opposite saturated polarity makes it useful as a voltage-comparer.
- To understand the working of an op-amp inverting amplifier, we rely on the idea of virtual ground at the − input terminal.
- A linear op-amp amplifier uses an external feedback resistor to close the loop.
- To understand the working of an op-amp noninverting amplifier, we rely on the idea of the difference voltage, $v_{\text{IN(dif)}}$, being virtually zero.
- Closed-loop op-amp amplifiers, both inverting and noninverting, have their voltage gain set by two external resistors, R_{f} and R_{i}.

● The voltage follower is a noninverting amplifier with zero ohms in the feedback path ($R_f = 0 \, \Omega$). It has $A_v = 1$. It takes virtually no current from the signal source.

● The op-amp offset problem may cause v_{OUT} to be a nonzero dc value when $V_{IN} = 0$ V. The problem can be corrected by installing resistors at certain locations in the circuit.

● An op amp's slew rate is the maximum rate at which it can change its output voltage.

● Because of the slew-rate limitation, the maximum frequency of an op-amp amplifier may be less than its small-signal cutoff (bandwidth) frequency.

● If an op-amp amplifier malfunctions, there are only a small number of possible problems with the external resistors. Each possible problem produces definite symptoms.

● A summing amplifier produces an output voltage V_{OUT} that is equal to the algebraic sum of several input voltages, multiplied by A_v.

● An integrator is an-op amp circuit whose output voltage v_{OUT} depends on the magnitude of V_{IN}, and on the amount of time that V_{IN} has been present.

● For an integrator, if V_{IN} is a steady dc value, V_{OUT} is a ramp of the opposite polarity. That is, a positive V_{IN} produces a negative-going ramp.

● Integrators are often used in measurement and control applications.

● An op-amp differentiator is a circuit whose output voltage v_{OUT} depends on how quickly the input voltage is changing.

● Active filters use op amps combined with external resistors and capacitors to discriminate among different frequencies — the same function performed by passive filters.

● The four basic filtering operations of low-pass, high-pass, band-pass, and band-stop can be produced by active filters, often with better selectivity and less signal loss than for passive filters.

QUESTIONS AND PROBLEMS

1. Draw the schematic symbol for an op amp, labeling the 5 main terminals. Show how the dc power supplies are connected and how the load is connected.

2. (T-F) For an op amp, it is a necessary requirement that one of the two input terminals, either –IN or +IN, be directly connected to ground.

3. An op amp's output voltage v_{OUT} goes positive when the +IN terminal is more _____ than the –IN terminal.

4. An op amp's output voltage v_{OUT} goes negative when the +IN terminal is more _____ than the –IN terminal.

5. We can summarize the answers to Questions 3 and 4 by saying that v_{OUT} goes _____ when the actual differential input voltage polarity matches the terminal schematic markings; and v_{OUT} goes _____ when the actual differential input polarity is opposite to the schematic markings.

6. The two common physical appearances of op amps are the _____ package and the _____ package.

7. (T-F) A plain IC op amp has a moderate voltage gain, $A_{v(OL)}$. It is usually between 10 and 50.

8. (T-F) A plain IC op amp has a moderate input resistance, $R_{IN(OL)}$. It is usually between 500 Ω and 10 kΩ.

9. If an op amp is powered by a ±15 V dual-polarity dc supply, the saturation voltages will be approximately ± _____ V.

The op amp comparators in Fig. 13-6 showed the reference voltage V_{REF} and the input voltage v_{IN} connected directly to the –IN and +IN terminals. Actually, v_{IN} and V_{REF} are applied through buffering resistors, as shown in Fig. 13-35 (resistors R_3 and R_4). Figure 13-46 also shows buffering resistors. In all those circuits, assume that V_S = ±15 V, giving a saturation value of ±13 V. Also assume that v_{IN} can vary between –15 V and +15 V.

10. In the comparator of Fig. 13-39(A), what range of v_{IN} causes v_{OUT} to go to +13 V? What range of v_{IN} causes v_{OUT} = –13 V?

11. In Fig. 13-39(B), suppose V_{REF} is adjusted to +4 V.
    a) What is v_{OUT} for v_{IN} = –6 V?
    b) What is v_{OUT} for v_{IN} = +2 V?
    c) What is v_{OUT} for v_{IN} = +7 V?

12. In Fig. 13-39(C), suppose V_{REF} is adjusted to +6 V.
    a) What is v_{OUT} for v_{IN} = –7 V?
    b) What is v_{OUT} for v_{IN} = –5 V?
    c) What is v_{OUT} for v_{IN} = +5 V?
    d) What is v_{OUT} for v_{IN} = +7 V?

13. In Fig. 13-39(C), suppose V_{REF} is adjusted to –6 V.
    a) What is v_{OUT} for v_{IN} = –7 V?
    b) What is v_{OUT} for v_{IN} = –5 V?
    c) What is v_{OUT} for v_{IN} = +5 V?
    d) What is v_{OUT} for v_{IN} = +7 V?

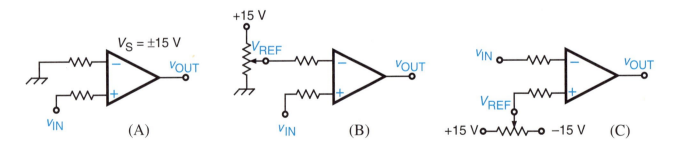

Figure 13-39 For Problems 10 through 13.

14. Draw the schematic diagram for an op amp inverting amplifier, and label the external resistors.

15. Explain the distinction between closed-loop voltage gain, $A_{v(CL)}$, and open-loop voltage gain, $A_{v(OL)}$. Refer to your schematic from Problem 14.

16. In your inverting amplifier, suppose $V_S = \pm 12$ V, $R_i = 4.7$ kΩ, $R_f = 33$ kΩ, and $R_{LD} = 5$ kΩ. If you included a bias resistor R_B, let it be $R_B = 4.3$ kΩ.
    a) Find $A_{v(CL)}$.
    b) If $V_{in(p-p)} = 1.6$ V, find $V_{out(p-p)}$.
    c) Sketch waveforms of V_{in} and V_{out}, showing their phase relationship.
17. In your inverting amplifier from Problem 16,
    a) Find input resistance $R_{in(CL)}$.
    b) Calculate the input current from the signal source.
    c) Calculate the output current to the load.
    d) Find closed-loop current gain, A_i.
18. In your inverting amplifier:
    a) The voltage at the +IN terminal is virtually _____ at all times.
    b) The voltage at the –IN terminal is virtually _____ at all times.
    c) The current flowing through the op amp between the +IN and –IN terminals is virtually _____ at all times.
19. For your inverting amplifier from Problem 16, consider the positive peak instant of the V_{in} waveform.
    a) Find the current through the 4.7-kΩ input resistor R_i. Specify direction.
    b) Find the current through the 33-kΩ feedback resistor R_f. Specify direction.
20. Draw the schematic diagram for an op-amp noninverting amplifier, and label the external resistors.
21. In your noninverting amplifier of Problem 20, suppose $V_S = \pm 12$ V, $R_i = 10$ kΩ, $R_f = 36$ kΩ, $R_{LD} = 5$ kΩ, and $R_B = 7.5$ kΩ.
    a) Find A_v.
    b) If $V_{in(p-p)} = 1.6$ V, find $V_{out(p-p)}$.
    c) Sketch waveforms of V_{in} and V_{out}, showing their phase relationship.
22. In your noninverting amplifier from Problem 21,
    a) Find input resistance R_{in}.
    b) Make a statement about the magnitude of the input current from the signal source.
    c) Calculate the output current to the load.
    d) Make a statement about the closed-loop current gain, A_i.
23. In your noninverting amplifier:
    a) (T-F) The voltage at the +IN terminal is virtually equal to v_{IN} at all times.
    b) The voltage at the –IN terminal is virtually _____ at all times.
    c) The current flowing through the op amp between the +IN and –IN terminals is virtually _____ at all times.
24. For your noninverting amplifier from Problem 21, consider the positive peak instant of the V_{in} waveform.
    a) Find the current through the 10-kΩ input resistor R_i. Specify direction.
    b) Find the current through 36-kΩ feedback resistor R_f. Specify direction.
25. For an op-amp voltage-follower, $A_v =$ _____ and current demand from the signal source is virtually _____ .
26. For an op-amp amplifier, if $V_{IN} = 0$ V, and V_{OUT} is not exactly 0 V, that is called the _____ problem.
27. Describe the two steps that can be taken to combat the offset problem.

28. Some op amps have an external frequency-compensation terminal. What is this terminal used for?

29. Draw the schematic diagram of a frequency-compensation circuit.

30. The slew-rate limitation of an op amp can limit the amplifier's maximum usable frequency when the signal magnitude is _____ . (large or small)

31. A certain sine wave has $f = 150$ kHz and $V_{pk} = 7$ V. Find the wave's maximum rate of change of voltage $(\Delta v / \Delta t\,|_{max})$.

32. For an op-amp amplifier to produce the sine wave of Problem 31 at its output, the op amp must have a slew rate of at least _____ .

33. A certain inverting amplifier has $A_v = 5$. It uses a fast op amp, with slew rate of $SR = 25$ V/μs. If $V_{in(pk)} = 2$ V, what is the highest frequency the amplifier can handle. (Considering slew rate only). Use. Eq. (13-20).

34. Suppose that the op amp in Problem 33 has a small-signal cutoff frequency (bandwidth) of $f_{high} = 1$ MHz. If the input signal voltage is lowered to $V_{in(pk)} = 0.5$ V,
    a) Calculate the amplifier's slew-rate limitation frequency.
    b) What is the amplifier's actual frequency limitation, the slew-rate limitation frequency from part (a), or the small-signal cutoff frequency?

35. The inverting amplifier of Fig. 13-40 has the waveforms shown. What are the possible causes of this trouble? Describe the steps you would take to fix the problem.

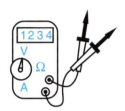

Figure 13-40

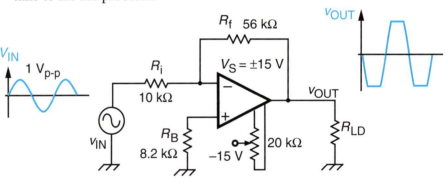

36. The inverting amplifier of Fig. 13-41 has the waveforms shown. What are the possible causes of this trouble? Describe what steps you would take to fix the problem.

Figure 13-41

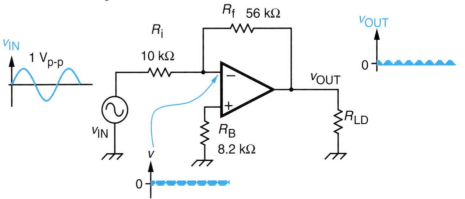

37. The noninverting amplifier of Fig. 13-42 has the waveforms shown. What are the possible causes of this trouble? Describe what steps you would take to fix the problem.

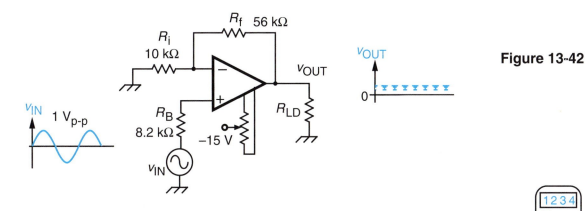

Figure 13-42

38. The noninverting amplifier of Fig. 13-43 has the waveforms shown. What are the possible causes of this trouble? Describe what steps you would take to fix the problem.

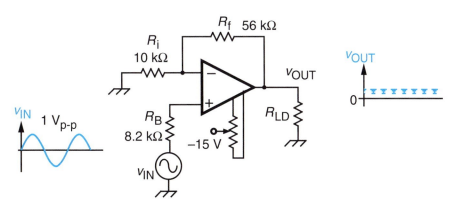

Figure 13-43

39. A certain summing amplifier has $R_1 = R_2 = R_3 = 10\ k\Omega$ and $R_f = 33\ k\Omega$. If $V_1 = -2.0\ V$, $V_2 = +3.6\ V$ and $V_3 = -2.9\ V$, find V_{OUT}.

40. Find V_{OUT} for the summing-type amplifier of Fig. 13-44.

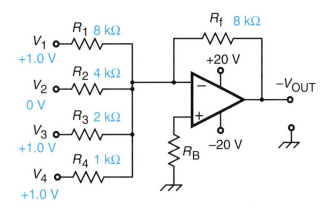

Figure 13-44

41. In Fig. 13-44, assume that the input voltages are digital. Their only possible values are 0.0 V and 1.0 V. Demonstrate that the presence of +1.0 V at V_2 means twice as much (is twice as important) as the presence of +1.0 V at V_1.

42. Extending the idea of Problem 41, V_3 is twice as important as _____, which makes it _____ times as important as V_1.

43. Voltage V_4 is twice as important as _____, making it _____ times as important as V_1.

44. The integrator of Fig. 13-45 receives a square-wave input waveform that switches back and forth between +8 V and –8 V. Use Eq. (13-32) to find the slope of the ramps at v_{OUT}. Then sketch the v_{OUT} waveform.

Figure 13-45

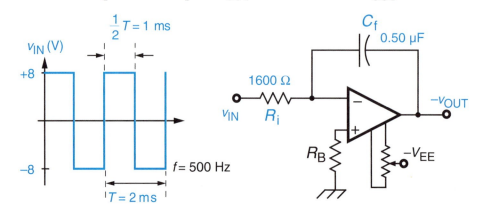

45. In the voltage-to-frequency converter of Fig. 13-35, suppose C_f is changed to 0.02 μF. V_{IN} remains +1 V.
   a) The integrator's downward ramps will become _____ (faster or slower).
   b) What is the new amount of time required for the downward ramps to go from +5 V to –5 V?
   c) For V_{IN} = +1 V, what will be the new output frequency?
   d) The frequency / voltage proportionality constant of this converter is now changed to _____ hertz per volt.

CHAPTER 14

ZENER DIODES – REGULATED DC

Hydroponics is the science of growing plants without soil. In this greenhouse, vegetable plants hang from an A-frame structure. Their roots are periodically sprayed with a nutrient-containing water solution, with nutrient infusion controlled electronically. Also shown on page 4 of the color section.

Courtesy of NASA

OUTLINE

Zener Diode Characteristics
Dc Voltage Regulation
Improved Dc Regulation Using a
    Series Transistor
Prepackaged IC Regulators
Adjustable Dc Regulators

NEW TERMS TO WATCH FOR

zener diode
reverse breakdown voltage
knee current
maximum reverse current
dynamic zener resistance
line regulation
regulated dc supply
load regulation
full-load condition
IC regulator

The standard silicon *p-n* junction diode is used for rectification and other related applications that depend on one-way conduction. We studied the standard diode in Chapters 2 and 3. Solid-state researchers have invented several variation on the standard diode. These diodes have special purposes not related to rectification. Of these special-purpose diodes, the most important is the zener diode.

✔

After studying this chapter, you should be able to:

1. Draw the schematic symbol for a zener diode. Show the current direction and voltage polarity when a zener is reverse-conducting.
2. Draw the overall characteristic graph of *I* versus *V* for a zener diode.
3. Apply Kirchhoff's voltage law and Ohm's law to a zener-resistor series circuit.
4. Explain the meaning of zener dynamic resistance r_z, with reference to a zener's characteristic curve.
5. Use the r_z idea to explain how a zener is able to reduce ripple dramatically.
6. Explain how a zener diode can provide line regulation. Given the circuit component values and specifications, calculate how good the line regulation is.
7. Define what is meant by a regulated dc power supply, as opposed to an unregulated power supply.
8. Explain how a zener diode can provide load regulation. Given the circuit component values and specifications, calculate the load regulation factor.
9. Draw the schematic diagram of a basic zener diode/series-pass transistor regulating circuit. Explain its principle of operation.
10. Given the circuit component values and specifications for the zener-transistor regulating circuit, calculate the load regulation factor.
11. Explain the advantage of using a Darlington pair in a zener/series-pass transistor regulating circuit.
12. Draw the basic schematic layout of a zener/op amp/series-pass transistor regulating circuit, and explain how it operates.
13. Draw the schematic symbol for a 3-terminal IC regulator, and show the external supporting capacitors that are required.
14. Draw the basic schematic layout of an adjustable-voltage regulator, and explain how it operates.
15. Draw the schematic symbol for a 4-terminal adjustable IC regulator. Show how the adjustment pot is connected, as well as the supporting capacitors.

14-1 ZENER DIODE CHARACTERISTICS

The schematic symbol for a **zener diode** is shown in Fig. 14-1. The diode terminals are called anode (A) and cathode (K), just like a standard silicon diode. To understand the operation of a zener, first recognize that it behaves just like a standard diode, for low values of applied voltage. As shown in Fig. 14-2(A), with the cathode pointing toward the more negative potential, the zener conducts. In Fig. 14-2(B), with the cathode pointing toward the more positive potential, the zener blocks.

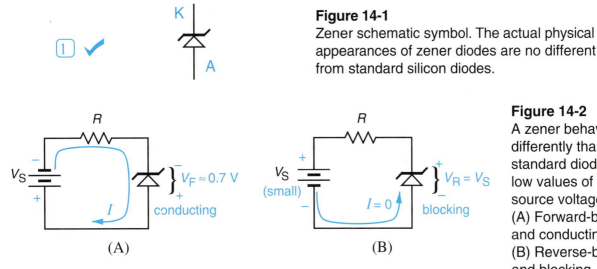

Figure 14-1
Zener schematic symbol. The actual physical appearances of zener diodes are no different from standard silicon diodes.

Figure 14-2
A zener behaves no differently than a standard diode, at low values of source voltage V_S.
(A) Forward-biased and conducting.
(B) Reverse-biased and blocking.

The usefulness of a zener diode is this:

> A zener diode is specially designed to operate successfully in the reverse breakdown mode, with its current limited by a series resistor.

Get This

This is unlike a standard *p-n* diode, which is never subjected to reverse breakdown in a normal rectifying application..

The reverse breakdown idea was mentioned in Sec. 2-3. Refer to Fig. 14-3 to get a more detailed explanation of this idea.

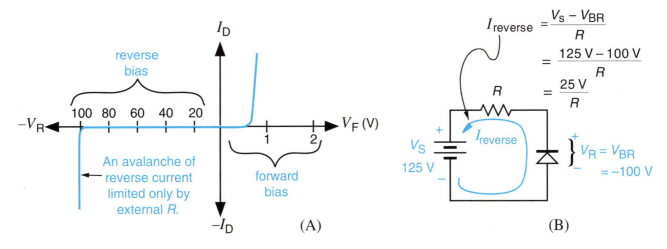

Figure 14-3 (A) Characteristic curve of a standard diode, with a V_{BR} rating of −100 V. The graph shows that reverse current surges out of control of the diode if the applied reverse voltage reaches −100 V. (B) In reverse breakdown, a diode's current is limited by whatever resistance R is present in the circuit.

A complete current-versus-voltage characteristic curve for a standard diode is shown in Fig. 14-3(A). In this case, the **reverse breakdown voltage**, V_{BR}, is −100 V. The diode will be able to block all reverse-bias voltages between 0 V and −100 V. But if a reverse voltage beyond −100 V is applied, the diode will start conducting heavily in the reverse direction, as the *I*-versus-*V* curve indicates. With the diode's voltage so large, current must be limited to quite a small value to avoid $P = IV$ failure. In Fig. 14-3(B), even if R is as high as 1000 Ω, we get $I_R = 25$ mA. Power P is given by $P = I_R V_R = 25$ mA $\times$ 100 V $= 2.5$ W, which is likely to be enough heat to destroy the diode.

> A zener diode is not harmed by reverse current, up to a certain limit.

A zener diode is specially manufactured to have a low reverse breakdown voltage. Figure 14-4(A) shows the characteristic I-vs-V curve for a zener diode with reverse breakdown voltage $= -10$ V. If a reverse voltage beyond -10 V is applied, the zener also conducts very easily in the reverse direction. As shown in Fig. 14-4(B), the reverse current is limited by series dropping resistance R_{ser}. By Ohm's law,

> The reverse breakdown voltage for a zener diode is called zener voltage, symbolized V_Z.

$$I_{reverse} = \frac{\text{difference between applied voltage and zener voltage}}{R_{ser}}$$

$$I_{reverse} = I_Z = \frac{V_S - V_Z}{R_{ser}} \qquad \text{Eq. (14-1)}$$

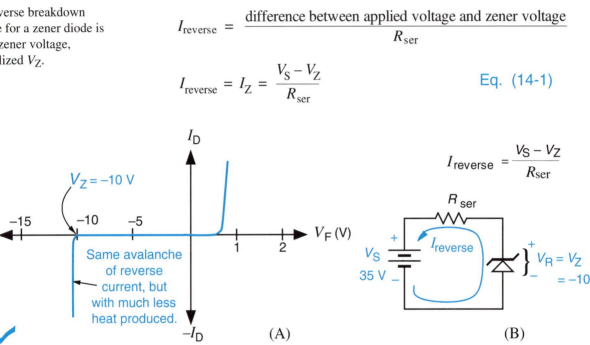

Figure 14-4 (A) Characteristic curve of a zener diode with $V_Z = -10$ V. (B) Reverse-biased zener circuit.

EXAMPLE 14-1

Suppose that a 10-V zener diode is driven by an applied voltage $V_S = 35$ V in series with $R_{ser} = 1000$ Ω, as shown in Fig. 14-4(B).
a) Find the reverse zener current I_Z, and state its direction.
b) Tell the magnitude and polarity of the voltage across the zener diode.
c) Find the power burned by the zener diode, and comment on the zener's ability to handle the generated heat.

SOLUTION

a) Equation (14-1) gives

$$I_Z = \frac{V_S - V_Z}{R_{ser}}$$

$$= \frac{35 \text{ V} - 10 \text{ V}}{1000 \text{ Ω}} = \frac{25 \text{ V}}{1000 \text{ Ω}} = \mathbf{25 \text{ mA}}$$

The current is in the reverse direction for a diode, namely **entering on the anode and exiting on the cathode**. Figure 14-4(B) makes this clear.

b) When a zener is conducting in the reverse direction, its voltage is the zener value V_Z. In this example, $V_Z = $ **10 V, negative on the anode**.

c) $P = I_R V_Z = 25$ mA (10 V) = **0.25 W**

Even the smallest standard electrical components can dissipate 1/4 watt, so the zener will dissipate the heat without harm.

Zener Dynamic Resistance

In the breakdown region, a zener curve has a very steep slope. This means that a large change in reverse current I_Z causes very little change in the actual voltage across the zener's terminals, V_Z. For example, the typical characteristic curve for a type 1N961 10-V zener is drawn in Fig. 14-5. It shows that the zener's actual voltage changes by only 0.3 V as the 1N961 zener goes from its **knee current**, 0.25 mA, to its **maximum reverse current**, 40 mA.

In a normal application, the zener's actual current would not change through the complete range from $I_{Z(K)}$ to $I_{Z(max)}$. Therefore the actual zener voltage V_Z would change by less than 0.3 V.

In Fig. 14-4, V_Z stood for the *nominal* zener voltage value. But in this context, V_Z means the *actual* voltage across the zener, which will be slightly greater.

Meaning of $I_{Z(K)}$ and $I_{Z(max)}$. Also, see figure caption.

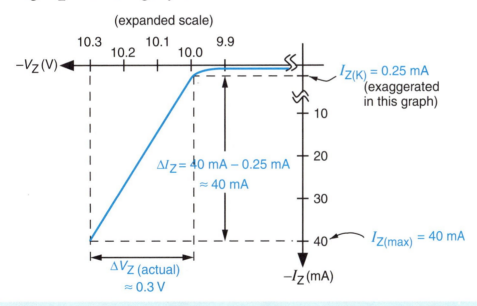

Figure 14-5
Expanded view of the reverse breakdown portion of the curve for a typical zener diode. The current value at which the zener diode starts its steep downward slope is called its knee current, symbolized $I_{Z(K)}$. The maximum current that the zener can carry is symbolized $I_{Z(max)}$.

EXAMPLE 14-2

Figure 14-6 shows a common zener application. A dc supply, which is not absolutely constant in its dc value, also contains some ac ripple. As shown in the schematic, such a supply is connected to a zener-resistor series combination.

a) Calculate the average (dc) reverse current through the zener diode. This is I_Z.

b) Locate this value of I_Z on the zener's characteristic curve. Mark it as the Q-point.

c) Superimpose the ac ripple on the 15-V dc value to find the total maximum and minimum instantaneous values of i_Z.

d) Mark these two points as P_1 and P_2 on the zener's characteristic curve.

e) Project up to the voltage axis from points P_1 and P_2, to find the actual instantaneous values of zener voltage, v_Z. Comment on what the zener circuit has done to the ripple.

SOLUTION

a) Ignoring the ripple for the time being, we apply Eq. (14-1) to Fig. 14-6(A).

$$I_Z = \frac{V_{S(dc)} - V_{Z(nominal)}}{R_{ser}}$$

$$= \frac{15\text{ V} - 10\text{ V}}{250\ \Omega} = \frac{5\text{ V}}{250\ \Omega} = \mathbf{20\ mA}\ (dc)$$

b) The 20-mA I_Z point is marked on the expanded zener curve of Fig. 14-6(B).

c) The ripple's positive peak is +1 V. Applying Eq. (14-1), we get

$$I_{Z(+\text{peak})} = \frac{(15\text{ V} + 1\text{ V}) - 10\text{ V}}{250\ \Omega} = \frac{16\text{ V} - 10\text{ V}}{250\ \Omega}$$

$$= \frac{6\text{ V}}{250\ \Omega} = \mathbf{24\ mA}\ (\text{at} + \text{peak instant})$$

The –1-V negative peak gives

$$I_{Z(-\text{peak})} = \frac{(15\text{ V} - 1\text{ V}) - 10\text{ V}}{250\ \Omega}$$

$$= \frac{4\text{ V}}{250\ \Omega} = \mathbf{16\ mA}\ (\text{at} - \text{peak instant})$$

d) The 24-mA and 16-mA values have been marked as points P_1 and P_2 on Fig. 14-6(B).

e) Projecting up from P_1 , we hit the voltage axis at about 10.16 V. A projection from P_2 hits at 10.10 V. This means that the peak-to-peak ripple across the zener is

$$V_{ripple(p\text{-}p)} = 10.16 - 10.10 = 0.06\text{ V}$$

and the average dc voltage across the zener is

$$V_{Z(avg)} = \frac{10.10\text{ V} + 10.16\text{ V}}{2} = 10.13\text{ V}$$

A zener diode is very effective at reducing ac ripple.

Using these values, the output waveform is sketched for comparison to the input waveform in Fig. 14-6(C). The output ripple has been reduced to only 3% of the input amount, a tremendous improvement.

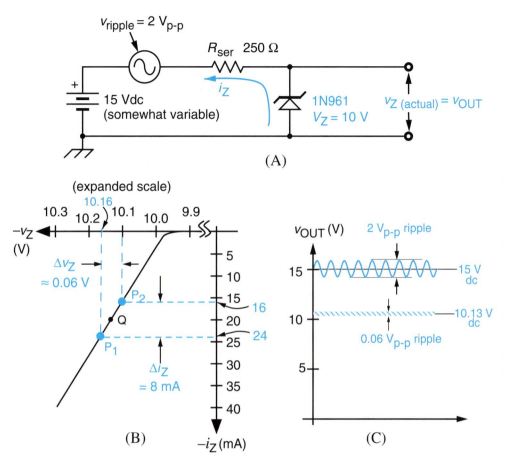

(A)

(B)

(C)

Figure 14-6

(A) Realistic rectified dc power supply has a slightly variable dc value with ac ripple superimposed. Here a realistic dc supply is driving a zener circuit.

(B) The zener swings between these two points on its curve.

(C) Waveform of v_{OUT}. Because the zener holds its terminal voltage nearly constant, it reduces ripple dramatically.

For any particular zener diode, the steepness of its characteristic curve is an indication of its ability to maintain a constant voltage. The ratio of change in voltage to change in current is called the **dynamic zener resistance**, symbolized r_z. That is,

$$r_z = \frac{\Delta v_Z}{\Delta i_Z}$$ Eq. (14-2)

The meaning of dynamic zener resistance, r_z.

r_z is a figure of merit. The lower the r_z value, the better the zener diode, all other things being equal.

EXAMPLE 14-3

From the typical 1N961 zener curve shown in Fig. 14-6(B), what is the value of dynamic zener resistance, r_z ?

SOLUTION

Equation (14-2) can be applied to the small piece of the curve between points P_1 and P_2, giving

$$r_z = \frac{\Delta v_Z}{\Delta i_Z}$$

$$= \frac{0.06 \text{ V}}{8 \text{ mA}} = \frac{0.06 \text{ V}}{8 \times 10^{-3} \text{ A}} = 7.5 \ \Omega$$

The straight-line character-istic is an idealization. A real zener would not be perfectly straight.

Since the characteristic curve is shown as a perfectly straight line below the knee, an alternative approach would be to take the entire reverse breakdown region. This gives

$$r_z = \frac{10.3 \text{ V} - 10.0 \text{ V}}{40 \text{ mA} - 0.25 \text{ mA}} = \frac{0.3 \text{ V}}{39.75 \text{ mA}} = \mathbf{7.55 \ \Omega}$$

If a zener's r_z value is known, then either Δi_z or Δv_z can be found from Eq. (14-2), if the other is given. We will apply this idea to the study of voltage regulation in the next section.

SELF-CHECK FOR SECTION 14-1

1. Draw a schematic diagram of a 12-V dc source driving reverse current through a 5-V zener diode in series with a 750-Ω resistor.
2. In the circuit of Problem 1, specify:
   a) The voltage across the zener, both magnitude and polarity.
   b) The voltage across the resistor, both magnitude and polarity.
   c) The current, both magnitude and direction.
3. In the reverse-conducting region of a zener characteristic curve, a steeper slope indicates a _____ dynamic resistance, r_z.
4. A certain zener diode has $V_Z = 14$ V and $r_z = 12 \ \Omega$. It can carry a maximum reverse current of 30 mA. Draw a carefully scaled graph of I_Z versus V_Z in the reverse conducting region. Assume a straight-line characteristic, as in Figs. 14-5 and 14-6.
5. On the graph that you drew for problem 4:
   a) Locate the operating point $I_Z = 25$ mA. What is the zener voltage V_Z at this point?
   b) Repeat part (a) for $I_Z = 15$ mA.
   c) If zener current I_Z varies between 25 mA and 15 mA, the voltage V_Z across the zener will vary by what amount?
6. For this zener diode, if $\Delta I_Z = 22$ mA, what will be the change in zener voltage, ΔV_Z.

14-2 DC VOLTAGE REGULATION

Line Regulation

In Section 14-1, we saw how a low dynamic zener resistance can eliminate a great deal of ripple in a dc power supply. Low r_z has another desirable effect:

Get This

In a dc power supply, a zener diode can minimize changes in the dc output voltage that result from variations in the ac supply voltage.

To understand why this is true, look at the dc power supply shown in Fig. 14-7. To analyze this circuit, let us assume that we have a 116-V rms ac supply voltage applied to the transformer primary winding. Then, by the transformer

voltage law,
$$V_S = n V_P$$
$$= (0.1)(116 \text{ V}) = 11.6 \text{ V rms}$$

We are assuming that the transformer is ideal – zero winding resistance.

The peak value of secondary voltage is
$$V_{S(p)} = (1.414)(11.6 \text{ V}) = 16.4 \text{ V}$$

Turning on two silicon diodes in the full-wave bridge causes a voltage loss of $2(0.7 \text{ V}) = 1.4 \text{ V}$. Therefore the filter capacitor charges to a voltage of
$$V_{dc} = V_{S(p)} - 1.4 \text{ V} = 16.4 \text{ V} - 1.4 \text{ V} = 15.0 \text{ V}$$

A standard 3-1/2 digit DVM would read this as 10.15 V or 10.16 V, as shown in Fig. 14-7(B).

which is the same dc voltage that we had in Fig. 14-6. Therefore the dc voltage across the supply's output terminals is about 10.13 V, the same as it was in Example 14-2.

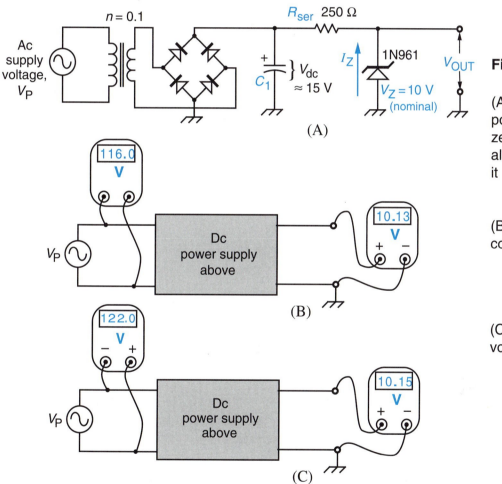

Figure 14-7

(A) Regulated dc power supply. The zener holds V_{OUT} almost constant; it regulates V_{OUT}.

(B) Initial line voltage condition.

(C) Changed line voltage condition.

Now suppose that the ac supply-line voltage changes by, say, 5%. Such a change, from 116 V to 122 V, is often seen on the ac line. In a plain unregulated dc supply (with no zener diode), the dc output voltage would rise by roughly the same percentage, 5%. But with the zener diode present in Fig. 14-7(A), here is what happens:

Solving the *line regulation* problem.

The increased secondary voltage is given by
$$V_S = (0.1)(122 \text{ V}) = 12.2 \text{ V}$$
with a peak value of about
$$V_{S(p)} \approx (1.414)(12.2 \text{ V}) = 17.2 \text{ V}$$

Subtracting 1.4 V for turning on the bridge diodes leaves approximately
$$V_{dc} = 17.2 - 1.4 = 15.8 \text{ V}$$
The new value of zener current is given by Eq. (14-1) as

$$I_Z = \frac{V_S - V_Z}{R_{ser}}$$

$$= \frac{15.8 \text{ V} - 10 \text{ V}}{250 \, \Omega} \approx 23 \text{ mA}$$

Previously, (in Example 14-2), I_Z was 20 mA. So the change in I_Z is

$$\Delta I_Z \approx 23 \text{ mA} - 20 \text{ mA} = 3 \text{ mA}$$

Therefore the zener's dc operating point moves down its characteristic curve by about 3 mA, producing a voltage change given by Eq. (14-2).

$$\Delta V_Z = (r_z)(\Delta I_Z)$$

$$= (7.5 \, \Omega)(3 \text{ mA}) \approx 0.02 \text{ V}$$

Thus, the dc output voltage changes from 10.13 V to

$$V_{OUT(new)} = V_{OUT(old)} + 0.02 \text{ V}$$

$$= 10.13 \text{ V} + 0.02 \text{ V} = 10.15 \text{ V}$$

In other words, dc V_{OUT} changes hardly at all. Compare Fig. 14-7(C) to 14-7(B).

Get This

When a dc supply has the ability to maintain a nearly constant value of V_{OUT} even when the ac supply-line voltage changes considerably, we say that the supply has good **line regulation**. A zener diode gives a dc supply good line regulation.

The meaning of the term regulated dc supply.

7 ✔

Any dc power supply that has some ability to hold V_{OUT} at a steady value, rather than simply allowing V_{OUT} to go through naturally occurring variations, is called a **regulated dc supply**. Installing the zener diode in Fig. 14-7(A) created a 10-V regulated dc supply. If the zener were not present, we would have a 15-V unregulated supply.

Regulated dc supplies must react to two kinds of variations that tend to change V_{OUT}. The two are line variations and load variations. We have seen the first; now let us look at the second.

Load Regulation

The Fig. 14-7 dc power supply has no load connected to its output terminals. To be useful, of course, the supply must be able to deliver current through a load resistance. This is shown in Fig. 14-8.

With R_{LD} connected, not all the current through R_{ser} passes through the zener. Now, some of I_{Rser} passes through R_{LD}. Applying Kirchhoff's current law to the right side of R_{ser}, we can say

$$I_Z = I_{Rser} - I_{LD} \qquad\qquad \text{Eq. (14-3)}$$

This is the same as saying that the zener current changes by an amount equal to I_{LD} when a load is added. That is,

$$\Delta I_Z = I_{LD}$$

The change in I_Z produces a slight change in V_Z as the zener moves along its characteristic curve. In other words,

Equation (14-2), $\Delta V_Z = r_z\,(\Delta I_Z)$, applies to changes in the load, just as it applies to changes in the line.

Get This

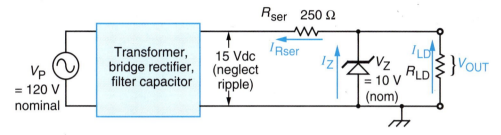

Figure 14-8
Connecting a load to a zener-regulated dc power supply. The three currents are related by Kirchhoff's current law:

$$I_{Rser} = I_Z + I_{LD}$$

EXAMPLE 14-4

In Fig. 14-8, suppose $R_{LD} = 750\ \Omega$.
a) Calculate the load current, I_{LD}, from Ohm's law.
b) Is there any change in the current I_{Rser} through the dropping resistor, from the no-load condition?
c) State the change in zener current, ΔI_Z, between the original no-load condition and the new loaded condition.
d) Using Eq. (14-2) with $r_z = 7.5\ \Omega$, find the approximate change in output voltage caused by connecting the load.
e) Give the approximate value of the new output voltage, under load.

SOLUTION

a) The dc output voltage is always *close* to the nominal zener voltage V_Z, so

$$I_{LD} = \frac{V_{OUT}}{R_{LD}} = \frac{V_Z}{R_{LD}}$$

$$= \frac{10\ V}{750\ \Omega} = \textbf{13.3 mA}$$

b) No. Ideally, the average dc current through the dropping resistor has not changed. It is still **the same 20 mA** that flowed in Example 14-2.
c) From Eq. (14-4),

$$\Delta I_Z = I_{LD} = \textbf{13.3 mA}$$

d) $\Delta V_Z = (\Delta I_Z)\,r_z$

$$= (13.3\ mA)\,(7.5\ \Omega) \approx \textbf{0.1 V}$$

e) The zener moves up its characteristic curve to the new operating point shown in Fig. 14-9(A). The approximate output voltage under load is

$$V_{OUT(LD)} = V_{OUT(NL)} - \Delta V_Z$$

$$= 10.13\ V - 0.1\ V \approx \textbf{10.03 V}$$

8 ✔

The difference between the no-load and loaded situations is pointed out in Figs. 14-9(B) and (C).

When a dc supply has the ability to maintain a nearly constant value of VOUT even when the load current changes considerably, we say that the supply has good **load regulation**. A zener diode gives a dc supply good load regulation.

Figure 14-9

The results of connecting a 750-Ω load to the zener-regulated supply.

(A) Change in zener operating point.

(B) Old (no-load) output situation.

(C) New (loaded) output situation. The change in V_{OUT} is slight.

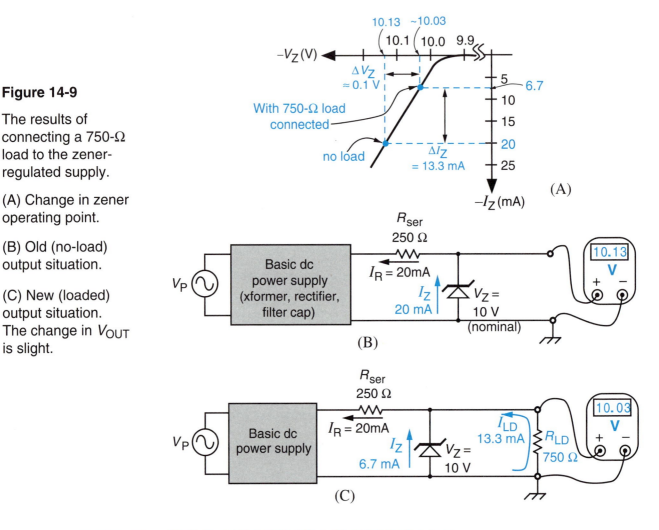

SELF-CHECK FOR SECTION 14-2

7. In a dc power supply that is *not* regulated by a zener diode, if the ac supply-line voltage decreases by 7%, the dc output voltage decreases by about _____ %.

8. (T-F) In a dc power supply that is regulated by a zener diode, if the ac supply-line voltage decreases by 7%, the dc output voltage decreases by a very small amount, probably a negligible amount.

9. For the dc power supply of Fig. 14-7(A), imagine that the ac line voltage decreases from 116 V to 105 V. Find the new value of V_{OUT}, by following the method that was used for the increase from 116 V to 122 V.

10. Define what is meant by a regulated dc power supply.

11. A well-regulated dc supply can compensate for both _____ variations and _____ variations.

12. Repeat Example 14-4 for $R_{LD} = 555\ \Omega$. Read the zener characteristic graph, Fig. 14-6, as best you can.
13. What is the absolute maximum amount of load current that can be taken from the dc supply of Fig. 14-8, before the voltage-regulation ability is lost? Explain this. What amount of load resistance R_{LD} will cause this?
14. Based on your answers to Problems 12 and 13, it might be reasonable for the supply's manufacturer to specify the full-load condition as $R_{LD} = 555\ \Omega$, $I_{LD} = 18$ mA (advising the user to connect *no less than* 555 Ω). Using your results from problem 12, the change in voltage from no-load to full-load condition can be calculated as a percentage of the nominal output voltage, 10 V. Perform this calculation. Your result is called *percent load regulation*. It is a figure of merit for the supply.

COMPACT FLUORESCENT LAMPS

Fluorescent lamps produce light by an altogether different principle than incandescent lamps. Instead of passing a large amount of current through a very-high-temperature filament, fluorescent lamps rely on smaller-magnitude current through low-temperaturemetallic vapor inside the sealed glass enclosure. Because of their lower current requirements for comparable light production, fluorescent lamps have always been much more energy-efficient than incandescent lamps.

Until recently, fluorescent lamps needed a large surface area of inside-coated glass in order to produce reasonable amounts of visible light. This is why fluorescent "tubes" tended to be at least 2 feet in length and at least 1 inch in diameter. But new inside-coating materials with better light-emission qualities have allowed manufacturers to shrink the size of their tubes dramatically.

Courtesy of General Electric Co.;
photo by Arnaud de Wildenberg.

This photo shows the manufacturing process for fluorescent lamps that are small enough to screw into a standard incandescent light socket. The equipment's flexible tubing serves to evacuate the air from the small glass enclosures and inject a mixture of low-pressure mercury vapor and inert argon gas.

In today's market, such lamps are approximately 30 times as expensive as equivalent-lumen incandescent lamps. But they last about 10 times longer and their electric energy consumption is only about 20% as great. So in the long run they are more economical.

14-3 IMPROVED DC REGULATION USING A SERIES TRANSISTOR

The simple zener diode installation of Figs. 14-6, 14-7 and 14-8 improves the performance of the dc supply greatly, especially in terms of line regulation. However, its load-regulating ability is rather limited, as you found in your answers to Problems 12, 13, and 14 in the previous **Self-Check**.

To expand the range of load currents that the zener can successfully regulate, we combine the zener with a transistor. The simplest circuit is shown in Fig. 14-10(A).

In that figure, we have changed the zener diode to a type No. 1N962A, which has a 1-V higher nominal V_Z value, namely 11 V. This is necessary to compensate for the approximate 0.7-V loss across the transistor's forward-biased B-E junction. The dc output voltage is then

$$V_{OUT} = V_Z - V_{BE}$$
$$\approx 11\ V - 0.7\ V = 10.3\ V$$

With the zener-regulating circuit connected to the base of the transistor, the zener's effects are amplified by a factor of $\beta = 50$. In other words:

Get This

The zener-transistor combination now has the ability to regulate a load current that is β times as large as the zener could handle by itself alone.

This idea is illustrated in Fig. 14-10(B).

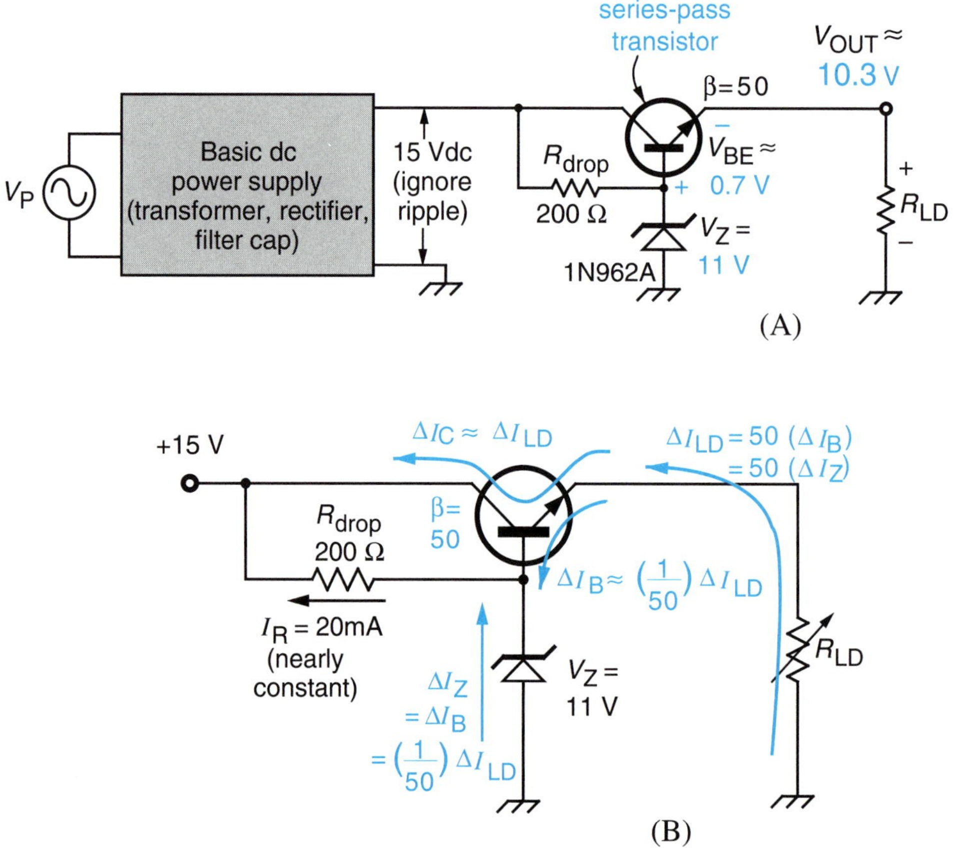

Figure 14-10

(A) Combining a series-pass transistor with a zener diode.

(B) Because of the transistor's current gain β, the change in zener current is only a small fraction $(1/\beta)$ of the change in load current. Looking at it the other way, the controllable variation in load current can be β times as large as the change in zener current.

9 ✔

In Fig. 14-10(B), the current through the dropping resistor R_{drop} is given by

$$I_{Rdrop} = \frac{15 \text{ V} - V_Z}{R_{drop}}$$

$$= \frac{15 \text{ V} - 11 \text{ V}}{200 \text{ } \Omega} = 20 \text{ mA}$$

This is the same value that we had throughout Section 14-2 in Figs. 14-6, 14-7 and 14-8.

If no load is connected to the dc supply this entire 20 mA passes through the zener diode, because the B-E path carries zero current. Therefore the zener's current can change by almost 20 mA before it loses its ability to regulate a nearly constant voltage at the transistor's base terminal. But with the load connected directly in the collector-emitter path, and with $\beta = 50$, the load current changes by 50 times as much as the base current. The change in base current is the same as the change in zener current, so

$$\Delta I_{LD} = \beta (\Delta I_Z) = 50 (\Delta I_Z) \qquad \text{Eq. (14-4)}$$

Therefore, if the maximum allowable change in I_Z is 18 mA, say, the maximum allowable change in I_{LD} is

$$\Delta I_{LD} = 50 (18 \text{ mA}) = 900 \text{ mA}$$

With the zener reverse-biased at $I_Z = 20$ mA, load current I_{LD} can be as large as 50×18 mA $= 0.9$ A.

EXAMPLE 14-5

In Fig. 14-10(A), assume that the type No. 1N962A zener's dynamic resistance is $r_z = 8 \text{ } \Omega$. This is slightly greater than the 7.5 Ω of the type 1N961A. If a drastic change in R_{LD} causes the load current to increase from 200 mA to 900 mA,

Higher-voltage zeners tend to have higher r_z values.

a) Calculate the change in base current, ΔI_B.
b) State the change in zener current, ΔI_Z. Does I_Z increase or decrease?
c) Using Eq. (14-2), calculate the change in V_Z.
d) Assuming that the transistor's base-emitter voltage V_{BE} holds perfectly constant at $V_{BE} = 0.70$ V, what is the change in V_{OUT} as I_{LD} changes from 200 to 900 mA?

SOLUTION

a)
$$\Delta I_B = \frac{\Delta I_C}{\beta} \approx \frac{\Delta I_{LD}}{\beta}$$

$$= \frac{900 \text{ mA} - 200 \text{ mA}}{50} = \frac{700 \text{ mA}}{50} = \textbf{14 mA}$$

b) The change in I_Z has the same magnitude as the change in I_B. This is because there is a nearly constant supply of current through R_{drop}. Whatever portion of I_{Rdrop} is not taken by the base of the transistor must flow through the zener [similar to Eq. (14-3)].

$$\Delta I_Z = \Delta I_B = \textbf{14 mA}$$

As I_B increases, I_Z **decreases**.

c) $\Delta V_Z = (r_z)(\Delta I_Z)$

$\quad = (8\ \Omega)(14\ mA) \approx \mathbf{0.11\ V}$

d) $V_{OUT} = V_Z - V_{BE}$. If V_{BE} is absolutely constant, then V_{OUT} changes by the same amount that V_Z changes, namely about **0.11 V**.

We conclude that the zener-transistor regulator has very good load-regulation ability, even for relatively large values of load current.

Power Burned by the Series-Pass Transistor: In many cases the series transistor in a regulated dc supply consumes quite a large amount of power. For example, the supply of Fig. 14-10 can deliver a maximum load current of about 900 mA and still maintain good zener-voltage regulation. The maximum load situation is shown in Fig. 14-11.

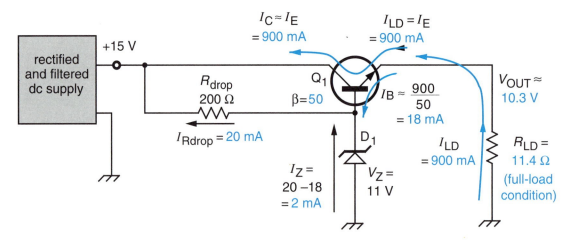

Figure 14-11 The zener-transistor series-regulated dc supply operating under full-load conditions. With $R_{LD} = 11.4\ \Omega$, $I_{LD} = 10.3\ V \div 11.4\ \Omega = 0.9\ A = 900\ mA$. The transistor's emitter current I_E is related to the base current I_B by the approximate formula $I_E \approx \beta I_B$. Therefore, $I_B \approx 900\ mA/50 = 18\ mA$. This leaves $I_Z = 20\ mA - 18\ mA = 2\ mA$ to keep the zener diode reverse-biased and still able to regulate voltage. The transistor's power dissipation is over 4 watts, requiring a high-power type. High-power transistors tend to have low betas. This is why we picked the value $\beta = 50$ for this example circuit.

Under this full-load condition, the power consumed by transistor Q_1 is given by the product of its current times its voltage, or

$$P_Q = I_C V_{CE}$$

$$= 0.9\ A\ (15\ V - 10.3\ V) = 0.9\ A\ (4.7\ V) = 4.2\ W$$

A small-signal transistor cannot dissipate this much power. The common plastic package and metal-can package transistors shown in Fig. 4-34 parts (a), (b) and (c) are generally limited to less than 1 watt maximum power dissipation, even when they are attached to a heat sink. Therefore, Q_1 in this example would need to be a high-power type, perhaps a type that comes in the plastic-and-metal package shown in part (d) of Fig. 4-34. It would be mounted on a large-surface heat-sink to dissipate its heat rapidly.

For even higher-current/higher-power applications, we would need one of the all-metal football-shaped transistor types, shown in part (e) of Fig. 4-34.

Darlington Series Regulation

Installing a single series-pass transistor with current gain of β provides great improvement in load regulation. So it is reasonable to take the next step, which is to boost β tremendously with a Darlington pair. This is shown in Fig. 14-12.

⓫ ✔

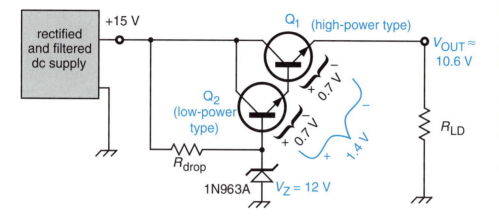

Figure 14-12
Zener/series-transistor regulation using a Darlington pair. The zener type has been changed again, to type No. 1N163A, with nominal V_Z = 12 V. V_{OUT} will be well-regulated at about 12 V – 1.4 V = 10.6 V.

Now the effective overall beta is equal to $β_1$ times $β_2$, giving perhaps $β_T$ = 50 x 100 = 5000, or greater. The change in zener current becomes very small [$ΔI_Z = ΔI_{LD} ÷ 5000$, from Eq. (14-4)]. Therefore the change in V_Z is also very small. Most of the variation in V_{OUT} will now be due to variations in the base-emitter voltages across Q_1 and Q_2 in Fig. 14-12.

Also, with $ΔI_Z$ so small, we can now get by with a no-load (quiescent) value of I_Z much less than the previous 20 mA. We would do this by increasing R_{drop}. The zener will now operate cooler with reduced self-heating from $P_Z = V_Z × I_{Z(quiescent)}$. This reduces temperature variation in the zener diode, which was a problem that we ignored all through Secs. 14-2 and 14-3.

SELF-CHECK FOR SECTION 14-3

15. (T-F) A dc supply that is regulated by a zener-transistor combination has a much greater range of output current than a simple zener regulator.
16. In a zener-transistor regulated supply with β = 70 for the series transistor, assume the load current can vary from 0 to 800 mA.
    a) What is the approximate variation in the transistor's base current ($ΔI_B$)?
    b) What is the approximate variation in the zener current ($ΔI_Z$)?
17. In Problem 16, if the zener diode has r_z = 10 Ω, calculate the variation in zener voltage ($ΔV_Z$).
18. (T-F) In Problem 16, if the voltage V_{BE} across the transistor's B-E junction is absolutely constant, the variation in V_{OUT} is just the same as $ΔV_Z$.
19. (T-F) In a transistor dc regulator, the power burned by the series-pass transistor increases if the load current increases.
20. (T-F) In a transistor dc regulator, the power burned by the series-pass transistor depends in part on the difference between the rectifier output voltage and the final dc output voltage, V_{OUT}.
21. (T-F) All other things being equal, a dc regulator with a Darlington pair of series-pass transistors will have better load regulation than a dc regulator with a single series-pass transistor.
22. Explain your answer to Question 21.

14-4 PREPACKAGED IC REGULATORS

Another high-gain method of regulating dc output voltage is shown in Fig. 14-13. Here, the R_{drop}-zener diode combination develops a voltage across the zener that is slightly larger than the nominal V_Z value. Let us suppose it to be $V_Z = 10.1$ V. This V_Z voltage is applied directly to the noninverting input (+IN) of an op amp. Because the op amp's input resistance is very high, its loading effect on the zener diode is negligible.

Therefore, the zener experiences virtually zero change in current and voltage, as a result of variations in the dc output load current. It still has some small amounts of Δi_Z and Δv_Z resulting from ac ripple though. Also, line variations can still produce nonzero ΔV_Z.

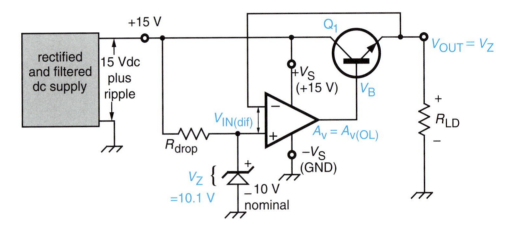

Figure 14-13 Dc regulator circuit with virtually perfect load regulation. If V_{OUT} tried to be a bit lower than V_Z, then $V_{\text{IN(dif)}}$ would not be virtually zero. Then the huge open-loop gain would keep pushing V_B higher, toward the op amp's $+V_{\text{sat}}$ value. At some point, V_B is bound to reach just the right value to make V_{OUT} virtually identical to V_Z.

The op amp-Q_1 combination makes V_{OUT} virtually equal to V_Z. It works because V_{OUT} is fed back directly to the –IN terminal of the op amp. Here is its explanation:

$\boxed{1}$ If V_{OUT} were a slight bit greater than V_Z, say $V_{\text{OUT}} = 10.15$ V, then the actual input voltage $V_{\text{IN(dif)}}$ to the op amp would mismatch the input terminal polarity marks. [$V_{\text{IN(dif)}}$ would be 0.05 V, more positive on the –IN terminal]. The op amp's voltage gain would then cause the transistor's V_B voltage to become less positive then the midpoint of the op amp's single-polarity power supply voltage. In this case, V_B would be brought below (less positive than) 7.5 V, half of 15 V. But such an action would reverse-bias the B-E junction of Q_1. The transistor's current flow path would be cut off, and V_{OUT} would immediately fall. Thus, we conclude that the regulation circuit of Fig. 14-13 will not allow V_{OUT} to be greater than (more positive than) V_Z.

$\boxed{2}$ On the other hand, if V_{OUT} were a bit lower than V_Z, say $V_{\text{OUT}} = 10.05$ V, the actual input voltage $V_{\text{IN(dif)}}$ would match the input terminal polarity marks. [$V_{\text{IN(dif)}}$ would be 0.05 V, more positive on the +IN terminal].

$V_{\text{IN(dif)}} = V_Z - V_{\text{OUT}}$
$= 10.1$ V $- 10.15$ V
$= -0.05$ V.

With $V_{\text{OUT}} = 10.15$ V applied to the emitter, and base voltage $V_B \approx 7$ V, the V_{BE} voltage would be the wrong polarity. Q_1 would shut off.

$V_{\text{IN(dif)}} = V_Z - V_{\text{OUT}}$
$= 10.1$ V $- 10.05$ V
$= +0.05$ V.

Now the op amp's $A_{v(OL)}$ voltage gain will drive V_B more positive than 7.5 V. The op amp will continue driving V_B more and more positive until Q_1 is turned on hard enough to make V_{OUT} virtually identical to V_Z. We know this must happen, because $V_{IN(dif)}$ must stabilize at an extremely small value, virtually zero, to keep the huge-A_v op amp in a non-saturated condition.

In summary, the Fig. 14-13 design will hold a constant value of V_{OUT} over a very wide range of I_{LD} variations. The I_{LD} variations are not limited by the need to maintain reverse current through the zener diode. I_{LD} is limited only by the current- and power-handling capabilities of transistor Q_1, and the rectifying and filtering circuit.

The components of Fig. 14-13 can be integrated onto a silicon chip and packaged as an integrated circuit. Such integrated circuits are called **IC regulators**. Depending on its output current specification, the IC can have the appearance of any one of the standard transistor packages shown in Fig. 4-34 parts (C) through (E).

An IC regulator is shown schematically as a three-terminal box. This is drawn in Fig. 14-14(A). A complete IC-regulated positive dc power supply is shown in Fig. 14-14(B). Capacitor C_1 is the main filter capacitor for the full-wave rectifying circuit. Small-to-medium-value capacitor C_2 may be connected to the input if the IC is located some distance away from main filter cap C_1, to stabilize V_{OUT} against sudden variations in R_{LD} and I_{LD}. Small-value capacitor C_3 is connected to the output to suppress high-frequency noise and eliminate the possibility of self-oscillation.

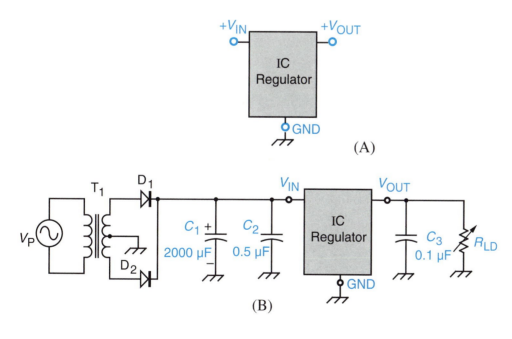

IC regulators are usually mounted on a heat sink.

Figure 14-14
IC voltage regulator.
(A) Three-terminal schematic symbol.
(B) Complete positive dc power supply, showing the external capacitors that are connected to an IC regulator. Capacitor C_1 must be sized according to the maximum load current and the desired ripple reduction. Capacitors C_2 and C_3 are sized according to advice given by the manufacturer of that particular type.

Negative IC Regulators: With some simple changes, the circuit of Fig. 14-13 can be converted into a negative dc voltage regulator. All we need to do is replace Q_1 with a *pnp* transistor and reverse the zener diode. Figure 14-15(A) shows the three-terminal box symbol for a negative IC regulator. A complete IC-regulated dual-polarity ±12-V dc power supply is drawn in Fig. 14-15(B).

Figure 14-15

(A) Three-terminal schematic symbol for negative IC regulator.

(B) Schematic of complete dual-polarity dc power supply. This circuit could be used to provide the dc supply for a collection of op amps.

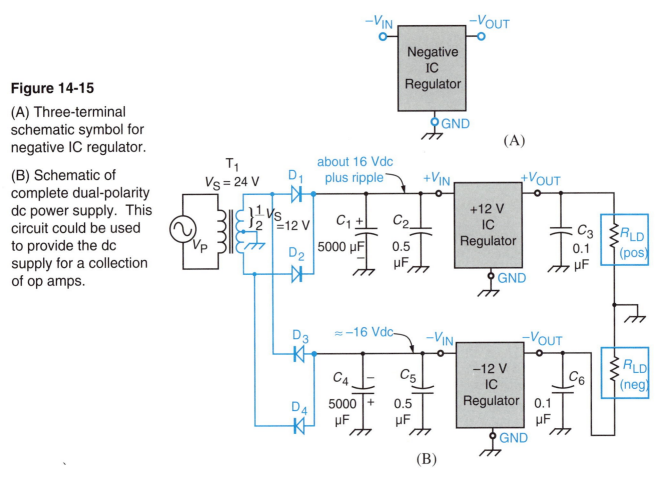

(A)

(B)

IC Regulator Specifications

The type No. LM7812C is a 12-volt positive IC regulator with maximum current capability greater than 1 ampere. Its integrated component count is 16 BJTs, no FETs, 2 zener diodes, 18 resistors and 1 capacitor. It comes in either the plastic-and-metal package shown in Fig. 4-34 part (D), or the large football package of Fig. 4-34(E). Table 14-1 gives its most important operating specifications.

For a particular application, it may be that the IC's V_{OUT} range of 12 V ±0.5 V is too wide a variation. In that case, you must test each IC individually, and choose only those individual units that are within your acceptable tolerance range. For example, if your application required that V_{OUT} be within the range of 11.9 V to 12.2 V, your would have to test and reject every individual IC that falls outside that range.

The actual load-regulation ability of the 7812C is not virtually perfect in Table 14-1 because of the IC's short-circuit protection circuitry. This circuitry does not appear in Fig. 14-13.

Robotic device for cleaning and applying coatings to solid-fuel rockets.

Courtesy of Hercules Aerospace, Vadeko International, and NASA

V_{OUT}	Guaranteed between 11.5 and 12.5 V
Range of V_{IN}	14.5 to 35 V
Load regulation ability	Typically, V_{OUT} changes by 0.012 V as I_{LD} varies from 5 to 1500 mA.
Line regulation ability	Typically, V_{OUT} changes by 0.004 V as V_{IN} varies from 14.5 to 30 V.
Ripple reduction	Typically, output ripple is lower than input ripple by a factor of 4000. So 1 V input ripple causes 0.25 mV output ripple.
Short circuit protection	If an accidental short circuit occurs across the output, V_{OUT} decreases to some low value. All internal currents are automatically limited to prevent damage to the IC.
Ambient air temperature range	0°C to 70°C (32°F to 158°F)

**Table 14-1
Performance
specifications for
the 7812C
IC regulator.**

14-5 ADJUSTABLE DC REGULATORS

A basic dc regulator with adjustable output is shown in Fig. 14-16. As the pot wiper is dialed toward the top, base voltage V_B approaches closer to the zener voltage V_Z. This causes the transistor to conduct harder, which increases V_{OUT}. As the pot wiper is moved toward the bottom, the pot wiper voltage (transistor base voltage V_B) becomes smaller, causing V_{OUT} to decrease.

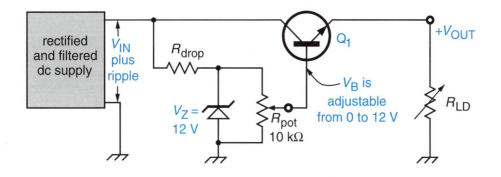

Figure 14-16 Zener/series-transistor dc regulator with adjustable V_{OUT}.

This simple design can be improved by using the op amp idea from Fig. 14-13. By adjusting the *portion* of V_{OUT} that is fed back to the –IN terminal, as shown in Fig. 14-17, we can adjust the final output voltage V_{OUT}.

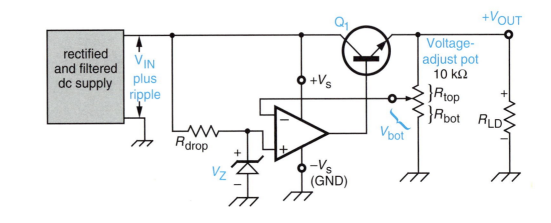

Figure 14-17 Improved adjustable dc regulator.

V_{bot} must become virtually identical to V_Z so that $V_{IN(dif)} \approx 0$ V, balancing the op amp in a non-saturated state.

This circuit works by making V_{OUT} become whatever value is necessary to cause $V_{bot} = V_Z$. For example, suppose $V_Z = 5$ V, and suppose the pot is adjusted right in the center, with $R_{bot} = R_{top} = 5$ kΩ. Then V_{bot} applied to the –IN terminal will be half of V_{OUT}. Looking at it the other way, $V_{OUT} = 2 \times V_{bot}$. Therefore V_{OUT} must go to 2×5 V $= 10$ V.

If we now adjust the pot so that $R_{bot} = 2.5$ kΩ and $R_{top} = 7.5$ kΩ, V_{bot} becomes only one fourth of V_{OUT}. Then V_{OUT} must increase to 20 V to make $V_{bot} = 5$ V, balancing V_Z.

In general, the voltage-division formula gives

$$\frac{V_{OUT}}{V_{bot}} = \frac{10 \text{ k}\Omega}{R_{bot}} = \frac{V_{OUT}}{5 \text{ V}}$$

$$V_{OUT} = \left(\frac{10 \text{ k}\Omega}{R_{bot}}\right) 5 \text{ V} \qquad \text{Eq. (14-5)}$$

EXAMPLE 14-6

In Fig. 14-17, with $V_Z = 5$ V and $V_{IN} = 30$ V:
a) Find V_{OUT} if the pot is adjusted to $R_{bot} = 4.0$ kΩ.
b) What value of R_{bot} is needed to make $V_{OUT} = 24$ V?

SOLUTION

a) From Eq. (14-5),

$$V_{OUT} = \left(\frac{10 \text{ k}\Omega}{4.0 \text{ k}\Omega}\right) 5 \text{ V} = (2.5 \text{ V})(5 \text{ V}) = \textbf{12.5 V}$$

b) Rearranging Eq. (14-5) to solve for R_{bot} gives

$$R_{bot} = 10 \text{ k}\Omega \left(\frac{5 \text{ V}}{V_{OUT}}\right)$$

$$= 10 \text{ k}\Omega \left(\frac{5 \text{ V}}{24 \text{ V}}\right) = \textbf{2.08 k}\Omega$$

An adjustable IC regulator is shown schematically in Fig. 14-18. The external adjustment pot has its end terminals connected to the $+V_{OUT}$ and GND terminals, and its wiper terminal connected to the IC's CONTROL terminal. Stabilizing capacitors C_2 and C_3 are required across the input and output, as usual.

15 ✔

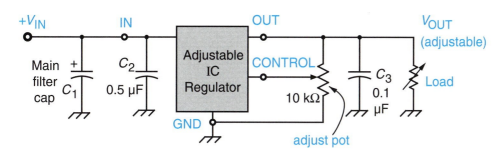

Figure 14-18
Type 78MG 4-terminal adjustable dc regulator with support components. Output voltage range is 5 to 30 V. Line regulation = 1%. Load regulation = 1%. Maximum output current = 0.8 A, typical. Ripple reduction factor = 10 000, typical.

FORMULAS

$$I_{\text{reverse}} = I_Z = \frac{V_S - V_Z}{R_{\text{ser}}} \qquad \text{Eq. (14-1)}$$

$$r_z = \frac{\Delta V_Z}{\Delta I_Z} \qquad \text{Eq. (14-2)}$$

$$I_Z = I_{R\text{ser}} - I_{LD} \qquad \text{Eq. (14-3)}$$

For a series-pass transistor regulator,

$$\Delta I_Z = \frac{\Delta I_{LD}}{\beta} \qquad \text{Eq. (14-4)}$$

SUMMARY OF IDEAS

● A zener diode operates in the reverse conducting mode, with electron current entering on the anode A and leaving on the cathode K.

● When a zener is conducting in the reverse direction, its voltage V_Z changes by only a small amount, even if its reverse current V_Z varies by a large amount.

● A zener diode's dynamic resistance r_z is the ratio of the zener's change in voltage to its change in current ($r_z = \Delta V_Z \div \Delta I_Z$).

● Standard low-power zener diodes have low dynamic resistance, r_z. Usually, r_z is only a few ohms.

● Because of its low r_z value, a zener diode can closely regulate the voltage across a load, even if the load varies. This is called its load-regulation ability.

● A zener can closely regulate V_{OUT} even if the supply-line voltage varies. This is called its line-regulation ability. For the same reason, a zener can greatly reduce ripple.

● By combining a series-pass transistor with a zener, we can regulate V_{OUT} over a much wider range of I_{LD}, compared to a zener alone.

● A regulating circuit containing a very high-gain amplifier, such as a Darlington pair or an op amp, can give excellent line and load regulation.

● An IC regulator is an integrated circuit containing the zener, and the high-gain amplifier and series-pass transistor all on one chip, in a package.

● Both positive and negative IC regulators are available.

● Both fixed-voltage and adjustable-voltage IC regulators are available.

● Many IC regulators must be heat-sunk to prevent their internal series-pass transistor from overheating at high load currents.

QUESTIONS AND PROBLEMS

1. Draw a schematic diagram of an ideal 10-V zener diode in series with a 500-Ω series dropping resistor, reverse-biased by a 15-V dc supply.

2. For the circuit you drew in Problem 1, find:
   a) The voltage across the zener. Mark the terminal polarity.
   b) The voltage across the resistor R_{ser}.
   c) The circuit's current.
   d) The power burned by the zener.

3. For the circuit that you drew in Problem 1, suppose the zener is not perfectly ideal ($r_z \neq 0\ \Omega$). Suppose that the actual r_z value is 9 Ω.
   a) If the dc supply is increased to 20 V, the circuit current will increase to what approximate value?
   b) This new current value will cause the actual voltage across the zener to increase by what amount? Use Eq. (14-2).

4. (T-F) A zener diode can be used to reduce the ac ripple content of a rectified, capacitor-filtered dc power supply.

5. When a zener diode is used with a dc power supply, if the power supply's ac input voltage changes by 10%, will the dc output voltage also change by 10%? Explain.

6. For the situation described in Question 5, in approximate terms how would you describe the change in dc output voltage?

7. Referring to your answers in Questions 5 and 6, we say that a zener diode gives a dc power supply good _____ regulation.

8. Draw a schematic diagram of a zener-regulated dc power supply with $V_Z = 12$ V and $r_z = 9\ \Omega$. With the load current $I_{LD} = 50$ mA, a DVM measures $V_{LD} = 12.14$ V.
   a) If I_{LD} increases to 60 mA, the zener current will _____ by _____ mA.
   b) Using Eq. (14-2), calculate the change in zener voltage, ΔV_Z.
   c) What will be the new value of load voltage measured by the DVM?

9. Referring to your answers to Problem 8, we say that a zener diode gives a dc power supply good _____ regulation.

10. Draw the schematic diagram of a dc-regulation circuit using a zener diode and a series-pass transistor.

11. (T-F) When a dc power supply must operate over a wide range of load currents, it must use a series-pass transistor in combination with a zener in order to achieve good load regulation.

12. Explain your answer to Question 11, specifically stating why the series-pass transistor must be used, rather than just the zener diode alone.

13. (T-F) Series-pass transistors used for dc power supply regulation are usually high-power types.

14. Draw the schematic diagram of a Darlington-zener regulating circuit.

15. Explain why the Darlington-zener regulating circuit provides even better load regulation than a single series-pass transistor circuit.

16. In the packaged IC regulator design of Fig. 14-13, the load current I_{LD} passes through _____ . (Which component ?)

17. (T-F) In Fig. 14-13, the voltage at the –IN terminal is virtually equal to V_Z.

18. In Fig. 14-13, explain why the current through the zener diode, I_Z, is virtually the same as the current through the dropping resistor.

19. In the dc regulator of Fig. 14-13, suppose $V_Z = 10.2$ V. If V_{OUT} were momentarily at 10.0 V, explain how the regulator would automatically correct this.

20. Draw the schematic symbol for an 8-V IC regulator.

21. Show all the support capacitors that are needed by the IC regulator of Problem 20. For each of these capacitors, give some approximation of its value.

22. (T-F) A type 7812 IC regulator will be destroyed if its V_{OUT} terminal is accidentally short-circuited to ground.

23. (T-F) The actual dc output voltage V_{OUT} from an IC regulator is guaranteed to be within ±0.1 V of its nominal voltage rating.

24. In Fig. 14-17, suppose $V_Z = 7$ V. If the 10-kΩ voltage-adjust pot is adjusted to $R_{bot} = 3$ kΩ and $R_{top} = 7$ kΩ, find V_{OUT}.

25. For the adjustable dc regulator of Problem 24, what adjusted value of R_{bot} will produce $V_{OUT} = 9$ V?

26. For the regulator of Problem 24, suppose $V_{IN} = 30$ V. Most IC regulators quit working when the difference between V_{IN} and V_{OUT} decreases to about 2 V or less. Therefore, this regulator will quit when V_{OUT} reaches about +28 V. What is the minimum adjusted value of R_{bot}, to keep the regulator working?

27. Based on your result from Problem 26, suggest an alteration to the adjustable regulator circuits (Figs. 14-17 and 18) to prevent the problem of regulator shut-down due to the output voltage V_{OUT} rising too close to V_{IN}.

CHAPTER 15

THYRISTORS AND OPTICAL DEVICES

Courtesy of Cox Sterile Products, Inc.

The sterilization unit shown at left uses high-velocity air (50 ft/s) at moderately high temperature (375°F) to achieve very rapid, completely reliable sterilization of medical instruments. Most instruments can be sterilized in 6 minutes, considerably less time than by other methods. Using electronically controlled heating, the unit is also very efficient, requiring much less power than other methods.

With its fast cycle time and low electric power consumption, it is perfect for disaster-relief vehicles, which are not large enough to have powerful generators or to carry an extensive backup supply of medical instruments.

OUTLINE

NEW TERMS TO WATCH FOR

thyristor

conducting

blocking

four-layer diode

forward breakover voltage

holding current

silicon controlled rectifier (SCR)

trigger

bidirectional

diac

reverse breakover voltage

unijunction transistor (UJT)

standoff ratio

peak voltage

light-emitting diode (LED)

seven-segment display

photo-detector

photo-diode

optical coupler

isolation

solid-state relay

In this book, you have been studying *analog* electronics — the branch of electronics dealing with signals that go through continuous gradual changes, such as a sine wave. The other branch of electronics is called *digital*. In digital electronics, the signals change abruptly. They make sudden jumps between one particular voltage, often +5 V, and ground, 0 V. The waveforms are square. We have already touched on one example of digital operation, in Section 12-5.

There is a class of electronic devices that are neither clearly analog nor clearly digital. They tend to have an analog-type input, but they produce sudden abrupt switching action at the output. This class of devices is called thyristors. They are important in the control of high-power electrical loads like motors and welders. We will study them in this chapter.

Often, thyristors are linked to low-power electronic circuits by optical (light-operated) electronic devices. We will also study opto-electronic devices at this time.

After studying this chapter, you should be able to: ✔

1. Describe the operation of a four-layer diode, including its characteristic graph of current versus voltage.
2. Draw the schematic symbol for a silicon controlled rectifier and describe its operation, including the conditions needed to drive it out of the blocking state into the conducting state.
3. Explain the condition necessary to force an SCR out of the conducting state and back into the blocking state.
4. Explain the operation of an SCR motor speed control circuit.
5. Draw the schematic symbol for a diac and describe its current-versus-voltage operation.
6. Draw and explain the operation of a triac. State the essential difference between a triac and an SCR.
7. Draw the schematic symbol of a unijunction transistor and describe its current-versus-voltage operation.
8. Show how a UJT is used to control the triggering instant for an SCR or triac.
9. Draw the schematic symbol for a light-emitting diode and explain its operation.

10. Draw the schematic symbols for a photo-diode and a photo-transistor, and describe the behavior of each device.
11. Show how a how-power electronic signal can be optically coupled to a high-power thyristor.

15-1 THYRISTORS: THE FOUR-LAYER DIODE

A **thyristor** is any electronic device that can operate in only two conditions — saturated and cutoff. There are numerous devices that are members of the thyristor family. They all share this feature: the inability to operate in a linear manner.

We usually don't apply the transistor-related words saturated and cutoff to thyristors. Instead we use the words **conducting** and **blocking**. In this context, consider the word conducting to mean conducting almost perfectly, like a closed switch.

The simplest thyristor to understand is the **four-layer diode**, also called the Shockley diode. Its schematic symbol is shown in Fig. 15-1.

Figure 15-1
Schematic symbol of four-layer diode, showing anode (A) and cathode (K) leads.

A four-layer diode behaves like a standard diode when it is reverse biased, with the cathode more positive than the anode. It carries virtually zero current, or blocks. When a four-layer diode is forward-biased, it does not begin conducting when the forward voltage reaches 0.6 V. Instead, the forward voltage must increase to the **forward breakover voltage** value, symbolized $V_{BR(F)}$, in order for the device to switch into conduction. This behavior is shown by the characteristic curve of current versus voltage in Fig. 15-2.

Figure 15-2
Characteristic curve of a four-layer diode. The forward-biasing voltage must reach the value $V_{BR(F)}$ in order to start the diode conducting. Once that value is reached though, V_{AK} immediately drops to a smaller value.

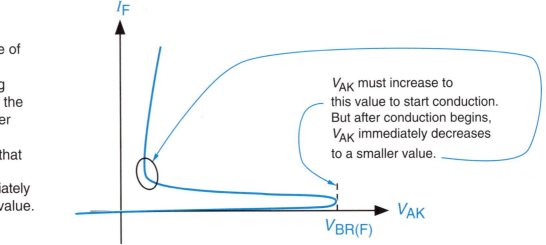

V_{AK} must increase to this value to start conduction. But after conduction begins, V_{AK} immediately decreases to a smaller value.

Notice that a conducting four-layer diode does not maintain a constant value of voltage across its terminals, as a zener diode does. The four-layer diode is a "surge-on" device, because its V_{AK} voltage breaks back to a reduced value once the critical breakover point is reached. A typical value of breakover voltage is 8 V.

Once a four-layer diode has been driven into the conducting state, it remains in that state as long as the forward current exceeds a certain critical value, called the **holding current**, symbolized I_H.

In Fig. 15-3, as long as the external circuit conditions cause the current I to be greater than the diode's I_H value, usually a few milliamps, the device will remain in its ON state.

For all thyristors, the conducting state is called the ON state.
The blocking state is called the OFF state.

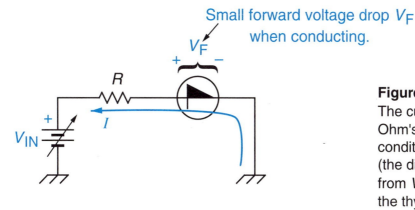

Small forward voltage drop V_F when conducting.

Figure 15-3
The current I is set mostly by Ohm's-law external circuit conditions V_{IN} and R (the diode's small V_F subtracts from V_{IN}). As long as $I > I_H$, the thyristor stays ON.

When we wish to turn OFF a four-layer diode, we could arrange for V_{IN} in Fig. 15-3 to decrease to a very small value. If V_{IN} gets small enough, Ohm's law will call for an I value that is less than I_H.

However, the more usual ways of turning OFF a thyristor are shown in Fig. 15-4.

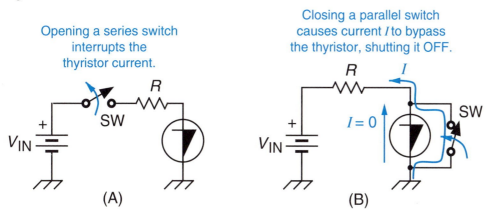

Figure 15-4 Standard methods for shutting OFF a conducting thyristor.
(A) Mechanical or electronic switch in series.
(B) Switch (usually electronic) in parallel.

Either method in Fig. 15-4 will place the thyristor into the blocking (OFF) state. When the switch is later returned to its original condition, the thyristor may or may not go back ON, depending on whether $V_{IN} > V_{BR(F)}$.

Another thyristor device that is closely related to the four-layer diode is the silicon unilateral switch, or SUS. It has a third terminal, called the gate. By connecting an external zener diode between the gate and cathode, the effective breakover voltage value can be made equal to the zener's voltage, V_Z.

Some people make no distinction between four-layer diodes and SUSs. Thus, you may hear a two-terminal four-layer diode *called* an SUS. This naming issue is not really very important, since both four-layer diodes and SUSs have become quite rare. We made this detailed discussion of the four-layer diode only because it is the simplest thyristor, and therefore gives us a good starting place for understanding all other thyristors.

15-2 THE SCR

The most common thyristor is the **silicon controlled rectifier**, or SCR. It has three terminals, as shown in the schematic diagram of Fig. 15-5.

Figure 15-5
SCR symbol. The main current flow path is from cathode to anode.

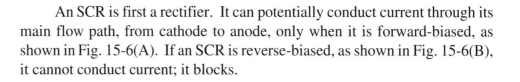

An SCR is first a rectifier. It can potentially conduct current through its main flow path, from cathode to anode, only when it is forward-biased, as shown in Fig. 15-6(A). If an SCR is reverse-biased, as shown in Fig. 15-6(B), it cannot conduct current; it blocks.

Figure 15-6
SCR rectifying behavior.
(A) With forward-bias V_{AK}, the SCR may permit current to flow.
(B) With reverse-bias V_{AK}, the SCR always remains OFF, blocking current.

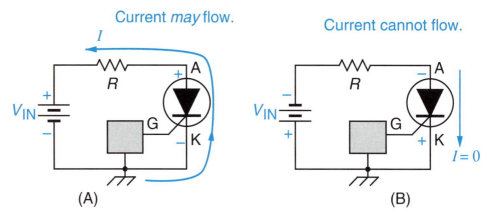

When an SCR is forward-biased, anode more positive than cathode, it would go into the conducting state if V_{AK} exceeded its forward breakover voltage rating, as suggested in Fig. 15-7. However, we almost never use SCRs this way.

Instead, with a smaller value of V_{AK} applied, an SCR can be **triggered** into the ON state by a momentary flow of current in the gate lead. If the gate current momentarily exceeds a certain critical value, symbolized I_{GT}, then the SCR breaks back from some point along the dashed line of Fig. 15-7. This places it in the forward conducting region, where it becomes a virtual short circuit, with V_{AK} near zero volts. The forward current I_F is limited only by the series circuit resistance, R in Fig. 15-6(A).

An SCR breaks back from any small V_{AK} value (as indicated by the left-most blue segment in Fig. 15-7) when its gate current I_G reaches the critical trigger value, I_{GT}.

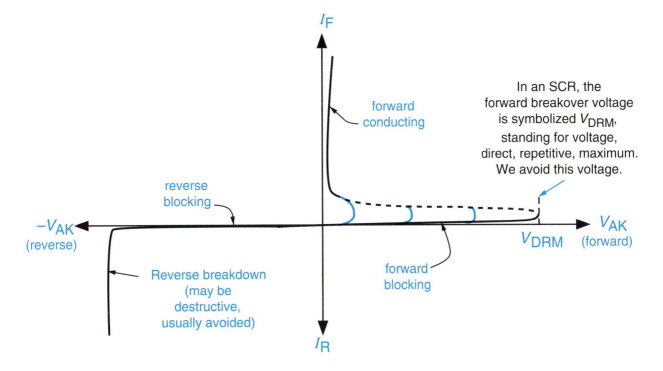

In an SCR, the forward breakover voltage is symbolized V_{DRM}, standing for voltage, direct, repetitive, maximum. We avoid this voltage.

Figure 15-7 Characteristic curve for an SCR .

Such gate-controlled triggering is the normal means of driving an SCR into the conducting state.

Once an SCR has been triggered into the ON state, it remains in that state as long as its main-terminal current stays above its holding value, I_H. There is no need to maintain a continuing flow of gate current I_G.

② ✔

Power-Controlling Ability: SCRs are commonly used to control high-power electric loads, such as lights, motors, heating coils, and welders. They are manufactured in the same package configurations as high-power transistors [see Figs. 4-34(D) and (E)]. For comparable package size and heat sinks, their maximum current capabilities tend to be greater than for transistors. This is because the SCR is always in a "saturated" condition, with V_{AK} at quite a low value, whenever it is conducting. This low value of voltage multiplied by the large cathode-anode current produces a non-threatening power dissipation $(P = VI)$.

Some very high-current units are manufactured in a package that looks like a giant stud diode. Or you may see a package that reminds you of a hockey puck — same shape and about the same size.

The voltage across an SCR's A-K terminals during conduction is symbolized V_T, with the T standing for triggered. Most SCRs have a V_T value less than 2 volts.

Switching back to Blocking State: SCRs can be used in either dc or ac circuits. In most applications, they are switched between conducting and blocking states (ON and OFF) at a fairly rapid rate. When used in a 60-Hz ac line circuit, the ON-OFF switching rate is the same as the line frequency — 60 Hz; or, in some full-wave circuits, 120 Hz. With ac , there is a natural turn-OFF mechanism at work. Every time a positive half-cycle ends, the ac source reverse-biases the SCR. This stops the holding current and puts the SCR back into blocking condition.

③ ✔

When used in a dc circuit, this natural mechanism is not available. Then the SCR must be turned OFF by one of the methods described in Fig. 15-4.

Ac Waveforms: Figure 15-8 shows an SCR in an ac circuit. Here is how it works.

During the negative half-cycle, the SCR is OFF, the current is zero, and the load voltage is zero. This is shown on the far left in Fig. 15-9. When the positive half-cycle begins, capacitor C begins charging + on top, − on bottom in Fig. 15-8. The rate at which C charges is set by the pot. If the pot is adjusted up toward the top, it picks off a larger portion of the rising source sine wave and charges C rapidly. If the pot is adjusted down toward the bottom, C charges more slowly.

At some point during the positive half-cycle, the capacitor voltage will rise to a large enough value to drive the critical amount of current, I_{GT}, through the G-K junction and resistor R_G. This gate current path is shown in Fig. 15-8. At that moment, the SCR is triggered into the ON state. Thyristor switching is so fast that we regard it as instantaneous. Therefore, the current (or v_{LD}) waveform instantly jumps to the instantaneous value of the ac source. Figure 15-9 shows this clearly.

Figure 15-8
Typical SCR power-control circuit. The load is essentially in series with the SCR. Relatively little current flows in the gate control circuitry.

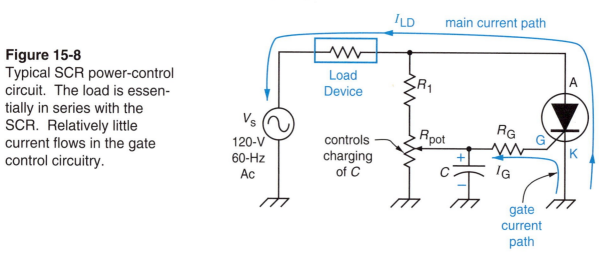

The load current and voltage waveforms then track the source sine wave for the rest of the positive half-cycle.

When the source passes through negative-going zero at 180°, the thyristor loses its forward bias, the main current slips below I_H, and we return to blocking state. This same switching sequence is repeated over and over, 60 times per second, indefinitely.

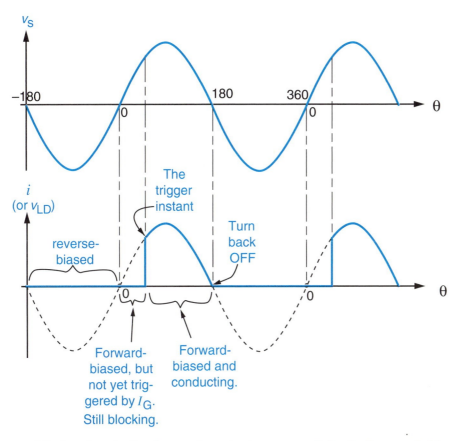

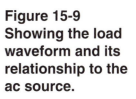

**Figure 15-9
Showing the load
waveform and its
relationship to the
ac source.**

Notice that the load waveform is dc, not ac. It is similar to a half-wave rectified waveform, but it contains less overall power, because a piece of the wave is missing.

Now here is how the SCR becomes so useful. We can change the amount of overall load power by changing the triggering instant. For instance, by adjusting the pot to a higher position, we can charge the gate capacitor faster and trigger the SCR earlier in the cycle. This is represented by the waveforms in Fig. 15-10(A).

It is not difficult to design SCR circuits that use both half cycles. They produce waveforms similar to a full-wave rectifier.

The triggering instant is also called the *firing* instant.

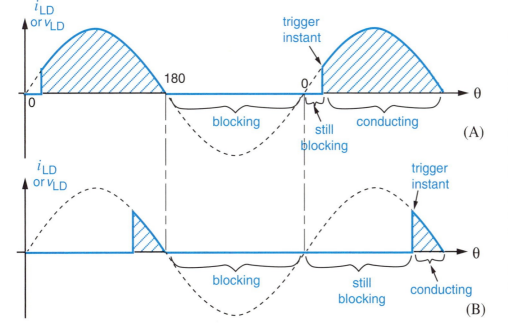

Figure 15-10
Showing the effect of changing the triggering instant.

(A) Earlier triggering gives more of the waveform, so load power increases.

(B) Later triggering gives less of the waveform, so the load is throttled down (less average current and less power).

Or, we can adjust the pot lower, charge *C* slower, and trigger our SCR later in the cycle. This is shown in Fig. 15-10(B). Under this condition, the load receives less power. If the load were a bank of lights, the light brilliance would decrease. If the load were a motor, it would develop less torque, causing it to slow down.

SELF-CHECK FOR SECTIONS 15-1 AND 15-2

1. The family of electronic devices that never operate in a middle range, but that always switch abruptly from completely OFF to completely ON is the _____ family.
2. A four-layer diode switches abruptly from OFF to ON when the applied anode-to-cathode voltage exceeds the _____ _____ voltage.
3. Give the symbol for the voltage in Question 3.
4. With thyristors, the formal term for turned-OFF is _____. The formal term for turned-ON is _____.
5. (T-F) An SCR *can* be turned ON by exceeding its forward breakover voltage, but that is rarely done.
6. Considering your answer to Question 5, what is the usual way of forcing an SCR into the ON state?
7. (T-F) When an SCR is used in an ac circuit, it rectifies the ac to produce dc across the load.
8. Draw the load current (or voltage) waveform for a triggering instant at 90°, exactly 1/4 of the way through the ac cycle.
9. Repeat Problem 8 for a 45° triggering instant.
10. In Problem 9, the earlier triggering instant causes the load waveform to deliver _____ power to the load. (Answer greater or lesser.)

15-3 BIDIRECTIONAL THYRISTORS: THE DIAC

The four-layer diode, SUS, and SCR are all unidirectional (only one direction of current) thyristors. There are also **bidirectional** thyristors. They are able to carry current in both directions.

The diac is also called the bidirectional trigger diode.

The simplest bidirectional thyristor is the **diac**, shown schematically in Figs. 15-11(A) and (B). Its characteristic curve is shown in Fig. 15-12.

Figure 15-11 Schematic symbols for a diac. The terminals are not called anode and cathode because the diac is not a rectifier. Instead, they are called anode 1 and anode 2.

If a diac is either forward- or reverse-biased by a low-magnitude voltage, it blocks. But when the biasing voltage exceeds the breakover value in either polarity, the diac surges on, as Fig. 15-12 shows.

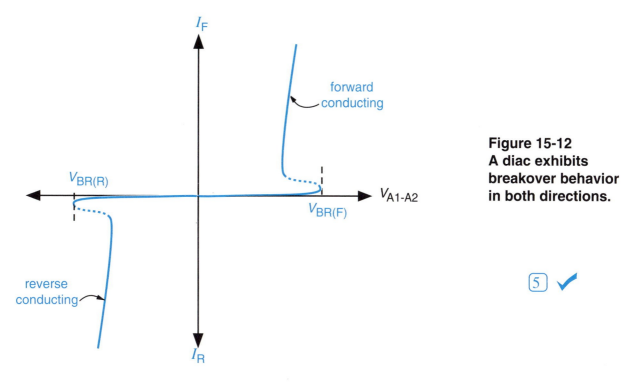

**Figure 15-12
A diac exhibits
breakover behavior
in both directions.**

Ideally, the forward breakover voltage $V_{BR(F)}$ and the **reverse breakover voltage** $V_{BR(R)}$ should have identical magnitudes. Thus, a 32-V-rated diac should break into the forward ON state at exactly 32.0 V positive, and it should break into the reverse ON state at exactly –32.0 V negative. In reality there may be a slight difference in the $V_{BR(F)}$ and $V_{BR(R)}$ magnitudes. The diac manufacturer will specify the worst-case difference that can occur between those two voltages. The smaller this difference is, the better the diac is considered to be. A high-quality diac will have a worst-case difference of about 2 V. Most units within a batch will have much less difference.

In order to remain in a conducting state, a diac is like all other thyristors. If its current stays above the critical I_H value (either direction), the diac will remain ON. If current I drops below the I_H magnitude, the diac reverts to the blocking state.

This technician is doing an initial visual inspection of newly manufactured circuit boards, to spot any obvious faults. After visual inspection, the boards will be subjected to electronic testing.

Courtesy of Enerpro, Inc.

The silicon bilateral switch (SBS) is a device that is related to the diac. It gives symmetrical bidirectional breakover action, with V_{BR} usually in the range of 8 to 15 V. An SBS has a third terminal, the gate G. The user can connect an external zener diode between the G terminal and either A1 or A2. In this way, you can tailor the SBS breakover voltage(s) to any desired value that is less than the natural V_{BR} rating.

15-4 THE TRIAC

The **triac** is a bidirectional version of the SCR. It is a high-current thyristor that is triggered into the conducting state in either direction, by a momentary surge of gate current. Its schematic symbol is shown in Fig. 15-13(A) and its characteristic curve in Fig. 15-13(B).

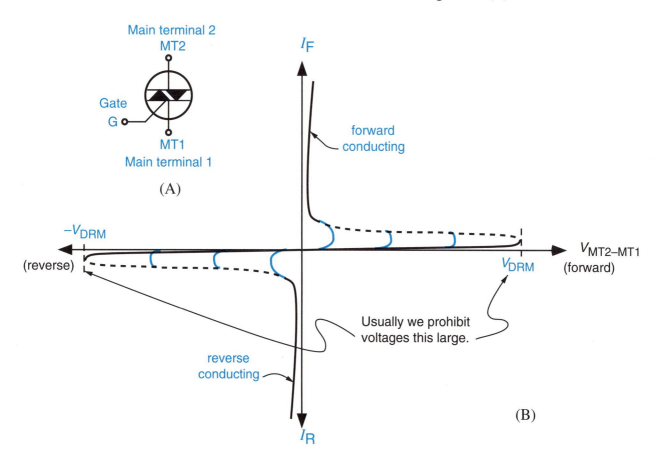

Figure 15-13 Triac (A) Schematic symbol. The main terminals are called just that, MT2 and MT1. The one closer to the gate is MT1. (B) Characteristic curve.

As with an SCR, a triac *could* be driven into conduction by applying a main terminal voltage greater than V_{DRM}, but that is almost never done. Instead, it is triggered into abrupt conduction by a surge of gate current I_G. The current surge must be greater than the triac's critical value, I_{GT}. It is possible to buy triacs with relatively small values of I_{GT}, a few tenths of a milliamp. These are called sensitive-gate triacs. Or you can obtain triacs with larger I_{GT} values, several milliamps or several tens of milliamps, for situations where there is danger of false triggering.

The same choice of sensitive-gate or non-sensitive-gate is available with SCRs.

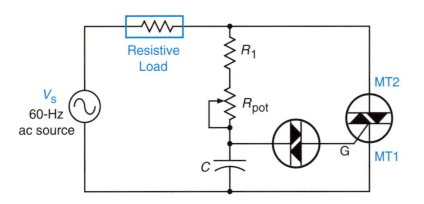

Figure 15-14
A triac is connected essentially in series with the load device it is controlling. Current through the *RC* gate-triggering circuit is negligible.

Figure 15-14 shows a triac controlling ac power to a load device, with a diac used in the gate trigger-control circuit. It works as follows.

When a positive half-cycle begins, it starts charging the capacitor in the *RC* circuit. The capacitor will eventually reach the breakover voltage $V_{BR(F)}$ of the diac. When that happens, the diac surges into conduction. This produces a surge of current in the triac's gate lead. The current surge is certain to exceed I_{GT}, since there is little or no resistance in the gate current path. Therefore the triac is reliably triggered into the ON state.

In Fig. 15-15, the triggering moment is shown at about 70°. As we know, that point can be adjusted earlier or later, by adjusting the pot resistance in the *RC* charging circuit.

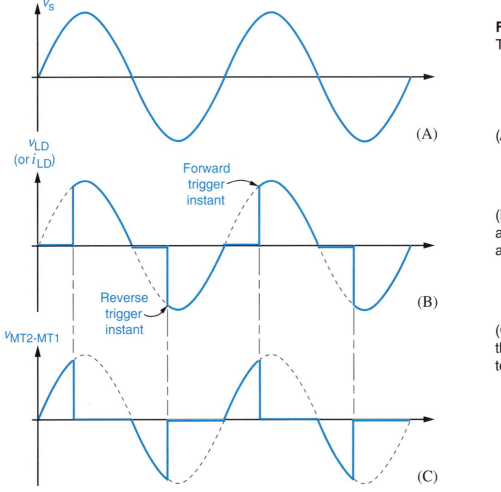

Figure 15-15
Triac waveforms.

(A) Ac source.

(B) Load voltage for a trigger instant at about 70°.

(C) Voltage across the triac's main terminals.

During the negative half-cycle, the same events occur, but with the opposite polarity. The capacitor now charges negative on top. When it reaches the diac's reverse breakover voltage $V_{BR(R)}$, it causes a burst of reverse current in the gate. (Electron current now flows into the gate terminal, through the G-MT1 junction, and out the MT1 terminal.) This triggers the triac into the reverse conducting state.

Because the diac has identical forward and reverse breakover voltages [$V_{BR(F)} = V_{BR(R)}$, ideally], it triggers the triac at identical points on both half-cycles. The triac could not do this on its own, because it has different I_{GT} values in the forward and reverse directions.

Also, triacs, being high-power devices, are subject to batch instability and temperature instability. Diacs are manufactured to be much more stable in both regards. Thus, placing a diac in the gate lead eliminates the triac's natural instability problems.

Notice that the load waveform in Fig. 15-15 is basically ac, not dc. Triacs are better adapted to driving ac loads. SCRs are better adapted to driving dc loads.

SELF-CHECK FOR SECTIONS 15-3 AND 15-4

11. Explain the difference between a bidirectional thyristor and a unidirectional thyristor.
12. (T-F) Ideally, a diac's forward breakover voltage, $V_{BR(F)}$, and its reverse breakover voltage, $V_{BR(R)}$, are identical in magnitude.
13. (T-F) In reality, $V_{BR(F)}$ and $V_{BR(R)}$ usually differ by a great deal, as much as 15%.
14. Draw the fundamental schematic relationship between an ac source, a load, and a triac.
15. (T-F) A triac's gate trigger-control circuitry is likely to contain an *RC* charging circuit.
16. (T-F) Including a diac in a triac's gate trigger-control circuitry improves the system's symmetry (same performance for both positive and negative).

15-5 UNIJUNCTION TRANSISTOR (UJT)

A **unijunction transistor** behaves very much like a low-power thyristor. It blocks current until a certain critical voltage is reached. Then it abruptly switches into the conducting state. What is different about a UJT is that its breakover voltage is not a definite fixed value. Rather, it is a certain percentage of an externally applied bias voltage. To understand this, refer to the UJT schematic symbol in Fig. 15-16(A) and the standard circuit arrangement in Fig. 15-16(B).

As shown in Fig. 15-16(A), the terminal names are base 2 , base 1 , and emitter. Do not try to make any sense out of these names. They do not make sense according to our usual understanding of the words base and emitter.

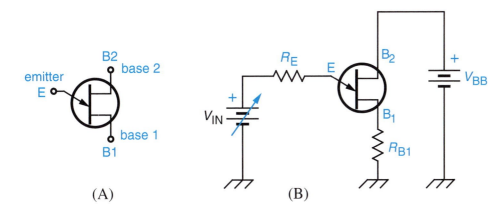

Figure 15-16

(A) UJT schematic symbol and terminal names.

(B) UJT connected into an operating circuit.

(A)   (B)

Figure 15-16(B) shows how a UJT is usually connected. Resistance R_{B1} is a few hundred ohms, typically. The V_{BB} bias voltage is generally in the 10 to 20-V range.

There is an internal current flow path from terminal B1 to terminal B2, having several thousand ohms of resistance. Therefore, a moderate amount of current flows through R_{B1}, through the B_1 to B_2 body of the UJT, and into the V_{BB} supply terminal, as shown in Fig. 15-17(A). Most of the V_{BB} voltage is dropped across the UJT itself; very little voltage appears across R_{B1}. This is because the internal interbase resistance is much larger than R_{B1}.

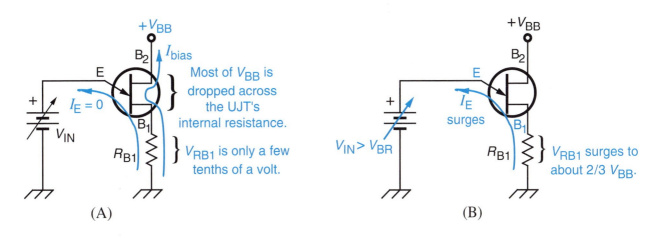

(A)   (B)

Figure 15-17 (A) With V_{IN} less than the breakover value, zero current flows through the E-B1 junction. (B) If V_{IN} exceeds the breakover value, a sudden surge of current flows through the E-B1 junction and through R_{B1}.

The breakover voltage value for the UJT is given by the formula

$$V_{BR} = \eta\, V_{BB} + 0.6\text{ V} \qquad \text{Eq. (15-1)}$$

where η is the **standoff ratio** for that particular type of UJT. Most UJTs have a standoff ratio in the range of 0.5 to 0.8. If the externally applied input voltage V_{IN} rises and exceeds V_{BR}, the UJT's internal resistance suddenly declines. This causes a sudden surge of current through the B1-E path, as shown in Fig. 15-17(B). The current surge produces an abrupt increase in v_{RB1}. This v_{RB1} voltage pulse is detected and used as the circuit's output.

7 ✔

EXAMPLE 15-1

In the circuit of Fig. 15-17, suppose $V_{BB} = 12$ V, and the UJT has $\eta = 0.75$. Find the breakover voltage for this circuit.

SOLUTION

From Eq. (15-1),

$$V_{BR} = \eta V_{BB} + 0.6 \text{ V}$$

$$= (0.75)\, 12 \text{ V} + 0.6 \text{ V} = 9.0 \text{ V} + 0.6 \text{ V} = \mathbf{9.6 \text{ V}}$$

The term peak voltage for a UJT is a confusing term, since peak voltage usually means the maximum value of a sine wave. It is unfortunate that this term and symbol have been widely accepted for UJTs.

In UJT terminology, the breakover voltage is usually called the **peak voltage**, symbolized V_P.

UJT Application

A popular application of the UJT is as a trigger-control device for high-power thyristors — SCRs and triacs. Figure 15-18 shows a schematic example.

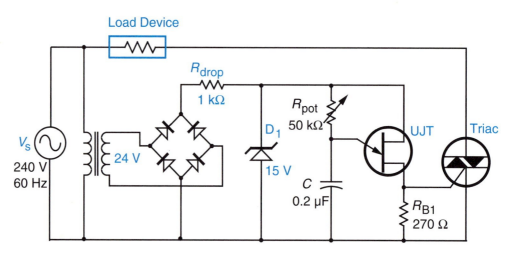

Figure 15-18
A UJT functions very well as a surge-type triggering device for a high-power thyristor.

The full-wave bridge rectifier, in combination with R_{drop} and the zener diode D_1, applies +15 V dc to the UJT timing circuit. This happens on every half-cycle, positive or negative, as shown in the Fig. 15-19(B) waveform.

Actually, there would be a small fixed resistor R_f installed just above R_{pot}.

Capacitor C starts charging through R_{pot} when the 15 V dc is applied. Depending on the pot adjustment of the charging time constant, the capacitor will raise the UJT's emitter voltage v_E to the 11-V breakover value (peak value), either early or late in the half-cycle. In Fig. 15-19(C), it is reaching the peak value at about 100°. At that moment the UJT fires, causing the capacitor's charge to surge through R_{B1} and the triac gate lead. This short-lived surge is represented by the v_{RB1} (and v_G) waveform of Fig. 15-19(D). The triac is triggered into the ON state and remains conducting for the rest of the half-cycle, as shown in Fig. 15-19(E).

Once the capacitor's charge has been exhausted and the UJT emitter current falls below its holding value, the UJT's E-B1 junction reverts to its blocking state. Therefore the capacitor begins charging a second time (within that half-cycle), as Fig. 15-19(C) makes clear. Near the end of the half-cycle, as the UJT's base-bias voltage falls toward zero, so does its peak (breakover) voltage, V_P. Therefore the UJT is bound to trigger a second time, when its V_P falls below whatever voltage is on the capacitor. This second burst, visible in the part (D) waveform, has no effect on the triac, since it is already conducting anyway at that moment. However, it does serve to discharge the capacitor so that it starts over fresh on the next half-cycle.

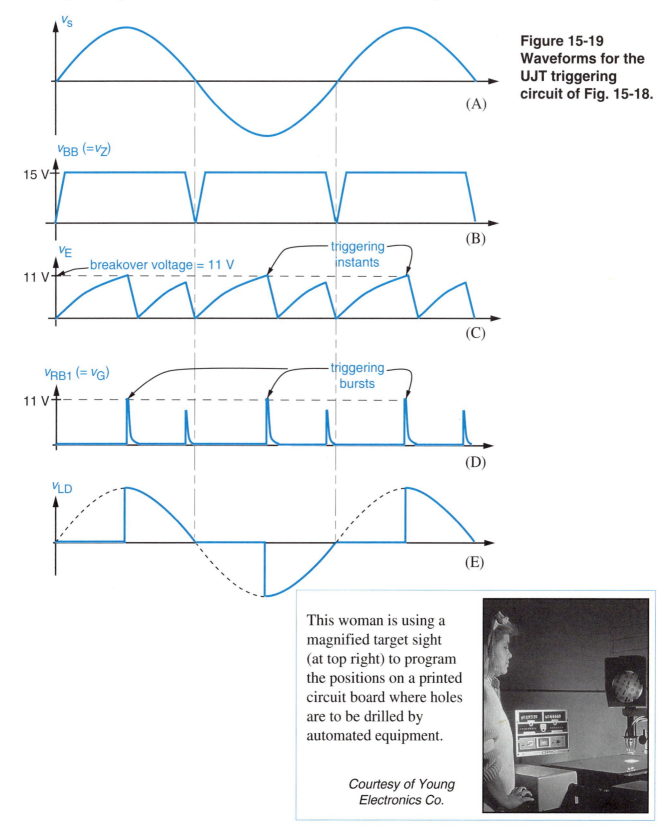

Figure 15-19
Waveforms for the
UJT triggering
circuit of Fig. 15-18.

This woman is using a magnified target sight (at top right) to program the positions on a printed circuit board where holes are to be drilled by automated equipment.

Courtesy of Young
Electronics Co.

15-6 LIGHT-EMITTING DIODE (LED)

The **light-emitting diode**, shown schematically in Fig. 15-20, has become a very popular device. When it is forward-biased by approximately 2 V or more, it starts conducting and gives visible light. This is indicated in Fig. 15-21(A).

Figure 15-20
Schematic symbol for an LED.
Sometimes, the arrows are
drawn with wavy lines.

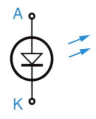

When an LED is reverse-biased, as in Fig. 15-21(B), it carries zero current and gives no light — it goes "dark".

This action is useful for giving a clear visible indication of the status of a certain electrical point. If the voltage at a point is above a certain value, we arrange for the LED to turn ON ; below that value, the LED reverts to OFF. Thus, at a glance, the LED tells us the condition at that point.

Figure 15-21
(A) Forward-biased LED
conducts current,
turns ON,
becoming "light".
(B) Reverse-biased LED
blocks current,
turns OFF,
going "dark".

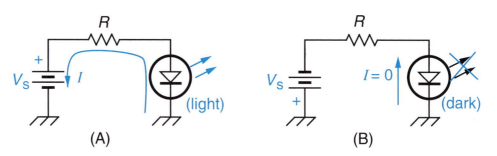

The advantage of an LED over a simple lamp bulb is its low power consumption. An LED can produce a visible amount of light with only a few tens of milliamperes current drain from the circuit. This is especially important in low-power digital circuits. Another LED advantage is its very long life, compared to an incandescent lamp.

LEDs are manufactured for several different colors. Red is the most common; but green, yellow, orange and blue are also available. The difference is truly a different light color, not just a different-colored lens cap, as in incandescent lamps.

Most LEDs can be tested with a standard ohmmeter. A good LED should measure very high resistance when reverse-biased by the ohmmeter, and lower resistance when forward-biased. However, the ohmmeter's applied voltage must be greater than about 3 V, since the forward conducting voltage for LEDs can range from about 1.5 V to 3 V. Also, the reverse breakdown voltage for LEDs is typically less than 15 V, and sometimes as low as 5V. To avoid being fooled by a low-resistance indication in reverse breakdown, you can place a signal diode in series with the red ohmmeter lead before touching it to the LED's cathode.

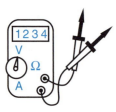

An LED may or may not give off enough light to be visible when it is forward-biased, since the ohmmeter's current-limiting resistor may keep the current very low.

Seven-Segment Displays: Light-emitting diodes are often used in seven-segment displays for forming numbers and letters. The physical appearance of a display module is shown in Fig. 15-22(A). The internal LED circuit diagram is shown in Fig. 15-22(B). In this arrangement, all seven of the LED cathodes are jumpered together and grounded. The anodes are brought out on seven separate leads, labeled **a** through **g**. A control circuit applies voltage to the proper combination of leads to form the desired numeral or letter. All ten numerals can be formed, and most of the letters of the alphabet.

Actually, there is another technology for making 7-segment displays, which is more popular than LEDs. That is liquid-crystal technology.

(A)

(B)

Figure 15-22
(A) Seven-segment display.
(B) Each segment is a lens-capped container enclosing an LED.

This is called *common-cathode* – the cathodes are tied together and all seven anodes are brought out separately. There is also a *common-anode* arrangement.

SELF-CHECK FOR SECTIONS 15-5 AND 15-6

17. Draw the schematic symbol and label the leads of a unijunction transistor.
18. (T-F) For a UJT, a dc bias voltage V_{BB} is applied positive on B2 (top) and negative on B1 (bottom).
19. (T-F) The breakover circuit path for a UJT is between emitter E and B1.
20. A UJT's breakover voltage is formally called _____ voltage. It is symbolized _____ .
21. (T-F) A UJT's breakover voltage is not a fixed value, but depends on the value of applied bias voltage V_{BB}.
22. Draw the schematic symbol for a light-emitting diode and explain what an LED does.
23. (T-F) Most LEDs require a forward current greater than 100 mA in order to produce a visible amount of light.

This cancer researcher is using a cytometer, an electronic instrument that measures the concentration of particular kinds of blood cells.

*Courtesy of
University of Miami*

ELECTRONIC SCOREBOARD

The electronic scoreboard in Los Angeles' Dodger Stadium is 10.25 meters wide by 7.68 meters high —about 34 feet by 25 feet. It works by illuminating small segments or "elements" of the screen, each one measuring 8 cm by 8 cm. There are 12 288 of these small screen segments, with each segment containing 16 colored "dots" in a 4× 4 array. The dots have a square shape, 1.3 cm on a side, spaced 2 cm between centers. The total number of dots on the screen is 196 608.

   The brightness of each dot is variable in 16 steps. Therefore the average brightness of each screen segment is variable in 256 steps (16 steps per dot multiplied by 16 dots per element = 256 steps.) The average color of each screen segment is controlled by the relative brightnesses of its 16 dots, eight of which are green, four blue, and four red.

   The board's electronic control circuit completely scans and alters all 196 608 dots in 1/60 second (60 complete screen updates per second). This enables the screen to display high-quality images of fast action as it occurs on the field.

   In the control room, the scoreboard operators can see the images from all the video cameras located throughout the stadium, on their console monitors. They can switch any one of the video images onto the scoreboard in an instant.

Photos by Andrew Semel, courtesy of Mitsubishi Electronics America, Inc.

Prerecorded messages and sports information can also be displayed when action on the field is slack.

15-7 OPTICAL COUPLING/ISOLATION

There are devices that work in the reverse way from an LED. Instead of emitting light when an electrical signal is applied, they produce an electrical signal when light shines upon them. Such devices are called **photo-detectors**. The most common one is the **photo-diode**, shown schematically in Figs. 15-23(A) and (B). Notice that the light-signifying arrows point toward the diode in part (A), rather than away from it, like an LED's arrows. Another way of symbolizing the presence of light is with the Greek letter λ, shown in part (B).

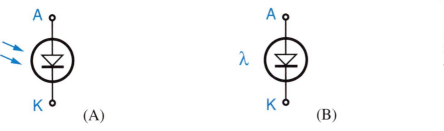

**Figure 15-23
Schematic symbols
for a photo-diode.**

(A)   (B)

Photo-diodes are always operated with reverse bias, as indicated in Fig. 15-24. They are never forward-biased. When a reverse-biased photo-diode is dark, as in Fig. 15-24(A), its reverse current is a small fraction of a microamp. When it is exposed to light, as in Fig. 15-24(B), its reverse current increases by a factor of perhaps 500:1.

Even though $I_{R(light)}$ is still quite small at $50\,\mu A$, it is noticeably different from $I_{R(dark)}$. This difference can be amplified to produce a final large output signal that tracks light intensity.

⑩ ✔

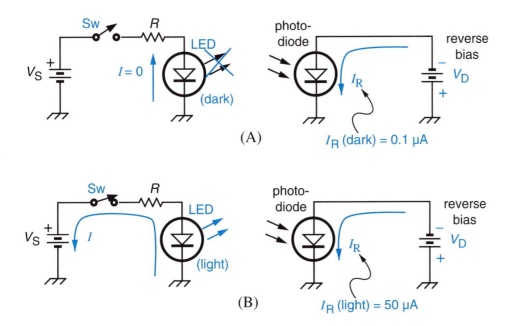

Figure 15-24
A photo-diode responds to the presence of light shining through the lens on its body.

(A) With no light present (LED is OFF) the reverse current, $I_{R(dark)}$, is very small.

(B) With LED light shining on the photo-diode, its reverse current increases by a large factor.

There are also photo-transistors and photo-SCRs. The latter are called LASCRs (pronounced lass$'$ – car), standing for <u>l</u>ight-<u>a</u>ctivated SCR.

It is common for an LED and a photo-detector circuit to be placed together in a sealed, light-tight enclosure. This enclosure is then referred to as an **optical coupler**. The arrangement is shown in Fig. 15-25.

Also called an optical isolator.

The major advantage of an optical coupler is that there is total electrical and magnetic **isolation** between the input circuit and the output circuit. Therefore, electromagnetic noise occurring in one circuit cannot be accidentally coupled into the other circuit. This is an important consideration in situations where a sensitive electronic input circuit must be protected from large noise spikes that can occur in the high-power load output circuit.

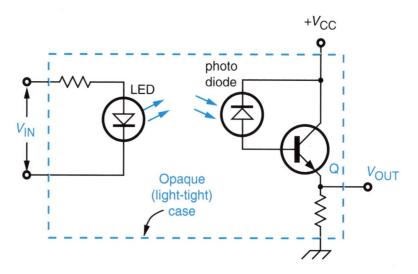

Figure 15-25
Optical coupler.
If V_{IN} turns the LED ON, the photo-diode carries a larger reverse current. This turns Q on harder, causing V_{OUT} to increase.

Optical couplers are manufactured that are especially adapted to triac triggering. They contain an LED and the equivalent of a photo-sensitive diac, as suggested in Fig. 15-26.

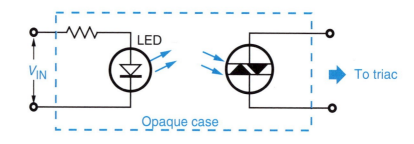

Figure 15-26
Schematic representation of an optical triac driver.

It is common to see an optical triac driver along with a triac and its gate supporting circuitry all enclosed in one package. This is shown in Fig. 15-27(A). Such a collection is called a **solid-state relay**, or SSR.

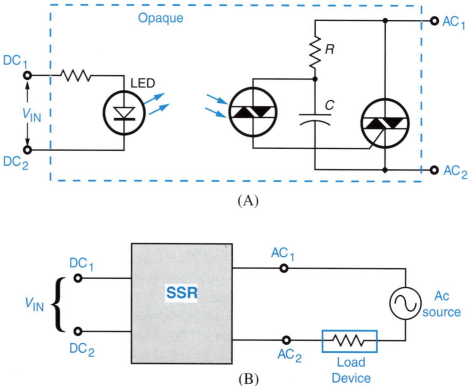

Figure 15-27
(A) Simplified schematic of a solid-state relay. The *RC* charging time constant is very fast, so it reaches the V_{BR} value of the photo-diac very shortly after a half-cycle begins. This triggers the triac almost immediately after the start of the half-cycle, if the photo-diac is enabled by light.

(B) Showing the connection arrangement for an SSR, the ac line, and the load.

An SSR is connected to the external power circuit as shown in Fig. 15-27(B). When a proper polarity V_{IN} is present, the triac essentially passes the entire ac wave to the load, like a closed relay contact. When V_{IN} is removed, the LED goes dark and the photo-diac cannot conduct. Therefore the triac never triggers. This blocks the ac current path to the load, like an open relay contact.

FORMULA

For a UJT

$$V_P = \eta \, V_{BB} + 0.6 \text{ V} \qquad\qquad \text{Eq. (15-1)}$$

SUMMARY OF IDEAS

● Thyristors are the class of devices that switch abruptly from nonconducting (blocking) to full conduction. They cannot operate in a partial conduction manner.

● Thyristors, for example the four-layer diode, have a certain critical voltage, called breakover voltage, at which they break into conduction.

● The silicon controlled rectifier (SCR) is a high-current thyristor. It conducts in one direction only (it rectifies).

● SCRs and other high-current thyristors are almost never driven into the conduction state by exceeding their main-terminal breakover voltage. Instead, they are triggered into the conducting state by a momentary flow of current in their gate lead.

● The diac is a low-current bidirectional (breaks over in both directions) thyristor. It is used as a stabilizing device in the trigger-control circuit of a high-current thyristor.

● Diacs have good symmetry. This means that their forward or positive breakover voltage and their reverse or negative breakover voltage are very close in magnitude.

● A triac is a bidirectional high-current thyristor.

● When a triac is placed in series with a load, it produces an ac current through the load whose average (rms) value is continuously variable by adjusting the trigger instant.

● A unijunction transistor (UJT) is a unidirectional thyristor whose breakover voltage V_P is not fixed. Instead, it depends on the applied bias voltage, V_{BB}.

● UJTs are useful as trigger-control devices for high-power thyristors, SCRs and triacs.

● The light-emitting diode (LED) is a diode that emits colored light when it is conducting. An LEDconsumes much less power than a standard lamp, and has much greater life expectancy.

● A photo-detector is a semiconductor device that produces an electrical signal change when light shines upon it. There are photo-diodes, photo-transistors, and photo-SCRs.

● By combining an electronic light source (LED) and an electronic light sensor (photo-detector) in a sealed enclosure, we can transmit electrical signals by light. This is called optical coupling.

● Optical coupling has the advantage that there is absolute electrical and magnetic isolation between the sending circuit and the receiving circuit. Therefore electric or magnetic noise produced in one circuit cannot affect the other circuit.

QUESTIONS AND PROBLEMS

1. What is the class of electronic devices whose input voltage can vary in a linear manner but whose output switches abruptly between blocking and full conduction?

2. Draw the I-versus-V characteristic curve for a four-layer diode with $V_{BR(F)} = 8$ V, and reverse breakdown voltage = 50 V.

3. Explain the meaning of holding current I_H.

4. The "R" in SCR stands for _____, so an SCR converts ac into ___ .

5. Draw the schematic diagram and label the terminals for an SCR.

6. For the diagram you drew in Problem 5, mark the main terminal polarity that forward-biases the device. Then mark the main terminal (anode-cathode) polarity that reverse-biases it.

7. (T-F) Once an SCR becomes forward-biased across its main terminals, it automatically switches from the blocking state (OFF) to full conduction (ON).

8. (T-F) It is a common practice to force an SCR into conduction by causing V_{AK} to exceed the forward breakover voltage, V_{DRM}.

9. What is the usual way of forcing an SCR into conduction, assuming that the anode-cathode main terminals are already forward-biased?

10. Draw the schematic diagram that shows the relation among an ac source, a load, and an SCR.

11. For the schematic diagram that you drew in Problem 10, draw the waveforms of the ac source v_s and the load voltage v_{LD}, assuming a trigger instant of 90 degrees.

12. Repeat Problem 11, assuming a trigger instant of 60°.

13. Which situation, the 90° trigger instant of Problem 11 or the 60° trigger instant of problem 12, will cause greater power delivery to the load? Explain why.

14. Once an SCR has been triggered into the ON state, is it necessary to maintain the gate current at a value greater than I_{GT}? Explain.

15. (T-F) In an ac circuit, a conducting SCR automatically reverts to the blocking (OFF) state when the positive half cycle ends.

16. When an SCR circuit is powered by a dc source voltage, it is switched out of conduction and back to blocking by somehow reducing the main terminal current below the _____ current value.

17. Draw the schematic symbol for a diac. Then draw the I-vs.-V ideal characteristic curve for a 35-V diac.

18. The high-current thyristor that can carry current in both directions (bidirectional) is called the _____ .

19. Draw the schematic symbol and label the terminals of the device in Question 18.

20. What design step is usually taken to insure good symmetry of the load waveform in a triac circuit? Explain why this works.

21. Draw the load waveform for a triac power control circuit with 120° triggering instant on both half-cycles.

22. In a triac (or SCR) gate trigger-control circuit, when the *RC* time constant is made shorter (faster), the triac is triggered _____ (answer earlier or later). This _____ the load power (answer increases or decreases).

23. Draw the schematic symbol and label the terminals of a unijunction transistor (UJT).

24. A certain UJT has an applied base bias voltage, $V_{BB} = 18$ V. Its standoff ratio is 0.7. What is its peak voltage (breakover voltage)?

25. If the same UJT had a different base voltage of 24 V, calculate its new V_P.

26. When a UJT triggers, its current surges from the _____ terminal to the _____ terminal.

27. Draw the schematic symbol for a light-emitting diode.

28. An LED emits light when it is _____-biased. It stays dark when it is _____-biased.

29. (T-F) LEDs give off white (all colors mixed together) light, like incandescent bulbs.

30. What is the approximate range of forward voltage for an LED that is conducting current and emitting light?

31. State the advantages of an LED over a plain incandescent lamp.

32. The class of devices that produce a change in electrical signal in response to light is called _____ .

33. An optical coupler is an opaque enclosure containing a(n) _____ and a(n) _____ .

34. The main advantage of optical coupling is the total electrical and magnetic _____ between the input and output circuits.

35. Explain the input/output relationship for a solid-state relay. If the dc input is active, the SSR _____ the ac line current to the load; but if the dc input is removed, the SSR _____ the ac line from the load.

36. (T-F) In an SSR, if the opto-coupler is active, the triac is triggered almost immediately after the beginning of each ac half-cycle.

Answers to Self-Checks and some End-Of-Chapter Problems

Chapter 1:
p. 7 1. fourteen 2. four 3. eight 4. covalent 5. F 6. F 7. F 8. Five outer-shell (valence) electrons 9. Three outer-shell electrons 10. Five- valent atoms 11. Three-valent atoms 12. free electron 13. hole 14. F **p. 11** 15. F 16. Zero net charge. The silicon atoms are neutral. The boron atoms have an equal number of protons and electrons, so they are also neutral. 17. F 18. positive 19. negative 20. Holes are so rare (roughly one per million silicon atoms) that it's very unlikely that an electron will fall into one.
p. 17 22. reverse 23. forward 24. T 25. T 26. about 0.6 V 27. A resistor 28. about 0.7 V 29. Some value between 0.6 and 0.7 V **EOC** 1. insulator 2. F 3. T 4. three, five 6. *n*-type: five-valent atoms; *p*-type: three-valent 7. *n*-type 8. *p*-type 9. F 10. F 11. F 12. junction 13. T 14. depletion 15. F 16. T 17. T 18. 0.6 19. greater 20. almost none 21. almost none 23. T 24. less

Chapter 2
p. 27 1. A and K 2. blocks 3. conducts 4. closed 5. open 6. conducting: A, D; blocking: B, C, E, F 7. $V_{diode} = 8$ V; $V_R = 0$ 8. $V_{diode} = 6$ V; $V_R = 0$ 9. $V_{diode} = 0.7$ V; $V_R = 9.3$ V 10. Pr. 7, $I = 0$: Pr. 8, $I = 0$: Pr. 9, $I = 9.3$ V/R
p. 32 11. 0.7 12. current, voltage 13. about 0.8 V 14. F 15. good 16. probably bad 17. bad, shorted
EOC 1. negative; positive 2. closed; open 3. a) conducting b) 0 V c) 24 V, + on top d) 240 mA, entering on bottom. 4. a) blocking b) 30 V, + on K c) 0 d) 0 5. $V_F \approx 0.7$ V for real; $V_F = 0$, ideally. 6. a) conducting b) 0.7 V, + on top c) 7.3 V, + on top d) 0.1 A 7. There would be nothing to limit the current, so diode would overheat and be destroyed. 8. F 9. F 10. Less than 1 μA 12. a) yes b) 0.7 V, + on right c) 0.7 V, + on bottom d) 8.6 V e) 86 mA 13. a) No, neither is conducting because D_2 is reverse-biased. b) 12 V, + on top c) 0 d) 0 14. forward breakdown 15. T 16. low, high 17. bad 18. bad 19. cathode 20. Diode failed open

Chapter 3
p. 40 1. one 2. peaks at +45 V 3. peaks at +5.5 V 4. Slightly less; A small portion of the source's positive half cycle time is spent moving between 0 V and +0.7 V 5. It never reverses polarity. 6. T **p. 49** 7. two 8. F 9. T 10. Full-wave pulsations, peaking at +70.7 V 11. Full-wave pulsations, peaking at +6.4 V 12. b, d or f 13. 120 14. four 15. blocking 16. Connect the load's top terminal to the anode tie-point of D_2–D_4. Bottom to cathode tie-point of D_1–D_3. **p. 56** 16. capacitor 17. smaller 18. 60 Hz; 16 ms 19. 120 Hz; 8 ms 20. Cap is – on top. 21. C_2 also – on top **p. 62** 22. 68 V 23. R_{LD} having a low value. Increasing I_{LD} always increases ripple. 24. + peak at +9 V; – clip at –4 V **EOC** 2. Half-wave pulsations peaking at +17 V 3. peaking at 16.3 V; No, less than 8.3 ms. 4. Peaks at –17 V 5. a) Full-wave, peaking at about 33 V b) It hardly matters. 0.7 V out of about 34 V is almost insignificant. 7. It goes to half-wave. 8. Overcurrent blows the fuse. 14. Output goes to half-wave. 15. Overcurrent blows the fuse. 16. halved 17. 0.2 V p-p

19. Negative on top, double the source peak. 20. T 21. + peak at +8 V; – clip at –5 V. 22. – peak at –9 V; as the waveform rises from –9 V, it is clipped at –3 V. The output tracks a sine wave between –3 V and –9 V; it holds at –3 V at all other times.

Chapter 4
p. 73 2. Entering at B, exiting at E. 3. Entering at C, exiting at E. 4. Exiting at E. 5. Because i_E is split internally between two separate currents, i_B and i_C 7. Entering at E, exiting at B (electron-flow). 8. Entering at E, exiting at C. 9. Entering at E. 10. T 11. BJT 12. β 13. F 14. T 15. current 16. T 17. current-controlled 18. Too much work. Very few. **p. 80** 19. forward, reverse 20. positive; even more positive 21. T 22. hole, electron 23. T 24. base — hole; collector — free electron 25. T 26. a) 60.8 μA b) 13.7 mA 27. stay the same **p. 86** 30. 120 μA p-p 31. 7 mA 32. current 33. equal 34. current-controlled 35. I_e 36. a) 42.3 μA b) 5.28 mA c) 6.34 V d) 8.66 V 37. a) 10 μA p-p b) 1.25 mA p-p c) 1.5 V p-p d) 1.5 V p-p 38. a) Centered on ground (0 V), oscillating between ±0.05 V. b) Centered on 6.34 V, oscillating between 7.09 V and 5.59 V. c) Centered on 8.66 V, oscillating between 9.41 V and 7.91 V. 39. a) 47.3 μA b) 9.45 mA c) 4.44 V, – on top. d) –7.56 V e) 25 μA p-p f) 5 mA p-p g) 2.35 V p-p 40. i_B swings between –34.8 and –59.8 μA, centered on –47.3 μA (25 μA p-p). i_C swings between –6.95 and –11.95 μA, centered on –9.45 μA; (5 mA p-p). v_{RC} swings between –3.27 and –5.62 V, centered on –4.44 V; (2.35 V p-p). v_C swings between –6.39 and –8.74 V, centered on –7.56 V; (2.35 V p-p). **p. 95** 41. v_B is less positive than 0.6 V. That includes being outright negative on B. 42. v_B is about 0.7 V, or slightly greater. 43. 0 44. 0 45. V_{CC} and R_C 46. $i_{C(sat)} \approx V_{CC}/R_C$ 47. V_{CC} 48. 0 V, or slightly greater. 49. 0 50. V_{CC}, or slightly less. 51. forward 52. Reverse (or possibly forward, but by an amount less than 0.6 V). 53. active (linear) 54. saturated 55. active (linear) 56. negative 57. –0.7 V **p. 100** 58. T 59. T 60. Curve turns sharply upward when v_{CE} reaches the breakdown value. 61. T 62. At $I_B = 20$ μA, $I_C = 3$ mA (flat). Every additional 20 μA of I_B causes I_C to increase by an additional 3 mA (flat). Thus, the I_C values are 3, 6, 9, 12, 15 and 18 mA. **EOC** 20. 8.6 kΩ target; 9.1 kΩ nominal. 21. T 22. Zero, or slightly above zero. 23. V_{CC} or slightly less. 24. At $V_{CE} = 30$ V, the 150-μA I_C curve shows $I_C \approx 20$ mA. β = $I_C/I_B \approx 20$ mA/150 μA = 133. 25. $I_C \approx 133 \times 75$ μA = 10 mA (half of 20 mA, since I_B is half of its initial 150-μA value). 26. 50 to 60 V 27. six 28. high, low 29. high, high

Chapter 5
p. 116 1. F 2. half 3. F 4. greater; less 5. about 110 μA 6. about 128 μA **p. 121** 7. T 8. T 9. 0 V 10. 0 11. Left: $I_C = 8$ mA, $V_{CE} = 0$; right: $V_{CE} = 12$ V, $I_C = 0$. 12. $I_C = 41.9$ μA; dc bias point slightly above the $I_C = 40$ μA line. 13. Left: $I_C = 23.5$ mA, $V_{CE} = 0$; right: $V_{CE} = $

20 V, $I_C = 0$. 14. a) $I_B = 118$ μA b) $I_B = 165$ μA c) $I_B = 71$ μA **p. 123** 15. a) 29.4 mA b) Left: $I_C = 29.4$ mA, $V_{CE} = 0$; right: $V_{CE} = 15$ V, $I_C = 0$. 16. a) 2.3 V b) 12.8 mA 17. $V_{CE} = 8.5$ V 18. a) 5 mA (between about 15.3 mA and 10.3 mA) b) 2.6 V (between about 7.2 V and 9.8 V) 19. a) $I_C = 15$ mA p-p (from about 20.3 mA to 5.3 mA) b) 7.6 V p-p (from about 4.7 V to about 12.3 V) **p. 134** 20. V_{out}, V_{in} 21. $A_v = R_C / (r_{Ej} + R_E)$ 22. a) stay the same b) decrease c) increase 23. 9.6 24. T 25. Make $R_{E1} = R_{E2} = 75$ Ω. **p. 137** 26. 6.49 27. 2.60 V p-p 28. 13.0 mA p-p 29. a) 107 μA p-p b) 85 and 333 μA p-p c) 525 μA p-p 30. 24.8 31. 161 **p. 140** 32. The output terminal of stage 1 is connected to the input terminal of stage 2. 33. The equivalent overall resistance that is seen by an ac source that is connected to its input. 34. T 35. increase 36. 27.5 **EOC** 1. T 2. T 3. batch 4. 984 μA 5. 2.2 V 6. 1.5 V 7. 6.8 mA 8. 5.6 V 9. +6.4 V 10. The voltage-divider / emitter resistor circuit forced the I_E value to be 6.8 mA by Ohm's law and KVL. 11. 11.5 mA 12. Left: $I_C = 11.5$ mA, $V_{CE} = 0$; right: $V_{CE} = 12$ V, $I_C = 0$. 13. $I_C = 6.8$ mA, $V_{CE} = 4.9$ V. 14. 45.3 μA 15. 4 Ω 16. 3.66 17. 2.93 V p-p. From C to ground. 18. 451 Ω 19. 2.0 20. 12 21. Make a careful visual inspection for faults. 22. F 23. ground 24. bad, good 25. T 26. Saturated or shorted 27. Cut off or open 28. Open

Chapter 6

p. 166 1. base, emitter 2. collector 3. F 4. T 5. F 6. T 7. a) 13.0 V b) 10.9 mA c) 0.99 8. a) 10.4 V b) 15.3 mA c) 0.99 d) 4.95 Vp-p e) 36.6 kΩ f) 3.77 kΩ g) 1.33 mA h) 19.8 mA p-p i) 14.9 j) 14.8 **p. 172** 9. emitter, collector 10. base 11. T 12. F 13. T 14. T 15. a) 3.6 V b) 2.9 V c) 3.9 mA d) 7.7 Ω e) 286 f) 5.7 V p-p 16. a) 5.9 V b) 5.2 V c) 0.77 mA d) 39 Ω e) 144 f) 5.77 Vp-p g) 39 Ω h) 1.0 mA i) 0.58 mA p-p j) 0.58 k) 83 17. Common-emitter; common-base. **EOC** 3. emitter-follower 4. less 5. in 28. Large; no (less than 1.0) 29. low 30. Bad loading effect (poor resistance match). 31. T 32. T

Chapter 7

p. 182 1. forward, reverse 2. zero 3. forward, reverse 4. a) –0.7 V b) 11.3 V c) 3.42 mA d) 3.42 mA e) 5.14 V f) 7.6 V 5. a) +0.7 V b) 5.3 V, + on bottom c) 1.96 mA d) 1.96 mA e) 5.9 V, + on bottom f) –4.8 V **p. 186** 6. T 7. T 8. a) 17.8 kHz b) 2.4 kΩ c) 109 d) 10.9 V p-p e) 2.73 V p-p f) 27.3 9. a) 0.69 mA b) –9.6 V c) 43 Ω d) 8 kΩ e) 186 f) 9.3 V p-p g) 0.93 V p-p h) 18.6 **p. 189** 10. T 11. $R_{in} = r_{b\,in}$ alone. 12. To provide an ac flow path for the secondary current through ground and r_{Ej}. 13. Again, to complete the ac flow path for secondary current. 14. a) 2.64 kΩ b) 16.5 kΩ 15. a) 0.62 V b) 0.25 V c) 6.23 V; 6.23 **p. 195** 16. direct 17. T 18. Dc drift, because a slight bias shift gets amplified, unlike an ac amp. 19. F 20. negative feedback 21. F 22. Stabilization of bias point; reduction of distortion. **p. 197** 26. T 27. 18 750 28. high **EOC** 28. i_C swings between about 0.57 mA and 0.82 mA. 29. v_{CE} swings between about 4.8 V and 14.4 V. 30. T 31. 11.0 V p-p 32. Yes it has, because the transistor has plenty of room for v_{CE} to *decrease* ($\Delta v_{CE} = 9.6$ V), but

not very much room for v_{CE} to *increase* ($\Delta v_{CE} = 15.1$ V – 9.6 V = only 5.5 V). 33. Shift to the left, to get the Q-point located closer to the horizontal center of the *ac* load-line. 34. a) $V_{CE} = 7.0$ V b) Q at $V_{CE} = 7.0$ V, $I_C = 0.87$ mA. 35. $V_{ce(max)} \approx 13.8$ V, compared to 11.0 V with the old Q-point.

Chapter 8

p. 219 1. tank 2. 14.5 kHz 3. Nonzero resistance in the inductor winding, connecting leads, and capacitor structure. 4. amplifier 5. positive 6. a) 4.5 kHz b) 9 (90% ÷ 10%) 7. a) 5.28 kHz b) 10 ($C_2 \div C_1$) **p. 225** 8. T 9. high 10. F 11. piezoelectric 12. expand 13. positive 14. T **EOC** 1. T 2. T 3. F 4. amplifier 5. tapped inductor 6. Colpitts — series capacitor combination; Armstrong — feedback transformer 7. F 8. 1.0 9. 0 10. F 11. F 12. 6.67 13. 5 14. T 15. a) 2.13 nF b) 345 kHz 16. a) 1.35 V b) 0.27 c) 3.7 17. 6.35 V 18. d 19. piezoelectric 20. T 21. T 22. T 23. F 24. Capable of low-frequency operation; more easily miniaturized. 25. Check integrity of V_{CC} connections to R_{B1} and R_C, and ground to R_{B2}, R_E, C_E, C_1, and C_2. 26. Measure dc bias conditions of the amplifier. 27. Transistor has a shorted C-B junction. 28. The amplifier's dc bias is proper, so the amplifier circuit has no faults. 29. Disconnect amplifier input, inject an ac signal to test amplifier's ac operation. 30. An open somewhere in the C_E circuit. 31. C_{out} is open. 32. The entire amplifier is functioning, dc and ac. 33. Try to ring the LC tank circuit with a scope. 34. F 35. 100 μs or 50 μs / cm. 36. The LC tank circuit is not functioning properly. 37. Measure C_1, C_2, and L with an ohmmeter. 38. The LC tank is OK. 39. Look for a break in the feedback path.

Chapter 9

p. 240 1. antennas 2. radio waves, or electromagnetic waves; electric, magnetic 3. horizontal, vertical 4. T 5. F 6. 20 kHz, 300 GHz 7. modulation 8. demodulation **p. 245** 9. sensitivity 10. selectivity 11. selectivity 12. sensitivity 13. audio (low-frequency) **p. 250** 14. F 15. T 16. The demodulated signal would contain ripple at the carrier frequency. 17. The demodulated signal could not respond to quick changes in the signal content (a loss of higher audio frequency message fidelity). 18. 20% 19. 16 V 20. 40%; 12 V 21. Number 20. A greater percentage of modulation is demodulated by the receiver as a larger-magnitude audio signal, which drives the speaker harder. **p. 257** 22. 997 kHz; 1003 kHz 24. time 25. frequency 26. 8 kHz 27. 8 kHz 28. stagger tuning **p. 262** 29. 5 30. 10 31. 575, 585 32. F. They could potentially garble each other since their sidebands touch. 33. T. They are so widely separated in frequency that they cannot interfere. 34. microphone **p. 268** 35. T 36. When we change the resonant tuned frequency of one LC filtering circuit, a second LC filtering circuit does not change by exactly the same amount. 37. Mixing, or heterodyning 38. 1725 39. 1270, 1725, 455, and 2995 kHz 40. They are filtered out by the 455-kHz IF amplifiers. **p. 273** 42. T 43. F 44. F 45. a) –2.5 V b) 9.86 V c) 2.14 V d) 1.44 V e) 2.82 mA f) 10.6 Ω g) $A_{v(2)} = 80$ h) $A_{v(total\ IF)} = 6400$

p. 282 (287) 46. amplitude, frequency 47. resting, or center 48. F 49. 150 kHz 50. f_c, $f_c + f_{mod}$, $f_c + 2f_{mod}$, $f_c - f_{mod}$, $f_c - 2f_{mod}$, $f_c - 3f_{mod}$. 51. 75 kHz **p. 286** 52. T 53. limiter 54. f_{local} is 10.7 MHz higher than the tuned carrier. 55. T **EOC** 1. radio. electric, magnetic 2. antenna 3.0×10^8 m/s; about 13 ms 5. selectivity 6. sensitivity 8. a 9. b and c 13. 3998, 4000, 4002 kHz 14. The 3997 and 4003 kHz sidebands are greater (taller) than the 3998 and 4002 kHz sidebands. 15. 4.000, 4.005; 3.995, 4.000. 16. 10 kHz 17. 10 kHz 18. 3995 kHz, 4005 kHz 19. Square. So that frequencies near the cutoffs are not penalized relative to frequencies near the midband. 20. piezoelectric effect 21. Faraday's law effect 22. b and d 23. 1905 kHz 24. 1450, 1905, 455, and 3355 kHz 25. 455 kHz; they are permanently tuned to that frequency 26. Automatic gain control (AGC) 27. a) Increase. More time elapsing between carrier cycles allows greater capacitor discharge. b) High. With the capacitor too large for optimum performance, it doesn't discharge quickly enough to respond to quick changes in the audio signal. 28. No. The demod capacitor sees only one frequency — 455 kHz. 29. F 30. a) –4 V b) $V_{RB1} = 10.9$ V c) $V_B = 1.12$ V d) 0.42 V e) 0.82 mA f) 36.4 Ω g) $A_{v(2)} = 23.3$ h) $A_{v(total\ IF)} = 544$ 34. F 35. Sideband frequencies at 4.001, 4.002, 4.003 MHz, etc., continually declining in magnitude. And likewise in the lower sideband. 36. Same spectrograph structure, but with higher-magnitude harmonic components. 37. 150 38. 150 39. 10.7 40. 10.625, 10.775 41. Noise bursts are removed by diode clipping action. 42. deviation 43. 0 44. a, d and e 45. a, b and e

Chapter 10

p. 304 1. a) 19.6% b) 205 W 2. A 3. B 4. increase, decrease 5. efficiency 6. Transformer push-pull **p. 311** 7. a) 5.55 Ω b) 5.55 V c) 6.67 V 8. There is not dc bias current *around which* we are oscillating. Therefore we cannot justify using any particular current value as the center or dominant value to plug into $r_{Ej} = 30$ mV / I_{dc}. 9. It's less than 78.5%. 10. Increase. Swinging over a greater portion of the ac load-line will produce quite a lot of additional load power with very little increase in power wasted in R_E or in the transistor. **p. 316** 11. *npn, pnp* 12. AB 13. collector 14. Slightly less than 1.0 15. It has no transformers, saving cost, weight, and low-frequency roll-off. 16. Just a triangle with input on the flat side and output at the apex. **p. 327** 17. –5, +5.6 18. F 19. F 20. T 21. F 22. 27 MHz 23. near zero **p. 339** 24. a) 18.5 dB b) 22.6 dB c) 27.8 dB d) 31.8 dB e) 6.99 dB f) 0 dB 25. 10, 20 26. 20, 30 27. a) 2.0 b) 39.8 c) 100 d) 1.26 e) 1000 f) 10 000 28. Smaller (lower), less 29. a) –1.74 dB b) –3.01 dB c) –10 dB d) –13.0 dB e) –20 dB f) –30 dB 30. adding 31. Add decibels for amplifiers (they're positive numbers) and subtract decibels for attenuators (they're negative). 32. a) +21 dB b) 126 33. 0 **p. 341** 34. Load resistance 35. a) Yes, because $R_{in} = 600$ Ω and it's driving a load with $R_{LD} = 600$ Ω. b) 42.9 dB 36. a) No, because the input resistance and the load resistance are unequal. b) Yes, because we can calculate both P_{in} and P_{out} directly,

knowing V_{in} and V_{out}. c) 59.3 dB **EOC** 1. 18.8% 2. class-A 4. In the same position on the load-line as for class-B, at $I_C = 0$. 5. a) 57% b) class-B or class-AB 6. a) 10 mA b) 5 mA c) $I_{out(pk)} = 1.25$ A d) 250 7. a) $P_{out} = 6.25$ W b) $P_{in} = 5.0$ mW c) $A_P = 1250$ or 31 dB 9. a) 7.1 Ω b) 7.67 Ω 10. The slope line goes from the left end-point at $i_C = 1.96$ mA to the right end-point at $v_{CE} = 15$ V. 12. a) 0.75 V b) 1.34 A 13. a) 9.51 V (pk) b) 7.13 V (pk) 14. $A_v = 3.57$ 15. a) 13.4 mA b) $I_P = 5.03$ mA (pk) c) $I_{ld(pk)} = 0.891$ A d) 177 16. $A_P = 633$ or 28 dB 18. 29 mA 19. 1.0 Ω 20 T 21. 7.6 V (pk) 22. 6.44 V (pk) 23. $A_v = 0.805$ 24. 326 Ω 25. 24.5 mA 26. $I_{ld(pk)} = 0.322$ A, $A_i = 13.1$ 27. $A_P = 10.6$ or 10.2 dB 28. increase 29. T 30. small 31. increase 32. 1.2 MHz 33. T 34. a) 20 dB b) 26.8 dB c) 18.0 dB d) 33.4 dB e) 8.5 dB f) 40 dB 35. a) 100 b) 316 c) 794 d) 1000 e) 2 f) 1 36. a) –2.2 dB b) –3.0 dB c) –10 dB d) –13.0 dB e) –20 dB f) –23.0 dB 37. a) 0.5 b) 0.25 c) 0.1 d) 0.01 38. a) +22 dB b) 158 39. 8 dB 40. Scale the vertical axis from –11 dB to + 30 dB. 41. load 42. $A_v = 35.0$ dB 43. a) 8.89 V b) 2.81 V

Chapter 11

p. 353 1. gate voltage 2. V_{DD} is positive; V_{GG} is negative. 3. a) V_{GS} is – on gate relative to + on source. b) I_D enters at source, exits at drain (electron flow). c) + on drain relative to – on source. 4. All polarities and directions are opposite. 5. 6.4, 8.1, 1.7 6. 3.6, 4.9, 1.3 7. F 8. decreases 9. T 10. F **p. 363** 11. gate, source; drain, source 12. a) End-points at 12.2 mA on left, and 22 V on right. b) Q-point at $I_D = 4.9$ mA, $V_{DS} = 13.2$ V c) $r_D = 1125$ Ω; So slope line has end-points at 5.3 mA on the left, 6 V on the right. d) End-points at: $i_d = 16.6$ mA on left, and $v_{ds} = 18.7$ V on right. e) P_1 is at $i_d = 6.4$ mA , $v_{ds} = 11.5$ V; P_2 is at $i_d = 3.6$ mA , $v_{ds} = 14.6$ V. f) $V_{ds} = 3.1$ V p-p g) $V_{ds} = 3.15$ V p-p h) $A_v = 3.1$ 13. T 14. High input resistance 15. F 17. Imagine $V_{GS} = 0$ V; on the right, we have $V_{GS} = 0$, $I_D = 1.77$ mA. Then choose $V_{GS} = -6$ V; on the left, we have $V_{GS} = -6$ V, $I_D = 2.74$ mA. **p. 370** 18. –4 V 19. 2.7 20. greater 21. Also greater 22. a) P_1: $V_{GS} = -2.0$ V, I_D is about 0.81 mA. b) P_2: $V_{GS} = -2.4$ V, I_D is about 0.53 mA. c) 700 μS 23. a) P_1: $V_{GS} = -2.6$ V, I_D is about 0.36 mA. b) P_2: $V_{GS} = -3.0$ V, I_D is about 0.17 mA. c) 475 μS 24. g_m varies dramatically as bias Q-point changes. 25. 2.36 26. 1.60 27. 1.39 28. 475, 1.09 **p. 373** 29. in– 30. current, voltage 31. No voltage drop across R_D because R_D isn't present. 32. a) No b) No c) 1.98 d) 1.1 MΩ 33. a) $V_S = 0$ V, $V_{RS} = 6$ V, $I_S = 0.8$ mA b) $V_S = +1$ V, $V_{RS} = 7$ V, $I_S = 0.93$ mA c) $V_{GS} = $ about –1.8 V, $I_D = $ about 1.03 mA 34. $V_{GS} = -1.6$ V, $I_D = 1.15$ mA; $V_{GS} = -2.0$ V, $I_D = 0.81$ mA; $g_m = 850$ μS 35. $A_v = 0.43$ **p. 378** 36. Shorted gate junction. 37. An open in the drain circuit. The *p-n* junction is forward-biased, but is surviving. **EOC** 27. Open gate junction. 28. Open R_S or any open condition in the source circuit.

Chapter 12

p. 390 1. Arrow points in, toward center. 2. T 3. Metal Oxide Semiconductor Field Effect Transistor 4. negative; positive 5. D / E-type 6. T 7. nonlinear; JFET, BJT

p. 392 8. T 9. T 10. T 11. increases 12. increases
p. 398 13. F 14. For an E-type, $V_{GS} = 0$ V produces $I_D = 0$, so the flow-path is OFF. For a D-type, $V_{GS} = 0$ V produces $I_D = I_{DSS}$, so the flow-path is ON. 16. $V_{GS(th)}$ 17. a) $V_{GS} = 6.7$ V b) I_D is about 3.2 mA c) P_1: $v_{GS} = 5.7$ V, I_D is about 2.0 mA. P_2: $v_{GS} = 7.7$ V, I_D is about 4.6 mA. d) 1.3 mS e) $r_D = 1.75$ kΩ; 2.28 **EOC** 1. V_{DD} is positive. 2. V_{DD} is negative. 3. F 4. T 5. T 6. $V_{GS(off)}$ is the x-axis intercept; I_{DSS} is the y-axis intercept. 7. The curve's slope at $V_{GS} = 0$ V, right on the y (I_D) axis. 8. greater, less 9. V_{DD} is positive. 10. V_{DD} is negative. 11. D-type has unbroken vertical line. E-type has broken vertical line. Unbroken vertical line suggests complete channel. Broken line suggests incomplete channel. 12. F 13. T 14. $V_{GS(th)}$ is the x-axis intercept.

Chapter 13

p. 414 2. inverting 3. noninverting 4. T 5. differential 6. noninverting (+IN); inverting (–IN); positive. 7. inverting (–IN); noninverting (+IN); negative. 8. F 9. 8-pin DIP; 14-pin DIP; round metal can, usually 8 pins. 10. CCW 11. Not Connected **p. 417** 12. F 13. T 14. 0 15. T 16. T 17. T 18. 83 μV 19. F; it will saturate almost immediately after a sine-wave crossover. 20. –12 V to +5.49 V; +5.51 V to +12 V **p. 423** 22. $A_{v(cl)}$ is the voltage gain of the complete op-amp circuit, with supporting resistors connected; $A_{v(ol)}$ is for the raw op amp. 23. $A_v = R_f/R_i$. 24. Inverted (opposite) in polarity. 25. a) 8.18 b) 2.2 kΩ c) V_{out} is an 8.18-V peak sine wave, 180° out of phase (inverted) with respect to V_{in}. d) V_{out} is a clipped sine wave, inverted relative to V_{in}. Its peaks are clipped at about ±13 V, about 50° after a crossover. 26. virtual **p. 432** 28. a) $A_v = 5.7$ b) $R_{in} \approx \infty$ 29. a) +6.84 V b) –8.55 V c) –10 V (saturated) d) $V_{out} = 11.4$ V p-p, in-phase with V_{in}. 30. T 31. voltage-follower **p. 437** 32. V_{OUT} is not exactly 0 V when V_{IN} is exactly 0 V. 33. 1) Place compensating resistor R_B in the +IN lead. 2) Install an offset-adjust pot if the op amp has provisions for one. 34. T 35. Install an external RC compensation circuit. 36. The op amp's fastest possible rate of output voltage change. 37. F 38. 10.1 kHz 39. a) 180 kHz b) 180 kHz c) 171 kHz **p. 442** 40. R_f is open. 41. R_i is open or R_f is shorted. 42. R_i is open or R_f is shorted. 43. R_B is open; nulling pot is open. **p. 452** 44. –2 V 45. +10 V 46. +13 V (+V_{sat}) 47. integrator 48. differentiator 49. integrator. **EOC** 2. F 3. positive 4. negative 5. positive; negative 6. DIP, metal can 7. F 8. F 9. 13 10. 0 V to +15 V; –15 V to 0 V 11. a) –13 V b) –13 V c) +13 V 12. +13 V for parts a, b and c; –13 V for part d. 13. a) +13 V; –13 V for parts b, c and d. 16. a) 7.02 b) 11.2 V p-p c) V_{out} is inverted. 17. a) 4.7 kΩ b) 340 μA p-p c) 2.24 mA p-p d) 6.58 18. a) zero b) zero c) zero 19. 170 μA, right to left; 170 μA, right to left. 21. a) 4.6 b) 7.36 V c) V_{out} is in-phase with V_{in}. 22. a) virtually ∞ b) virtually zero c) 1.47 mA p-p d) A_i is virtually infinite. 23. a) T b) equal to v_{IN} c) zero 24. a) 80 μA, left to right. b) 80 μA, left to right. 25. 1.0, zero 26. offset 30. large 31. 6.6 V/μs 32. 6.6 V/μs 33. 398 kHz 34. a) 1.59 MHz b) Small-signal cutoff, 1 MHz. 35. R_f open or R_i shorted. Turn off power,

eliminate signal source, measure resistances with an ohmmeter. 36. R_i open or R_f shorted. Same procedure as Problem 35. 37. R_i open or R_f shorted. Same. 38. R_B open. Same. 39. +4.3 V 40. –13 V 41. $A_{v(2)} = R_f/R_2 = 8$ kΩ / 4 kΩ = 2; $A_{v(1)} = R_f/R_1 = 8$ kΩ / 8 kΩ = 1. Therefore V_2 gets 2 times as much gain as V_1. 42. V_2, four. 43. V_3, eight. 44. slope = 10 V/ms; v_{OUT} ramps between ±5 V. 45. a) slower b) 2 ms c) 500 Hz d) 500

Chapter 14

p. 468 2. a) 5 V, + on cathode. b) 7 V c) 9.3 mA, electron current entering on the anode. 3. lower 5. a) –14.30 V b) –14.18 V c) 0.12 V 6. 0.26 V **p. 472** 7. 7% 8. T 9. 10.08 V 10. One that maintains nearly constant output voltage, regardless of changes in operating conditions. 11. line, load 12. a) 18.18 mA b) No, $I_{Rser} = 20$ mA. c) 18.18 mA d) $\Delta V_Z \approx 0.14$ V e) $V_{OUT(LD)} \approx 9.99$ V. 13. 20 mA; 500 Ω 14. Percent load regulation = 1.4%. **p. 477** 15. T 16. a) 11.4 mA b) 11.4 mA, the same. 17. 0.11 V 18. T 19. T 20. T 21. T 22. ΔI_Z is smaller because of the extremely large β value. **EOC** 2. a) 10 V b) 5 V c) 10 mA d) 0.1 W 3. a) 20 mA b) 0.09 V 4. T 5. No. It changes hardly at all. 6. Probably negligible. 7. line 8. a) decrease, 10 mA b) 0.09V c) 12.05 V 9. load 11. T 12. The zener diode can tolerate a current change of $(1/\beta) \times I_{LD}$ but it can't tolerate a current change of *all* of I_{LD}. 13. T 15. ΔI_Z is even smaller because β_T is so much greater. 16. Q_1 17. T 18. The op amp's differential input current is virtually zero. 19. Difference voltage of 0.2 V applied to input terminals with schematic-matching polarity will drive the op-amp output higher, toward +V_{CC} (+15 V). 22. F 23. F 24. $V_{OUT} = 23.3$ V 25. $R_{bot} = 7.78$ kΩ 26. $R_{bot(min)} = 2.5$ kΩ 27. Install a 2.7-kΩ resistor below the pot, connected to ground.

Chapter 15

p. 494 1. thyristor 2. forward breakover 3. $V_{BR(F)}$ 4. blocking; conducting 5. T 6. Making gate current exceed the critical gate trigger current, I_{GT}. 7. T 10. greater **p. 498** 11. A bidirectional thyristor can carry current in either direction, not just one direction. 12. T 13. F 14. They're in series. 15. T 16. T **p. 503** 18. T 19. T 20. peak; V_P 21. T 22. Produces light when forward-biased and conducting; goes dark when unbiased or reverse-biased. 23. F **EOC** 1. thyristors 3. The minimum amount of main-terminal current that is required to maintain a thyristor in the ON state. 4. rectifier, dc 6. Forward is + on A, – on K. 7. F 8. F 9. Raising i_G above the trigger level, I_{GT}. 13. The 60° instant. A greater portion of the line voltage waveform is applied to the load. 14. No. Gate current has nothing to do with maintaining an SCR in the ON state, after it has been triggered. 15. T 16. holding 18. triac 19. Terminals MT2, MT1, G. 20. Installing a diac in the gate lead. The diac has good symmetry itself. 22. earlier; increases 23. Terminals B2, B1, E. 24. about 13.2 V 25. about 17.4 V 26. B1, E 28. forward; zero or reverse 29. F 30. 1.5 to 3 V 31. Low power consumption; very long life. 32 photo-detector 33. LED, photo-detector 34. isolation 35. conducts; blocks 36. T

Index